Change, Transformation and Development

J. Stan Metcalfe · Uwe Cantner
Editors

Change, Transformation and Development

With 41 Figures
and 61 Tables

Springer-Verlag Berlin Heidelberg GmbH

Professor John Stan Metcalfe
University of Manchester
Centre for Research on Innovation and Competition
and School of Economic Studies
Manchester M13 9QH
United Kingdom

Professor Dr. Uwe Cantner
Friedrich-Schiller-University Jena
Department of Economics and Business Administration
Chair of Economics/Microeconomics
Carl-Zeiss-Straße 3
07743 Jena
Germany

Some of the contributions have been published in
"Journal of Evolutionary Economics", Vol. 12, No 1–2, 2002

ISBN 978-3-642-62410-0 ISBN 978-3-7908-2720-0 (eBook)
DOI 10.1007/978-3-7908-2720-0

Bibliographic information published by Die Deutsche Bibliothek
Die Deutsche Bibliothek lists this publication in the Deutsche Nationalbibliografie; detailed bibliographic data available in the internet at *http.//dnb.ddb.de*

http://www.springer.de

Originally published by Physica-Verlag Heidelberg New York in 2003
Softcover reprint of the hardcover 1st edition 2003

Cover design: Erich Kirchner, Heidelberg

SPIN 10893730 88/3130-5 4 3 2 1 0 – Printed on acid-free paper

Table of Contents

iii. Transforming capabilities and organisations

iv. Schumpeter and the transformation of economic thought

v. Empirical studies

Change, transformation and development

Introduction to the volume

The general theme of the 8th International Joseph A. Schumpeter Society Conference, held at the University of Manchester was the exploration of economic and social dynamics in relation to the innovation process and its outcomes, broadly defined. This theme is very firmly rooted in the Schumpeterian tradition in which an economic perspective is mutually embedded in a wider awareness of the role of other disciplines. Indeed since Schumpeter's time, the degree of specialisation within the social sciences has risen many fold, new sub disciplines continue to emerge, highly specialised theoretical tools and empirical methods continue to be developed, and new fields for the study of management and business overlap with the more traditional social sciences. Consequently, there is a need for connecting principles to offset the dangers of intellectual fragmentation. Evolutionary economics and evolutionary analysis more generally, certainly provide some of these connecting principles although much of this field remains to be developed. The central ideas of variation, selection and development, applied within the instituted frame of modern capitalism, provide a powerful set of concepts to consider the interaction between economic and other forces and to focus attention on its propensity to change and transform itself from within.

The various contributions to this volume reflect upon the above mentioned general themes in a number of ways. In the following we want to characterise each paper very briefly. For this purpose we have classified them under six different sections starting with section i titled *evolutionary processes*. Here, Stan Metcalfe's (University of Manchester, Manchester, UK) Presidential Address 'Knowledge of Growth and the Growth of Knowledge' explores the idea of restless capitalism, the notion that economic transformation is inseparable from the transformation of knowledge in general and practical knowledge in particular. The conference opened with Dick Nelson's (Columbia University, New York, US) paper on 'Bringing Institutions into Economic Theory', in which a knowledge of social technologies is given a comparable weight to the more formal technological and scientific knowledge upon which so much attention has been lavished by innovation scholars. This section concludes with Pavel Pelikan (Royal Institute of Technology, Stockholm, Sweden) who discusses in 'Choice, Chance, and Necessity in the Evolution of Forms of Economies' the search of humans for forms of society as a higher-level Darwinian evolution and looks for regularities and principles in such error driven processes.

The section *ii*, titled *demand, innovation and evolution*, draws attention to the demand side in evolutionary economics, a field still to be developed and explored. In 'Designing Clunkers: Demand-Side Innovation and the Early History of the Mountain Bike', Guido Buenstorf (Max Planck Institute, Jena, Germany), highlights the dominant role of consumers in consumer good innovations. It seems to be an important general feature of this learning based evolutionary process that the principle design of the consumer good under consideration was established before commercial interests entered the scene.

Section *iii*, brings together papers oriented towards the discussion of *transforming capabilities and organisations*. John Matthews (Macquarie University, Sydney, Australia) opens the section with 'A Resource–Based View of Schumpeterian Economic Dynamics'. He develops a discussion of entrepreneurship and industrial dynamics on the basis of the concept of the 'resource economy', in which the focus is on the productive resources produced and exchanged between firms – and thus on the dynamical capital structure of the economy. John Finch (University of Aberdeen, Aberdeen, Scotland) delivers an application of Richardson's concept of capabilities understood as tacit, personal, subjective and context specific knowledge shared across small groups of actors. His empirical inquiry is concerned with the development of those capabilities within an industry (the UK upstream oil and gas industry) promoted by organisational change in the activities across firms in the sector and contracting and supply companies. Keith Jakee and Heath Spong (Royal Melbourne Institute of Technology, Melbourne, Australia) conclude this section with a theoretical discussion of endogenous change driven by entrepreneurial activities and their effects on institutions.

Section iv, *Schumpeter and the transformation of economic thought*, elaborates on evolutionary thinking in relation to economic thought. Maria Brouwer (University of Amsterdam, Amsterdam, The Netherlands) demonstrates how Weber, Schumpeter and Knight influenced each other's thinking about entrepreneurship and how their debate is still vivid in contemporary thinking about governance and organisation. With 'Connecting Principles, New Combinations, and Routines: Reflections inspired by Schumpeter and Smith', Brian Loasby (University of Sterling, UK) discusses evolutionary processes of knowledge creation in not fully connected systems and relates this to the work of Smith and of Schumpeter.

Section v, *empirical studies*, contains a number of papers with a variety of empirical approaches to the study of innovation and evolution. Innovating routines, a core element in evolutionary thinking are identified and classified in the paper 'Innovating Routines in the Business Firm: What Matters, What's Staying the Same and What's Changing?' by Keith Pavitt (University of Sussex/SPRU, Brighton, UK). The sources of success in recently established and growing industries are investigated for the case of the wind turbine industry in "The Emergence of a Growth Industry: A Comparative Analysis of the German, Dutch and Swedish Wind Turbine Industries" by Anna Johnson and Staffan Jacobsson (Chalmers University of Technology, Göteborg, Sweden). Two new taxonomies of manufacturing industries, one based on intangible investment and one based on human resources, are suggested by Michael Peneder (WIFO, Vienna, Austria), 'Intangible investment and Human Resources', who then investigates the presumed complementarity between the taxonomies. An investigation into the joint dynamics of market concentration, the number of firms and the range of products is provided in the paper 'The Long-Term Evolution of Vertically-Related Industries' by Andrea Bonnaccorsi and Paoloa Giuri (Sant' Anna School of Advanced Studies, Pisa, Italy). They test their model with data on the commercial jet and turboprop aircraft and engine industries for the period 1948–1998. In 'The Role of Innovation and Quality Change in Japanese Economic Growth' Derek Bosworth, Silvia Massini and Masako Nakayama (Manchester School of Management, Manchester, UK) are concerned with qualitative change and the important question of the separation in price indices of quality change and pure inflation. The types of entrepreneurial action which placed the Swedish mobile phone industry into a first-mover position, are investigated in 'Entrepreneurs, Innovations and Market Processes in the Evolution of the Swedish Mobile Telecommunications Industry' by Staffan Hultén (Stockholm School of Economics, Stockholm, Sweden and Ecole Centrale Paris, Paris, France) and Bengt Mölleryd (Stockholm School of Economics, Stockholm, Sweden). Finally, in their paper, 'The New Geography of Corporate Research in Information and Communication Technology', John Cantwell (University of Reading, Reading, UK) and Gracia Santangelo (University of Reading, Reading, UK and Università degli Studi di Catania, Catania, Italy), investigate the regional dispersion of research of multinational firms in information and communications technology and find differences herein between intra-industry competition and inter-industry co-operation.

Our final section *vi* is concerned with papers oriented towards *policy, transformation and development*. Bo Carlsson (Case Western Reserve University, Ohio, USA) and Ann-Charlotte Fridh (Royal Institute of Technology, Stockholm, Sweden) begin with 'Technology Transfer in United States Industries', where they examine the roles of technology transfer offices in commercialising research results of US universities. The systems approach to innovation policy is taken up by Morris Teubal's (The Hebrew University, Jerusalem, Israel) paper 'What is the systems perspective to Innovation and Technology Policy (ITP) and how can we apply it to developing and newly industrialised economies'. Last, not least, in their paper 'Knowledge Production and distribution and the economics of high-tech consortia' Maurice Cassier (CNRS, CERMESD, Paris, France) and Dominique Foray (CNRS, IMRI, Dauphine University, Paris, France) discuss collective invention and how for this purpose research consortia run the management of knowledge and intellectual property.

Collectively and individually, these papers provide a good guide to the range of issues discussed at the conference and the rich collation of ideas and methods that constitute a modern Schumpeterian agenda. What they do not do is capture one if the most inspiring sessions of the conference in which, Wolfgang Stolper, Dick

Nelson and Giovanni Dosi presented the views of three different generations of scholars on the Schumpeterian enterprise. Sadly this was to be Wolgang Stolper's last Schumpeter conference and we all shall miss his wise council and engaging conversation. It is perhaps fitting that we conclude with the words with which he summed up his assessment of the current state of the Schumpeterian research programme. "The fundamental ideas of Schumpeter have proven to be developable, and that we can meet here in Manchester is a joyful realisation of this fact". We will do the memory of Wolfgang Stolper, and indeed Joseph Schumpeter, no little service by continuing with the development of these profound ideas.

Manchester and Jena, August 2002

Stanley J. Metcalfe

Uwe Cantner

Knowledge of growth and the growth of knowledge

J. S. Metcalfe

Centre for Research on Innovation and Competition and School of Economic Studies, University of Manchester, M13 9QH, UK (e-mail: stan.metcalfe@man.ac.uk)

Abstract. The central theme of this address is the complicated relationship between the growth of the economy and the growth of knowledge. This theme is explored with the help of a single concept "restless capitalism" which is used to capture the idea that capitalism in equilibrium is a contradiction in terms precisely because the growth of knowledge cannot be meaningfully formulated as the outcome of a constellation of equilibrating forces. This theme is explored through a discussion of growth accounting, the relationship between innovation, markets and institutions and, as an example, the development of innovation in the field of ophthalmology. We also discuss some pioneering contributions made by Simon Kuznets and Arthur Burns to the discussion of evolutionary growth. From this Schumpeterian perspective we see the economy as an ensemble not an aggregate entity and so see more clearly the importance of microdiversity in the relationship between growth of knowledge and growth of the economy.

Key words: Growth – Knowledge – Evolution – Innovation – Institutions

JEL Classification: B25, E11, O30, O40

1 Knowledge of growth and the growth of knowledge

The central theme of this address is the double-sided relation between the growth of an economy and the growth of knowledge in general and practical knowledge in particular. This relation is at the core of Schumpeterian perspectives on economic transformation even though in Schumpeter's work it was subsumed within his general concept of entrepreneurship. It is equally central to modern evolutionary accounts of economic growth and to neoclassical theories of endogenous growth, although, in each case, from a quite different standpoint with regard to the nature of knowledge and the processes by which it is accumulated and directed to economic ends. I shall explore this theme with the help of a single concept, "restless capitalism", which I use to capture the idea that capitalism in equilibrium is a contradiction in terms precisely because the growth of knowledge cannot be formulated meaningfully as the outcome of a constellation of equilibrating forces. I shall begin with a discussion of how growth accounting underestimates the role of technical progress in economic growth, move on to outline the nature of growth in relation to innovation, markets and institutions more generally, and then I introduce some growth theory from the 1930s to capture the relevant features of restless capitalism and economic transformation. I conclude with a discussion of innovation in relation to the growth of medical knowledge to highlight the theme of ongoing transformation and structural change.

2 Growth and technical progress

Modern interest in the growth-knowledge relationship dates from the development of growth accounting and the discovery that the rate of tangible capital accumulation could only account for a small proportion of measured economic expansion. The initial studies of Schmookler and Abramovitz, subsequently refined by Solow, Kendrick, Dennison and others all pointed in the same direction. To put it in Solow's terms, measured *shifts in* the production function were far more important than *factor substitution around* a given production function as an accounting explanation of economic growth. Following these initial studies a great deal of effort has been devoted to refining measures of input and output with the purpose of eliminating as far as is possible the residual element in economic growth, and for a good reason. To the extent that the residual reflects the growth and application of a multiplicity of kinds of new knowledge then we have to admit that our understanding of it lies beyond the scope of existing methods of analysis. There is no doubt that this motivated Jorgensen and Griliches to conduct their pioneering attempts to tame the residual, effectively but wrongly, claiming that it could be reduced to a series of measurement errors. Notice though that they did not claim that the growth of knowledge was not central to the growth process, rather that its application to the economy depended on processes of investment in equipment and human skills – the so-called embodiment hypothesis. Now the attempt to shift the balance of explanatory power between "shifts in" and "movements around" an aggregate production function, is not simply an empirical problem. More fundamentally, the method of treating the two contributions as independent and additive (Nelson, 1982) is open to an obvious objection. Namely, that measured capital deepening is not necessarily an independent contributor to growth, rather it is contingent upon prior shifts in the production function, that is to say the rate of technical progress. Consequently, to deduct the capital deepening effect from the growth in labour productivity, however measured, is to systematically underestimate the contribution of technical progress to economic growth. This is the return of a "measure of ignorance" with a vengeance, or as Usher (1980) put it so effectively, no progress means no growth. All measured growth in this view involves an alteration in input:output relationships, that is to say it requires changes in the application of knowledge broadly defined. The point is very simple and involves nothing more than realisation that the answer to the question, "How much has technical progress shifted the production function?" is not the same as the answer to the question, "How much has technical progress contributed to economic growth?". To see this, consider the conventional model of an aggregate economy in steady growth according to the stylised facts with a constant saving ratio, constant distributive shares and a constant capital:output ratio. Suspend for present purposes any deeper objections to the idea of an aggregate production function and the notion that factor prices measure marginal products.

In such a configuration it is easy to see that the rate of capital deepening is not independent of the shift in the production function. We can express this idea either in terms of the extra saving per head, which is made possible by the shift in the production function, or, if we recognise that capital goods are necessarily produced means of production, then technical progress means that a given rate of saving commands a greater effective labour force employed in the production of capital goods (Rymes, 1971). At the risk of belabouring the point let $\hat{q}$, $\hat{k}$, $\hat{A}$ be respectively the logarithmic rates of increase of output per head, capital per head, and the shift factor for the production function (technical progress), and let β be the share of capital income in total product. Then the formula

$$\hat{A} = \hat{q} - \beta\hat{k}$$

in principle is the growth accounting measure of the shift in the production function. However this shift understates the contribution of technical progress to growth. Let this be measured by $\hat{T}$ then we should write

$$\hat{T} = \hat{q} - \beta\left(\hat{k} - \hat{T}\right)$$

to reflect the fact that only that amount of capital deepening that is independent of technical progress should be given an explanatory role. In steady growth it follows immediately that

$$\hat{T} = \hat{q} = \hat{k}$$

and that

$$\hat{A} = (1 - \beta)\hat{T}$$

In short, all measured growth, in these conditions, is due to shifts in the production function, and the conventional residual underestimates the contribution of technical progress to economic growth. Moreover, the greater is the measured capital share in final output, the greater is the degree of understatement. The rate of growth of labour productivity is the proper measure of technical progress, as it should be when labour is the only primary factor of production. Consequently, there are grounds for claiming that the growth of knowledge is more important to economic growth than the growth accounting approach recognises. Out of steady growth, with a varying saving ratio, the story would be different but the evidence apparently does not support the idea of strong independent saving effects (Prescott, 1998).

Side by side with the development in productivity accounting, growth theorists of various persuasions had assembled a theory of economic growth but not a theory of the growth rate. Whether we take the Meade, Solow, Swann approach or that associated with Kaldor, Pasinetti and Robinson we find models of the properties of steadily growing economies, not models of the growth rate. The determinants of the latter, population growth and Harrod-neutral technical progress, were left unexplained. It is to fill this vacuum that a range of theories of endogenous growth have been developed (Aghion and Howitt, 1999).

Schumpeterian evolutionary theories are certainly of this kind but they are fundamentally different from the theories developed by endogenous growth economists of whom I take the work of Jones (1995) as representative. What is central to his approach is the idea of a production function for ideas, in which the rate of increase in the stock of ideas is dependent upon current research effort and the existing stock of ideas. Now this is not a new approach, Machlup sketched out the principle elements of it in 1962. Put briefly, there are diminishing returns, after a point, to current research effort, with the existing stock of ideas acting as the 'fixed factor' in the knowledge production process. New ideas in the output of research change the stock of solved problems, so shifting the marginal relations between research input and the output of knowledge. Machlup suggested that this could happen in one of two broad ways, either by reducing or enhancing the research agenda that is derived from the existing stock of ideas. It is this model of knowledge accumulation that Jones deploys in his analysis, and in so doing he is right to try and capture the autocatalytic nature of knowledge accumulation. However, he does so in a way that, to my view, stands in the way of understanding the knowledge-growth relationship. In particular, he works with an aggregate stock of ideas and looks for the conditions under which this stock, and of necessity all its component elements, are growing at the same steady rate. It is this double condition, required to service the idea of equilibrium growth, that Schumpeterian and evolutionary proponents of growth will find so unacceptable as a basis for understanding capitalism as a knowledge-based economy. The steady state accumulation of knowledge is for them an idea too far.

I now want to explore this claim in a number of ways. First, in terms of the internal problems with this approach and how it hides the central element of knowledge-based growth, namely, pervasive structural change, or flux and economic transformation. Secondly, and following the theme of structural change, the dominant role of service activities in modern economies suggests that our understanding of growth should give a prominent role to innovation and knowledge accumulation in relation to services. This does not mean that we downplay manufacturing innovation but rather focus on the interaction between services and manufacturing in the process of productivity growth (Metcalfe and Miles, 2000). I will illustrate this second theme with a brief outline of a particular medical innovation, the intra-ocular lens that has transformed the delivery of ophthalmic services for cataract patients. It is through detailed comparative and historical studies of this kind that I claim we will better understand the knowledge-growth relationship.

3 The conceptual problems

What kinds of analytical frameworks will help to unravel the knowledge-growth relationship? It seems clear that the answer to this question will not be found in economics alone. Rather the most useful frameworks will enable effective dialogue with the work of historians, sociologists, political scientists, and management scholars, reflecting the fact that knowledge advances more rapidly when the connecting principles between a variety of perspectives are strong. However, from an economic view the relevant frameworks are likely to display several defining characteristics.

Firstly, they will encompass ongoing structural change; that is to say, they will provide explanations for the diversity of growth rates in the economy and not assume their uniformity. Equilibrium growth, every activity expanding at the same geometric rate, is a device for avoiding this most pervasive of all the stylized facts of modern growth. Growth involves transformation, transformation leads to development and capitalism in steady growth is a contradiction in terms. Whatever steady state growth theory is about it is not obviously knowledge-based capitalism. Secondly, while we can measure at the macro economic level we cannot comprehend the growth process in these terms. It is not simply the statistical truism that macro aggregates are constructed by averaging away the diversity in patterns of economic change. It is a much deeper point; aggregation hides the evolutionary process that generated the aggregates in question. Evolutionary growth processes depend on the existence of variety, on the relative spread or diffusion of rival products and methods of production and these essential elements are written out of the aggregate picture. Only when the elements in an economy grow at the same rate do macro aggregates have a clear meaning, but when the micro growth rates are different the interpretation of the change in any aggregate is far more opaque. Quite what does it mean to say that measured GDP in country X is $Y\%$ higher today than 10, 20, 50 years ago, when the bundle of economic activities differs structurally and qualitatively between the various dates? Thus a knowledge-growth framework not only begins at the micro level and formulates appropriate rules for aggregation, it also explains how the composition of those aggregates changes in a systematic fashion in the course of economic development.

Thirdly, the problem of aggregation cannot be solved unless we understand the process of interaction between micro agents and this takes us directly to the problem of co-ordination and the role of market processes. Markets are the central instituted form by which economic order and changes in order are generated within capitalism. Order produces pattern, order is not equilibrium in the sense of a state of rest from which there is no internally generated tendency to depart. In this sense equilibrium, which can only mean an equilibrium of beliefs, surely means that history has come to a stop since there cannot be internally generated reasons for beliefs to change. In equilibrium time passes but nothing happens. As Schumpeter remarked, to understand capitalism you have to understand its capacity to transform itself from within and this requires an understanding of why the economy is "far from equilibrium' as a modern physicist might put it. In this regard knowledge is like energy it defies equilibrium by maintaining a potential for change that is ever present.

Of course, not all the relevant co-ordinating institutions fall into the market category. In relation to the production of knowledge in particular, networks and communities of practitioners and practice play very important roles. Indeed, one of the major conceptual and empirical challenges is to integrate more closely these differently instituted forms of co-ordination to explain how each is embedded in the other.

From the perspective of the growth-knowledge relation, markets take on a new light. We see them not as devices to optimally allocate given resources to given ends but as institutions to facilitate change, to permit entrepreneurship, to encourage challenges to the established order. Thus they are devices for keeping the economy ordered but out of equilibrium, they are the frameworks which shape ongoing structural change. Nor are market institutions given. They have to be established and their establishment, growth, stabilisation and decline involve the investment of real resources in market making activity. It is surely somewhat shocking that the detailed understanding of markets and their formation plays so little role in the modern theory of growth of any persuasion.

Fourthly, it is clear that a serious attempt to link growth and transformation means abandoning the device of the representative agent. Rather more accurately it involves abandoning any attempt to theorise in terms of the uniform agent. Indeed, the Marshallian notion of a representative unit of behaviour, the firm in his case, was a device for coping with economic evolution, not economic equilibrium. In evolutionary theory a representative agent need not correspond to any actual agent in the relevant population nor can its attributes be determined a priori. What is representative is the emergent outcome of the economic process not a precondition of it. What is representative depends on the manner of co-ordination of the relevant behaviours, and hence it will change with the economic process even when the individual behaviours of the "real" agents are fixed. I need hardly mention the difficulties that arise in introducing innovation and change of knowledge into the world of the uniform agent. Again it is the macro perspective that has led us astray. For the device of the uniform agent is merely a way of having the micro conform to a prior conception of the macro.

What kinds of frameworks will help us to meet these difficulties? They will be evolutionary and adaptive simply because evolutionary theory is naturally about growth rates and is premised on the micro diversity that characterises a knowledge-based economy. They will emphasise market co-ordination as a selective process because it is through co-ordination that growth is created. Growth rates appear as emergent phenomena, that is to say, they are not intrinsic properties of agents such as firms; rather they are properties that emerge as a result of the interaction between firms in the market selection process. However, selection is not enough because, left to themselves, selection processes destroy the microdiversity in economic attributes upon which evolution depends. Variety needs to be replenished for growth and transformation to continue and this requires that development processes be given equal weight with selection processes (Foster and Metcalfe, 2001). Now, development processes are deeply connected to the growth of knowledge and by this I do not only mean the formal knowledge associated with science and technology. Practically useful knowledge cannot be so circumscribed, knowledge of organisation, knowledge of market and social knowledge of how to interact must be given due weight. Clearly these different kinds of knowledge are accumulated and diffused by very different kinds of development processes. Moreover, in so far as market experience is a key element in the growth of knowledge, this implies that the process of selection and the process of development are inseparable and mutually determining (Dosi, 1997). The competitive process depends on microdiversity and the generation of microdiversity reflects the competitive process. Change the working of market processes and you change the way practical knowledge is accumulated. All of this puts the modus operandi of markets at the centre of the growth process while recognising that the way markets work is contingent upon wider sets of institutional factors. It is surely not an accident that many of the scholars who have contributed to the innovation systems literature are also the scholars who have developed a Schumpeterian perspective on the market process. For Schumpeter's theory of development joins innovation and markets, development and selection, in the medium of the entrepreneur; and I need not remind this audience that entrepreneurship is *impossible* in equilibrium.

Another aspect of this approach is that the treatment of co-ordination processes means that the demand side of the innovation process needs more attention than it has thus far received. Perhaps because of Schumpeter's view on the passivity of consumers in the innovation process we have ended with a perspective dominated by supply side considerations. However, there is a growing body of evidence, much of it from the pen of technology historians, which gives the consumer/user a very active role in shaping innovation. Not only through Lundvall's supplier-user interaction, for example, but also in the conceiving of new applications for products quite different from those conceived of by the original designers and entrepreneurs. Susan Douglas's study of the radio industry (1987) provides a fine example of this kind of influence showing how the radio evolved from a means of maritime communication to a means of mass entertainment. Thus, a knowledge-based theory of growth will emphasise the link between the micro diversity of behaviours and processes of creativity and the formulation of novelty by consumers as much as by firms. Indeed it is the continual generation of novelty on both sides of the market relationship that underpins the idea of restless capitalism and keeps capitalism "far from equilibrium". Consequently the dynamics of the growth process cannot be governed by a process of convergence to equilibrium states, for the states of rest are continually being redefined by the accumulation of consumer and producer knowledge that occurs in the market process (Kaldor, 1954).

The technicalities of this market-based approach are, in my view, best handled by a general class of processes under the broad label of replicator dynamics. They have a potentially important property, namely that the dynamics of the process are governed by the distribution of behaviours around population averages (representative behaviours!) not by the distance of the system from some long-period attractor (Iwai, 1974; Metcalfe, 1998; Dosi, 1997). It is this feature, which corresponds to the idea of open-ended, adaptive, evolutionary development. The system evolves and in the process grows but not by chasing states of rest; so avoiding the embarrassment of multiple equilibria or equilibria that are changing more quickly than the process of convergence can accommodate to. The fundamental reason for this historical indeterminacy is to be found in the nature of knowledge and its accumulation processes. It is the characteristic of knowledge that one idea leads to another in typically unpredictable ways (Popper, 1996) reflecting the immense possibilities for the recombination of ideas and the use of ideas. It is because knowledge is used but not used up, that ideas feed inexorably on ideas, which make increasing returns in the production of ideas to be of far greater importance than increasing returns in the production of goods and services. Whether this can be encompassed in the idea of a knowledge production function I have very severe doubts. How is the stock of ideas to be defined? Is it the simple sum, the product, the union, the intersection or the combinatorial combination of individual ideas? Since ideas are not obviously commensurable, by what weights are we to combine one idea with another? If these are value weights, how are the relevant rates of exchange to be determined? Surely not by market prices. There is a very real prospect that the idea of a stock of knowledge capital will fall prey to exactly the same logical difficulties that destroyed the idea of a productive input called aggregate capital. The specification of inventive labour will prove just as problematic for it is obviously not homogeneous as Machlup (1962) rightly insisted. The production function route is perhaps best avoided unless we are prepared to emphasise the extreme micro heterogeneity of the underlying inventive processes.

4 Alternative foundations for the knowledge/growth relationship

If we step back to the growth economists of the 1920s and 1930s, not only Schumpeter but also scholars such as Arthur Burns, Simon Kuznets and Allyn Young, we find the beginnings of an empirically grounded non-aggregative growth theory that meets many of the requirements outlined above. These contributions were swept aside by the Keynesian revolution when growth theory went macro but they fit extremely well with our growth-knowledge perspective and with the idea of growth as transformation.

In his comprehensive review of modern economic growth Abramovitz (1989) identified structural change and a tendency towards retardation in the growth of output as two salient empirical generalisations about the process of economic growth. Structural change is, of course, a necessary reflection of diversity in the growth rates of different activities. Retardation, however, is a different phenomenon, the systematic tendency for rates of growth of specific entities or their ensemble to decline with the passage of time. To anyone brought up on the economics of uniformly expanding economies, whose structure cannot change over time and whose rate of growth is constant, neither of these propositions will have much resonance. Yet they are central to the literature to which Abramovitz is referring, in particular to the work of the two principal retardation theorists, namely A.F. Burns and S. Kuznets. Both are concerned with the measurement and explanation of secular or long time movements in the volume of economic activity. Accepting that the modern economic system is 'characterised by ceaseless change', neither could proceed with an aggregate analysis of growth nor accept the idea of uniform progress in all branches of activity.

Let us begin with A.F. Burns's detailed study of American economic growth in the period 1880-1937. Burns gathered a great deal of evidence to establish that a central feature of modern economic development is the diversity of growth rates of output across different sub-sectors and industries in the economy. What might appear to be smooth progress of production and trade in the aggregate hides a considerable diversity of experience. His list of diversity creating factors has a thoroughly modern ring to it: new commodities; new raw materials; changes in methods of production; new methods for the recovery of waste products; changes in forms of industrial organisation; increases in the number of uses of given materials and in the number of materials put to a given use; and, finally, the emergence of what he calls learning products and style goods. In sum, Burns claims that "These changes have resulted in an increasing divergence of production trends for they have served to stimulate or depress but to an unequal extent, the development of various industries" (p. 63). Furthermore, what makes an economy progressive is not diversity *per se* but a positive skew to the distribution of growth rates. Thus the focus was on the micro diversity of industrial growth experience and what he called his "law of industrial growth", that individual growth rates are subject to retardation and ultimate decadence. That is to say, for any industry its percentage growth rate declines with time eventually becoming negative.

> Apparently, just as the forces making for growth of individual industries have dominated in the American system over the forces making for decline, so have the forces making for retardation in the growth of individual industries dominated in the system over the forces making for acceleration" (p.122).

This he claimed was a characteristic feature of a progressive economy, an economy in a perpetual state of flux qualitatively as well as quantitatively.

Simon Kuznets too had explored independently the same themes (1929, 1954) and from a broadly similar perspective, and he stated the problem clearly as follows,

> As we observe various industries within a given national economy, we see that the lead in development shifts from one branch to another. A rapidly developing industry does not retain its vigorous growth forever but slackens and is overtaken by others whose period of rapid development is beginning. Within one country we can observe a succession of different branches of activity in the vanguard of the country's economic development, and within each industry we can notice a conspicuous slackening in the rate of increase' (Kuznets, 1929/1954, p. 254).

Of course, the long secular movements of the shares of agriculture, industry and service sectors in total output provide confirmation at higher levels of aggregation of the enduring presence of growth rate diversity and structural change. As do the shifting rural-urban balance of the population, changes in working hours and changes in the pattern of household consumption. Indeed the long swing of development must have been marked by as much change of consumption behaviour and pattern of demand as change of industry.

Alongside Burns and Kuznts equal attention should also be paid to Allyn Young (1928). His central concern was the link between productivity growth and the *extension* of the division of labour within and between industries. Crucially, this led him to emphasise the role of demand and how demand considerations lead to the interdependence of productivity growth rates between industries. What he didn't say, for he had no need to, was that this interdependence precludes any simple adding up of sectoral productivity growth rates to explain the economy wide rate of productivity increase

The point is that, Young along with Burns and Kuznets, articulated a theory of growth that is non-aggregative in character, that depends on the details of market co-ordination and that emphasises the dynamics of knowledge accumulation at the level of individual sectors and firms. Schumpeter surely would have found this agreeable and evolutionary growth theorists can find much of value in this literature to underpin their interest in agent- based models of growth. Perhaps there is no better way to summarise this theme than with Eliasson's (1990) notion of an experimentally organised economy.

5 The IOL story: competition and the institution of innovation systems

I turn now to my final theme in this discussion of economic transformation, that of the growth of the service economy and its relative neglect by economists interested in innovation and the knowledge growth dynamic. This neglect is problematic for we all recognise the predominant role of service production in the modern economy. If we are to make sense of the relation between knowledge and growth then the process of innovation in the service economy needs far more attention. This is a large topic (Metcalfe and Miles, 2000) and so I shall use a particular case of innovation in medicine to capture some of the main themes. I shall be brief because I want to get to the main point, which is the interaction between selection processes and development processes in relation to economic growth and the complementarities between services and manufacturing, between intangibles and artefacts.

The medical service innovation in question is the removal of cataract and the restoration of functional sight in individuals afflicted with this condition. The new service depends on a particular artefact innovation, the intra-ocular lens. The intra-ocular lens (IOL) is the solution to a pressing medical condition, age related cataracts, that affect over half the population of people over fifty – the fastest growing population cohort in the OECD countries. The traditional treatment involved removal of the cataract-damaged lens from the eye, so leaving the patient severely disabled visually. In 1949 a British clinician, Harold Ridley, set in motion a train of events that by 1980 had resolved this problem. His innovation was the insertion of a plastic lens within the eye, "where nature intended", in the place formerly occupied by the defective natural lens. What was for Ridley, and the patient, a hazardous procedure has now become a routine operation, the most frequently performed operation in the over-50s age cohort. This is not the place to dwell on the ins and outs of this fascinating case of knowledge accumulation and application (Metcalfe and James 2001). It involved hostility to the original innovation by the established ophthalmic profession and a "swarm" of following imitators in true Schumpeterian fashion. It involved extended sequences of innovation along a design trajectory, as Dosi suggested, with important complementary innovations in materials, lens design and operative technique along the way. Moreover, not only has the treatment of patients been transformed, there is a new division of labour being created in this medical service activity. A treatment that formerly involved extended hospitalisation is now carried out on an ambulatory basis, in many settings the procedure is organised as an effective production line.

Important though these aspects are, they are not what I want to emphasise. For Ridley and his immediate followers and imitators were "hero-surgeons" working in clinical contexts, governed by the selection processes of instituted practices and resource allocation in publicly funded health services in Europe. By the 1980s the locus of innovation and development had shifted to a medical-industrial complex dominated by five or six transnational companies located in the USA. These companies, as a matter of policy, build very close working relations with the present generation of ophthalmic clinicians and fund a great deal of the R&D activity. What they have done is to have created a new kind of innovation system in the search for competitive advantage. In so doing they join together the processes of selection and development on which the growth of this particular activity depends. This takes us to an interesting perspective on the emergence and development of systems of innovation, namely that innovation systems are not natural givens, they have to be constructed and they are constructed around specific innovation problems. In the process the system and the problem co-evolve.

Here is a sketch of the argument. At the national level there are sets of knowledge accumulating, storing and transmitting capabilities, in universities, hospitals and research institutes. In medicine, these capabilities are connected by a range of informal and formal, national and international practitioner networks. However, these constellations of capabilities do not constitute an innovation system, they are at best a science system or a technological system in Carlsson's sense (1995). To translate latent capabilities into an innovation system requires the activities of for profit firms, focussed upon specific classes of innovation problem, for example, IOL devices and surgical techniques, for a specific purpose, to gain competitive advantage in the market process. Firms play the key role in constructing an innovation system, making the connections between different actors to focus attention on the solution to problems they define, and articulating and combining together the multiple bands of knowledge required for innovation. In its combinatorial role the firm is a unique organisation within innovation systems. Thus the development system for a particular class of problems is not there naturally, it is assembled within the competitive process and competition leads to connection and connection to collaboration. It is not simply that the innovation processes are distributed across multiple agencies and actors, it is much more that they are embedded in market selection processes and that the associated, very specific innovation system constitutes the external organisation of the firms. By virtue of this link with the competitive process we are dealing with rival innovation systems as fluid as the competitive processes that underpin them. Connections are made and broken as commercial advantage dictates. Such systems are certainly not monolithic, they are created, grow, stabilise and decline, and they involve, as in the IOL case, a subtle and changing interaction between public capability and private action. But that is the exactly the point, the link between competition and innovation is multi-faceted. Innovation creates diversity and diversity, in true evolutionary fashion, makes competition feasible. Competition in turn stimulates the search for innovation based advantage and in the process, we suggest, creates innovation systems from general capabilities. Thus the relation between the knowledge of growth and the growth of knowledge really is double-sided. But so then is the relation between the service and the manufacturing elements in this story, in this extended division of labour where the one ends and the other begins is not obvious. This particular service economy is not separable from the associated manufacturing economy and neither are the associated innovation processes.

6 Conclusion

I have suggested that the Schumpeterian perspective on the knowledge growth relationship cannot usefully be treated in macroeconomic terms, that the economy is an ensemble of connected elements not an aggregate entity. Rather an emphasis on the micro diversity of behaviours and their co-ordination by market and other instituted processes is the route to understanding the knowledge-growth relation and the transformations it implies. For this reason, I have argued that capitalist economies are restless, they never are, indeed never can be in equilibrium, and they are driven at root by experiments in novelty creation. Markets and other instituted arrangements provide the connections that influence the ongoing growth of knowledge. To understand their workings requires much more than knowledge of economic relationships narrowly defined and it requires greater recognition of the dominant role of service activities in the modern economy. This is a co-disciplinary endeavour, and the frameworks we deploy must be open to the cross fertilisation of ideas. Schumpeter might, I think, have approved, he was after all as much a sociologist, as he was an economist. He warned continually of the link between progress and the break with existing instituted patterns in economy and society. Our task is neither to copy, nor to imitate but to develop this idea.

References

Abramovitz M (1989) Thinking about growth. Cambridge University Press, Cambridge, MA

Aghion P, Howitt P (1999) Endogenous growth theory. Boston, MIT Press

Burns AF (1934) Production trends in the united states since 1870. New York, NBER

Carlsson B (1995) Technological systems and economic performance: the case of factory automation. Dordrecht, Kluwer

Dosi G (1997) Opportunities, incentives and the collective pattern of technological change. Economic Journal 107: 1530–1547

Douglas S (1987) Inventing american broadcasting 1899–1922. Johns Hopkins University Press, Baltimore

Eliasson G (1990) The knowledge based information economy. Industrial Institute for Economic and Social Research, Stockholm

Foster J, Metcalfe JS (eds) (2001) Frontiers of evolutionary economics. Edward Elgar, Cheltenham

Iwai K (1974) Schumpeterian dynamics Part II. Journal of Economic Behaviour and Organisation 5: 321–351

Jones C (1995) R&D-based models of economic growth. Journal of Political Economy 103: 759–804

Kaldor N (1954) A classificatory note on the determinateness of equilibrium. Review of Economic Studies 1: 122–136

Kuznets S (1929) Secular movements in production and prices. Boston, NBER

Kuznets S (1954) Economic change. Heinemann, London

Machlup F (1962) The Supply of Inventors and Inventions. In: Nelson R (ed) The rate and direction of inventive activity. NBER, Princeton University Press
Metcalfe JS (1998) Evolutionary economics and creative destruction Routledge, London
Metcalfe JS, James A (2000) Emergent innovation systems and the delivery of clinical services: the case of intra-ocular lenses. CRIC Discussion Paper, University of Manchester
Metcalfe JS, Miles I (2000) Innovation systems in the service economy. Dordrecht, Kluwer
Nelson R (1982) Research on productivity growth and productivity differences: dead ends and new departures. Journal of Economic Literature 12: 1029–1064
Popper KR (1996) A world of propensities. Thoemmes, Bristol
Prescott EC (1998) Needed: a theory of total factor productivity. International Economic Review 39: 525–551
Rymes TK (1971) On concepts of capital and technical change. Cambridge University Press
Usher D (1980) The measurement of economic growth. Blackwell, Oxford
Young AA (1928) Increasing returns and economic progress. Economic Journal 38: 527–542

Bringing institutions into evolutionary growth theory

Richard R. Nelson

School of International and Public Affairs, Columbia University, 420 W. 118th St., New York, NY 10027, USA (e-mail: rrn2@columbia.edu)

Abstract. Classical economics was both evolutionary and institutional. With the rise of neoclassical economics, both the evolutionary and the institutional aspects were squeezed out of main line economic theory. The last quarter century has seen a rebirth of both traditions, but as minority intellectual positions, and to a considerable extent separate ones. This essay argues the need for a rejoining of evolutionary and institutional economics, and suggests a way to bring the two strands together in a coherent way.

Key words: Evolution – Institutions – Technology – Growth

JEL Classification: B52, B15, N01, A10

1 Introduction

In this essay I lay out a way to bring economic institutions into evolutionary economic growth theory, that I think has considerable promise. Those of us who have been developing evolutionary growth theory have known for a long time that this needed to be done. The question was how to do it. I have been working on this question for a number of years, and now think I see a natural way to do it. This essay is basically a report on that work.

The economists who have been active in the development of evolutionary growth theory over the last twenty years have been motivated in large part by their perception that neoclassical economic growth theory, while assigning technological change a central role in economic growth, is totally inadequate as an abstract characterization of economic growth fuelled by technological change (Nelson and Winter, 1982) In particular, that theory represses the fact that efforts to advance technology are to a considerable extent "blind". This proposition does not deny the purpose, the intelligence, and the often powerful body of understanding and technique, that those aiming to advance technology bring to their work. But it seems always to be the case that different inventors and R and D teams lay their bets in different ways, and what will work best is virtually impossible to predict in advance. Hence industries and eras where technological advance has been rapid and cumulative almost always have been marked by a number of competing efforts and actors, with ex-post selection rather than forward looking planning determining the winners and losers. The broad notion that technological advance proceeds through an evolutionary process has been developed independently by scholars of technological advance operating in a variety of different disciplines including sociologists (Constant, 1980; Bijker, 1995), technological historians (Rosenberg, 1976; Vincenti, 1990; Petroski, 1992; Mokyr, 1990), as well as economists interested in modeling (Nelson and Winter, 1982; Metcalfe, 1998; Saviotti, 1996).

Needless to say, explicit recognition that technological advance proceeds through an evolutionary process leads one to formulate a growth theory that has a very different structure than that contained in neoclassical growth theory, new or old. However, for the most part evolutionary growth theory, like neoclassical growth theory, has not as yet taken on board the complex institutional structures that are characteristic of modern economies. For a discussion see Nelson, 1998.

On the other hand, sophisticated empirical scholars of technological advance always have understood that the rate and character of technological advance was influenced by the institutional structures supporting it, and that institutions also strongly condition whether and how effectively new technology was accepted and absorbed into the economic system. These themes are clear, for example. in David Landes' magisterial *Unbound Prometheus* (1970) and in Christopher Freeman's *The Economics of Industrial innovation* (1982). And recently, of course, the notion of a national or a sectoral innovation system, which clearly is an institutional concept, has played a significant role in theorizing about technological advance (see e.g. Lundvall, 1992; Nelson, 1993; Mowery and Nelson, 1999).

However, it seems fair to say that by and large modern evolutionary economists writing about technological change, and modern economists who have been stressing the role of institutions in economic development, have had little interchange. The principal purpose of this and kindred essays is to build a bridge between the two intellectual traditions, and to suggest a way they may be joined.

2 Institutional analysis and evolutionary economic theory: the historical connections

I want to begin by proposing that, before modern neoclassical theory gained its present preponderant position in economics, much of economic analysis was both evolutionary and institutional. Thus, Adam Smith's analysis [Smith, 1937 (1776)] concerned with how "the division of labor is limited by the extent of the market" and, in particular, his famous pinmaking example, certainly fits the mold of what I would call evolutionary theorizing about economic change. Indeed, his analysis is very much one about the co-evolution of physical technologies and the organization of work, with the latter, I would argue, very much a notion about "institutions". In many other places in *The Wealth of Nations*, Smith is expressly concerned with the broader institutional structure of nations, in a way that certainly is consonant with the perspectives of modern institutional economics. Karl Marx of course was both an evolutionary theorist, and an institutional theorist. If you consider the broad scan of his writing, so too was Alfred Marshall. Thus, evolutionary growth theorizing that encompasses institutions in an essential way has a long and honorable tradition in economics.

As neoclassical economic theory became dominant in economics, and increasingly narrowed its intellectual scope, both the institutional and the evolutionary strands of economic analysis came to become "counter-cultures". In some cases, they were intertwined. Thus, they certainly were in Veblen, and Hayek.

However, there was a tendency for the dissonant strains of institutional economic theorizing, and evolutionary economic theorizing, to take their own separate paths. Thus, in the United States, Commons (1924, 1934) helped to define the American institutional school. However, his analysis was not very evolutionary. Nor was the perspective of Coase (1937, 1960) who, later, had a major shaping role on "the new institutional economics". On the other hand, Schumpeter (1942), whose work arguably has provided the starting point for modern evolutionary economics, is seldom footnoted by self-professed institutionalists, despite the fact that Schumpeter was very much concerned with economic institutions. And Schumpeter's institutional orientation was ignored, as well, in the early writings of the evolutionary economists who cited Schumpeter as their inspiration.

Thus, what has been called the "new institutional economics," and the new evolutionary economics, have different immediate sources. And their focal orientations have been different. The orientation of institutional economics is toward the set of factors that mold and define human interaction, both within organizations, and between them. In contrast, much of modern evolutionary economic theorizing is focused on the processes of technological advance.

However, in my view at least, recent developments have seen the two strands coming together again, as Hodgson (1988, 1993) and Langlois (1989) long have argued should be the case. Thus, Douglass North (1990), perhaps today's best known economic "institutionalist," gradually has adopted an evolutionary perspective regarding how institutions form and change. And, as I noted earlier, many of the scholars who did the early work on the new evolutionary economics recently have become focused on such subjects as "national innovation systems", which is an institutional concept par excellence.

There certainly are strong natural affinities, in the form of common core assumptions and perceptions, between institutional economists, at least those in the school of North, and modern evolutionary economists. There also are very strong reasons more generally why they should join forces.

Both camps share a central behavioral premise that human action and interaction needs to be understood as largely the result of shared habits of action and thought. In both there is a deep-cutting rejection of "maximization" as a process characterization of what humans do. There also is a rejection of the Friedmanian notion that, while humans do not go through actual maximizing calculations, they behave "as if" they did, and, therefore, that behavior can be predicted by an analyst who calculates the best possible behavior for humans operating in a particular context. Thus, for scholars in both camps, patterns of action need to be understood in behavioral terms, with improvements over time being explained as occurring through processes of individual and collective learning. For economic evolutionary theorists, this exactly defines the nature of an evolutionary process.

Scholars in both camps increasingly share a central interest in understanding the determinants of economic performance, and how economic performance differs across nations, and over time. Modern evolutionary theorists focus centrally on what they tend to call "technologies." For evolutionary theorists, a country's level of technological competence is seen as the basic factor constraining it's productivity, with technological advance the central driving force behind economic growth. As noted, increasingly evolutionary economists are coming to see "institutions" as molding the technologies used by a society, and technological change itself. However, institutions have not as yet been incorporated into their formal analysis.

On the other hand, institutional economists tend to focus exactly on these institutions. Many would be happy to admit that the influence of a countries institutions on it's ability to master and advance technology is a central way that institutions affect economic performance. However, institutionalists have yet to include technology and technological change explicitly into their formulation.

The arguments for a marriage I think are strong. Below I map out what a marriage might look like.

3 Routines as a unifying concept

I begin by noting the essential function the notion of a „routine", or an equivalent concept, plays in modern economic evolutionary theory. As Sidney Winter and I have developed the concept (1982), the carrying out of a routine is "programmatic" in nature, and like a program tends largely to be carried out automatically. Like a computer program, our routine concept admits choice within a limited range of alternatives, but channelled choice.

Thus the routines built into a business firm, or another kind of organization that undertakes economic activity, largely determine what it does under the particular circumstances it faces. The performance of that firm or organization will be determined by the routines it possesses and the routines possessed by other firms and economic units with which the firm interacts, including competitors, suppliers, and customers. At any given period of time, many of the routines are largely common to firms in the same line of business, but some are not, and these latter, therefore, provide the stuff that determines how firms do relative to their competitors. The distribution of routines in an economy at any time determines overall economic performance. Under evolutionary economic theory, economic growth is caused by changes in the distribution of operative routines, associated both with the creation of superior new routines, and the increasingly widespread use of superior routines and the abandonment of inferior ones. The latter can occur through the relative expansion of organizations that do well, or the adoption of better techniques by organizations that had been using less good ones, or both.

As noted, most of the writing by evolutionary economists has focused on "physical" technologies as routines. However, the notion of a routine fits very well with the conceptualization of many institutional economists, if the concept is turned to characterize standardized patterns of human transaction and interaction more generally. Indeed, if one defines institutions as widely employed "social" technologies, in the sense I will develop shortly, it is easy to take institutions on board as a component of an evolutionary theory of economic growth.

In order to see what I am suggesting here, it is useful to reflect a bit on some important characteristics of productive routines. A routine involves a collection of procedures which, taken together, result in a predictable and specifiable outcome. Complex routines almost always can be analytically broken down into a collection of subroutines. Thus, the routine for making a cake involves subroutines like pour, mix, and bake. These operations often will require particular inputs, like flour, and sugar, and a stove. In turn, virtually all complex routines are linked with other routines that must be effected in order to make them possible, or to enable them to create value. Thus, a cakemaking routine presupposes that the necessary ingredients and equipment are at hand, and the acquisition of these at some prior date requires its own "shopping" routines. And still further back in the chain of activity, the inputs themselves needed to be produced, in a form that met the requirements of cake makers.

A key aspect of productive routines that I want to highlight here is that, while the operation of a particular routine by a competent individual or organization generally involves certain idiosyncratic elements, at its core almost always are elements that are broadly similar to what other competent parties would do in the same context. By and large, the ingredients and the equipment used by reasonably skilled bakers are basically the same as those used by other skilled cakemakers. And the broad outline of the steps generally can be recognized by someone skilled in the art as being roughly those described in *The Joy of Cooking*, or some comparable reference.

There are two basic reasons why productive routines tend to be widely used by those who are skilled in the art. The first is that great cake recipes, or effective ways of organizing bakeries, or for producing steel or semiconductor, tend to be the result of the cumulative contributions of many parties, often operating over many generations. This is a central reason why they are as effective as they are. Widely used routines are widely used because they are effective, and they are effective because over the years they have been widely used. To deviate from them in significant ways is risky, and while the payoffs may be considerable, there also is a major chance of failure.

The second reason is that particular routines tend to be a part of systems of routines. This systemic aspect forces a certain basic commonality of ways of doing particular things. The needed inputs tend to be available, routinely, for widely known and used routines. If help is needed, it generally is easy to get help from someone who already knows a lot about what is needed, and to explain the particulars in common language, if the routine involved is widely known and practiced. In contrast, idiosyncratic routines tend to lack good fit with complementary routines, and may require their users to build their own support systems.

4 Social technologies and institutions

In an earlier paper (Nelson and Sampat, 2000) where Bhaven Sampat and I developed many of these notions, we proposed that, if one reflects on the matter, the program built into a routine generally involves two different aspects: a recipe that is anonymous regarding any division of labor, and a division of labor plus a mode of coordination. We proposed that the former is what scholars often have in mind when they think of "physical technologies." The latter we called a "social technology," and proposed that social technologies are what many scholars have in mind when they use the terms "institutions." North and Wallis (1994) have proposed a similar distinction between physical and social technologies.

Widely employed social technologies certainly are defined by and define "the rules of the game," the concept of institutions employed by many scholars. Social technologies also can be viewed as widely employed "modes of governance," which is Williamson's notion (Williamson, 1985) of what institutions are about. And in the language of transaction costs, which is widely employed in the institutional literature, generally used "social technologies" provide low transaction cost ways of getting something done. As this discussion indicates, the concept of social technology is broad enough to encompass both ways of organizing activity within particular organizations – that is, the M form of organization is a social technology – and ways of transacting across organizational borders. Thus, markets define and are defined by "social technologies." So too are widely used procedures for collective choice and action.

This formulation naturally induces one to see prevailing institutions not so much as "constraints" on behavior, as do some analysts, but rather as defining the effective ways to get things done when human cooperation is needed. To view institutions as "constraints" on behavior is analogous to seeing prevailing physical technologies as constraints. A productive social technology (an institution) or a physical technology is like a paved road across a swamp. To say that the location of the prevailing road is a constraint on getting across is basically to miss the point. Without a road, getting across would be impossible, or at least much harder.

5 Institutions in an evolutionary theory of economic growth

The question of how institutions fit into a theory of economic growth of course depends not only on what one means by institutions, but also on the other aspects of that theory. I suggest that the concept of institutions as social technologies fits into evolutionary theories of economic growth very nicely.

5.1 Technological advance as the driving force

While these days almost all scholars studying economic growth see technological advance as a large part of the story, evolutionary theorists put special weight on technological advance. The reason is that, while neo-classical theory sees economic actors as facing a spacious choice set, including possible actions that they never have taken before, within which they can choose with confidence and competence, evolutionary theory sees economic actors as at any time bound by the limited range of routines they have mastered. Each of these has only a small range of choice. Further, the learning of new routines by actors is a time consuming, costly, and risky thing. Thus while neo-classical growth theory sees considerable economic growth as possible simply by "moving along the production function", in evolutionary theory there are no easy ways to come to master new things.

Put more positively, from the perspective of evolutionary theory, the economic growth we have experienced needs to be understood as the result of the progressive introduction of new technologies which were associated with increasingly higher levels of worker productivity, and the ability to produce new or improved goods and services. As a broad trend, they also were progressively capital using. Elsewhere (Nelson, 1998) I have developed the varied reasons for the capital using nature of technological change. Rising human capital intensity also has been a handmaiden to that process, being associated both with the changing inputs that have generated technological advance, and with the changing skill requirements of new technologies.

Within this formulation, new "institutions" and social technologies come into the picture as changes in the modes of interaction-new modes of organizing work, new kinds of markets, new laws, new forms of collective action– that are called for as the new technologies are brought into economic use. In turn, the institutional structure at any time has a profound effect on, and reflects, the technologies that are in use, and which are being developed.

I believe that the concept of institutions as social technologies, the routines language for describing them, and the theory sketched above of how institutions and institutional change are bound up with the advance of physical technologies in the process of economic growth, becomes more powerful, the closer the analysis gets to describing actual social technologies in action. Thus I turn now to two important particular developments in the history of experienced modern economic growth: the rise of mass production industry in the United States in the late 19th century, and the rise of the first science based industry-synthetic dyestuffs-in Germany at about the same time. Given space constraints, the discussion must be very sketchy, but I hope to provide enough detail so that one can see the proposed conceptualization in action.

5.2 The rise of mass production

As Alfred Chandler (1962) and other business historians tell the story, during the last parts of the nineteenth century and the first half of the twentieth, manufacturing industry in the United States experienced rapid productivity growth, associated with the bringing into operation of methods of production-new technologies or routines-that came to be called "mass production". These methods were accompanied by growing scale of plants and firms, rising capital intensity of production, and the development of professional management, often with education beyond the secondary level. However, these latter increases in "physical and human capital per worker", and in the scale of output, should not be considered as independent sources of growth, in the sense of growth accounting; they were productive only because they were needed by the new technologies.

At the same time, it would be a conceptual mistake to try to calculate how much productivity increase the new technologies would have allowed, had physical and human capital per worker, and the scale of output, remained constant. The new production routines involved new physical technologies which incorporated higher levels of physical and human capital per worker than the older routines they replaced. To operate the new routines efficiently required much larger scales of output than previously.

And they also involved new "social technologies". Chandler's great studies are largely about the new modes of organizing business that were required to take advantage of the new opportunities for "scale and scope". The scale of the new firms exceeded that which owner-managers plus their relatives and close friends could deal with, either in terms of governance or finance. The growing importance of hired professional management, and the diminished willingness of the original family owners to provide all the financial capital, called for the development of new financial institutions and associated markets. The need for professional managers also pulled Business Schools into being. More generally, the new industrial organization profoundly reshaped shared beliefs of how the economy worked, and came to define the concept of modern capitalism.

The development of mass production proceeded especially rapidly in the United States, in part at least because of the large size of the American market, but also because the associated new institutions grew up rapidly in the new world. In general Europe lagged. On the other hand, the rise of new institutions to support science based industry occurred first in Europe.

5.3 Synthetic dyestuffs

I turn now to consider another example: the rise of the first science based industry, in Germany, that occurred over roughly the same time period as did the rise of mass production in the U.S.

The basic story has been told by several scholars, but the account I draw most from here is that contained in the thesis by Peter Murmann (Murmann , 1998). Murmann's account is presented in standard language. The account presented here is "semi-formal" in the sense that it makes explicit use of the concept of routines, and the physical and social technologies involved in routines.

Several new routines play the key roles in the story. The first is a new "physical technology" for creating new dyestuffs, with university-trained chemists as the key inputs. This new physical technology came into existence in the late 1860s and early 1870s as a result of improved scientific understanding of the structure of organic compounds. The second key element in the story was the development of the "social technology" for organizing chemists to work in a coordinated way for their employer-the invention of the modern industrial research laboratory. The third element in our story is another social technology, the system of training young chemists in the understandings and research methods of organic chemistry. This social technology was university based, and funded by national governments. Finally, there are new markets with their own particular rules and norms. One links the firms interested in hiring chemists with the supply of chemists. Another market links dyestuff firms with users of the new dyestuffs.

Several different kinds of "institutionalized" organizations play key roles in our theoretical story. First, there are chemical products firms, of two types. The old type does not possess an industrial research laboratory, and achieves new dyestuffs slowly through processes that involve only small levels of investment. The other kind of firms, the "new" type, invests in industrial research laboratories, and because of those investments achieves new dyestuffs at a much faster rate than do old firms. There are two other kinds of organization in this story as well. One is national chemical products industry associations, who lobby government for support of university training. The other is national universities who train young chemists. National political processes and government funding agencies also are part of our story, but they will be treated implicitly rather than explicitly.

As noted, this account also involves specification of certain "institutionalized" markets, and the recognition that these markets differ somewhat from nation to nation. In particular, chemists have a national identity, and the firms do also. German chemists (I will assume that these all are trained in German universities) require a significantly higher salary to work in a British firm than in a German one, and British-trained scientists require more to work in Germany than in Britain. (Alternatively, the best of the national graduates would rather work in a national firm.) This means that, other things being equal, it advantages national firms if their national universities are training as many chemists as they want to hire.

There also are national markets for dyestuffs. The British market is significantly larger than the German market throughout the period under analysis. Other things being equal, British firms have an advantage selling in the British market, and German firms in the German market. However, the advantage of national firms can be offset if a foreign firm is offering a richer menu of dyestuffs. Under our specification, if a foreign firm does more R and D than a national firm, it can take away the latter's market, at least partially.

There are several key dynamic processes, and factors influencing them, in our story. To a first approximation, the profits of a firm, gross of its R&D spending, are an increasing function of the level of its technology, defined in terms of the quality of the dyestuffs it offers, and the volume of its sales. This first approximation, however, needs to be modified by two factors. One is that the profits of a firm that does R and D depend on whether the chemists it hires are national or not. The other is that, for a given level of the other variables, British firms earn somewhat more reflecting their advantaged location regarding the market.

R and D is funded out of profits, but not all firms invest in R and D. Firms can spend nothing on R&D (as do "old-style" firms), or they can invest a fraction of their profits in R and D (as do "new-style" firms). Initially, all firms are "profitable enough" to be able to afford a small-scale R&D facility. Some (the "new style" firms) choose to do so, and others don't. If the profits of a new style firm grow, they spend more on R and D.

Given the availability of the new R and D technology, it is profitable to invest in R and D and, given the competition from "new style firms", firms that do not do R and D lose money. This is so both in Germany and the U.K. In both countries a certain fraction of firms start to invest in R and D when the new technology arrives. These profitable firms expand, and the unprofitable ones contract. As firms that do R and D expand, their demand for trained chemists grows too. National firms hire nationally trained chemists first, and then (at higher cost) foreign trained chemists.

The supply of chemists provided to industry by universities is a function of the funding those universities receive from government. For a variety of reasons the supply of German chemists initially is much greater than the supply of British chemists. This initial cost advantage to German firms that do R and D is sufficient to compensate for the disadvantage regarding the location of the product market. And over time, the political strength of the national industry association, and the amount of money they can induce government to provide national universities, is proportional to the size of that part of the national industry that undertakes organized research.

Start the dynamics just before the advent of the new scientific understanding that creates a new technique for creating new dyestuffs. There are more (and bigger) British firms than German firms in this initial condition, reflecting their closeness to the large part of the market. No firm has an industrial research laboratory. The supply of chemists being trained at German universities is more than sufficient to meet the limited demands of German firms, and British firms, for chemists.

Now, along comes the new scientific technique for creating new dyestuffs. Some British firms and some German firms start doing industrial R&D on a small scale. They do well, and grow. The demand for university-trained chemists grows. Since most of the existing supply of chemists, and the augmentations to that supply, are German-trained, German firms are able to hire them at a lower price than can British firms. The German firms who invest in R&D do well, on average, relative to British firms, and their German competitors who have not invested in R&D. They grow, and as they do their R and D grows. The effectiveness of German university lobbying for government support of training of chemists increases as the German industry grows. You can run out the rest of the scenario.

6 Promise and challenges

I believe the conception of institutions as defining or shaping standard social technologies is coherent, broad enough to square with the concepts of insitutions proposed by other scholars, and particularly well suited to be brought into evolutionary economic growth theory. The concept of social technologies as standard ways that things are done when the doing involves interactions among different people or organizations pairs up nicely with the concepetion of physical technologies as recipe-like, but which is mute regarding the organization of labor.

In my view at least, the advance of physical technologies continues to play the leading role in the process of economic growth. In the example of the rise of mass production, social technologies enter the story in terms of how they enable the implementation of physical technologies. In the case of the rise of the industrial

R and D laboratory, new social technologies are needed to support activities that create new physical technologies. Perhaps a useful way of looking at this obvious interdependence is to posit, or recognize, that physical and social technologies coevolve. And this coevolutionary process is the driving force behind economic growth.

References

Basalla G (1988) The evolution of technology. Cambridge University Press, Cambridge
Bijker W (1995) Of bicycles, bakelites, and bulbs. Cambridge University Press, Cambridge
Chandler A (1962) Strategy and structure: chapters in the history of the industrial enterprise. MIT Press, Cambridge
Chandler A (1977) The visible hand: the managerial revolution in American business. Harvard University Press, Cambridge
Coase R (1937) The nature of the firm. Economica 4: 386–405
Coase R (1960) The problem of social cost. Journal of Law and Economics 3: 1–44
Commons JR (1924) Legal foundations of capitalism. Macmillan, New York
Commons JR (1934) Institutional economics. University of Wisconsin Press, Madison, WI
Constant E (1980) The origens of the turbojet revolution. Johns Hopkins, Baltimore
Freeman C (1982) The economics of industrial innovation. Pinter, London
Hayek F (1967) Studies in philosophy, politics, and economics. Routledge and Kegan Paul, London
Hayek F (1973) Law, legislation, and liberty, Vol.I Rules and order. Routledge and Kegan Paul, London
Hodgson G (1988) Economics and institutions. Polity Press, Cambridge
Hodgson G (1993) Economics and evolution: bringing life back into economics. Polity Press, Cambridge
Landes D (1970) The unbound prometheus. Cambridge University Press, Cambridge
Langlois R (1989) What was wrong with the old institutional economics (and what is still wrong with the new)? Review of Political Economy 1: 270–298
Lundvall BA (1992) National systems of innovation. Pinter, London
Metcalfe JS (1998) Evolutionary economics and creative destruction. Routledge, London
Mokyr J (1990) The lever of riches. Oxford University Press, New York
Mowery D, Nelson R (1999) The sources of industrial leadership. Cambridge, New York
Murmann P (1998) Knowledge and competitive advantage in the synthetic dye industry: 1850-1914. Columbia University Business School, New York
Nelson RR (1993) National innovation systems: a comparative analysis. Oxford University Press, New York
Nelson RR (1998) An agenda for growth theory: a different point of view. Cambridge Journal of Economics 22: 497–520
Nelson RR, Sampat B (2000) Making sense of instituions as a factor shaping economic performance. Journal of Economic Behavior and Organization (forthcoming)
Nelson RR, Winter SG (1982) An evolutionary theory of economic change. Harvard University Press, Cambridge
North D (1990) Institutions, institutional change, and economic performance. Cambridge University Press, Cambridge
North D, Wallis J (1994) Integrating institutional change and technological change in economic history: a transaction cost approach. Journal of Institutional and Theoretical Economics: 609–624
Petroski H (1992) The evolution of useful things. Knopf, New York
Rosenberg N (1976) Perspectives on technology. Cambridge University Press, Cambridge
Saviotti S (1996) Technological evolution, variety, and the economy. Edward Elgar, Cheltenham
Schumpeter J (1942) Capitalism, socialism, and democracy. Harper and Row, New York
Silverberg G, Dosi G, Orsenigo L (1988) Innovation, diversity, and diffusion: a self organizing model. Economic Journal 98: 1032–1054
Smith A (1937 [1776] The wealth of nations. The Modern Library, New York

Soete L, Turner R (1984) Technology diffusion and the rate of technical change. Economic Journal 94: 612–623
Veblen T (1915) Imperial germany and the industrial revolution. Macmillan, New York
Veblen T (1919) The place of science in modern civilization and other essays. Huebsch, New York
Vincenti W (1990) What engineers know and how they know it. Johns Hopkins Press, Baltimore
Williamson O (1985) The economic institutions of capitalism. Free Press, New York

Choice, chance, and necessity in the evolution of forms of economies

Pavel Pelikan

Royal Institute of Technology, Department of Industrial Economics and Management, S-100 44 Stockholm, Sweden (e-mail: p.pelikan@lector.kth.se)

Abstract. Whereas ants have the constitution of their anthills prescribed by their genes, humans appear free to choose the form of their societies themselves. Ex post, however, this freedom turns out strongly limited by severe performance tests that human societies, to avoid crises and disintegration, must be able to pass. As humans knew nothing and still know only little about which forms of their society have this ability, they have been forced to search for such forms by imperfectly informed trials and possibly costly errors, and thus run, without fully realizing it, a kind of higher-level Darwinian evolution. Narrowing attention from the forms of human societies to the forms of their economies, this essay searches for principles and regularities of this evolution the knowledge of which, if understood and applied, could avoid at least some otherwise likely errors in the future.

Acknowledgements. I thank Luciano Andreozzi, Leonard Dudley, Dan Johansson, Geoffrey Hodgson, J.S. Metcalfe, Bedrich Velicky, and participants of a seminar at the Center for Theoretical Study, Charles University, Prague, for valuable comments on earlier drafts. The usual caveat applies. The work on this paper was partly supported by CTS Research Project MSM 110000001.

1 Introduction

Like ants, humans need to live in societies, but unlike ants, they are not told by nature which ones. Whereas each ant is given the constitution of its anthill as part of its genetic endowment, humans appear free to choose the form of their societies themselves. The much greater richness of their genetic endowment includes indeed the abilities to create and to adapt to a great variety of societies. Although one may speculate that this variety may be genetically constrained – much like the variety of human languages is seen constrained, following Chomsky (1967), by a genetically determined universal grammar – historical evidence demonstrates that, at least for limited periods of time, it is enormous.

In the long run, however, much of this freedom turns out not to be real, as many of the forms of societies which humans could create and live in for more or less long periods in the past have proved unable to last. More than by genetic constraints ex ante, nature thus limits the variety of forms of human societies ex post, by various performance tests, which each society, to avoid crises and disintegration, must be able to pass. A decisive role is played by the tests of economic performance, which the current demographic growth, globalization, and increasing scarcity of basic resources in general make increasingly severe, and thus the freedom increasingly limited.

For a long time, however, humans knew nothing, and even now they know only little, about which forms of their societies may, and which ones cannot, lastingly succeed. They have therefore been forced to search by poorly informed trials and often costly errors for the scarce forms able to so. They have thus extended, without fully realizing it, the Darwinian evolution of forms of life to the forms of their economies and societies. It is, in other words, as if nature had in mind a rather narrow range of forms of human society, but instead of informing us directly (as it does with ants), it forces us to guess, and hits us with crises and misery if we guess wrong.

While nothing can yet be considered certain, some societies now clearly appear much closer to guessing right – at least as far as the form of their economy is concerned – than others. The wrong guesses, many of which can still be found at work in Third World economies, also turned out to include several ingredients of the First and Second Worlds economies that many people, including many theoretical economists, used to believe promising – in particular national planning, extensive welfare states, and selective industrial policies. What now appear to be the main ingredients of the right guesses – as indicated by the long period of growing prosperity of those economies which have used them most – are private enterprise and market competition, including competition in financial markets. But much uncertainty still remains about other ingredients – in particular about the extent and the contents of the role of government, if it is to solve those problems that markets cannot solve, without hindering markets in solving all the problems which they can solve.

In several earlier papers I used a simple evolutionary analysis to try to foresee, at least roughly, which forms of economy may be successful, and which ones cannot.[1] In this paper I wish to examine, by similarly simple means, how such forms evolve over time. This evolution can be seen to depend upon three main

[1] See, e.g., Pelikan (1988, 1989, 1992, 1993). My results were only verbal, and thus not very precise (a more recent attempt to increase their precision is in Pelikan, 1999). Nevertheless, they proved well centered on what turned out later, after the collapse of real socialism and the Japanese financial crisis, to be the right answers. The simple evolutionary comparison of alternative economic institutions for the scope of entrepreneurial trials that they allow and encourage, and for the rigor and speed of the correction of errors that they enforce, foresaw quite well what now empirically turn out to be the main difference between the right and the wrong guesses, mentioned in the previous paragraph. Let me add that little of this difference was seen by the much more precise standard analysis of that time; many of its mathematical models were on the contrary proving the optimality of the wrong guesses.

factors: (i) human cognitive abilities, including abilities to create and to learn; (ii) the cultural heritage of different societies; and (iii) the ultimate sanctions of nature in the broadest sense, which also includes human nature and all the actually evolved and currently competing human societies. The first two factors supply the inputs for the guessing, while the third is an ex post feedback through which the actually implemented guesses are tested, and either more or less temporarily accepted – which allows the guessing to take a pause – or rejected – which causes a crisis and makes a new guess (reform, transition) necessary. While not all new guessing need be enforced by crises – people often try to improve even upon economies which work reasonably well – the feedback is always there, ready to hit whenever a guess, be it enforced or voluntary, proves to cause serious malfunctions.

In existing theories of societal change, the three factors appear to attract unequal attention. While the first two are increasingly discussed,[2] the ultimate sanctions of nature are often underestimated or neglected. One reason may be that most authors of these theories come from prosperous capitalist economies which have not been hit by serious sanctions of this type for a long time, so that they may tend to forget how devastating these sanctions can be. In this paper, I wish to decrease this inequality by putting these sanctions in the center and showing that, in the evolution of human societies, the last word is always theirs.

This will have interesting implications for the other two factors. When the importance of sanctions of nature is correctly appreciated, our cognitive abilities will turn out to have limits which, in the evolution of forms of economies, make them often comparable to the sources of random (uninformed) mutations in the evolution of life – however informed and rational we might believe them ourselves. Although cultures will still be recognized to constrain both the actual form and the possible reforms of economies, and thus make the evolution of each economy path-dependent, this path-dependency will turn out to be weaker and less irrevocable than usually believed. Namely, many cases of path-dependency will be found inevitably and often brutally interrupted by requirements of nature's performance tests: a culture which prevents its economy from evolving into and preserving a reasonably efficient form, able to cope with both scarcity of resources and competing economies, will itself prove unable to last – unless subsidized and thus artificially kept alive by more efficient friends, at *their* risk and expense.

The ultimate sanctions of nature will thus be found to imply severe *objective* constraints upon the evolution of forms of economies, which, in the long run, decrease the importance of chance and of culturally conditioned *subjective* interests and beliefs. In other words, the process of societal change will be found to be based on objective fundamentals, which determine a more or less rich, but nevertheless limited, variety of sustainable forms of economies. That what will be determined is a variety, and not a single form, deserves emphasis, to make it clear that historical determinism will not be advocated.[3]

[2] See, e.g., North (1990, 1999, 2000); Grief (1994); Denzau and North (1994); Vanberg (1994); and Knight and North (1997).

[3] Note that the argument that the future of societies depends upon the performance of their economies is indeed also the basis of the Marxist historical determinism. But, while Karl Marx deserves credit for being one of the first students of the evolution of economies and societies, he was grossly mistaken in at

The unpleasant problem is that there are no corresponding ex ante constraints to prevent policies driven by culturally conditioned subjective interests and/or false beliefs from implementing forms from outside this variety. Although such forms cannot last – the constraints of nature (in the present broad sense) will sooner or later force humans to change them – the social costs of such changes may be very high. This raises the important but so far little examined questions of what the constraints of nature are, and how they can best be understood by theory and respected by policy, to avoid or minimize experiments with unsustainable forms of economies. Without aspiring to answer these questions in any detailed and definite way, the main purpose of the following pages is to call attention to them and consider how they might fruitfully be addressed.

The paper is organized as follows. Section 2 clarifies the main concepts used and finds good reasons why, in an evolutionary perspective, the form of economies is best characterized by their institutions. Section 3 outlines how these concepts relate to each other. Section 4 discusses the conditions of sustainability of institutions, considering both their objective evaluation by nature, according to the real economic performance to which they may lead, and their subjective evaluation by the human agents concerned, expressed by the political support they may obtain. As nature is found to put a growing premium on adaptability, Section 5 examines which parts of an economy may have to be adaptable, and finds out that they need not include its institutions, provided that these have a sufficiently high adaptation potential, in the sense that they provide for sufficient adaptability of organizations and outputs. Section 6 shows that the institutions with the highest adaptation potential, which in severe and variable environments are also the only sustainable ones, belong to the category that can globally be denoted as 'modern capitalism.' Section 7 considers the cultural prerequisites for the forming and maintaining of such institutions, and draws from it conclusions about the limits of cultural relativism and the first task of development policies.

2 Concepts

The present argument is built around three main concepts: *culture*, *institutions*, and *organizations*. All are parameters which condition and constrain human actions in the short run, but may themselves change as a result of some of these actions in a longer run. In agreement with North (1990) and following my earlier uses of these terms (Pelikan, 1987, 1992, 1995), they are understood as follows.

Institutions are humanly devised rules that constrain human decisions sets. They can be said to shape human actions and interactions, and can be compared to 'the rules of a game.' In addition to rules that concern all agents, such as property rights, the institutions of an economy also define the instruments of economic policy that government is allowed or required to use.

least two respects: (i) the claim that this evolution is a deterministic process with a unique outcome, and (ii) the prediction, based on a distorted and distorting labor theory of value, that capitalism will perish and the outcome will inevitably be a form of socialism or communism – which now turns out to be the least likely thing to happen. For a brief assessment of Marx's story, fully compatible with the present view, see North (1990: 132–133).

According to their origins, institutions can be divided into two types: *formal*, consisting of politically determined written laws; and *informal*, consisting of culturally evolved and often unwritten custom, such as moral norms or religious taboos. An important difference between the two is in the speed at which they can change: while a determined policy-maker (legislator, reformer) can change formal institutions overnight, informal ones cannot but evolve relatively slowly. In consequence, they often constitute a binding constraint on how fast effective institutional change can proceed.

A *culture* includes, in addition to its specific informal institutions, specific *values* and *beliefs*. It thus not only determines an important part of the rules of the game, but moreover strongly influences how the game is actually played: its values influence which objectives (preferences) the players pursue, and its beliefs influence which part of the information available they are able and willing to employ.

A set of interconnected beliefs – possibly expressed in terms of mental models, religious dogma, or theories – can be referred to as an *ideology* (cf. North 1990:23). Ideologies can thus be seen as substitutes for, or competitors with, knowledge, providing answers to questions that science cannot answer, or has not yet answered. They may be flexible (open) or rigid (closed), depending on whether or not they allow beliefs to be qualified or replaced by newly gained knowledge. A culture may be tolerant or intolerant, depending on whether it admits several parallel ideologies, or only one. A tolerant culture is a prerequisite for political democracy, which includes the task of organizing a civilized competition among different ideologies.

All these concepts can be seen at work in *organizations*. An organization is defined as a set of interacting agents, whose actions are constrained (shaped) by its institutions, and whose choices of the specific actions actually taken depend on their preferences and cognitive abilities, conditioned by the values and beliefs (ideologies) of its culture. In other words, if institutions are seen as the rules of a game, an organization is an interrelated collection of the players playing this game.[4]

As noted, the present focus is on *economic* organizations, defined, as usual, to consist of *economic* agents – such as producers, consumers, investors, and savers – engaged in production, allocation, and the use of scarce resources. But, and this is perhaps less usual, economic organizations are understood here broadly to include not only relatively tightly and deliberately organized firms and government agencies (economic organizations in the narrow sense), but also less tightly and more spontaneously organized markets and entire economies.[5]

This means that a national economy is seen as an economic organization the agents of which are constrained by national institutions, and which contains a number of smaller economic organizations – such as firms, markets, and government

---

[4] To make such a sharp difference between institutions and organizations is the first necessary step to making institutional analysis operationally clear. In most common languages and older institutional economics, 'institutions' may mean both 'rules' and certain lasting 'organizations' - such as central banks, ministries, or universities – which cannot but lead to analytical confusion.

[5] In many economic problems, of course, distinguishing firms, government agencies, and markets from each other is of prime importance. Here, however, we need to begin with a more elementary distinction between the level of actors, where things actually happen, and where all the three can be seen to belong, from the level of rules which constrain, and thus shape, this happening.

agencies. Each of these organizations has internal institutions of its own, but is constrained, both in its actions and in the choice of its institutions, by certain national institutions – such as the corporate law which constrains the choice of the rules of corporate governance within a firm.

All the (smaller) organizations that an economy actually contains, together with their internal institutions and their market and/or non-market interrelationships, will be referred to as the economy's *organizational structure*. Intuitively, this structure can be visualized as the 'working body' of the economy. While in standard analysis, organizational structures are usually assumed constant – e.g., all the markets and firms studied are assumed once for all given – in the study of societal change they are important variables. It is the actual state of this variable – in particular the quality and the quantity of the actually existing firms, the extent and the transaction costs of the actually existing markets, and the abilities of the actual government agencies – that determines the range of possible outputs, or in other words, implies the economy's performance.

In the study of the evolution of forms of economies, the first question is, which concept can best characterize such forms. To concentrate on the most important evolutionary changes, without being distracted by less important short-term fluctuations, we need a concept that does not change very often, and may thus remain stable for relatively long periods, during which other economic variables may keep changing. In static analysis, in which organizational structures are assumed constant, the form of an economy is usually characterized by this structure, often called an 'economic system' – such as a given set of firms and households interconnected by a given set of markets, or by a given mechanism of planning. But evolutionary analysis, in which organizational structures turn out often to change – e.g., by entry and exit of firms, rise and decline of industries, and opening, enlarging, or closing of markets – needs another concept, for which the institutions of economies appear to be the best candidate.[6] By determining the forms of property rights and the instruments of economic policy, institutions indeed determine all the features by which forms of economies are usually characterized – such as 'liberal capitalism' with a minimum of government intervention, or various forms of 'mixed economies' with different types and degrees of government control, and industrial and/or welfare policies. They may also remain stable for relatively long periods of time, while the organizational structure under them, within the more or less broad limits implied by their rules, may keep changing and developing, and thus more or less successfully adapting to changes in technologies and/or in relative prices. Although institutions also may, and from time to time indeed do, change, these changes, beside being in average slower that those of organizational structures, also appear to contain principles and regularities which can be more easily extracted and comprehended

[6] To characterize the form of an economy by its institutions is in agreement with North (1990) and Hayek (1967, 1973), who refers to what is called here 'institutions' by the terms 'general rules' and 'the order of rules' (cf. also North, 1999). I advocated this characterization in Pelikan (1988, 1992), where I also showed that characterizing the form of economies by their institutions instead of organizational structures formally corresponds to characterizing living organisms by their genotypes instead of the more variable phenotypes, as modern biology is indeed doing.

by theoretical analysis. The evolution of the forms of economies is thus understood to be the story of *their* changes.

3 Interrelations

An economy's culture, institutions, and organizational structure are interrelated in multiple ways. As noted, the culture directly determines all the informal institutions, and strongly influences, by its values and beliefs, the preferences and the cognition (rationality) of human individuals, and thereby the actual behavior of the economy's agents under both the formal and informal institutions. The culture thus determines an important part of the rules of the economic game, and strongly influences the ways in which the rules are actually exploited and the game is played.

In addition to the roles of economic agents, human individuals may, and usually do, assume the roles of political and cultural agents. Among the actions they may take in these roles, the present focus is on the innovations by which they contribute to changing the institutions and/or the culture. Legislating changes of formal institutions is an example of political innovation, while introducing or imitating new informal rules of conduct, and modifying or refuting received beliefs are examples of cultural innovations. Such innovations are the micro-causes of societal change; it is by them that societal change is driven.

Although the three roles are often considered separately, each by its specialized social scientists, they must be recognized as being linked by the prevailing culture, which conditions the preferences and the cognitive abilities of every individual in all these roles. This means that in addition to influencing how people as economic agents play the economic game, the culture also influences both how they as political agents try to change formal institutions and how they as cultural agents try to change the culture itself.

It is instructive to regard culture, institutions, and organizational structure as a chain of successively hardening constraints. This may perhaps best be seen by starting from the end of the chain, with the economy's output. As noted, this is directly constrained by the economy's organizational structure: each specific structure implies a certain limited range of feasible outputs. To obtain more or better outputs, the structure must first itself be changed – e.g., by the entry of new firms and the reorganization or exit of some of the incumbent ones.

The ways in which the structure can change, and thus develop or decay, are in turn constrained by the prevailing institutions – such as the institutional conditions for entrepreneurship, which include corporate, labor and competition laws, and the extent of allowed or required industrial subsidies. While some institutions allow or even encourage extensive structural changes, others make many such changes difficult or impossible. If the performance cannot be improved without a substantial structural change, and if this is prohibited by the prevailing institutions, the performance cannot be improved without first changing these institutions.

In the next link of the chain, the variety of institutions with which an economy can be endowed is constrained by the prevailing culture; for instance, this can exclude certain forms of property rights or economic competition for religious or ideological reasons. Cultures also differ in the severity of these constraints: some allow a wide variety of alternative institutions, while others are tightly linked to institutions of a specific type.

The final link is self-enclosed: the variety of feasible changes of a culture is constrained by the culture itself. More precisely, each culture, to preserve itself over time, must contain an essential core which it must effectively protect – otherwise it could not last long as an identifiable entity. Usually one of its core beliefs is that all of its core beliefs are eternal truths which must never be put in doubt. In consequence, cultures may differ not only in their actual contents, but also in how these contents are partitioned between the protected core and the modifiable parts. Depending on how tolerant a culture is – in other words, how many different ideologies and types of scientific knowledge it may accommodate and allow to compete – its protected core may be relatively small and many of its parts may therefore be modifiable by learning, or the core may be large, and most of it may thus be dogmatically kept unchanged.

In addition to this chain of influences leading from culture through institutions and individual behavior to the economy's organizational structure and performance, there are also important feedback influences leading in the opposite direction. The main feedback is the impact of the economy's performance upon the psychical and physical health and well-being of the members of the society. It is this feedback which is seen to constitute the 'ultimate sanctions of nature.'

This real-term feedback is accompanied by several informational feedbacks, through which its actual and future states are more or less incompletely signalled, and whose signals are interpreted and acted upon by the agents – again depending on their culturally conditioned values and beliefs. The signalling is often based on different conventional, more or less aggregated performance criteria or indicators – such as the growth of national product, the income per capita, the level of literacy, or the expected length of life – which usually represent a compromise between what is valued and what is possible to measure.

Many innovative actions can indeed be understood as deliberate responses to such signals – such as policies or economic reforms trying to cure an actual crisis or to prevent a threatening one. But societal change may also be driven by innovative actions based on arbitrary fantasies – such as beliefs in various utopias – and the difference between the two may not be very sharp. If, as is usually the case, the signalling is incomplete and/or the innovators' cognitive abilities are limited or culturally biased, what the innovators may subjectively believe to be deliberate rational responses may objectively be close to arbitrary fantasies, whose role in the evolution of economies is not very different from the role of random mutations in the evolution of species – i.e., they are necessary to keep the evolution going, but most of them turn out to lead in the wrong direction.

There is a subtle relationship between the subjective, socially chosen performance criteria, which are part of the information feedback and thus guide the members of society in their endeavors, and the objective criteria of nature, which determine the ultimate sanctions. The basic point is that the former criteria must be reasonably close proxies of the latter, if the economy is not to be hit surprisingly by sanctions of the latter while apparently performing well according to the former. For instance, while its national product and the income per capita may be growing, a large part of its population may fall sick or start dying, because of stress, lack of exercise, or an unhealthy diet. This point is important to realize, to avoid the widespread fallacy that socially chosen performance criteria are the only ones that matter, and that the valuation of performance of economies is therefore purely a matter of culturally conditioned social conventions.

A qualification is in order. To the extent that socially chosen performance criteria may become internalized in human minds, they may also influence real feelings of satisfaction, and thereby also the objective criteria of nature: as is well known from elementary psychology, if some subjective criteria are believed important, the failure to meet them can objectively affect people's physical and mental health. But these influences have strict limits: nature imposes many criteria which no cultural conditioning can change. Thus, most people cannot be convinced to abstain from demanding food, shelter, medical care, and probably also a minimum of personal freedom; the few deviants who might be convinced to do so would be unlikely to survive for long and could definitely not provide the basis for a successful human economy.

Unfortunately, however, nature declines to inform us not only about the form of a successful human society, but even about the success criteria that it requires such a society to meet. Our situation is thus very much like the one of a student who has to pass an examination without fully knowing what it may be about. Our quest for a successful form of economy cannot be only about the means, but much of it must also concern the ends.

4 The double feedback in the selection of human institutions

To understand the evolution of institutions, we need to know when an economy's institutions may remain stable, and when a more or less radical institutional change (reform, transition) becomes inevitable.

The answer involves a double feedback: (i) economic, or external, which consists of the real-term economic performance to which the institutions lead; and (ii) political, or internal, which transmits the approval or disapproval of the institutions by the members of the society. Thus, to be sustainable, an economy's institutions must both provide for a minimum objective economic performance, which would allow the members of the society to keep in reasonably good physical and mental health, *and* satisfy the members' subjective, culturally conditioned values and preferences, so that they will not feel frustrated and consequently try to reject the institutions by political means.

It is again instructive to recall the example of anthills and note that the sustainability of their institutions depends only on feedback (i): as the institutions of anthills are genetically given, ants cannot reject them and not even put them in doubt. In contrast, the strong dependence of human institutions upon feedback (ii) substantially complicates their evolution, opening it to the risk of repeated conflicts and crises.

One reason is that feedback (ii) alone is complex. Depending on their culturally conditioned values and cognitive abilities, people may judge, and possibly feel dissatisfied with, the performance of their economy, or its institutions, or both. As the ignorance of the links between the two is still widespread, even among theoretical economists, the two judgements may clash: desired institutions may lead to undesired outcomes and vice versa. Moreover, because of the above-mentioned imperfect knowledge of nature's ultimate criteria of economic performance, feedback (ii) may clash with feedback (i): not only may desired institutions fail to deliver desired performance, but even if delivered, such performance may not be very wise from the point of view of feedback (i) – for instance, it may turn out to harm individual health and/or undermine the economy's future performance.

The possibility of such clashes can be illustrated by Schumpeter's (1942/76) argument that capitalism will fall not because of an economic crisis, as Marx claimed, but because it will be increasingly disliked, in particular by influential intellectuals, and therefore, sooner or later, politically rejected. But now when all forms of socialist economies have proved, contrary to what Marx believed and Schumpeter argued, prohibitively inefficient, and their transition to forms of capitalism appears to be the only effective remedy, it must be concluded that new socially disastrous clashes of the two feedbacks can be prevented only if Schumpeter could be proven wrong also about capitalism's unpopularity, and some of its economically successful forms could gain sufficient intellectual and political support.[7]

Another illustration is Hayek's (1976) argument that the social justice that should be valued is the 'procedural' one of rules, and not the 'substantial' one of outcomes. This argument implies that societies can succeed economically only if their institutions offer efficient incentives and just rewards for contributions to production and productivity increases. Such institutions should therefore be positively valued, even if they turn out to lead – because individuals in all societies inevitably differ in their abilities and effort – to a highly unequal income and wealth distribution. But socially damaging clashes of the two feedbacks are again likely, for large wealth inequalities are more or less negatively valued in all known human cultures. The justice of outcomes can thus hardly escape valuation and, if this is sufficiently negative, the institutions that lead to it – even if they were the only ones able to save the society from economic collapse – can hardly resist political rejection.[8]

In feedback (i), the main ingredients are efficiency and adaptability. With a slight extension of the usual meaning, the former refers to the ability of an economy to obtain all the private and public goods needed for keeping its members in reasonably good physical and mental health by exploiting its current environments – in particular the available natural resources and the terms of trade with other economies. The latter means the ability to maintain the former ability when the environments change.

[7] Another argument of Schumpter's was that capitalism would erode the moral norms (informal institutions) on which it is based. But even this argument may be considered disproved: in the light of the strongly negative influences of socialism and extensive welfare states on working morale and civic honesty, capitalism proves superior even in this respect.

[8] When arguing in favor of procedural justice and against substantive justice, Hayek can be said to have abandoned the position of a positive scientist, who must recognize that people of all cultures value the distribution of outcomes, to turn into a social reformer trying to convince them not to do so.

The working of this feedback depends on both properties of an economy and properties of its environments. How efficient and adaptable an economy must be, in order to avoid being forced by crises and threats of disintegration into a radical institutional change, depends on two properties of its environments: their *severity*, meaning the costs of obtaining the needed goods, and their *variability*, which can be expressed in terms of the frequency, the amplitude, and the regularities (or irregularities) with which these costs may keep changing. The severity puts demands on the economy's efficiency: the higher the costs, the less inefficiency feedback (i) admits. The variability puts demands on the economy's adaptability: the more the environments change, and the more they do so in irregular and thus unpredictable ways, the more adaptability is required.

While economists have mostly been concerned with an economy's efficiency in assumedly constant environments, the present argument is that, in the long run, the main criterion of feedback (i) is the requirement of adaptability. At first sight, there may seem to be both historical and logical reasons to doubt this argument. Historically, several highly rigid economies proved able to exist for millennia. Logically, adaptability is only required in one of four types of environments: those that are both severe and variable. If the environments are generous, inefficiency due to poor adaptation is only weakly penalized, and if they are severe but stable, it suffices to obtain the required efficiency only once, after which it can be routinely repeated without further changes. But both doubts have a simple answer: the growing world population, the increasing scarcity of vital resources, and the consequent need to keep producing and adapting to technological innovations make the other three types of environments increasingly unlikely to be encountered, which makes rigid economies increasingly unlikely to last.

5 Conditions of adaptability of economies

The requirement of adaptability of an economy leads to the question of what exactly must keep adapting when environments are changing. As opposed to the popular view that changing environments require changing institutions (see, e.g., North 1990, 2000), I shall argue that the economy's institutions may remain stable, and thus spare the economy the high costs and risks of forced institutional changes, provided that they have certain suitable properties. My starting point is the elementary but often forgotten principle that can roughly be put as follows: *if something is regularly changing, there must be something else which realizes the regularity and itself remains stable.*[9]

More precisely, the ability of a variable or a parameter suitably to change in response to changes of another variable or parameter requires at least temporary

[9] In slightly different words, this principle is discussed by Hofstadter (1979: 686), who speaks of 'modifiable software' which must always be based on 'inviolate hardware.' A clear empirical example is the human brain, whose high and multilevel flexibility and adaptability – such as the ability to learn to learn – is ultimately due to the genetic recipes for building it, contained in the genetic endowment of homo sapiens (the brain of apes is substantially less adaptable!), which have remained quite stable during several tens of thousands of years, and which remain constant for each individual during his entire life.

stability of yet another parameter, namely the one realizing the response relationship (function, routine, regularity) which makes such suitable changes feasible. If furthermore this relationship is to change and adapt, there must again be another, higher level parameter which must remain relatively stable, to make such a higher level adaptability feasible. Formally: if y is suitably to change in response to changes of x, say $y = f(x)$, the suitable $f(.)$ must remain relatively stable; and if, in a longer run, $f(.)$ is suitably to change in response to changes of z, say $f(.) = G(z)$, the suitable $G(.)$ must again remain relatively stable.

This calls for a sharp distinction between the ability of a parameter suitably to change its own state, and its ability to provide for suitable changes of other variable(s) or parameter(s). Let me reserve the term 'adaptability' for the former ability, and denote the latter as 'adaptation potential.' Thus, adaptability is the ability of variable parameters to vary themselves, while adaptation potential is the ability of actually stable parameters to provide for adaptability of other variable(s) or parameter(s).

The case of an economy's institutions is both important and instructive. While considered in more detail below, to see clearly the difference between the two abilities, it may be helpful to realize already now that *the higher the institutions' adaptation potential, the less they need to be adaptable.* [10]

More generally, the present framework distinguishes four levels of parameters or variables within an economy that, to allow it to cope with changing environments, may need to be adaptable, or have a high adaptation potential, or both: (1) the qualities and quantities of the economy's output; (2) the organizational structure producing the output; (3) the institutions, which shape both the working and the forming of the structure; (4) the culture, which includes the informal part of the institutions and the values and beliefs that guide individuals in their economic and non-economic behaviors.

The main relationships can be summarized as follows: the adaptability of the output depends on the adaptation potential of the organizational structure, the adaptability of the structure depends on the adaptation potential of the institutions, and the adaptability of the institutions depends on the adaptation potential of the culture. As the constraints upon the adaptability of a culture are also features of this culture, the adaptability of a culture depends on its adaptation potential vis-a-vis itself.

[10] It is this difference that appears overlooked in North (2000). He correctly notes that the changes of an economy's environments may be both important and unpredictable, but appears to believe that, to avoid a crisis, such changes require correspondingly important changes of the economy's institutions. Although some environmental changes may indeed be so important that even previously successful institutions must be changed, the link between environmental changes and institutional changes is less direct. The present point is that much depends on the institutions' adaptation potential: highly adaptive institutions with a high adaptation potential such as the US ones, allow their economy to adapt to a broad range of environmental changes, within which they may thus themselves remain stable, whereas institutions with a low adaptation poential, such as those of many less developed economies, prevent adaptations to much more limited environmental changes. Such institution must indeed be adapted (reformed, transformed) if a crisis is to be avoided, as soon as an actual environmental change exceeds these narrow limits.

What an economy always needs to keep adapting is its output. Whether it also needs to adapt its organizational structure depends on how much the outputs are required to vary compared to the adaptation potential of the structure. If the required output changes stay within the limits of the structure's adaptation potential, the structure need not be adaptable. This case, however, appears increasingly rare. Because of the limited flexibility of many firms to change their products and methods of production, it is unlikely that all the changes of outputs required by rapid technological progress and important changes in relative prices could be achieved without some changes of the structure, including continuous entry and exit of firms ('creative destruction'). In other words, the required adaptability of outputs often exceeds the adaptation potential of existing organizational structures – including the Japanese one, which many economists for a long time believed to be the miraculous exception – so that some structural adaptability is always required.

This requirement, in turn, raises the question of whether the economy's institutions need to be adaptable. Much as in the previous case, this answer again depends on how the adaptability required compares to the adaptation potential providing for it. But the result appears less universal. In contrast to all the known organizational structures, of which none appears to have a sufficiently high adaptation potential to provide for sufficient output adaptability, institutions appear more promising, in the sense that the adaptation potential of some of their forms may provide for sufficient structural adaptability even in widely and irregularly changing environments. The problem is that such institutions are still far from widespread, but remain limited to advanced capitalist economies. For example, the adaptation potential of US institutions has been so high that they have allowed the structure of US economy to keep adapting successfully to widely changing terms of trade, while themselves remaining basically stable for over a century (counting from what may be considered the last substantial changes: income tax and antitrust law). By contrast, the institutions of less developed economies have a much lower adaptation potential, which causes structural rigidities and, as argued below, is indeed the main obstacle to their development.

The requirement of adaptability of institutions is thus selective, limited to those with an insufficient adaptation potential. For them, the inquiry continues to the prevailing culture, raising the question of whether its adaptation potential is high enough to allow them to increase this potential by a suitable change. If yes, the culture may be maintained, if not, it must itself change, and is therefore itself required to be adaptable. But here the story comes to its end: as a culture is defined to include the constraints on its own adaptability, if it is to be adaptable, it must have a sufficiently high self-adaptation potential. This means that those of its parts that are required to adapt must belong to its periphery modifiable by learning, and not to its stubbornly self-protected core. Otherwise the culture cannot adapt, and neither can, in consequence, the economy's institutions, structure, and outputs. It then helps little to complain that the environments are no longer what they used to be.

To recapitulate, the requirement of adaptability, which stems from properties of an economy's environment and begins with its outputs, may be relayed, through a domino-like chain of required changes, through the economy's organizational structure and institutions, all the way up to its culture. How high the relaying must actually reach – in other words, how many of these parameters need to change and therefore need to be adaptable – depends in part on the importance of the environmental changes and in part on the adaptation potential of the parameters. A parameter with a sufficiently high adaptation potential, which allows the variable(s)-parameters(s) under its influence to be sufficiently adaptable, can remain stable and thus stop the relaying.[11]

An important property of this chain is that both the difficulties and the social and individual costs of adaptations appear to increase steeply with the level of the required changes – from adaptations of quantities and/or qualities of outputs, which, if they can be realized within a stable structure, are often easy and cheap, through the more difficult and more expensive structural and institutional adaptations, to the cultural ones which appear to be the costliest and the most troublesome.

One reason for these rising costs is subjective: be it due to genetically given instincts or cultural conditioning, people often tend to value changes negatively. The more fundamental the changes are and the more uncertain their consequences, the more painful they are usually perceived. Another reason is the lack of knowledge about what specific adaptations are required, which usually also increases with the levels. The less knowledge is available, the higher the expected costs of the errors in the inevitable trial-and-error search for a suitable adaptation.

The need for cultural adaptation and the social costs of searching for it are strikingly illustrated by some of the least advanced post-socialist economies, where the incumbent values and beliefs, culturally evolved through a mixture of feudal and socialist conditioning, are among the main reasons why the economically needed institutional and structural changes cannot find sufficient political initiative and support. This situation may also be understood as a particularly severe clash of feedback (ii) with feedback (i).

6 Which economic institutions can be sustainable?

Until now, our attention has mostly been limited to general conditions that sustainable institutions of economies must meet. Now it is time to identify these conditions in more detail and ask which specific institutions can effectively meet them. Let me start broadly by considering an entire politically organized society (such as a nation) with the two basic conditions that it must meet to avoid misery and crises: (1) its economy must be sufficiently efficient (non-wasteful) in terms of its socially chosen performance criteria and in relation to the severity and variability of its environments; (2) the criteria must be sufficiently wise not to deviate too much from the objective criteria of nature (in the broad sense). This leads to the following series of questions: What can the members of the society do to meet these conditions? What role must be played by the economy's institutions? Which specific institutions can play this role best, or at least sufficiently well?

[11] It may be instructive to think of our genetic endowment as so highly adaptive, and thus providing for such a high adaptation potential (learning abilities) of human brains, that it could itself remain basically constant for tens of thousands of years, while the environments to which humans kept successfully adapting varied widely in highly unpredictable fashions.

To meet Condition (2), the members must overcome the above-mentioned difficulties with nature which does not inform them in advance of its objective criteria (cf. the end of Section 3). Thus, if they want to minimize the probability of unpleasant surprises, they must search for knowledge about these criteria themselves, including learning from past mistakes, and then wisely use whatever knowledge they may acquire in the choice of their social criteria. This is certainly a difficult and risky task: as the knowledge will hardly ever be complete, sound intuition and above all good luck are also necessary if the socially chosen criteria are not to be unwise, directing even the most efficient economy towards social misfortunes. A simple illustration is the current controversy about the effects of humanly caused CO_2 emissions on global warming and resulting policy objectives. It is indeed still difficult to know how the causes of the currently observed global warming are divided between natural fluctuations and human industrial activities, and the risks include failing to avoid natural disasters at one extreme, and unnecessarily causing economic difficulties at the other.

Most of this, however, is not of a direct concern to economists. According to the standard division of labor among sciences, the search for knowledge about nature's criteria belongs to the natural sciences and, to the extent that these criteria also depend on human nature, to such social sciences as anthropology and social psychology. How the knowledge found can be used, under influences of the prevailing values and ideological beliefs, in the political choice of the social criteria of economic performance (policy objectives, social preferences) is mostly the subject for political scientists. Economists can certainly do a few things, such as checking these processes for logical consistency and warning about possible distortions due to imperfect incentives and insufficient competence of politicians and government officials. Basically, however, the politically chosen criteria belong to the givens from which they have to start.[12]

In consequence, the role of economic institutions may be considered limited to helping the society to meet Condition (1). But this is not a simple task. In environments that are increasingly severe and variable, the society needs an economy which, as noted, is not only highly efficient in relation to their current state, but also sufficiently adaptable, to be able to regain and keep its efficiency even after the environments have possibly changed. The economy's institutions, in addition to allowing and inducing the economy to form an efficient organizational structure for current environmental conditions, must therefore also have a sufficient adaptation

[12] The welfare economists who take the postulate of consumer sovereignty for granted may claim that they have much more to say about these criteria. Libertarians may object against all kinds of political choice and claim that each individual consumer must have the exclusive right to decide what is best for him. Two comments are in order. First, the division of such rights between individual and political decisions must itself result from a political decision: libertarians would have to gain political power to be able to abolish all political valuations of economic performance. Second, regardless of our personal tastes for either consumer sovereignty or political paternalism, such tastes are only parts of our culturally evolved ideologies, which undoubtedly determine much of our currently chosen performance criteria, but without any guarantee of compatibility with the ultimate criteria of nature. Which specific mixtures of consumer sovereignty and political paternalism may prove sustainable is thus less a matter of subjective tastes than another question for an objective evolutionary analysis – which, however, will not be entered here.

potential, to allow this structure to be adaptable over the whole range of possible environmental changes.

This task, which is difficult by itself, is compounded by the need to make the required efficiency and adaptability compatible with the politically chosen performance criteria. This means that, to the extent that these criteria are more than market aggregations of individual consumers' preferences, the institutions must define policy instruments by which the economy could be steered to do well also according to the additional, politically chosen parts of these criteria. There are several reasons why also this task is difficult. First, the extent of these parts – which start with the traditional public goods and spillover effects – is likely to be quite large if Condition (2) is to be met in today's world, where spillovers among both individuals and economies appear to grow in both intensity and density. A second reason is the interference of most of *feasible* policy instruments with both efficiency and adaptability, which leads to difficult trade-offs – such as the well-known one between efficiency and equity.[13] A third reason is the imperfection of real-world policy-makers, who cannot be expected to be either perfectly benevolent or perfectly competent (unboundedly rational). The needed policy instruments must thus not only have the intended positive effects – which is often difficult to achieve by itself – but these must not be overridden by their negative side-effects, including the possible social losses that they might cause in the hands of imperfectly benevolent and/or imperfectly competent policy-makers. In other words, the institutions must define only such policy instruments that are needed to avoid greater social losses than those that their use in non-idealized real-world conditions could be expected to cause.

The crucial question thus is: which specific features must an economy's institutions have to meet reasonably well all these requirements? While this question is of prime importance for both theory and policy, it appears to attract little attention of today's economists. Interesting parts of the answer can be found in the literature on law and economics, public choice, and new institutional economics, but they are mostly limited to the effects of institutions on efficiency, while leaving aside the in the long run much more important problem of adaptability.

To deal with this problem, I used a simple evolutionary analysis (cf. footnote 1), which can be summarized as follows. As it is never fully known in advance which structural change will be successful, structural adaptation cannot proceed without imperfectly informed entrepreneurial trials, most of which may turn out to be errors. The adaptation potential of institutions therefore depends on how large the variety of such trials they allow and encourage, and how fast and rigorous the correction or elimination of errors they enforce. This makes it possible to conclude that the

[13] The emphasis on 'feasible' is to warn against the old bad habit of theoretical economists of playing with unfeasible policy instruments – such as lump-sum taxes. To rank feasible policy instruments according to the efficiency losses caused, adaptive efficiency provides sharper criteria than the usually considered allocative efficiency. Thus, as bases for taxation, final consumption appears relatively benign, whereas capital, reinvested capital gains, and financial market transactions prove by far the most harmful. As ways of responding to spillover effects, policies building on market exchanges and competition of private enterprises – such as tax financed vouchers for boosting merit consumption and tradeable emission rights for limiting air and water pollution – prove adaptively less inefficient than direct government control (Pelikan 1993, 1999).

institutions with the highest adaptation potential must be variants of what can be denoted as 'modern capitalism,' the basic property of which is to provide for the formation, development and preservation of competitive markets, including financial markets, and for private and tradable ownership of firms, including commercial and investment banks. In contrast, institutions which allow or prescribe national planning, and/or government ownership of firms and/or industrial subsidies and/or government controlled allocation of productive investment, can be shown to reduce the variety of trials, or the speed and the rigor of the elimination of errors, or both, which causes their adaptation potential to be significantly lower.

That the main comparative advantage of capitalism is in its adaptability, and not in the usually studied static efficiency, and that it was necessary to employ evolutionary analysis to show it, deserves emphasis. Evolutionary analysis appears indeed to be the only way to justify fully in theory the policy advice that most practical economists now give to countries in economic difficulties, advice such as privatization, deregulation, and replacement of government allocation or control of productive investment by development of civilized financial markets, including risk-capital markets. This policy advice can find only half-hearted support in standard analysis: because of the great freedom in the choice of its simplifying assumptions, this analysis proved equally able in showing the optimality of the opposite policies.

On the other hand, my evolutionary analysis was only very rough. While it showed the futility of looking for adaptive institutions elsewhere than among variants of modern capitalism, it said little about their possibly significant details. Much structural adaptability may indeed depend on institutional details – such as details in the formulation of bankruptcy law, patent law, or antitrust law – which remain a largely uncharted territory.[14] It is also in such details that even the institutions with the highest adaptation potential may still have to be adaptable, ready to change in face of new developments, such as important technological and organizational innovations.

A recent example is the development of information technologies and the spread of products with high information content in general. To deal with such fundamentally novel types of products and technologies in ways that facilitate suitable structural adaptations and prevent most of the unsuitable ones, even the most adaptable capitalist economies turned out to require institutional changes, such as modifications of and additions to the legislation of property rights.

This example also illustrates what is often viewed as a causal arrow between technological and institutional changes: the former are seen to cause the latter. This view has a long history, as it already appears in the old Marxist thesis that technological development (which is roughly what Marxists call 'the development of the forces of production') is exogenous, as if automatically falling from the sky,

---

[14] It is such important nuances that the recent developments of law and economics and modern institutional economics is often about. As noted, however, most attention is paid to the effects on static efficiency, and much less to those on adaptability. While sometimes the two types of effects work in the same direction, at other times they may conflict: for example, duplication of research efforts or investment in unsuccessful ventures may appear allocatively inefficient – yet both may be the necessary price to pay for finding the initially missing information about what projects of which agents can be part of a successful structural change, and thus be invaluable for structural adaptability.

and constitutes the prime cause to which the prevailing institutions (included in the marxist term 'superstructure') are forced, possibly by means of revolution, to keep adapting. What appears less often noted is that this is only a half of the story. The other half was pointed out by North and Thomas (1973), who found that a not less important causal arrow leads in the very opposite direction: their study of the property rights that define the freedoms and incentives of inventors and innovators made it possible to conclude that it is the form of the prevailing institutions that determines whether technologies will develop or stagnate.

The present view leads to a synthesis of the two arrows with a strong qualification of the Marxist one. Since technological innovations can be counted as special parts of the economy's output, their production is seen to depend, along the North and Thomas arrow, on the prevailing institutions: the institutions strongly influence the forming and the working of the economy's organizational structure, by which the output is produced. The qualification of the marxist arrow is that the changes caused by the actually produced innovations are in the first place structural (e.g., some new markets may have to open, some incumbent firms may have to reorganize or exit, and some new firms may have to enter) and not necessarily institutional. Institutions are required to change only if their adaptation potential is not sufficiently high, so that the structure, to accommodate the innovations, would have to be more adaptable than the institutions allow.

This implies that two arrows form a feedback loop, which may seem to allow several institutional equilibria, states in which an economy's institutions would not be forced to change. A necessary condition is that the institutions must not allow a greater variety of innovations to be generated than the one to which they allow the economy's structure to adapt. It may thus seem that both institutions with a low adaptation potential and those with a high adaptation potential may be sustainable, provided that they both satisfy this condition. This, however, may only be true for isolated economies. Otherwise, when different economies enter into economic or military competition, the more adaptable ones will prove to have a decisive comparative advantage over the less adaptable ones. This will force the latter to increase the adaptation potential of their institutions by a possibly radical reform, or to suffer a lasting decline.[15]

7 Some implications for cultural relativism and development policies

The economically relevant findings of the previous section may be summarized as follows: (1) the institutions of economies, to be sustainable in severe and variable environmental conditions, must have a high adaptation potential; (2) to have this potential, the institutions must contain many components of what is usually called 'modern capitalism.' A serious problem arises if the actual environmental conditions

[15] This effect of competition among institutions can indeed explain why Western Europe, which was during centuries divided into many relatively small competing states, could evolve institutions with a high adaptation potential, which have allowed it to prosper, while much larger but relatively isolated empires – such as China and Mongolia – could preserve for a long time institutions with a low adaptation potential, but at the price of a subsequent long-lasting decline. For this explanation, or reasoning leading to it, see, e.g., Bernholz (1995); Kerber and Vanberg (1995); and Rosenberg and Birdzel (1986).

of an economy are indeed severe and variable, but its institutions, instead of the needed components, contain some detrimental rules – for instance, they hinder establishment or smooth functioning of markets, or require extensive government ownership of firms, or let market competition be damaged by private cartels and/or corruption of the government bureaucracy.

The general strategy for solving this problem is clear: the institutions need to be adapted (reformed, transformed) to satisfy point (2). But how actually to implement this strategy is much less so. As follows from Section 5, and in somewhat different words from Vanberg (1992), Denzau and North (1994), and Knight and North (1997), the adaptability of institutions is constrained by the adaptation potential of the prevailing culture. What makes these constraints particularly strong is that they act through several parallel, mutually reinforcing channels. The incumbent culture not only constrains people as cultural agents in their abilities and willingness to modify informal institutions, but it also constrains, through its more or less strongly self-protected ideological beliefs, the cognitive abilities of social scientists needed to produce or import the knowledge about what institutional changes are necessary and of politicians needed to understand this knowledge and use it to suitably change formal institutions.

To be sure, as has often been pointed out, institutional change is also constrained by the vested interests of currently privileged agents. Ultimately, however, even this constraint depends on the culture. While initially much of it is determined by the resources with which the privileged agents can defend their position, the relative importance of the culture is bound to grow over time: as the economy continues to decline, these resources inevitably dwindle. What then matters most is the actual understanding of the situation and the learning abilities by which this understanding can be adapted to new facts, which are both, as noted, strongly culturally conditioned. What happens next depends on how these abilities are divided between the incumbent elite and the other agents: for instance, if its learning abilities are high, the elite may start reforming the economic institutions itself, and thus save, at least for some time, its privileged position. The Chinese political leaders, who still call themselves 'communists' but are in fact transforming Chinese economic institutions into highly capitalist ones, illustrate this alternative.

The crucial role of culture in institutional change, and the fact that different cultures differ in how they help or hinder institutions in evolving toward and remaining close to favorable forms, have important implications for cultural relativism and development policies. Cultural relativism turns out to be strongly limited by certain absolute conditions. Although the notion of truth may indeed be relative to cultures – different cultures can be found to contain different beliefs (ideologies, hypotheses, theories, mental models) – these beliefs turn out to be subject to the ultimate sanctions of nature (in the above broad sense), which are absolute. The beliefs that favor efficiency and adaptability of economies thus endow their cultures with important evolutionary advantages over cultures whose beliefs consist of less relevant fantasies.

Why these advantages have often been overlooked may be due to the fact that they were of low importance in the generous and stable environments which surrounded for a long time many primitive human societies, especially those living in tropical climate, and which some social scientists still appear, more or less implicitly, to assume. As noted, such environments tolerate both low efficiency and low adaptability, and thus allow a wide variety of cultures with a wide variety of beliefs to prosper. This may then give the impression that 'anything goes' and nothing is absolute. It is only when the environments of economies grow severe and variable, as they now appear to do virtually everywhere, that these advantages become decisive – whether social scientists see them or not.

But the limitation of cultural relativism also has its limits. It concerns only those parts of cultures that determine informal economic institutions and influence the behavior of humans as economic agents and as authors of formal economic institutions. The other parts of cultures – such as music, dances, fashion and diet – are not concerned. Although even they may be connected to efficiency and adaptability of economies – e.g., the diet of some cultures may be healthier than the diet of other cultures, which may cause significant differences in labor productivity and the costs of medical care – these connections are relatively weak. For these parts, cultural relativism can fully flourish.

The finding that cultures are economically unequal, in the sense that some of them provide for economic development and prosperity while others cause economic stagnation and misery, is not very new.[16] But its implications for development policies are still little developed. The present one is that in the frequent cases in which economic underdevelopment is caused by lack of adaptability of structures and institutions, the dominant culture is crucial: unless it has, or can be supplied with, a sufficiently high adaptation potential, all other efforts to promote economic development are bound to fail. While North (2000) is certainly right that the favorable cultural conditions of the developed Western economies took several centuries to form, and that we do not know how to form them more rapidly, to try to learn it and to use the knowledge acquired both for assistance to legislation and for educational campaigns is the only promising strategy of helping poor countries to become richer. A necessary condition for this strategy to work is that the developed 'West,' instead of suffering from guilt complexes, becomes fully aware of the economic advantages of its culture and institutions, without which the educational campaigns could hardly be convincing.

Emphatically, however, that the poor 'South,' to become richer, must be helped to adopt economically favorable cultures and institutions by the prosperous 'West' is no reason for the 'West' to become pretentious. On the contrary, it must be modest enough to recognize that most of its present success is due to past chance: it has simply been extremely lucky that its cultural development, which cannot be claimed to have been driven by knowledge of the conditions of economic success, happened to coincide so well with these conditions. Moreover, its success is still far from definitive. As mentioned in the beginning of Section 6, if economic efficiency is not to work in the wrong direction, it needs more knowledge about the ultimate criteria of nature, and another large portion of good luck, to choose sufficiently wise criteria of economic performance, which would allow it to steer clear of both

---

[16] For similar results, see, e.g., Banfield (1958); North (1990); Grief (1994); Knight and North (1997); and Landes (1998).

the Scylla of exaggerated environmentalism and the Charybdis of neglecting vital spillover effects which could fatally damage its natural environment or its moral basis.

References

Banfield EC (1958) The moral basis of a backward society. The Free Press: Glencoe, Illinois
Bernholz P (1995) Causes of changes in politico-economic regimes. In: Gerken L (ed.) Competition among Institutions, Macmillan: London, and St. Martin's Press: New York
Chomsky N (1976) Reflections on languages. Fontana Books: New York
Denzau AT, North DC (1994) Shared mental models: ideologies and institutions. Kyklos 47: 3–30
Grief A (1994) Cultural beliefs and the organization of society. Journal of Political Economy 102: 912–950
Hayek FA (1967) Studies in philosophy, politics, and economics. University of Chicago Press: Chicago
Hayek FA (1973) Law, legislation and liberty, vol. 1. rules and order. University of Chicago Press: Chicago and London
Hayek FA (1976) Law, legislation and liberty, vol. 2. the mirage of social justice. University of Chicago Press: Chicago and London
Hofstadter DR (1979) Gödel, Escher, Bach: An eternal golden braid. Basic Books: New York
Kerber W, Vanberg V (1995) Competition among institutions: evolution within constraints. In: Gerken Z (ed.) Competition among Institutions, Macmillan: London, and St. Martin's Press: New York
Knight J, North DC (1997) Explaining economic change: the interplay between cognition and institutions. Legal Theory 3: 211–226
Landes DS (1998) The wealth and poverty of nations: why some are so rich and some so poor. Norton: New York
Marris R, Mueller DC (1980) The corporation, competition, and the invisible hand. Journal of Economic Literature 18: 32–63
North DC (1990) Institutions, institutional change and economic performance. Cambridge University Press: Cambridge, UK
North DC (2000) Economic evolution and the process of change: a research agenda. Workshop on Evolutionary Analysis of Economic Policy, May 4–6, Ruhr University of Bochum
North DC, Thomas RP (1973) The rise of the western world: A new economic history. Cambridge University Press: Cambridge, UK
Pelikan P (1987) The formation of incentive mechanisms in different economic systems. In: Incentives and Economic Systems. Hedlund S (ed.) London, Sidney: Croom and Helm
Pelikan P (1988) Can the imperfect innovation system of capitalism be outperformed? In: Dosi G et al. (ed.) Technical Change and Economic Theory, Pinter Publishers: London
Pelikan P (1989) Evolution, economic competence, and the market for corporate control. Journal of Economic Behavior and Organization 12, 279–303
Pelikan P (1992) The dynamics of economic systems, or how to transform a failed socialist economy. Journal of Evolutionary Economics 2, 39–63; reprinted. In: Wagener HJ (ed.) On the Theory and Policy of Systemic Change, Physica-Verlag: Heidelberg, and Springer-Verlag: New York
Pelikan P (1993) Ownership and efficiency: the competence argument. Constitutional Political Economy 4: 349–392
Pelikan P (1995) Competitions of socio-economic institutions: in search of the winners. In: Gerken L (ed.) Competition among Institutions, Macmillan: London, and St. Martin's Press: New York
Pelikan P (1999) Institutions for the selection of entrepreneurs: implications for economic growth and financial crises, WP 510. The Research Institute of Industrial Economics: Stockholm
Rosenberg N, Birdzell LE (1986) How the west grew rich: the economic transformation of the industrial world. Basic Books: New York
Schumpeter JA (1942/76) Capitalism, socialism, and democracy. Harper & Row: New York
Vanberg V (1992) Innovation, cultural evolution, and economic growth. In: Witt U (ed.) Explaining Process and Change: Approaches to Evolutionary Economics. The University of Michigan Press: Ann Arbor
Vanberg V (1994) Rules and choice in economics. Routledge: London and New York

Designing clunkers: demand-side innovation and the early history of the mountain bike

Guido Buenstorf

Max Planck Institute for Research into Economic Systems, Evolutionary Economics Group, Kahlaische Strasse 10, D-07745 Jena, Germany (e-mail: buenstorf@mpiew-jena.mpg.de)

Abstract. Innovation studies in economics tend to focus on the supply side and to assign a passive role to consumers. This passivity of consumers in innovation processes is questioned by the present paper. It is suggested that supply-side considerations alone may be insufficient to account for innovations in consumer good industries. Based on the example of the early development of the mountain bike, the paper shows how consumers may be the dominant actors in consumer good innovations. It is demonstrated that the basic features of mountain bikes had already been established when commercial interests entered the industry, and that development of the mountain bike cannot be understood unless the group setting of its origins is taken into account. To explain the development of the mountain bike, economic, sociological, and psychological concepts are integrated. The dynamics identified in the present case study may hold for a broader class of commodities.

Key words: Consumption – Innovation – Learning – Communication – Social groups

JEL Classification: D11, D71, O33

1 Introduction

Following Schumpeter's path-breaking contributions, economists have in the past decades shown an extensive interest in technological innovation, and they have been able to establish a number of findings on determinants as well as effects of innovation. These findings may not be of universal applicability, however. Most research has focused on high-technology sectors characterized by large R&D budgets, and a supply-side bias is observable in the innovation literature. As with the interest in innovation itself, this supply-side bias can be traced back to Schumpeter, who considered spontaneous change of consumer preferences as exceptional and, accordingly, abstracted from the activity of consumers in the innovation process. Schumpeter argued that economic change normally is initiated by producers, whereas consumers "are, as it were, taught to want new things " (1934/1983, p.65). The passive role in the innovation process thus assigned to consumers is questioned in the present paper. It is suggested here that supply-side considerations alone may be insufficient to account for innovations in consumer good industries. Based on the example of the early development of the mountain bike, this paper shows how consumers may play a dominant role in consumer good innovations. It is shown that the basic characteristics and design features of the mountain bike had already been established when the first commercial interests entered the emerging industry. And although the enthusiasts who in the early 1970s created the mountain bike can be seen as Schumpeterian entrepreneurs, insofar as they introduced a new combination, their motivations were far more hedonistic than Schumpeter's characterization of the entrepreneur would allow.

Two critical features of the history of the mountain bike industry are, first, the active role of consumers in innovation and, second, the speed of technological development which distinguishes it from purely fashion-driven sectors such as apparel, where technology is essentially invariant. It appears as though both characteristics are shared by a number of industries, such as sports and leisure industries, but also parts of the computer and software industries. To explain the emergence and diffusion of innovations in these industries, a concept is required which can account for the consumers' role in commodity (re-) definition and (re-) design, and which is dynamic to accommodate the rapid pace of change. In section 2 of the present paper, theoretical approaches to tackle this task are discussed. Section 3 gives a narrative of the early history of the mountain bike. In section 4, this history is discussed in light of the concepts presented before. Section 5 concludes with an outlook on the potential scope of applicability of the explanation developed for the mountain bike case.

2 Accounting for the role of demand in innovation

Various studies have tried to identify the role of demand-side factors in the innovation process. In the early history of innovation studies, two strands of literature can be distinguished. First, triggered by some skepticism about the role of basic research in fostering innovativeness, a number of studies based on survey and interview techniques were conducted in the 1960s and 1970s in order to identify the driving forces of innovativeness. The findings of these studies were interpreted to demonstrate the important role of demand-side factors in innovation, and they gave rise to a controversy which centered around "demand-pull " versus "technology-push" models. In its strong form, the demand-pull model argues that price signals provide sufficient information about consumer preferences to direct the development efforts of producers. However, in a critical survey of the empirical literature on the demand-pull hypothesis, Mowery and Rosenberg (1979) found that the empirical studies do not in fact warrant such far-reaching conclusions. They argue that the surveyed studies suffer from a number of methodological problems, of which the most troubling seem to be the underspecified concept of demand and a bias toward innovations in producer rather than consumer goods.

The second strand of literature uses data on sectoral demand patterns and patenting activities. It was initiated by Schmookler's (1966) analysis of patent statistics suggesting that the sectoral distribution of innovations is affected by the relative demand for the products of the sectors, measured by investment levels in user industries. This finding was, with some important qualifications, confirmed for a larger data base by Scherer (1982). Some more recent empirical papers also find that innovation is affected by aggregate demand conditions at the sectoral or national level (Geroski and Walters, 1995; Brouwer and Kleinknecht, 1999). Although these results indicate a potential significance of demand factors in innovation, they differ from the argument to be developed below because of their more aggregate perspective. Moreover, while in both strands of the demand-pull literature demand is seen as inducing innovation, the actual act of innovation is nonetheless attributed to the supply side. Thus, according to this line of research, the role of consumers in the innovation process remains a passive one.

Dosi's (1982) influential contribution aimed at overcoming the dichotomy between demand-pull and technology-push accounts of innovation. It is based on the concepts of "technological paradigms" and "technological trajectories," which both guide and restrict technological change. Dosi argued that the market is weak as an *ex ante* selection mechanism between innovations, particularly in the initial stages of a technology, and that social and institutional factors operate as additional selection criteria; for example, industrial conflict between capital and labor is seen as a factor favoring mechanization. Under this concept, the potential impact of social factors on the specific form that technological innovations may take is acknowledged. There is, however, no direct involvement of the demand side in the innovation process. By contrast, a more active role of demand-side factors in innovation has been suggested by von Hippel (1978) who reported empirical evidence for a "user-dominated innovation pattern" in some industries producing specialized capital goods. In these industries, both major innovations and minor improvements to existing technologies tended to be generated, tested and utilized by the user industries before the specialized producers began to adopt and commercially market them. According to von Hippel, user-dominant innovation is a phenomenon restricted to specific industries, while innovations in other, seemingly similar industries were exclusively forthcoming from the producers themselves. The key difference between von Hippel's results and the argument developed below concerns the product of the respective industries. Von Hippel's results are derived from high-tech industries producing for a small number of specialized users. By contrast, the mountain bike is a consumer good, and its origins are markedly low-tech.

Outside of economics, there are approaches giving more emphasis to the role of consumers in innovation.[1] In particular, demand-side factors are emphasized by some sociologists and historians of technology (cf. Metcalfe, 2001). Among the

[1] It should be noted that the passive role assigned to consumers in economics more generally has recently been questioned by numerous authors. However, these criticisms (e.g., Bianchi, 1999; Swann, 1999; see also the contributions in Bianchi, 1998) focus on change in individual consumption patterns,

various frameworks proposed in the New Sociology of Technology (cf. Hughes, 1986; Bijker and Law, 1992), an approach congenial to the present purposes is the Social Construction of Technology concept (Pinch and Bijker, 1984; Bijker, 1995), which is derived from social constructivist concepts originally developed in the sociology of science. Central to the social constructivist view of technological change is the notion of "interpretative flexibility" of commodities: it is argued that the uses and the meaning of commodities are not determined technologically, but are constituted by the interpretations of potential users and other social groups. What meaning a specific group of users actually attaches to a commodity depends on the problem they attempt to solve by using the commodity. Commodities may be used in ways other than the ones for which they were originally designed, different groups may attach different meanings to the same commodity, and interpretations of commodities are subject to change over time. Competing and changing interpretations of a commodity may give rise to new technological developments and thus increasing design variety, which is subsequently reduced in processes of "closure" selecting between competing designs. Closure is achieved if a consensus is realized among the relevant groups on the meaning of an artifact. It may be caused by a new design variant able to solve all the problems of the various groups, or if groups with incompatible design requirements lose their interest in the commodity (for example, because a substitute becomes available which can accommodate their requirements). As closure does not necessarily require actual solutions to problems, but rather that the relevant social groups perceive the problem as being solved, it may also be achieved by rhetorical means alone (for example, by advertising claiming that the problem has indeed been solved) or by a redefinition of the relevant problems.

Social constructivist authors (Pinch and Bijker, 1984; Oddy, 1994; Bijker, 1995) have used the early history of the bicycle – which bears some parallels to the mountain bike case to be discussed below – to exemplify their approach. In doing so, they have identified several groups of cyclists with different interpretations of the bicycle, and accordingly with different design requirements. The authors argue that in the 19th century the bicycle was primarily used by young upper-class males as an item of sports equipment. For this group of users, races were an important use, and accordingly speed figured prominently among the design criteria for new models. The emphasis on speed motivated the design of the "ordinary" bicycle with an increasingly large front wheel (as "ordinaries" were powered by pedals directly driving the front wheel, wheel size determined the transmission ratio and thus the maximum speed). For these sportive users, the fact that high-wheeled "ordinaries" were inherently unsafe and difficult to ride was not a crucial problem; frequent falls and injuries were accepted as a normal part of the cycling activity. But the safety lacking in the high-wheeled bicycle limited the use of bicycles as a means of transportation. The desire for increased safety by older and less athletic cyclists provided a motivation for the development of various alternative designs such as tricycles and "safety" bicycles (chain-driven bicycles with smaller wheels as are

i.e. subjective novelty, rather than on the contribution of consumers to innovation in the sense of objective novelty.

still used today). The bicycle as a means of transportation represented a commodity different from the bicycle used for sports, and the different interpretations implied different directions in design.

The history of the air tire is another illustration of the social constructivist concept of interpretative flexibility. The air tire had originally been invented as an anti-vibration device competing with various shock-absorbing frame, seat, and handlebar designs. Contemporary eyes perceived air tires as ugly (they were ridiculed as "sausage tires"), and a majority of engineers did not consider them practical. For the sports cyclists on their high-wheeled "ordinaries" , vibration did not even represent a pressing problem. There was thus widespread opposition to the air tire among relevant user groups. That the air tire nonetheless became a commercial success was, according to Pinch and Bijker (1984), due to a reinterpretation of its use. When low-wheeled safety bicycles equipped with air tires outperformed high-wheeled ordinaries in races, air tires became seen not as a means to absorb shocks, but to increase speed. In this way, they became acceptable also for the sportive group of cyclists. The air tire's impact had on the history of the bicycle is more profound, though. It made the safety bicycle suitable for sports purposes and thus ensured its universal success (implying " closure" of the basic design controversy). The design requirements both for speed and for safety could now be matched by the safety bicycle, so there was no longer a reason to stick to the ordinary bicycle.

As the example of the safety bicycle indicates, the social constructive approach is able to enrich purely technological accounts of innovation processes by introducing social factors. In particular, it can capture changing uses of commodities, even if concomitant technological change is absent, and it can identify non-technological obstacles to the adoption of innovations. The active role of users in technological change can thus be included in the analysis, which helps to make sense of developments which might otherwise appear as detours in retrospect. The social constructive approach nonetheless has some disturbing shortcomings. First, from an individualistic perspective, the exclusive focus on social groups and their impact on product development is unsatisfactory. It remains unclear what individual factors, for example cognitive representations and processes, correspond to the macro-level phenomena discussed.[2] Second, social constructivism is not a dynamic theory. While it can be used to describe the constellation of relevant social groups in the innovation process, and also the meanings they attach to commodities, neither causal factors nor transmission mechanisms of change are identified. The internal dynamics governing the emergence, growth, and development of groups remain as much unspecified as the processes leading to changes in the constellation and relative weight of various groups.

By portraying consumption as a learning process, Witt (2001) has suggested a framework which appears useful as a point of departure in order to overcome the shortcomings of social constructivism. As it starts from the individual level and is dynamic in nature, this framework is able to capture both individual changes and changes within groups, and it can provide a micro-level explanation for the

[2] The analysis of relevant groups in the social constructivist approach has moreover been criticized for neglecting the wider social context (Rosen, 1993; Winner, 1993).

dynamics of innovative consumer goods. According to Witt, consumption serves to fulfill wants that are developed in a chain of learning processes starting from a set of innate basic wants. He argues that humans are born with a set of invariant wants corresponding to basic physical (food, sleep etc.) as well as psychic needs (affection, cognitive arousal, social recognition etc.), and also with innate learning propensities. Some wants can be satisfied directly by the consumption of inputs (e.g., food intake satisfies hunger), whereas others require the use of consumption "tools" (e.g., clothing to keep up body temperature). Witt differentiates between two levels of learning in consumption. First, new wants are learned in conditioning processes that lie outside the agent's awareness. New wants develop when neutral activities are repeatedly performed simultaneously with activities fulfilling existing wants so that the neutral activity becomes satisfying itself. Second, agents consciously and often deliberately learn how to satisfy existing wants in new ways, i.e. they identify new direct inputs or consumption tools capable of satisfying their wants, either separately or in varying combinations. In this way they acquire consumption knowledge about which direct inputs and consumption tools are suitable for satisfying particular wants.

Learning how to satisfy wants (i.e., acquisition of consumption knowledge) results from various activities. It may be based on personal experience: people deliberately experiment with new consumption activities, e.g., they try new items on restaurant menus or in supermarket grocery departments. It may also exploit the experiences made by other agents for adaptations of one's own consumption behavior. Additional learning results from communication within user groups and from the deliberate acquisition of information about consumption possibilities. Myriads of highly specialized magazines on automobiles, stereos, computers, and even teddy bears help in collecting this information, as does commercial advertising (or at least it pretends to do so). Witt (1996) argues that, at the behavioral level, learning of wants and of ways to satisfy them may result, in addition to shifts in the objects of wants, in a "refinement effect". Objects of wants are perceived in increasingly more detail as consumption knowledge related to the wants is accumulated. Pre-existing wants themselves are further reinforced and refined, which in turn tends to increase both the attention and the physical resources allocated to the want.

At the individual level, acquisition of wants and learning how to satisfy them are interrelated by positive feedback relations which give rise to non-linear dynamics of mutual reinforcement; wants and consumption knowledge therefore tend to co-evolve. These dynamics result from the restricted capacity of the working memory which causes human attention and information processing to be highly selective (Simon, 1986; Witt, 1996). Selective individual attention and information processing is affected by several factors. First, attention to information is shaped by the individual structure of wants and their current state of deprivation (Witt, 2001). The probability that a piece of information gains attention increases if it is associated with pre-existing wants that are not currently being satisfied. In turn, if attention is allocated to some piece of information, activities corresponding to the information are more likely to be performed. If these activities are experienced as satisfying, the want is further developed. In this way, a positive feedback loop between wants and attention is established. Additional guidance to the allocation of attention is provided by cognitive representations based on knowledge previously acquired. Prior knowledge affects both the attention to incoming stimuli and the meaning attached to them, i.e. the way in which they are "framed" by the agent (Tversky and Kahneman, 1981; Devetag, 1999). Another positive feedback loop is present here which links cognitive representations and selective attention. Information that can be made sense of within existing cognitive structures is more likely to be attended to and to be memorized, which in turn elaborates these structures and thus increases the probability of attending to future information of a similar kind. If undisturbed, the feedback process may cause representations to become increasingly detailed, but also increasingly rigid. In addition to attention, prior knowledge held by the agent also strongly affects human problem-solving (cf. Eysenck and Keane, 1995; Kellogg, 1995, and the literature cited therein). Both problem representation and the approaches taken to problem solution reflect the experience of the agent. With growing experience with a class of problems, problem-solving is increasingly based on domain-specific strategies that are partially automatic, i.e. the agent is not fully aware of her adoption of solution methods. While these effects are part of the explanation of the superior problem-solving performances of experts, they may also become dysfunctional if the situation changes. Expertise may thus come with inflexibility in reaction to environmental changes.

In addition to these individual determinants of the learning of wants and the acquisition of consumption knowledge, individual consumption behavior is affected by the social level. It has for a long time been argued that consumers' preferences are shaped by the social context (classic references are Veblen, 1899, and Leibenstein, 1950). Psychological evidence indicates that this influence has several dimensions (Aversi et al., 1999). First, the way in which individuals evaluate experiences is influenced by the reactions of their social context. Second, individual cognitive representations are affected by the representations of interacting agents and by their interpretation of information and situations. Another important link between the individual and the social level is established by the human capacity for observational learning, enabling them to learn from the experience of others. Observational learning, which figures prominently in social cognitive learning theory (Bandura, 1986), may be based on the actual observation of others (so-called models), but also on communication of knowledge in personal interaction and via mass media. Who gains attention and becomes a model depends on the characteristics of both the observer (again, effects of prior knowledge on selective attention play a role here) and the observed. Visibility, status and power increase an individual's potential to become a model for others. Within groups, prestigious members are more influential as models than are peripheral members. Moreover, the likelihood of attention and imitation by observers is enhanced by the success of the observed behavior, i.e. it increases with the rewards realized by the model.

According to social cognitive learning theory, observational learning from models can modify observers' attitudes, emotions, thought patterns, and behavior. It is another important determinant of attentional processes, as models draw observers' attention to particular objects in their environment. Patterns of behavior and communication within groups thus have an agenda-setting effect on group members' individual information processing; objects related to these patterns are more likely to be perceived and memorized than those which are not. Here as well, a positive feedback relation is present, as it can be expected that the activities attended to are in turn more likely to be performed (which may lead to further individual learning) and communicated to other group members. Observational learning therefore helps socialize individuals into groups; it may result in the homogenization both of behaviors and of cognitive representations within groups. Accordingly, Witt (2001) suggests that "wants and knowledge within intensely communicating groups tend to develop in much the same direction and may give rise to sub-cultural commonalities in consumption patterns". Due to the selective choice of models based on their personal traits and their position within the group, a few individuals may be able to affect strongly the thought and behavior of the entire group.

The impacts of direct modeling and communication within groups are further reinforced by the uni-directional information flow from mass media. Mass media are able to transmit information to a large number of individuals simultaneously. They therefore have an enormous potential for affecting individuals by selectively supplying information and providing models for observational learning. As with other forms of information, however, modeling through mass media is not necessarily effective, because the disseminated information is subject to the selection mechanisms described above. Communication groups and mass media coevolve: specialized media emerge to serve developing subcultural communication circles which results, first, in a growing variety of different publications, and, second, in increased communication flow within the group (and probably, in turn, less communication *between* groups). The spread of information through media increases the speed of learning within the group, and tends further to assimilate the wants and knowledge stocks of group members. Beyond this passive role as a disseminator of information, media construct reality as much as they reflect it (Harris, 1999). In particular, they may assume an active role in the development of fashions and trends when they differentially promote alternative developments. Because of their commercial interest in keeping the attention level high for their subject matter, media are likely to try and assume such an active role.

The effects of observational learning, communication and mass media on the development of individual wants imply that the assumption of autonomous individual decision-making commonly made in economics is at best questionable. On the other hand, by acknowledging social interaction, the emergence of collective phenomena such as fashions becomes explicable in a framework which starts from the individual level. Regularities in the patterns of change (but not in its specific contents, which depends on the unpredictable attribution of meaning in the individual minds) may in principle be identifiable.

Witt (2001) uses the framework of consumer learning to explain why consumer demand has adapted to the historical increases in consumer good variety, and why no satiation has occurred at the aggregate level. The learning processes outlined above do in no way presuppose that consumers merely react to changes on the supply side, however. In the following section, it will be argued that in the case of the mountain bike, a novel use for an existing commodity was first developed by consumers. In processes of individual and social learning, it was able to diffuse through a group of users and to motivate substantial technological development. Technological change and innovation originating from the users themselves resulted in the stabilization of the basic design before commercial interests and thus more clearly identifiable supply-side factors began to enter the market.

3 Evidence: the origins of the mountain bike[3]

The roots of the mountain bike can be traced back to the early 1970s, when some cyclists in Marin County (California) started riding down steep, unpaved mountain tracks on what were then called "clunkers" : sturdy dirt bikes built around pre-war frames from junkyards. These bicycles involved no sophisticated design, and riders equipped the second-hand frames with whatever suitable components were available (such as hubs and drum brakes from tandems, balloon tires, brake levers from motorcycles etc.). Off-road clunker riding developed into a counter-cultural activity in Marin County, with a distinct "hippie" flavor; for example, it is reported that among the first prizes awarded to race winners were envelopes of marihuana. By the end of 1973, clunker riding had attracted 20 to 30 followers in Marin County. Independently, a second group of some 10 dirt riders had assembled farther south in the Cupertino area of the Santa Cruz mountains. Two years later, the Marin County group had grown to some 50 riders.

The hippies of Marin County were not the first to cycle off-road. There are a number of documented earlier developments of cross-country bicycles (and probably a far higher number of undocumented ones). Predecessors of the mountain bike include Vernon Blake, a British cycling magazine editor who in 1930 constructed a sturdy bike equipped with balloon tires and low gears, as well as John Finley Scott, a Californian student assembling fat-tire off-road bikes equipped with derailleur gearshifts in the 1950s. In 1972, shortly before the mountain bike became popular, Tim DuPertuis constructed the first derailleur-geared fat tire bike in Marin County. He left the region shortly afterwards, and apparently had no influence on later developments. In contrast to the later Marin County clunker riders, all these predecessors remained isolated. Their bicycle designs fell into oblivion before they were able to spark further developments.

From the beginning, downhill racing played a central role in the clunker community. Races influenced the development of the mountain bike in several important ways. First, racing attracted a number of serious road cyclists to off-road cycling, some of whom later became key players in the development of the mountain bike. Second, races spurred the development of mountain bike design by providing incentives for improvement. Enormous mechanical stress on the components caused a large number of bicycles to break down in races. Some types of frames and forks proved better able than others to withstand the mechanical stress, and the various types of brakes available turned out to overheat to different extents during the long downhill stretches. Races were thus the primary "research labs" of early mountain bike development. The times needed by riders to finish the race, as well as whether they finished at all or whether their bikes broke down on the course, provided ob-

[3] This section draws heavily on the accounts of mountain bike history given by Kelly (1979, 1985, 1988, 1999) and Berto (1999).

jective data on the performance of various designs and components. Based on race experience, clunkers were improved in a trial-and-error process. This innovation-spurring effect was closely related to racing regulations, which essentially consisted of only a single rule: cyclists had to reach the finish line without external help. The reliability of the bicycle was therefore most important.[4] Besides the self-sufficiency requirement, no technical restrictions to bike designs existed.[5]

In addition to these direct effects on mountain bike technology, racing also had a more indirect impact on the development of the mountain bike. As races brought together a number of people, any successful new improvement quickly diffused within the community. If people had just gone cycling on their own, the activity would have been more dispersed geographically, and communication would have been less intense. Finally, racing spurred media interest in mountain bikes. A newspaper article in 1978 and a US-wide TV broadcast in 1979 helped make them popular outside Marin County.

Communication among mountain bike riders was increased not only by races, but also by specialized media which appeared rather early in the development (Penning, 1998). The first mountain bike magazine called *Fat Tire Flyer* was issued in 1980 by Charles Kelly. It originated from a typewritten and xeroxed newsletter featuring dates of mountain bike races and lists of stolen bicycles, supplemented by jokes and cartoons. The magazine cultivated a laconic and self-ironic "underground" style of writing which apparently reflected the attitude of the clunker community.

Early Marin County clunkers did not yet have derailleur gearing to make them suitable for uphill cycling. Nor was that of much significance then, because of their predominant use in downhill racing. The group of cyclists from Cupertino, by contrast, had started equipping their bikes with derailleurs, and their innovation diffused to Marin County. In December 1974, both groups met at a race, and the Marin County bikers were exposed to the gearshift-equipped Cupertino bikes. According to Berto (1999), the Cupertino bikes provided the inspiration to Gary Fisher, who afterwards became the first in Marin County to use a derailleur on a clunker. The December 1974 race was crucial for the development of the mountain bike, because mountain biking was abandoned in the Cupertino area shortly afterwards.

After the introduction of the gearshift, the use of mountain bikes underwent a major change, as downhill cycling and downhill races became less dominant. Without gears, cycling uphill had only been done in order to reach the point of departure for downhill cycling; it had been a substitute to motorized uphill transport, and had been restricted to strong riders. By contrast, the derailleur made cross-country cycling with uphill stretches more enjoyable, so that cross-country riding and also cross-country racing soon became increasingly popular. The first cross-country race took place in 1977. One year later a group of cyclists from Marin County drove to Crested Butte, Colorado, to participate in a long-distance race

[4] This is a key difference to cyclo-cross, an older off-road cycling discipline. In cyclo-cross, riders are allowed to change bicycles during the race, and therefore reliability is less of an issue.

[5] By contrast, design specifications in road cycling have been rather restrictive from the very beginning, which may have been one cause of the rather stagnant bicycle design over most of the 20th century (Rosen, 1995).

to Aspen over Pearl Pass. The race turned out to be more of a long-distance tour than a sportive event, which reflects a major difference between the Marin County clunker group and downhill cyclists elsewhere. The riders from Crested Butte were primarily interested in the thrill of downhill cycling, and they had organized the tour over Pearl Pass more for the spectacle than for physical exercise or competitive motives. Accordingly, they had given far less attention to improving their bicycles than had the cyclists from California. In the larger perspective, the spread of cross-country riding fundamentally changed the direction of further development of the mountain bike. Frame geometries were adapted, the number of gears increased, and more emphasis was placed on weight reductions.

Parallel to the development of mountain bike technology and mountain bike uses, a mountain bike *industry* gradually emerged. In the beginning, users and producers of mountain bikes had been identical. All through the 1970s, most clunker riders assembled their own bicycles from second-hand frames and standard components available in junkyards and in stores. After 1976, some riders began to assemble clunkers for sale within the local market as well as for their own use, but this did not represent any large-scale production, and apparently no one could make a living from it. A decisive step in the development of a mountain bike industry was the manufacturing of frames specifically designed for mountain bikes. In part, this was motivated by a diminishing supply of used parts, in particular sturdy frames, which was due to the increasing numbers of clunker riders, and the heavy toll taken by the downhill races. Moreover, even the best second-hand frames were not ideal, because they were heavy and often failed during races. The first custom-made frame was built in 1976, but it was a failure. Another attempt was made by Joe Breeze, who was both a clunker rider and an experienced frame builder. He needed two years to design and assemble ten all-new mountain bikes. They had the same frame geometry as the pre-war frames, but their design was different and they were made from better materials; in 1978, they sold for $ 750 apiece. The next step toward what would become the dominant mountain bike design, a classic diamond frame with oversized downtube, was made in 1979 by Tom Ritchey, one of the best clunker racers in Marin County and also an experienced builder of race bike frames. His first series consisted of a total of thirteen bicycles.

These first all-new bicycles proved that there was a market extending beyond that for recycled bicycle junk. Various producers started designing and producing mountain bike frames and whole bicycles. About 200 custom-made mountain bikes were sold in 1979 and about 300 in 1980. Custom-made bikes dominated the industry until 1982, when three large bicycle companies introduced mountain bikes, and total sales of the industry were an estimated 5,000 units. In 1983, industry sales increased to 50,000. Traditional bicycle producers, which had reacted late to the mountain bike boom, now started to enter the market. Similarly, most producers of bicycle components had not taken the mountain bike seriously in its early days. This changed when in 1983 two Japanese component makers, Suntour and Shimano, introduced complete groups of components consisting of cranksets, gearshifts, hubs, and bottom brackets specifically designed for off-road use.

Mountain bikes were a U.S. innovation. Like many other U.S. fashions, they came to Europe with a delay of several years. Nonetheless, early European mountain bike producers met with extreme skepticism from dealers who sometimes would not even stock the bicycles at the producer's risk. In Germany, for example, it was generally expected that mountain bikes would not sell, because the climate was less favorable than in California, traffic was more dense, and riding in the forest was subject to regulation and to potential conflict with other users. Dealers, trade associations and most of the bicycle industry apparently were unable to perceive the market potential of the mountain bike. Penning (1998) reports that when confronted with a market forecast predicting a market share of some 40 per cent for high-quality mountain bikes in the DM 2,000 price range, industry representatives were certain that there was a mistake in the position of the decimal point involved. Consequently, it was not before the mid-1990s that the German bicycle industry had made up its late start disadvantage in the mountain bike market and was able to compete successfully with imports.

A further differentiation of the mountain bike could be observed in the second half of the 1990s, after full-suspension mountain bikes had been introduced. A new group of younger users entered the stage, and downhill cycling again emerged as a discipline of its own. Specific bike designs for downhill use were developed, where weight was rather unimportant and suspension and brakes figured most prominently, and spectacular races were organized. A new 'free riding' subculture using its own symbols, fashions, and favorite brand names emerged. In contrast to the origins of the mountain bike, however, this new development cannot easily be characterized as a demand-side innovation. Commercial interests in the now established bike industry seem to have fostered this trend, as did mountain bike magazines and skiing resorts which found that downhill riders were profitable summer customers for ski lifts.

4 Discussion

The mountain bike was not invented by a single producer in a grand design to open new markets for bicycles. Rather, it was developed in a series of steps by an increasing number of users themselves. In this process, users did not only discover new uses for an existing technology, as in the cases referred to by Swann (1999) and Metcalfe (2001), but they also were innovators in the technological sense. Essentially all the major ingredients of what would become a mountain bike, including frame geometry, fat tires, derailleur shifting, cantilever brakes, a quick-release seat post, and thumb shifters, had been introduced through a process of trial and error, and a marketable product had been created, before the first mountain bikes were built for commercial purposes. The user-innovators were motivated by their personal enjoyment rather than by economic interests.[6]

In the development of the mountain bike, novelty in use largely preceded technological novelty. When the first clunkers were assembled from pre-existing components, there was not much constructive novelty involved, although of course the

[6] This is nicely illustrated by a 1979 *Bicycling* article written by Charles Kelly which characterizes mountain biking as an "edge sport" and which states: "This sport may never catch on with the American public, but its originators couldn't care less."

mixture of components was new. What turned clunkers into a new category of commodity, however, was their use for downhill racing on dirt and gravel tracks. It seems straightforward to describe the earliest development of the mountain bike as an act of social construction in the sense of Pinch and Bijker (1987), in which the earlier "closure" of bicycle design was questioned and the bicycle acquired a novel interpretation. Underlying the reinterpretation was a process of individual and social learning of wants and ways to satisfy them. At the individual level, the mountain bike clearly satisfies different wants than the road bike from which it was derived. Through their experiences with the first clunkers, cyclists learned that bicycles could be used to satisfy a want for excitement and thrill, something that would not necessarily have been suspected before as a function of the bicycle. A new want – 'cycling for thrill' – was acquired individually and spread through observational learning and communication. Once the new want had been learned, it gave rise to new demands on the bicycle's design, as improvements of the tool required to satisfy it. These demands were not met well by the available bicycles, so that a motivation was provided for experimenting with new designs. Thus, the technological development of the mountain bike, which was motivated by the attempt to satisfy better the newly acquired want, over time resulted in the accumulation of an increasing stock of consumption knowledge.[7]

This first discovery of a new use did not remain the only reinterpretation of the bicycle. When they started to equip clunkers with derailleur gearshifts, mountain bikers found that not only downhill riding could be rewarding, but also cross-country riding, which is quite a different activity. Being far more strenuous and less exciting, cross-country riding may be seen to satisfy a want for physical exercise rather than a desire for thrill. The popularization of cross-country riding thus meant another reinterpretation of the bicycle, which now became used as a tool for physical workout in an outdoor environment. Again, the new use of the bicycle was diffused in social learning and communication processes. Given that the mountain bike community at the time was rather small, single events such as the Pearl Pass tour, as the first long-distance mountain bike tour and as an event which brought together two different groups of mountain bike riders, may have been of significance in this learning process.

In turn, cross-country cycling affected the further technological development of the mountain bike by increasing the emphasis on weight reduction and on the number of gears. Consumption knowledge of how to use a mountain bike and technological knowledge of how to design a mountain bike thus coevolved in a process consistent with the theoretical notion of a refinement effect introduced above. Another interesting aspect of the introduction of derailleurs is the fact that it apparently was not an obvious thing to do. As has been outlined above, it took several years and an inspiration from outside the Marin County group before clunkers were equipped with derailleurs. In retrospect, the lagged introduction of derailleurs is

[7] This individualistic account of the popularity of the mountain bike is compatible with the sociological interpretations proposed by Rosen (1993) and Mills (1995). Rosen stresses the role of nostalgia and relates the mountain bike to a social context of "postmodern" lifestyles and "post-Fordist" production. According to Mills, mountain bikes can be seen as a cultural statement questioning the authority of generally accepted bicycle design.

surprising as no insurmountable problems were involved. The derailleur did not have to be invented anew for use on mountain bikes; it was a well-known technology used in road bikes. Many of the Marin County riders had backgrounds in road racing and some worked as mechanics in bike shops, so they were familiar with that technology. That it nonetheless required an inspiration from outside to combine the idea of the clunker with the derailleur can be interpreted as evidence for the impact of framing effects: road bikes and clunkers seem to have been seen as rather distinct commodities so that the association was not easily made. More generally, the lagged introduction of the derailleur indicates that even presumably straightforward technological improvements cannot be expected to happen "automatically", which strengthens the case for arguments based on cognitive and social theories.

With hindsight, one may argue that the mountain bike caused a reinterpretation of the bicycle in general. Before the introduction of the mountain bike, bicycles tended to be perceived as low-priced means of transportation.[8] There had hardly been any major innovation in bicycle technology for decades. Prices were low, as was quality. Bicycles were sold in department stores or (as was frequently the case in Germany) in small shops which also stocked sewing machines and which tended to be run by elderly men in grey overalls. Mountain bikes, by contrast, represented a high-quality, innovative, and fashionable product. In spite of some major drawbacks in day-to-day use, they were frequently preferred for city use and commuting both because of the superior quality of their components and because of their higher prestige. Over time, many of these components, e.g. improved gearshifts and brakes, found their way into regular bicycles. Average quality and bicycle prices increased, and bicycles even became capable of fulfilling status and prestige desires of their owners. People became accustomed to bicycle prices that had been inconceivable before.[9]

The early history of the mountain bike suggests a crucial role for social learning and communication processes. It has been noted above that there had been earlier developments of off-road bicycles which did not have any effect on the cycle industry and which soon fell into oblivion. People like Vernon Blake and John Finley Scott had been isolated eccentrics. By contrast, in Marin county there was a group of enthusiasts who kept the development going. Apparently, all early mountain bikers shared the same counter-cultural "hippie" background, which added to the group's coherence. Within the group, learning experiences of individuals could easily be communicated to others. Moreover, after 1980 the *Fat Tire Flyer* magazine not only spread information among mountain bike riders, but it may also have had an agenda-setting effect by reinforcing the hippie image through its laconic style of writing and its appearance.[10]

---

[8] This does not hold for specific kind of bicycles such as race bikes and BMX bikes. However, the development of these kinds of bicycles had little effect on both the design and the perception of regular bicycles.

[9] For example, the average bicycle sold in Germany in 1999 was priced at DM 650. Multiplied by total sales of almost 5 million bicycles per year, this figure implies a total industry revenue close to 5 billion DM (numbers reported in the *Ostthueringer Zeitung* of March 27, 2000).

[10] In a similar way, the emergence of the free riding subdiscipline in the 1990's seems to have been related to a "freestyle" subculture which is also identifiable as driving the development of snow boarding.

The group setting in a homogeneous cultural background by itself, however, seems insufficient to explain why the mountain bike boom started in Marin County and not elsewhere. It has to be recalled that in the early 1970s, there were groups of off-road cyclists elsewhere, such as the group in Cupertino and also the cyclists in Crested Butte. It may have been by chance that the mountain bike boom emerged from the Marin County developments, but a plausible case can also be made for a more systematic explanation based on the effects of racing. The Marin County group differed from other off-road cyclists in the sportive focus of their activities. Although situated in a hippie atmosphere, key figures in the group were active bicycle racers. These individuals set a competitive tone and, accordingly, the races organized by the group were taken seriously enough to time racers and document the results. Sufficient ambition was present in the races for competition to provide incentives for serious improvement efforts. As has been argued above, race regulations favored innovations, results provided data on the relative success of innovations, and race events were well suited both for reinforcing the bonds within the group and for the rapid diffusion of innovations.

The setting of the early days of the mountain bike in Marin County was therefore ideal for an extended process of social learning which helped establish basic design features and uses of the mountain bike, and which allowed the process, in contrast to similar developments at other locations and at other points in time, to gain a critical momentum. These boundary conditions, however, were shaped by the mountain bike riders themselves, and a few individuals played important roles in the process. Besides their direct impacts, these individuals served as models for observational learning in the sense of the social cognitive learning theory referred to above. Gary Fisher was one of these prominent persons. He was the first to use a derailleur on a Marin County clunker, one of the first to sell clunkers assembled from used parts, and co-founder of one of the earliest mountain bike producer firms. In addition, Fisher won a number of important races, and when he started to sell mountain bikes, he was still involved in road racing – he trained and raced in Europe, and attempted (without success) to enter the U.S. Olympic cycling team. Although Fisher seems to have realized earlier than others that there might be a market for mountain bikes, his relation to the mountain bike was more multi-faceted. Charles Kelly is another of the key figures. He organized many of the races, issued the *Fat Tire Flyer* and published articles on the mountain bike in cycling and outdoor magazines. The first custom-made mountain bike was produced for him, and he later founded a mountain bike firm with Fisher. However, Kelly, who is frequently portrayed as a 'missionary' for the mountain bike, did not exhibit much commercial interest. He retreated from the firm in 1983 and concentrated his further efforts on publishing.

In the early development of the mountain bike, producers and consumers were not yet clearly separate entities. Commercial production activities evolved as a by-product of leisure activities. The first custom-made mountain bike frames were produced by active mountain bike riders, both for their own use and for small-scale sale. In this phase of the industry, most established producers failed to appreciate the opportunities related to the mountain bike, a failure which cannot be explained by a lack of technological capabilities. Rather, it seems as though their framing of the bicycle as a means of inexpensive transportation precluded established producers from appreciating the ongoing development, and may even have caused hostility (characterized by Grant (1988) as "fear and loathing") toward the mountain bike. After 1982, when the viability of the mountain bike market had been demonstrated by increasing sales, more and more firms, both new ones and established bicycle producers, began to produce and sell mountain bikes. In addition to development activities, they also started marketing activities. A steady flow of innovations was generated to stay ahead of competitors and to keep demand high. Trends were increasingly made by producers, helped by a media scene which itself became more professional and profit-oriented over time. With increasing age and size, the mountain bike industry thus developed into a more regular consumer-good industry.

5 Concluding thoughts: a generalizable pattern of development?

The present paper aimed at demonstrating that the innovation and spread of the mountain bike cannot be understood unless the group setting of its clunker origins is taken into account. The mountain bike was not invented by a single individual, but gradually developed within a group of users. Its basic design and uses had been developed and stabilized before commercial interests entered the industry. I suspect that stories quite similar to that of the mountain bike might be written for a number of consumer goods that are used by groups of enthusiasts. Obvious parallels seem to exist to other sporting goods and outdoor equipment, such as wind surfboards, hanggliders, and snowboards. At the same time, commodities such as the mountain bike differ from ordinary fashion phenomena in that they may evolve into increasingly high-technology products with technological change coming from firms with fully-fledged R&D departments. In the case of the mountain bike, this development manifests itself in innovations such as carbon-based frame materials and electronically controlled suspension systems which are available today. There seems to exist a class of industries, then, the innovative dynamics of which are not sufficiently understood. It would therefore seem to be of importance to sort out the crucial characteristics for the success of "demand-side innovations". For example, an interesting issue is to ask whether, in consumer good innovations of the kind described here, the early, demand-driven phase of innovation can be substituted by commercial activities or if, in contrast, attempts commercially to generate fashions 'out of the blue' are bound to fail.

Moreover, the importance of demand-side innovative activities may extend far beyond this class of consumer goods between fashion and technology. Parallels also exist to the computer and software industries. For example, hobbyists played an important role in the early microcomputer industry (Langlois and Cosgel, 1999). Usenet, one of the predecessors of the Internet, was developed by a group of students who linked their computers to set up e-mail and newsletter services (Rosenzweig, 1998) in a deliberate attempt to circumvent access restrictions of a computer network sponsored by the U.S. Department of Defense (Arpanet). Their activities were motivated by idealistic motives and an interest in computer technology rather than by commercial motives, and they were part of a community of graduate students and young faculty of computer science. Another case in point is the Linux computer operating system which has for a long time been developed and continually improved by a community of "users=developers" (McKelvey, 2001) linked via the Internet. Its source code is freely available to everyone, and all additions to the program are made by volunteers who in turn make their programs freely available. In the Linux case, again, community values and identifications seem to be crucial for the motivation of software developers, and commercial interests came in only after the basic product had been established and had begun to diffuse. Demand-side innovation may thus be a pervasive phenomenon which could, if McKelvey's suggestions about emerging "Internet Entrepreneurship " (ibid.) are correct, be of increasing significance in the future.

Acknowledgements. Helpful comments on an earlier version by Per Ove Eikeland, Wilhelm Ruprecht, and Ulrich Witt are gratefully acknowledged, as is guidance to the mountain bike literature provided by Mike Lanza and Brion O'Connor. The usual disclaimer applies.

References

Aversi R, Dosi G, Fagiolo G, Meacci M, Olivetti C (1999) Demand dynamics with socially evolving preferences. Industrial and Corporate Change 8: 353–408
Bandura A (1986) Social Foundations of Thought and Action. A Social Cognitive Theory. Englewood Cliffs, NJ: Prentice-Hall
Berto FJ (1999) The Birth of Dirt. Origins of Mountain Biking. San Francisco: Cycling Resources
Bianchi M (ed.) (1998) The Active Consumer. Novelty and Surprise in Consumer Choice. London and New York: Routledge
Bianchi M (1999) Design and efficiency: new capabilities embedded in new products. In: Dow SC, Earl PE (eds.) Economic Organization and Economic Knowledge. Cheltenham and Northampton: Edward Elgar, pp. 119–138
Bijker WE (1995) King of the road: the social construction of the safety bicycle. In: Bijker WE Of Bicycles, Bakelites, and Bulbs. Toward a Theory of Sociotechnical Change. Cambridge, MA and London: MIT Press, pp. 19–100
Bijker WE, Law J (1992) General introduction. In: Bijker WE, Law J (eds.): Shaping Technology/Building Society. Studies in Sociotechnical Change. Cambridge, MA and London: MIT Press, pp. 1–14
Brouwer E, Kleinknecht A (1999) Keynes-plus? Effective demand and changes in firm-level R&D: an empirical note. Cambridge Journal of Economics 23: 385–391
Devetag MG (1999) From utilities to mental models: a critical survey on decision rules and cognition in consumer choice. Industrial and Corporate Change 8: 289–351
Dosi G (1982) Technological paradigms and technological trajectories. A suggested interpretation of the determinants and directions of technical change. Research Policy 11: 147–162
Eysenck MW, Keane MT (1995) Cognitive Psychology. A Student's Handbook. 3rd ed. Hove: Psychology Press
Geroski PA,Walters CF (1995) Innovative activity over the business cycle. Economic Journal 105: 916–928
Grant R (1988) Foreword. In: Kelly CR, Crane N: Richard's Mountain Bike Book London: Richard's Bicycle Books, pp. 9–12
Harris RJ (1999) A Cognitive Psychology of Mass Communication. Mahwah NJ; London: Lawrence Erlbaum Associates
Hughes TP (1986) The seamless web: technology, science, etcetera, etcetera. Social Studies of Science 16: 281–292
Kellogg RT (1995) Cognitive Psychology. Thousand Oaks: Sage
Kelly CR (1979) Clunkers among the hills. Bicycling (January): 40–42
Kelly CR (1985) The vanguard. ATBs and their dedicated, imaginative band of California builders. Bicycling (March): 126–139

Kelly CR (1988) The mountain bike. In: Kelly CR, Crane N: Richard's Mountain Bike Book London: Richard's Bicycle Books, pp. 19–156
Kelly CR (1999) Repack reunion. In: Berto FJ: The Birth of Dirt. Origins of Mountain Biking. San Francisco, CA: Cycling Resources, pp. 105–118
Langlois RN, Cosgel MM (1998) The organization of consumption. In: Bianchi M. (ed.): The Active Consumer. Novelty and Surprise in Consumer Choice. London and New York: Routledge, pp. 107–121
Leibenstein H (1950) Bandwagon, snob, and Veblen effects in the theory of consumer demand. Quarterly Journal of Economics 64: 183–207
McKelvey M (2001) Internet entrepreneurship: Linux and the dynamics of open source software. University of Manchester: CRIC Discussion Paper no. 44
Metcalfe JS (2001) Consumption, preferences, and the evolutionary agenda. Journal of Evolutionary Economics 11: 37–58
Mills R (1995) Thinking about thinking about cycles. In: Cycle History: Proceedings of the 5th International Cycle History Conference. San Francisco, CA: Bicycle Books, pp. 11–20
Mowery DC, Rosenberg N (1979) The influence of market demand upon innovation: a critical review of some recent empirical studies. Research Policy 8: 103–153
Oddy N (1994) Non-technological factors in early cycle design. In: Cycle History: Proceedings of the 4th International Cycle History Conference. San Francisco, CA: Bicycle Books, pp. 63–67
Penning C (1998) Bike History: Die Erfolgsstory des Mountainbikes. Bielefeld: Delius Klasing
Pinch TJ, Bijker WE (1984) The social construction of facts and artefacts: or how the sociology of science and the sociology of technology might benefit each other. Social Studies of Science 14: 399–441
Rosen P (1993) The social construction of mountain bikes: technology and postmodernity in the cycle industry. Social Studies of Science 23: 479–513
Rosen P (1995) Diamonds are forever: the socio-technical shaping of bicycle design. In: Cycle History: Proceedings of the 5th International Cycle History Conference. San Francisco, CA: Bicycle Books, pp. 51–58
Rosenzweig R (1998) Wizards, bureaucrats, warriors, and hackers: writing the history of the Internet. American Historical Review 103: 1530–1552
Scherer FM (1982) Demand-pull and technological innovation: Schmookler revisited. Journal of Industrial Economics 30: 225–237
Schmookler J (1966) Invention and Economic Growth. Cambridge, MA: Harvard University Press
Schumpeter JA (1934/1983) The Theory of Economic Development. Cambridge, MA: Harvard University Press. (Transaction edition)
Simon H (1986) Rationality in psychology and economics. Journal of Business 59: S209–S224
Swann GMP (1999) Marshall's consumer as an innovator. In: Dow SC, Earl PE (eds.) Economic Organization and Economic Knowledge. Cheltenham and Northampton: Edward Elgar, pp. 98–118
Tversky A, Kahneman D (1981) The framing of decisions and the psychology of choice. Science 211: 453–458
Veblen T (1899) The Theory of the Leisure Class. New York: Macmillan
Von Hippel EA (1978) Users as innovators. Technology Review (January): 31–39
Winner L (1993) Upon opening the black box and finding it empty: social constructivism and the philosophy of technology. Science, Technology, and Human Values 18: 362–378
Witt U (1996) The political economy of mass media societies. Max Planck Institute for Research into Economic Systems: Papers on Economics and Evolution #9601
Witt U (2001) Learning to consume – a theory of wants and the growth of demand. Journal of Evolutionary Economics 11: 23–36

A resource-based view of Schumpeterian economic dynamics*

John A. Mathews

Macquarie Graduate School of Management, Macquarie University, Sydney, NSW 2109, Australia
(e-mail: john.mathews@mq.edu.au,
URL http://www.gsm.mq.edu.au/faculty/home/john.mathews/index.html)

Abstract. This paper seeks to offer a theoretical platform where the modern "resource-based view" of the firm might meet with evolutionary economics and the study of entrepreneurship, and with the economics of industrial organization. It does so by proposing the concept of the "resource economy" within which productive resources are produced and exchanged between firms. This is presented as the dual of the mainstream goods and services economy – where the "resource economy" captures the dynamic capital structure of the economy. The paper is concerned to bring out the distinctive principles governing resource dynamics in the resource economy, capturing competitive dynamics in such categories as resource creation, replication, propagation, exchange and leverage; evolutionary dynamics in terms of resource variation, selection and retention; entrepreneurial dynamics in terms of resource recombination and resource imitation, transfer and substitution; and industrial organizational dynamics in terms of resource configuration, resource complementarities and resource trajectories.

Key words: Schumpeterian dynamics – Resource-based View – Resource economy – Evolutionary dynamics – Entrepreneurial dynamics

JEL Classification: L0, L1, O0, P0

* Comments by Bengt-Åke Lundvall, Stan Metcalfe, George Richardson, Lars-Gunnar Mattsson, Linda Weiss and Ivo Zander on earlier drafts of this paper, are gratefully acknowledged.

1 Introduction

For as long as one can remember, the edifice of the neoclassical economic synthesis has been under attack. Critiques have focused on the extreme unreality of the assumptions that underpin the Arrow-Debreu theorems of welfare economics. They have queried the excessive formalism of the edifice, and the lack of practical significance of many of the results. They have castigated the neoclassical synthesis for its internal incoherence (lacking an independent theory of capital, for example, one of the favorite topics of the Cambridge school), its lack of a dynamic element, its non-evolutionary character, its lack of any conception of "market process" – and so the list could be continued (Blaug, 1997). Through all this, the neoclassical synthesis remains as strong as ever, impervious it seems to these or any other attacks.

In this paper a different tack is taken. The neoclassical edifice is left alone, standing as a representation of what goes on *in a certain kind of economy* – namely the economy where goods and services are produced and exchanged. The paper then introduces another kind of economy, namely an economy of productive entities – called "resources" – that are needed to produce the economy of goods and services. Resources also are produced and exchanged. As "capital goods" they are used by firms to transform inputs into outputs, but the resources themselves are not consumed. In their totality they can be said to constitute the "resource economy." Now it turns out that the rules of production and exchange of resources are rather different from the rules governing the goods and services economy. Resources are not in general free-standing entities; they are tightly bound to firms. They can be accessed and exchanged – but usually through complex interfirm transactions. Resources are bundled together into firms – with the prime challenge for the firm being to build synergies between resources to ensure distinctiveness and generate entrepreneurial profits. The resource bundles within firms change over time, as firms adapt to changing circumstances, or as sudden new resource combinations are created – as described so clearly by Schumpeter in his *Theorie der Wirtschaftliches Entwicklung* (1912/1934/1996).

Thus the resource economy is best approached from a dynamic and evolutionary perspective, where path dependence and trajectories are paramount, rather than in terms of the static framework favoured by the neoclassical synthesis. It also turns out that the resource economy can only be approached from the perspective of disequilibrium, since resource dynamics are under the control of entrepreneurial production plans that are always inconsistent in any real economy – as understood and defended by economists such as Hayek (1941) and Lachmann (1956/1978) as well as Schumpeter. Thus the "resource economy" takes us to the heart of capitalist dynamics.

The idea of "resources" of course comes from the "resource-based view" (RBV) of the firm, which is a school of thought that has risen to prominence in the strategic management literature. It is otherwise known as the "capabilities view" in which guise it has been taken up and discussed by numerous economists (see Foss (1997), Foss and Knudsen (1996) or Foss and Loasby (1998) for representative discussions). Calls have been made to develop a synthesis of the RBV of the firm with evolutionary economics (Montgomery, 1995) and with entrepreneurship studies (Hitt and Ireland, 2000). Yet the disciplines remain stubbornly apart. One of the goals of this paper is to explore the possibility that the notion of the "resource economy" might provide common ground where these different discourses might meet, and where fresh insights might be generated.

To date, these various approaches have held back from developing a common language and set of concepts. The RBV of the firm, while making some welcome progress in accounting for the heterogeneity of firms (in terms of the distinctiveness of their resources and associated capabilities), has nevertheless stopped short of taking its insights into the wider economy. The RBV of the firm remains anchored to a view that sees firms developing their resources internally, ignoring the wider aspects of resource exchange.[1] The conventional RBV has made little use of any notion of shared resources or "extended" capabilities (Coombs and Metcalfe, 1999); moreover it appears to be wedded to an incumbent's view of competitive dynamics, ignoring the challenger's perspective and the strategies that challengers use to acquire or leverage resources externally. The RBV has also made little contact with the literature on the "market" for corporate control, which is concerned with the terms on which corporate assets are bought and sold. A focus on resources themselves, as fundamental entities, and their production and exchange in the wider economy, is needed to clear away these conceptual obstacles. The evolutionary view, for its part, discusses the evolution of firms, technologies, markets, or organizational "routines" – but rarely discusses the evolution of "resources" in general, with some important exceptions (Winter, 1995). Yet resources can be taken as a generalization in a certain sense of all these entities. Likewise the literature on entrepreneurship, with important exceptions, has remained detached from strategic concerns and has instead focused on equilibrium-returning features of the process. It is only recently that scholars have sought common ground (McGrath and MacMillan, 2000).

Thus an examination of the notion of a "resource economy" where firms trade productive resources with each other in order to enhance their competitive prospects, and build new combinations through entrepreneurial activities, and where these resources follow cumulative pathways exhibiting evolutionary dynamics, offers some hope of bringing these various fields together, and enriching them, in a way that it is independent of the concerns of the neoclassical synthesis.

2 The resource economy

Consider then, as an exercise in imagination, an entity to be called the "resource economy." By this is meant the totality of productive entities that make the production of goods and services possible. Resources are the fundamental units of value generation. They do not exist independently, but are contained within firms. Resources can be specialized and bundled together in highly distinctive configurations, to lend firms special competitive advantages. Resources can be built by firms

[1] Two of the influential contributions to the field, Barney (1986, 1991) and Dierickx and Cool (1989) discuss this issue (in terms of imperfections and incompleteness of "strategic factor markets"), but dismiss external sourcing since it is theoretically available to all firms and therefore cannot serve as a source of distinctive advantage.

internally, and they can be traded – as described every day in the business pages of the newspaper. The production and exchange of resources is what we shall describe as the "resource economy." Resources are the productive assets of firms, the means through which activities are accomplished. The basic insight that separates the RBV of the firm, and evolutionary economics generally, from conventional economic and industrial organization analysis, is that resources are seen as lending distinctiveness to firms, i.e. generating heterogeneity. There is no "representative" firm in the resource economy; the point is to model firms in all their heterogeneity, starting with their different resource endowments, and moving on to the dynamics of the processes through which these resource endowments may be changed (extended, contracted) through the development of routines and the inter-relations between firms (Eliasson, 1986). The firm's resources set limits to what the company can do.[2] As such, resources include tangible entities such as production systems, technologies, machinery, as well as intangibles like brands, or property rights such as landing rights for an airline or bandwidth for a telecoms company.[3]

Resources then are the basic constitutive elements out of which firms transform inputs into outputs, or generate services. To provide an airline service, for example, a firm needs to acquire and consolidate resources such as aircraft with crews to fly them; landing slots at airports and the administrative capacities to operate them; passenger booking systems and the skills to operate them; aircraft maintenance facilities and skills, and so on. Building a distinctive "competence" out of these consolidated resources, the firm will enter into the passenger transport industry, and equipped with a certain strategy (e.g., targeting the business traveller) it will either succeed or not. Its strategic capabilities and the competitive advantages generated over rival firms, rest on the distinctiveness of the resources at its command. These will not necessarily have to be owned; indeed the firm may lease its aircraft, subcontract its maintenance operations, and even its ground operations, reserving only the core functions of actually booking passengers and flying them as its distinctive competencies. An airline building a national air service can build its routes one at a time, or it can accelerate the process by acquiring routes from one or more regional operators (or acquiring the regional airlines themselves). Thus resources can be assembled by firms to reflect their current strategic imperatives.

The resource economy as defined may be characterized as the dual of the mainstream, neoclassical "goods and services" economy. As pointed out by Wernerfelt (1984) products (activities) and resources are two sides of the same coin.[4] By this is meant that they describe two facets of the same reality, namely a process of production of goods and services. In the economy as a whole, there are on the one hand

---

[2] Rumelt (1984) was one of the first to link strategic direction with resources; he argued that the firm's strategic significance is "characterized by a bundle of linked and idiosyncratic resources and resource conversion activities" (1984, pp. 561).

[3] Teece, Pisano and Shuen (1997, p. 521) prefer the term "specific assets" by which they mean, the firm's specialized plant and equipment, its "difficult-to-trade knowledge assets and assets complementary to them, such as its reputational and relational assets."

[4] As Wernerfelt (1984, p. 171) put it: "Most products require the services of several resources and most resources can be used in several products. By specifying the size of the firm's activity in different product markets, it is possible to infer the minimum necessary resource commitments. Conversely, by specifying a resource profile for a firm, it is possible to find the optimal product-market activities."

the activities conducted by firms, which transform inputs into outputs; the terms used to describe these processes are the familiar ones of inputs, outputs, costs and prices. The object of analysis is to determine for any given set of inputs and outputs a set of prices which will clear all markets, i.e. produce an equilibrium balance between supply and demand.

In the resource economy, on the other hand, the object of interest is the configuration of resources, i.e. their distribution in heterogeneous and distinctive bundles, within and between firms. What is of interest is the adaptive capacity of such an economy, in terms of its abilities to generate new resource configurations, and the evolutionary pathways along which such resource configurations develop. These resources, in totality, account for the production of the goods and services that are described in mainstream economics. It is the same economy we are dealing with, but we are viewing it from a fresh perspective.

Resources are very real and very expensive. They are bid for, won and lost every day, as reported in the business page of the newspaper. The price that productive resources fetch (eg a division of a company, a cellular telephone license, a group of media titles and their editorial staffs) is usually much greater than their asset "book value" and is determined by corporate valuations, such as stock market valuations if the company is listed. So in one sense the answer to the question: what is the total worth of a nation's "resource economy" is provided by the total stock market valuation of the economy's firms. In another sense, the "value" of a resource is entirely a matter of strategic judgment – the same resource can have very different value for different firms, depending on the use to which it is put, or to the same firm at different points in time.[5]

The resources in a real economy are in a constant state of flux, accounting for observed phenomena of competitive and evolutionary dynamics. Resources are being developed by firms and being exchanged between firms, through open-market deals (eg as in the sale of a division of one firm to another) or more commonly through various kinds of contractual arrangements (eg technology transfer agreements, subcontracting/OEM agreements, licensing arrangements) or through resource transfers effected as a result of mergers or acquisitions. It is through these contacts that *resources are exchanged and shared* between firms, either voluntarily or involuntarily. These can be identified as cases of resource propagation, resource replication, resource exchange, resource redeployment, resource sharing and resource leverage.[6] All are involved in the dynamics of the resource economy.

As emphasized in the "Austrian" theory of capital, the processes of resource exchange and dynamics operate in a non-equilibrium or disequilibrium frame-

[5] Resource valuation is a topic rarely tackled in the conventional RBV of the firm, which is perhaps one of the reasons it has not become a mainstream economic discipline. Valuation involves processes such as discounted cash flow and, where available, the valuation provided by stock markets. Insofar as resources are exchanged commercially between firms, the resource economy may be identified with the market for corporate control. However many more such processes of resource movement are encompassed in the concept of the resource economy.

[6] On resource exchange, see Moran and Ghoshal (1999); on resource redeployment, as a result of horizontal mergers and acquisitions, see Capron and Mitchell (1998); on resource leverage, see Prahalad and Hamel (1990). On resource leverage as a resource-focused catchup strategy, see Mathews (1997a, b; 1998).

work – for the simple reason that equilibrium would entail perfect congruity of entrepreneurs' business plans or business models. While equilibrium is at least plausible (if unrealistic) in the markets for goods and services, it is neither plausible nor realistic in the case of markets for resources.[7]

It bears repeating that the restlessness of the resource economy is quite distinct from the activities of the firms embodying these resources – their production activities. Of course there could be a great deal of production and other economic activity without much resource exchange – and vice versa, there could be a great deal of "resource churning" (eg huge numbers of mergers and acquisitions) without much effect on the level of productive activity. But in general, one would expect to find in a productive economy a reasonable degree of resource exchange activity. The extent of this depends on the development of specialized markets for resources. Resource exchange takes place largely through bilateral contractual arrangements, without the mediation of a "market" at all. But some firms show exceptional enterprise and actually create "markets" for resources through their brokerage activities. Merchant banks in particular take the lead role in this.[8] It is certainly an indicator of a very sophisticated economy when organized markets for resources start to appear – as "capital markets" made their early appearance and brought capitalist economies to a new level of sophistication.

3 Firms and resource dynamics

The first substantial issue to consider is how resources may be encapsulated within firms, and how firms may derive profitable opportunities from this bundling.[9] To be plausible, our account of the resource economy must translate into an account of firms and their capabilities that is consistent with the Penrose view, and with the insights of the RBV. This leads to questions such as what determines the rate of growth of firms as resource bundles, the limits to this growth, and how these matters are translated into entrepreneurial and management practice.

The disposition of resources within firms is the outcome of entrepreneurial action, or it is bequeathed from earlier resource combinations (Galunic and Rodan 1998). It is the task of the entrepreneur to assemble a bundle of resources and to capture synergies so that revenues generated (returns from sale of outputs net of costs of inputs) exceed rents paid on the resources utilized; this is the task

[7] See Hayek (1941) and Lachmann (1956/1978; 1973) for statements of this point of view; Lewin (1997) and Foss (1994) as well as Lewin and Phelan (1999) provide links between these views of capital and modern resource-based discussions of firms' competitive advantage.

[8] Merchant banks frequently recombine resources and launch them as new companies. For example Deutsche Bank in 1999 was assembling wireless communications licenses covering different parts of Europe in order to bundle them into a new corporate venture.

[9] It was Edith Penrose in *The Theory of the Growth of the Firm* (1959) who developed the first clear expression of a "resource-based view" of the firm. She considered firms to be "bundles of resources" and saw the specialization of these resources as fundamentally accounting for the variations between firms. As Penrose put it (1959/1995: 24): "... a firm is more than an administrative unit; it is also a collection of productive resources the disposal of which between different uses and over time is determined by administrative [management] decision." For a recent discussion of Penrose's contribution, see Pitelis and Wahl (1998) as well as Foss (2000).

of producing positive entrepreneurial profits.[10] It is the task of management to utilize such a resource stock and extract the most productive services from it in transforming inputs into outputs. The range of goods and services to be produced with the services provided by such a resource stock cannot be known in advance; it is a matter of discovery, a process of learning, where the outcome depends on the management's knowledge, experience and capacity for imaginative experiment. Management seeks to capture synergies between resources (utilizing a resource bundle for more than one kind of activity, or to produce goods for more than one kind of market). The capture of such synergies is the resource-economy equivalent of co-specialization of assets and the capture of cost-based economies of scope in the goods and services economy.[11]

What accounts for the growth of firms is their propensity to develop management or organizational "routines" which then liberate management attention to investigate and discover further development and diversification opportunities. Penrose (1959) puts this in terms of management capturing the services of an "excess" of resources that call for diversification into production of new products or entry into new markets. What then limits the size of firms, or their rate of growth, is the managerial burdens of keeping track of these diversifications. In the end, the firm can "pay attention" only to so many different kinds of activities. In the end, it can forge an effective union out of only so many resources; beyond that limit, at any time, the firm functions as no more than a conglomerate, where its resources have no synergistic interaction (and the whole is therefore no more than the sum of the parts). Such a disaggregated firm is a prime candidate for a hostile takeover.

This then is a behavioural account of the process through which managements are led to seek diversification and new market entry, based on an existing stock of resources, and why they are led to seek to combine those resources with others (eg through mergers and acquisitions) to further enlarge their "strategic options" (Itami, 1987). It works on the basis of a notion of "disequilibrium" within the firm, where the potential services rendered by a stock of resources is out of balance with the actual services being secured through the firm's existing organizational routines.[12]

[10] Note in this treatment I am not making assumptions as to whether resources are available at less than full cost on "imperfect" factor markets – as done within the RBV approach by, e.g. Barney (1986). I am making the conservative assumption, along with Schumpeter (1912/1934/1996) and earlier capital theorists such as Clark (1888), or Fetter (1904; 1927) that entrepreneurs pay full costs for the capital goods (resources) utilized. In this way I bypass completely all the complications associated with Ricardian treatment of rents as stemming from resources that are "rare" or "fixed." It is much more straightforward to assume that resources are available and can be secured, and that what counts is the packaging or bundling of the resources into a distinctive whole within the firm. Clark (1888) made the distinction between "capital" (as a fund) and capital goods, which included land and all productive factors; while Fetter simplified notions of rent to argue that rents are the earnings of any factor of production, irrespective of their rarity or fixity.

[11] Teece (1986; 1992) has developed an account of the dynamics of firm diversification in terms of the co-specialization of the assets involved and the capacity of managements to appropriate the services of these assets.

[12] Loasby (1991), building on the work of Hahn, has developed a formulation of this process in terms of attaining an "equilibrium" within the firm between the services provided by the current resource base and the services required by the current range of goods and services produced. This is a striking way of expressing the core of Penrose's argument. But of course it is a completely different use of the term

We shall discuss below the analogue to this intra-firm resource disequilibrium in the form of an economy-wide resource disequilibrium, which generates the motive for entrepreneurship.

In passing, it is worth noting that this provides a plausible foundation for a theory of management. Given a stock of resources within a firm (assembled through entrepreneurial action or bequeathed) it is management's task to develop the "organizational routines" needed to capture as many of the services from these resources as possible. Management has the task of rationalizing the resource base, in order to capture synergies. Yardsticks to measure management performance are then its effectiveness and efficiency in developing, and adapting, the routines needed to put in motion the firm's resource stock. This is a theory of management which is concerned with maximizing the creation of value through discovering new activities, rather than appropriating as much value as possible (through cost cutting) from a given set of activities – in keeping with the best of current treatments of the management function (Ghoshal, Bartlett and Moran, 1999).

To summarize the discussion so far, what we have is a picture of the economy where firms are built from encapsulated resources, and operated [managed] with a view to building and capturing resource synergies. Firms are involved in actively accumulating resources to enhance their dynamic distinctiveness and capabilities (Teece, Pisano and Shuen, 1997). As firms translate their newly discovered activities into "routines" so management attention is liberated for further discovery, and they are led to grow and diversify, building on their "excess" resource base, ie on a disequilibrium in their resources. Successful diversification is based on co-specialization of resources that act synergistically with each other.[13] Firms seek complementary resources from other firms with which they have direct dealings, through the dynamics of resource propagation, replication, leverage and transfer. These constitute the exchange dynamics of the resource economy, driven by disequilibrium considerations (rather than the equilibrium considerations which govern neoclassical analysis of the goods and services economy).

What drives firms in these patterns of behaviour is the competitive dynamics of an industry – the role played by rival firms, as well as by potential partners and other kinds of organizations. So we turn next to the analysis of competitive dynamics from the resource perspective, to see what added insights may be gleaned from this approach.

4 Competitive dynamics: incumbents and challengers

Firms are in unrelenting competition with each other, in terms of their products and services. Price competition is the primary vehicle through which these dynamics

'equilibrium' from its use in neoclassical economics, and it goes against the grain of the 'disequilibrium' tone of reasoning adopted in this paper – and therefore it is not pursued here.

[13] Substantive predictions follow from this account, such as that the "value" of firms will reflect the degree to which managements have succeeded in capturing resource synergies. Empirical work designed to test such predictions would have to utilize a value parameter such as *Tobin's q*, and proxies for the firm's resources – as is done in studies which seek to capture the effects on firm value of diversification. For a recent review of the issues involved, see Steiner (1996).

are expressed, as well as qualitative attributes like time to market, product quality, customer responsiveness and innovation – as described in the analysis of the goods and services economy. In the 1980s a vision of firms locked in competitive struggle within an "industry forces" framework was developed (Porter 1980; 1985). The basic assumption, in keeping with the neoclassical synthesis, was that firms are more or less uniform; what distinguishes their performance (and their potential sustainable competitive advantage) is the industrial setting in which they find themselves. Industrial pressures are transmitted through processes such as barriers to mobility that keep firms locked in (and out of) industries.

This "industry forces" view of competitive processes, based on a view of firms as homogeneous, has given way in the 1990s to an approach that sees firms as heterogeneous, and looks inside firms, to their resources, for an account of competitive performance. The essential insight of the RBV of the firm as developed in strategic management has been that underpinning these competitive struggles in product markets lie the attempts by firms to secure sustainable competitive advantages through the distinctiveness of their resource base (Wernerfelt, 1984; Barney, 1986; Dierickx and Cool, 1989; Peteraf, 1993, Amit and Shoemaker, 1993; and for a critical perspective, Foss, 1998, or Foss and Knudsen, 2001). Thus there are multiple levels to competitive dynamics. The most obvious and superficial level is that of product competition. Beneath that there is competition over product ranges and families, eg brand loyalties from one product to another, and product architectures (eg the Intel Pentium series of microprocessors). And beneath this level is the most fundamental of all, namely the underlying resources (assets and capabilities) that enable firms to consistently bring out new competitive products and thereby circumvent their rivals.

This is the insight that has generated a new perspective on the competitive dynamics of the resource economy. Firms are competing with each other, at the most basic level, through emulation, variation and substitution of each other's resources. *It is the competitive struggle over resources that may be viewed as the fundamental driving force of the capitalist economy.*

There is a Marshallian and a Schumpeterian dimension to these resource-based competitive dynamics. Marshall's conception of competitive dynamics involved firms with varying strategies and programs each implementing their different approaches; the market then "selected" the most appropriate strategic arrangement in line with current demand and industrial preferences. The Marshallian processes of competitive dynamics are observed every day as firms compete not just in terms of prices but in terms of their complementary offerings, involving technologies, or products connected together in value-chains. In industrial districts, the Marshallian forces may be seen in terms of the sharp competition between suppliers of similar goods or services, and the collaboration between complementary suppliers linked in a value chain. These are the origins of increasing returns in a manufacturing district – as discussed with perfect clarity by Young (1928).

Marshall captured an essential feature of these processes in the distinction between the economies which a firm could introduce for itself (internal economies) and those introduced by other firms but which are of benefit to the focal firm (external economies). In doing so, Marshall was able to reconcile the phenomena of increasing returns and inter-firm competition: the firm's activities are subject to diminishing returns, but the benefits it derives from other firms (externalities) enable increasing returns to be secured.[14] Translating across to the terms of the resource economy, it may be observed that firms derive advantages not just from the resources they embody themselves, but also from resources *external to the firm* to which the firm can secure access. Following Richardson (1972) we may call these complementary resources.[15] These are the critical insights, traceable to Penrose and Richardson, which have remained under-utilized in the RBV of the firm as developed in the strategic management literature.

4.1 Schumpeterian competitive dynamics

The more fundamental and sweeping kind of competition that drives capitalist dynamics is captured by Schumpeter's conception of the "creative gales of destruction" that regularly sweep through the capitalist system, initiated by entrepreneurs who break with existing arrangements in order to try out new combinations. From a resource perspective, such entrepreneurs are accomplishing *resource recombination* – one of the most powerful factors driving competitive dynamics.[16]

Of course Schumpeter did not use the language of resources, which has only come into vogue in the 1980s and 1990s – but it is easy to translate Schumpeter's insights into the language of competitive resource imitation and substitution. From the resource perspective, the Schumpeterian dynamics may be captured in terms of *resource imitation*, *resource transfer* and *resource substitution*. (These are the terms used in the RBV of the firm.) It bears repeating that we are talking here of processes at the resource level, not at the level of the goods and services produced from the resources.

The RBV of the firm emphasizes the sustainability of competitive advantages due to resource endowments. To do so, it is focused almost exclusively on the extent to which firms can capture resources that are difficult to imitate and not easily

[14] Prendergast (1992, p. 460) puts the matter in these terms: "By the time he published the first edition of his *Principles*, Marshall had formulated an ingenious theoretical solution to the problem of reconciling increasing returns and competition within the framework devised by Cournot. The solution involved the introduction of the concept of external economies which were viewed as the sole cause of increasing returns within a regime of competition. Interpreted as a perturbation of a firm's unit-homogeneous production function caused by changes in the output of the industry as a whole, external economies are a device of considerable power and elegance ..." For a critique of this position, see Hart (1996).

[15] Richardson (1972) referred to firms' activities and capabilities; he used capabilities very nearly in the sense referred to here by resources. Complementary activities are those which bind firms together in contractual arrangements, thus forming larger aggregates which constitute the "organization of industry." These issues, which have been ignored in the conventional RBV of the firm, are taken up below.

[16] As Schumpeter himself put it (1942/1975, p. 84): "... in capitalist reality as distinguished from its textbook picture, it is not (price-guided) competition which counts but the competition from the new commodity, the new technology, the new source of supply, the new type of organization (the largest-scale unit of control for instance) – competition which commands a decisive cost or quality advantage and which strikes not at the margins of the profits and the outputs of the existing firms but at their foundations and their very lives. This kind of competition is as much more effective than the other as a bombardment is in comparison with forcing a door ..."

transferred or substituted. This has always struck me as extremely odd. It takes an incumbent's perspective – whereas economics should, and normally would, be more concerned with promoting competition, and would therefore take a challenger's perspective. In the competitive dynamics as developed in this article, we are concerned with neither incumbent nor challenger advantage, but with how both incumbents and challengers drive each other to higher and higher levels of economic performance.

4.2 The incumbent's perspective: Uncertain imitability of resources

It is through uncertain imitability of resources that incumbents are able to establish sustainable competitive advantages. The more that incumbents are able to create (resource-based) isolating mechanisms, the more sustainable their advantages. Lippman and Rumelt (1982) and Rumelt (1984) introduced these ideas in the explicit context of a resource-based view of strategic competitive dynamics. They demonstrated how an analysis at the level of resources would shed light on the sources of sustainability, ie through uncertain imitability; Rumelt introduced the idea of an "isolating mechanism" as the (resource-based) firm-level analogue of mobility barriers (Caves and Porter, 1977) at the industry level.

For our purposes, where we are concerned as much with a challenger perspective as with incumbents, the Lippman and Rumelt theorem tells only half the story. The complementary proposition concerns how challengers successfully confront incumbents, even when they have built a resource base on causal ambiguity and strewn the competitive landscape with as many "isolating mechanisms" as they can devise.

4.3 The challenger's perspective: reliable imitability

It is through the fundamental imitability and transferability of resources that challengers are able to invade industry segments occupied by incumbents. Challengers acquire the requisite resources through internal development and through external leverage, where they are guided in their choice of which industry segment to attack by the availability of resources that are most easily imitated and transferred. We may coin the expression "reliable imitability" for such an approach, to bring out the complementarity with the uncertain imitability of Lippman and Rumelt.

Reliable imitability depends for its plausibility on such features of the resource economy as path dependence. Technologies, for example, are known to evolve along "trajectories" that reflect the path dependence of cumulative design and utilization decisions (Dosi et al. 1988, 1997). In resource terms, this may be described as a case of predictable resource evolution (as discussed below). Now a challenger can "read" a technological or resource trajectory as well as an incumbent – in fact, it can probably read the trajectory better, because it is unencumbered with the prior commitments that create inertia for firms, and make it so difficult for them to swing into new technological trajectories.[17]

If resources were non-transferable and non-imitable, then incumbents' competitive advantages would be sustainable forever. But firms are able to diversify and challenge incumbents' positions. They are able to do so because they adjust their resources to their strategic needs. A goal of entering a new market needs to be thought through, from the resources perspective, with an analysis of the resources required to support such a shift. This is what Itami (1987) calls "dynamic resource fit" and he gives numerous Japanese examples of firms building their resource base, or acquiring new resources, in order to support their new strategic thrust. Firms are able to draw on multiple connections, from industrial networks or supplier networks, in effecting these resource transfers. East Asian firms in Korea and Taiwan and Singapore have all learned much from these Japanese examples, and have applied the lessons in their own attempts to "leverage" resources from advanced firms in advanced countries. The case of the creation of a semiconductor industry in East Asia, entirely through strategies of resource leverage (knowledge, technology, market access) is one of the best examples of this process at work. The strategies pursued by the firms involved sought to make up their initial disadvantages in terms of their "latecomer" advantages – such as being able to read technological trajectories, and take advantage of the availability of process technology equipment from third party vendors. These are ways in which the imitation of a given resource base may be made more "reliable" (Mathews, 1997, 1998, 2001b; Mathews and Cho, 1999, 2000).

Competitive dynamics shape the rise and fall of firms within an industry setting at any moment in time. Incumbents seek to defend their position, through the uncertain imitability of their distinctive resource base. Challengers are constantly seeking ways to evade this resource base, or to appropriate it, through imitation, transfer and substitution of resources. Their success can be grounded in the sources of reliable imitability, such as the tendency of resources to evolve along certain well-defined trajectories. So we turn next to consider the evolutionary dynamics of the resource economy.

5 Evolutionary dynamics

The ingredients of an evolutionary approach in economics are now reasonably well-defined.[18] It is clear that a consistent and coherent account must identify some

[17] Henderson and Clark (1990) and Henderson (1993) give the graphic example of semiconductor equipment supply firms, where in each successive generation of the technology, the previous leading firm was unable to make the transition; this is plausibly interpreted by Henderson and Clark as a case of organizational failure to accommodate new technological architectures. The argument clearly carries over to the resource economy, where firms committed to a particular resource trajectory will find it difficult to accommodate new resource variations. This is the challenger's advantage, and the source of "reliable imitability."

[18] For an introduction, see Dosi and Nelson (1994); the definitive treatment is by Nelson and Winter (1982). Langlois and Everett (1994) provide an illuminating discussion informed by a reading of the current evolutionary debates in the biological sciences. Andersen (1994), Hodgson (1993), Witt (1992),

category or categories as unit of variation and something else (or the same) as unit of selection, together with an account of the actual processes involved in generating variations and selecting entities according to some designated "fitness" criterion. Furthermore there has to be some kind of "inheritance" function, or entity which accounts for retention. The point of course is that such a theoretical structure has nothing in common with the comparative statics of the neoclassical synthesis; it represents a completely different way of visualizing the workings of the economic system.

It was Nelson and Winter who first formulated a clear evolutionary account, as an alternative to the static, optimizing account of mainstream neoclassical economics.[19] They did so in terms of firms (as "phenotype") and their "organizational routines" as "genes" (or genotype) seeing these as lending continuity to economic life, as opposed to the random fluctuations and optimizing responses to prices envisaged by the neoclassical view. The resource-based view as extended in this paper can take over this description provided by Nelson and Winter, and subsequently elaborated, with the proviso that it is not "routines" but resources which are acting as the units of variation, selection and retention. The resource based view of the economy thereby provides a unifying account of the processes of economic evolution, via the dynamics of resource variation, selection and retention.

The distinction between variation and selection of "resources" as opposed to that of "routines" (Nelson and Winter) is subtle but important. Nelson and Winter argue that organizational routines are "sticky" in the sense that they vary slowly, and are "inherited" by successful firms as they grow and develop. Exactly the same arguments carry over to resources, but with even greater force. Resources as defined here are clearly good candidates for vehicles of variation and selection, in that they are explicitly exchanged between firms, as part of a process of adaptive learning.[20]

It was argued above that routines are the behavioural expression of resources. Managements utilize the firm's distinctive resources by creating routines; this is the origin of the firm's propensity to grow and expand, as managers look to extract enhanced services from "routinized" resources. Now the argument is transposed to an evolutionary context. If it is the underlying resources that are varying, then they are creating selection pressures which are experienced in terms of successful routines of competitively successful firms. This is interpreted by managements as "best practice."

---

and Metcalfe (1998a,b) provide expositions of the evolutionary approach to economics from different perspectives, while Vromen (1995) provides an extended comparison of evolutionary schools of thought.

19 See discussions by Nelson and Winter (1974, 1982) and the individual contributions of each author, such as Winter (1964) and Nelson (1994).

20 See Hutter (1994) for a discussion of the issue of what is the unit that evolves. Mathews (2001a) elaborates on this, introducing resources, routines and firms' relations as three fundamental categories of a simplified economic framework termed an "industrial market system." Evolutionary pressures operate then on resources, routines and interfirm relations, that are the "units that evolve" – they are changing through firms' adaptations and through sudden recombinations.

5.1 Co-evolutionary resource dynamics

In biological evolution, the phenomenon of species co-adapting to changes in their environment is frequently observed, so that they become co-specialized with respect to each other. This is termed co-evolution. Numerous examples include the microorganisms that evolve in the guts of certain mammalian species, or the ants that co-evolve with certain kinds of acacia to provide mutual advantages. Now it is coming to be observed that business works also according to co-evolutionary principles. Firms for example encourage business units to evolve in different but complementary directions, allowing them to seize opportunities for collaboration where they present themselves – rather than imposing predetermined patterns of divisionalized operation on them.[21] From a resource perspective, the notion of co-specialization of resources both within and between firms can be interpreted as the expression of co-evolutionary dynamics.

If resources can be described in terms of their evolutionary and co-evolutionary dynamics, what then is the significance of this perspective for economic performance? Variety is the driver of evolutionary dynamics, whether we are talking about technologies, firms or resources. This is the core of the Fisher principle, the "fundamental theorem" of systems in evolutionary motion. It states, when applied to competitive economic systems, in the words used by Metcalfe (1994, pp. 328) that "the rate of change of average behavior within a population of competing firms is governed by the degree of variety in behavior within that population."[22] The key issue then is how resource creation can exceed resource destruction to enhance the resource variety and diversity that drives economic learning and adaptation, i.e. evolutionary success.

Resource variety is generated by new combinations and, sometimes, by genuinely new resources, as in the case of a new technological standard emerging and driving the spawning of a new industry. This brings us to the consideration of entrepreneurship, innovation and technological dynamics, involving issues such as path dependencies, lock-in, adaptive learning and technological trajectories.

6 Innovation and entrepreneurship: Schumpeterian resource dynamics

From the resource perspective, novelty in the economy is generated principally through resource recombination, and the principal agents who accomplish these recombinations are entrepreneurs.[23] Schumpeter had the clearest possible conception that it was entrepreneurship which created new lines of development within

[21] See Eisenhardt and Galunic (2000) for a recent exposition of this perspective.

[22] Metcalfe (1994, pp. 328–329) notes that: "Implicit in this view are the four central themes of the evolutionary perspective: that it is differences in behaviour between firms which drive the evolutionary process; that these differences are evaluated economically within a population of competing behaviours; that this evaluation generates selective pressure to change the relative performance of each distinct form of behaviour in the population; and, that these behaviours are subject to inertia, changing slowly relative to the changes imposed by selection."

[23] Schumpeter developed such a theory of entrepreneurship, in the sense of initiating new lines of economic development, in his *Theory of Economic Development*. This classic was first published in German in 1912 (frequently erroneously cited as 1911 – yet the 1912 date of publication is clearly

an economy, in ways that could not be anticipated through analysis of the "circular flow" economy. Entrepreneurial initiative created new activities, whose profitability then attracted imitators, and so the resource distribution in the economy as a whole is shifted.

There is an uncanny resemblance between this Schumpeterian conception of economic dynamics and Kant's great theory of moral action; perhaps Kant was present in Schumpeter's mind as he was writing. Kant created a conception of a universe of causality in which scientific laws rule supreme, but in which new sources of causal chains can be created by acts of the will, i.e. a willed action by a human creates a chain of events whose links can be explained by science, but whose origin can only be accounted for in terms of moral values and free will. This is what Kant meant by his being transfixed by the two great phenomena of our existence – the starry heavens above (the universe and its regularity explicable by scientific laws) and the moral law within. Schumpeter's analogue is the entrepreneur who creates a new "line of business" which redistributes resources in the circular flow, and which once it is up and running, is amenable to traditional economic analysis.

Like Schumpeter, we keep a firm dividing line between "entrepreneurship" and "innovation." Sometimes the two coincide, as when a technologist develops a completely new product or process concept and starts a new company to exploit it. But usually the two are best treated separately.

6.1 Entrepreneurship: new resource combinations

From the resource perspective developed here, there is virtually nothing to be changed in Schumpeter's account. The resource economy is where resource recombination occurs. (Schumpeter: new enterprise formation occurs outside the "circular flow" of normal economic events.) New enterprises are created through new combinations of existing resources, adapted to new perceived needs or opportunities. (Schumpeter: new enterprises are created through recombination.) The new combinations are assembled not by managers but by entrepreneurs or other corporate promoters (eg a merchant bank); it is the entrepreneur/promoter who establishes the firm's initial business strategy on the basis of the particular combination of resources assembled. (Schumpeter: It is the entrepreneur who initiates a new sequence of economic operations – as the "new employment of existing production goods" (1012/1934/1996, pp. 136)).

Day (1986) has provided an intriguing reinterpretation of Schumpeter's entrepreneurial function in terms of disequilibrium dynamics of the goods and services economy. He takes the position that in any real setting the "circular flow"

shown on the first German edition, *Theorie der Wirtschaftliches Entwicklung*, published in Leipzig by Verlag von Duncker & Humblot. This first edition carried seven chapters, with the seventh chapter treating "the economy as a whole." A second German edition was published in 1924, dropping the long seventh chapter, and an English translation of this second edition was published by Harvard University Press in 1934. The Transaction Publishers edition was published in 1996; hence the bibliographic reference to TED as Schumpeter (1912/1934/1996). Much interest surrounds the "lost" seventh chapter of Schumpeter's great work, since it provides an overview of his method and approach. For a useful introduction to the text, in the context of Schumpeter's early career, see Swedberg (1991).

would rapidly swing wildly into disequilibrium, with one cycle of price formation and production decisions feeding off another to produce unstable swings – exactly as are observed in reality. He maintains that entrepreneurs actually bring stability to this unstable system, by identifying the sources of disequilibrium and initiating new actions that are then embodied in corrective fashion in a new round of the "circular flow." From our perspective, it is disequilibria in the goods and services economy which provide the stimulus for entrepreneurial action, combining and recombining resources in order to produce a new set of goods or services. Entrepreneurs have the capacity to translate the disequilibria into resource terms, and to visualize how a new resource combination can be effected to "correct" the disequilibrium identified. This is a nice way of illustrating the duality of the goods and services economy and the resource economy: entrepreneurs are guided by signals from the goods and services economy but their actions are conducted in the resource economy, which in turn change the dynamics of the goods and services economy.

6.2 Innovation: new resource creation

There is hardly a term in economics that has attracted as much confusion as "innovation." Is it the appearance of totally new forms, or their uptake in the economy? The term is frequently taken to imply much more than technological novelty; it can span the appearance of new marketing forms, or new organizational arrangements, or any other economic activity that shows signs of novelty. From the resource perspective, these ambiguities may be dispelled: innovation may be identified *tout court* with the creation of new resources – resources that have not existed before, as distinct from resource combinations achieved through entrepreneurial action.

An example of a completely new resource is a *technological standard.* Suppose that we regard a technological standard as a "resource" – since it becomes widely available as such, to a variety of firms, and not just to the originator.[24] Resource creation in this sense is needed to drive the formation of new industries and their diffusion. This is generally beneficial in its effects. But standardization can also lead to perverse outcomes, as for example where one resource is *created* and then *propagated* on such a scale that it precludes the creation of another, perhaps superior resource. This is a case of "lock-in" where the success of the inferior resource is generated through increasing returns.[25]

While interesting, lock-in effects are simply an extreme form of the more general phenomenon which may be described as *resource trajectories*, or path dependence (Antonelli, 1997). It is another way of saying that "history matters." From a resource perspective, resource accumulation, within firms and in the wider economy, can

[24] Standards can be interpreted as equilibria where users are agents with multiple technical choices (Cowan and Miller, 1998). But such game-theoretic formulations, while illuminating, miss the essential dynamic features of standardization. Often it is not foresight and calculation on the part of agents which leads to the emergence of a standard, but the outcome of unforeseen technological dynamics.

[25] Such lock-in effects are discussed by Arthur (1989), where the wide adoption of the perhaps inferior technology in itself generates "network externalities" that preclude the other, perhaps superior, technology from being started. The case of the QWERTY typewriter keyboard is the most famous such case (David, 1985).

clearly be expected to follow certain trajectories, or pathways, given that firms tend to develop their resource stock based on what they already have. There is nothing counter-intuitive in resource accumulation following a trajectory. Moreover, whole systems of firms may generate resource configurations that become "locked-in" in inferior economic performance. Resource pathways in this manner become of fundamental significance for economic performance, which in turn is linked to the issues of industrial organization.

7 Industrial organization and economic performance

The resource economy perspective is concerned not primarily with individual firm development, but above all with the interactions between firms – or with the "organization of industry" itself. The fundamental feature of an economy is the patterns through which the actions of economic agents are coordinated with a view to enhancing overall economic performance. I shall refer to *economic performance* as opposed to the performance of individual firms which populate the economy. I wish to demonstrate that this is critically linked to the way that resources are distributed within the economy, both within and between firms – in other words, to paraphrase Adam Smith, that economic performance is limited by the organizational configuration of resources within the economy.[26]

7.1 Organizational configuration of resources

Enhanced performance at the economic level, as at the organizational level, can be captured through specialization and the emergence of intermediate input suppliers, which in turn is associated with decomposing a process into a finer division of labour. Consider the case of a group of firms, each specializing in a particular range of products and overlapping with each other in terms of their resource. As the market expands, some firms can specialize in intermediate subassemblies, to create more complex value-adding pathways within the industry. Standardization of subassembly modules enables potential economies of scale to be captured, and an organizational reconfiguration of resources to be effected. It is the possibility of intermediate specialist activities emerging, as the scale of the market expands, that drives specialization of resources – as anticipated by Adam Smith.[27] If these activities are conducted by new, specialist firms, it is a case of horizontal division

[26] It was Richardson (1972) who drew attention to these issues, by introducing a range of firm interactions laid out across a spectrum whose endpoints were the integrated firm at one end and the open, anonymous market at the other. Utilizing a classification of activities as "similar" and "complementary" he argued that similar activities would be carried out within a single firm (under unified management) while dissimilar activities would be coordinated through the market. Complementary (but dissimilar) activities would be coordinated by direct negotiation between firms (as in various kinds of contractual arrangement). Without damage to Richardson's argument we may translate the terms across to resources. It is thus complementarity of activities which induces firms to act together, in order to pool resources, or to find ways to service the activities from a common resource.

[27] As expressed by Richardson (1996/1998: 168): "where the scale of an economic activity increases, it will be practicable for component processes within it to be separated out. In general, the cost savings

of labour (Langlois, 1989). If the activities are conducted within the same firm, it is a case of vertical division of labour (Stiglitz, 1951). We thus have a resource interpretation of the process first alluded to by Adam Smith, in his theorem proposing that the division of labour and its beneficial effects is limited by the extent of the market.

Sometimes the required further specialization is not achieved, and the economic performance of a group of firms is thereby degraded. This has occurred over and over again as industrial districts wax and wane. The district of Okayama, in western Japan, for example, became a flourishing centre of production of varied kinds of farm engines in the 1950s and 1960s, as Japan's farmers moved en mass to mechanize their operations. They needed one engine only per farm, to drive pumps, tractors, or threshing machines. Over 30 manufacturing firms arose in the Okayama district to service this need, producing small, light engines of variable but low horsepower to a variety of end-specifications, for distribution by specialized distributors throughout Japan. But nothing remains of this district today. It was wiped out by the rise of mass producing firms in Tokyo and other metropolitan centres, who were much more vertically integrated and connected to lengthy subcontracting chains than were the small Okayama producers who encapsulated all the technical capabilities needed to produce an engine in one small firm. As new kinds of engines appeared, such as faster and lighter machines, the small self-contained producers of Okayama found themselves unable to switch from being self-sufficient producers to specialized parts of a longer production chain. The longer metropolitan production chains, which encouraged specialized mass producers, therefore wiped them out (Tokumaru, 1998).

From the resource perspective, these Okayama producers were not able to make the breakthrough from self-sufficiency in resources to a new configuration where some resources are shared between firms. There was apparently no mechanism in this case to shift the cluster of firms to a new configuration. Successful clusters of firms, such as in a Silicon Valley, are able to make these configuration shifts; others stay "locked in" to a particular configuration and decline. The issue is how such shifts are accomplished, and whether they call for specific institutional interventions, or are accomplished by the actors themselves.

7.2 Clusters

One obvious way to impose an organizational configuration on economic activities, beyond encapsulating them within individual firms, is to cluster them, in local communities of firms specializing in closely complementary activities. These entities all entail an organizational structure between firms as opposed to one that holds within firms. Clusters of this form are well recognized and indeed are becoming the object of increasing attention – due to the outstanding success of such high tech clusters as Silicon Valley in the USA, and other science-driven clusters like

made available by an increase in the scale of a particular economic activity [lead] ... to a change in industrial structure, those stages exhibiting the greatest scale economies becoming the business of specialist suppliers."

Research Triangle Park in North Carolina, or the Hsinchu district in Taiwan where all the country's major IT and semiconductor activities are co-located.[28] It is widely recognized that the success of a Silicon Valley owes much to highly specialized complementarities that are closely co-located – something that cannot be accounted for in simple capital and labour terms in a production function.

Now from a resource perspective there is a clear interpretation to be offered for the phenomenon of clustering, which is that clusters constitute a form of economic organization where resources are shared between firms locally. The two operative words are *shared*, and *local*. Resources can be utilized by more than one firm – this is the very point of adopting a resource perspective on the economy (as opposed to the usual perspective which treats the firm on its own). Resources such as specialized manufacturing knowledge and technical capabilities can be shared in the form of a common "culture" of excellence and leading edge technical intelligence – where the latest developments are exchanged in cafes and meeting points, in workshops and seminars, and through rapid job-hopping, as in Silicon Valley. These are all ways in which one might describe resources as being "in the air" to adapt Marshall's telling phrase. But they are also local. Other forms of shared resource do not have to be local – as in worldwide R&D collaborative structures for example. But the point of the cluster is that it draws benefits from resources shared between firms that are closely co-located.[29]

Local sharing of resources in clusters can then be expected to improve economic performance, as numerous historical and contemporary examples attest. But again organizational configuration of resources holds the key. Not all locally clustered firms thrive economically. There are many examples of industrial districts, for example, which have declined, not because of poor management or technical capabilities, but because of their inability to adjust to changing external economic circumstances.[30] They were "locked in" to one particular kind of organizational configuration (of resources). And when economic circumstances changed, and this proved to be a sub-optimal configuration, they were unable to pull themselves spontaneously into a new configuration. This has happened on countless occasions as industrial districts have flourished for a time but have eventually declined as external economic circumstances changed – as in the Japanese case of Okayama.

---

28 See Porter (1998) for a recent discussion, Dyer and Singh (1998) for sources of "relational advantage" between firms, and Martin (1999) for a review of cluster-focused geographical economic studies. On clustering as a source of success in Taiwan and Singapore, see Mathews (1997a, 1999, 2001c) and Mathews and Cho (2000).

29 See Foss and Eriksen (1995), Foss (1996, 1999) or Best (1999) for a discussion of this phenomenon in an explicitly resource-based context, and Lawson (1999) for a similar argument extending the "competence perspective" from the individual firm to the region. Schmitz (1999) adds the point that firms in industrial districts develop collective action through conscious intervention, as in the formation of consortia.

30 See for example the study of the Italian footwear industrial districts of Fusignano and San Mauro Pascoli by Nuti and Cainelli (1996).

7.3 Non-local resource sharing

Non-local forms of organization, where again resources are shared, tend to be more successful in adapting to new circumstances and changing their form – or rather, they organize for shorter periods, and break up and re-organize as circumstances and opportunities change. Consider the case of R&D consortia, fashioned through private initiative or through public policy. Again from a resource perspective, the rationale and source of success is clear: it is through managed sharing of resources. Firms participate in such consortia in order to acquire access to knowledge and techniques which would be too difficult or expensive for each to acquire individually. But the consortium can allow Smith's division of labour to operate. Each firm or group of firms can specialize in certain aspects of a problem, while the consortium as a whole pools the results for the benefit of all – as in the case of Taiwan's R&D consortia (Mathews, 2001c).

It is important to stress that these resource configurations usually span firms – in "development blocks" or "technological systems" or "systems of tight linkages" or "national systems of innovation" - and call for supra-firm modes of organization that facilitate the sharing of resources.[31] There is a recursive feature to this process of resource encapsulation – from small groups of resources encapsulated within a small firm to capture synergies, to larger encapsulations within larger or divisionalized firms, and culminating in encapsulations in clusters, networks, alliances, or national systems. In each case the driving factor is encapsulation into a resource agglomeration that has an "identity" and a capacity for self-action, or adaptation.[32] They can be agglomerated through the expression of "market forces" or through deliberate, policy-guided action, as in the formation of numerous consortia and alliances. It is the heterogeneity of such resource aggregations that lies at the heart of national competitive systems, just as it is the heterogeneity of resource clusters within firms which accounts for their firm-level competitive advantage. And it is the capacity of an economy to form such resource configurations, and to adapt them as circumstances change, that constitutes what might be called "economic learning" – a notion that makes no sense in mainstream equilibrium analysis.[33]

The key organizational insight is that economic performance is not optimized by simply looking to optimize the performance of each productive resource, on its own. The organizational dimension is essential in order to deal with the issue of coordination. The organizational dimension operates at several different levels –

[31] On development blocks, see Dahmen (1989); on technological systems, see Carlsson and Stankiewicz (1991) and the contributions to Carlsson (1997); on "systems of tight linkages" see Cohen and Zysman (1987). Foss (1996) refers to all these forms of industrial organization as operating at the meso level – between the firm and the national industry. On national systems of innovation, see Lundvall (1988, 1992); this concept spans firms as well as supporting institutions such as public R&D laboratories. From the resource perspective, these concepts all embody the notion of resources held in common and shared within a specified group of firms and institutions.

[32] A useful analogy is an Object-Oriented software system, where the software "objects" are the elemental units, and larger programs are built through encapsulated systems of interacting objects. Such analogies are discussed in Mathews (1996a,b).

[33] See Lundvall and Johnson (1994) for a discussion of the concept of the learning economy; Mathews (1996c, 2000) gives an account of the organizational underpinnings of economic learning.

bundling resources in firms to capture synergies, and then connecting firms with each other to capture further synergies, and groups of firms with other groups of firms to capture further synergies again. These are what may be called the "organizational" sources of performance enhancement in the resource economy. This is the starting point for a resource-based approach to value creation and wealth generation in wider economic systems.

8 Concluding remarks

The claim to novelty in the preceding exposition lies not so much in the parts (where existing ideas are taken over and transmuted into resource equivalents) as in the whole. It represents a synthesis of non-neoclassical concerns, ranging from firm growth through competitive and evolutionary dynamics to the organization of industry, where the connecting thread is provided by the elemental category of "resource." But this synthesis would be of no more than passing interest, if it did not generate fresh insights. I have sought to demonstrate these, such as the account of incumbent-challenger competitive interactions, and the translation of Schumpeter's insights into resource-based terms. The advantages of such a synthesis can be enumerated briefly. First, it provides a unifying thread linking various disparate areas of economic and management investigation, such as in evolutionary economics, the RBV, and entrepreneurship; second, it offers a coherent interpretation of the totality of resources as the "resource economy"; third, it links with the processes of corporate acquisition and divestment that are described every day in the business pages of the newspaper; fourth, it provides fresh empirical challenges for economics; and fifth, it opens up new and challenging fields such as explaining the continuing relevance of clusters (like Silicon Valley) and even the very latest Internet phenomena.[34] There is therefore suggestive evidence that it may be worthwhile to pursue the notion of the "resource economy" as a unifying and empirically rewarding field of inquiry.

What is novel is the category of the "resource economy" as something distinct from the mainstream goods and services economy and labour economy of neoclassical analysis. Everything covered in the mainstream analysis of the goods and services economy continues as before; it is undisturbed by the present analysis. It is complemented by the analysis of the RBV, directed towards the hitherto unnamed "resource economy." The approach of the RBV is quite distinctive in that it is dynamic and evolutionary; it is descriptive rather than analytical, and concerned with accounting for outcomes in terms of processes. It is concerned with differences, and with how these drive evolutionary dynamics and overall economic performance. It is the private opinion of this author that in time these approaches will come to dominate the analysis of the "goods and services" economy as well, by which time the obsession with formalistic static equilibrium analysis will just be a bad memory. But for the moment all we need claim is that these evolutionary-oriented approaches are "tailor-made" for the resource economy.

[34] Amit and Zott (2001) provide an original adaptation of the RBV and entrepreneurial studies to the case of Internet-based firms, that is close in spirit to the account presented here.

The resource economy as a totality provides the setting for observing numerous parallels and analogies that would otherwise remain obscure. Note the strong analogy for example between the disequilibria in resource combinations that can be held to drive entrepreneurial interventions, at the level of the economy, with the resource disequilibria that can be held to drive managerial interventions at the level of the firm. In the one case, it is entrepreneurs recognizing a disequilibrium in the goods and services market (such as a clear shortage of some kind of intermediate good) and responding to reconfigure resources, while in the other case it is managers recognizing a "slackness" in the resource base of their firm and responding with a strategy of diversification or new market entry to take up the slack. Entrepreneurs create new resource combinations, while managers within firms can utilize the resource combinations they are given to create new sets of goods and services. They are both working within resource "constraints" and opportunities that derive their rationality from the resource economy – a rationality that is hidden from view if the goods and services economy is the sole object of attention.

This paper seeks only to sketch what an analysis of the "resource economy" might look like, and how it might complement the kinds of analysis subsumed under the rubric of the neoclassical synthesis. The case is made that the project is at least plausible and probably feasible. It has the merit that it is empirically oriented, and if taken up, will encourage empirical investigations of competitive resource dynamics, evolutionary resource dynamics, pathways and adaptations, and many other phenomena that the neoclassical synthesis ignores. This goes to the heart of the critique of the neoclassical synthesis, which is not so much that it is wrong, as that it discourages any kind of empirical inquiry – given that all interesting questions are settled in advance. In the resource economy, everything has to be settled by testing claims against reality. This might be a good foundation for an economics suited to the 21st century.

References

Amit R, Schoemaker PJ (1993) Strategic assets and organizational rent. Strategic Management Journal 14: 33–46

Amit R, Zott C (2001) Value creation in e-business. Strategic Management Journal 22: 493–520

Andersen ES (1994) Evolutionary economics: post-Schumpeterian contributions. Pinter, London New York

Antonelli C (1997) The economics of path-dependence in industrial organization. International Journal of Industrial Organization 15 (6): 643–675

Arthur WB (1989) Competing technologies, increasing returns, and lock-in by historical events. The Economic Journal 99 (March): 116–131

Barney JB (1986) Strategic factor markets: Expectations, luck and business strategy. Management Science 32: 1231–1241

Barney JB (1991) Firm resources and sustained competitive advantage, Journal of Management 17 (1): 99–120

Best MH (1999) Regional growth dynamics: A capabilities perspective. Contributions to Political Economy 18: 105–119

Blaug M (1997) Disturbing currents in modern economics. Challenge 41 (3): 11–34

Capron L Mitchell W (1998) Bilateral resource deployment and capabilities improvement following horizontal acquisitions. Industrial and Corporate Change 7 (3): 453–483

Carlsson B Stankiewicz R (1991) On the nature, function and composition of technological systems. Journal of Evolutionary Economics 1: 93–118

Carlsson BAV (ed) (1997) Technological systems and industrial dynamics. Kluwer, Boston Dordrecht

Caves RE Porter ME (1977) From entry barriers to mobility barriers. Conjectural variations and contrived deterrence to new competition. Quarterly Journal of Economics 91: 241–262

Clark JB (1888) Capital and its earnings. Publications of the American Economic Association, Vol 3, No 2

Coombs R Metcalfe JS (1999) Organising for innovation: Coordinating distributed innovation capabilities. ESRC Centre for Research on Innovation and Competition, University of Manchester

Cowan R Miller JH (1998) Technological standards with local externalities and decentralized behaviour. Journal of Evolutionary Economics 8: 285–296

Dahmén E (1989) Development blocks in industrial economics. In: Carlsson B (ed) Industrial dynamics. Kluwer, Boston

David P (1995) Clio the economics of QWERTY. American Economic Review Proceedings 75: 332–337

Day RH (1986) Disequilibrium economic dynamics: A post-Schumpeterian contribution. In: Day RH, Eliasson G (eds) The dynamics of market economies. North Holland, Amsterdam

Dierickx I, Cool K (1989) Asset stock accumulation and sustainability of competitive advantage. Management Science 35 (12): 1504–1511

Dosi G (1997) Opportunities, incentives and the collective patterns of technological change. The Economic Journal 107 (Sep): 1530–1547

Dosi G, Nelson RR (1994) An introduction to evolutionary theories in economics. Journal of Evolutionary Economics 4: 153–172

Dosi G, Freeman C, Nelson R, Silverberg G, Soete L (eds) (1988) Technical change and economic theory. Pinter, London New York

Dyer JH,Singh H (1998) The relational view: Cooperative strategy and sources of interorganizational competitive advantage. Academy of Management Review 23 (4): 660–679

Eisenhardt K Galunic D C (2000) Coevolving: At last, a way to make synergies work. Harvard Business Review Jan–Feb: 91–101

Eliasson G (1986) Micro heterogeneity of firms and the stability of industrial growth. In: Day RH, Eliasson G (eds) The dynamics of market economies. North Holland, Amsterdam

Fetter FA (1904) The relations between rent and interest. Proceedings of the 16th annual meeting American Economic Association, with commentaries by nine discussants, pp 199–227 (reprinted in: Rothbard (1977), pp 192–221)

Fetter FA (1927) Clark's reformulation of the capital concept. In: Hollander JH (ed) Economic essays contributed in honor of John Bates Clark. Macmillan, New York (reprinted in: Rothbard (1977), pp 119–142)

Foss NJ (1994) The sustrian school and modern economics: essays in reassessment. Copenhagen Business School Press, Copenhagen

Foss NJ (1996) Higher-order industrial capabilities and competitive advantage. Journal of Industry Studies 3 (1): 1–20

Foss NJ (1998) The resource-based perspective: An assessment and diagnosis of problems. Scandinavian Journal of Management 14 (3): 133–150

Foss NJ (1999a) Networks, capabilities, and competitive advantage. Scandinavian Journal of Management 15 (1): 1–16

Foss NJ (1999b) Edith Penrose, economics and strategic management. Contributions to Political Economy 18: 121–150

Foss NJ, Eriksen B (1995) Competitive advantage and industry capabilities. In: Montgomery C (ed) Resource-based and evolutionary theories of the firm: towards a synthesis. Kluwer, Boston, MA

Foss NJ, Knudsen C (eds) (1996) Towards a competence theory of the firm. Routledge, London New York

Foss NJ, Knudsen T (2001) The resource-based tangle: Towards a sustainable explanation of competitive advantage. Managerial and Decision Economics (forthcoming)

Foss NJ, Loasby BJ (eds) (1998)Economic organization, capabilities and co-ordination. Routledge, London New York

Freeman C (1987) Technology policy and economic performance: lessons from Japan. Pinter, London

Freeman C (1994) The economics of technical change. Cambridge Journal of Economics 18: 463–514

Galunic DC, Rodan S (1998) Resource recombinations in the firm: Knowledge structures and the potential for Schumpeterian innovation. Strategic Management Journal 19: 1193–1201

Ghoshal S, Hahn M, Moran P (1999) Management competence, firm growth and economic progress. Contributions to Political Economy 18: 121–150

Hayek FA (1941) The pure theory of capital. Routledge & Kegan Paul, London

Henderson R (1993) Underinvestment and incompetence as responses to radical innovation: evidence from the photolithographic alignment equipment industry. RAND Journal of Economics 24 (2): 248–270

Henderson R, Clark KB (1990) Architectural innovation: The reconfiguration of existing product technologies and the failure of established firms. Administrative Science Quarterly 35: 9–30

Hitt M, Ireland RD (2000) The intersection of entrepreneurship and strategic management research. In: Sexton DL, Landstrom H (eds) The Blackwell handbook of entrepreneurship. Blackwell, Oxford

Hodgson G (1993) Economics and evolution. Polity Press, Cambridge

Hutter M (1994) The unit that evolves: Linking self-reproduction and self-interest. In: Magnusson L (ed) Evolutionary and neo-Schumpeterian approaches to economics. Kluwer, Boston Dordrecht

Itami H (with Roehl TW) (1987) Mobilizing invisible assets. Harvard University Press, Cambridge, MA

Lachmann LM (1956/1978) Capital and its structure. Bell, London (1978, 2nd edn, reprinted by: Sheed, Andrews McMeel Inc, Kansas City)

Lachmann LM (1973) Macro-economic thinking and the market economy. The Institute of Economic Affairs, London

Langlois RN (1989) Economic change and the boundaries of the firm. In: Carlsson B (ed) Industrial dynamics. Kluwer, Boston

Langlois R, Everett M (1994) What is evolutionary economics? In: Magnusson L (ed) Evolutionary and neo-Schumpeterian approaches to economics. Kluwer, Boston Dordrecht

Lawson C (1999) Towards a competence theory of the region. Cambridge Journal of Economics 23 (2): 151–166

Lewin P (1997) Capital in disequilibrium: A re-examination of the capital theory of Ludwig M Lachmann. History of Political Economy 29 (3): 523–548

Lewin P, Phelan SE (1999) Firms, strategies, and resources: Contributions from Austrian economics. The Quarterly Journal of Austrian Economics 2 (2): 3–18

Lippman SA, Rumelt RP (1982) Uncertain imitability: An analysis of inter-firm differences in efficiency under competition. Bell Journal of Economics 13: 418–438

Loasby B (1991) Equilibrium and evolution. Manchester University Press, Manchester, UK

Lundvall B-A (1988) Innovation as an interactive process: from user-producer interaction to the national system of innovation. In: Dosi G et al (eds) Technical change and economic theory. Pinter, London New York

Lundvall B-A (ed) (1992) National systems of innovation: towards a theory of innovation and interactive learning. Pinter, London

Lundvall B-A, Johnson B (1994) The learning economy. Journal of Industry Studies 1 (2): 23–42

McGrath R, MacMillan IC (2000) The entrepreneurial mindset. Harvard Business School Press, Boston, MA

Marshall A (1890/1920) Principles of economics, 1st/8th edn. Macmillan, London

Mathews JA (1996a) Organizational foundations of economic learning. Human Systems Management 15 (2): 113–124

Mathews JA (1996b) Holonic organizational architectures. Human Systems Management 15 (1): 27–54

Mathews JA (1996c) Organizational foundations of economic learning. Human Systems Management 15 (2): 113–124

Mathews JA (1997a) Silicon Valley of the East: How Taiwan created a semiconductor industry. California Management Review 39 (4): 26–54

Mathews JA (1997b) The competitive advantages of the latecomer firm. Paper presented at Academy of Management Anual Meting (Organization and Management Theory), Boston, August

Mathews JA (1998) Jack the Beanstalk: The creation of dynamic capabilities through knowledge leverage by latecomer firms. Paper presented at Academy of Management annual meeting (Business Policy and Strategy), San Diego, August

Mathews J A (1999) A Silicon Island of the East: Creating a semiconductor industry in Singapore. California Management Review 41 (2): 55–78
Mathews JA (2001a) Interfirm competitive dynamics within an industrial market system. Industry and Innovation 8 (1): 79–107
Mathews JA (2001b) National systems of economic learning: The case of technology diffusion management in East Asia. International Journal of Technology Management 22 (5/6): 455–479
Mathews JA (2001c) The origins and dynamics of Taiwan's R&D consortia. Research Policy (forthcoming)
Mathews JA (2001d) Dragon multinational: a new model of global growth. Oxford University Press, New York (forthcoming)
Mathews JA, Cho DS (1999 Combinative capabilities and organizational learning in latecomer firms: The case of the Korean semiconductor industry. Journal of World Business 34 (2): 139–156
Mathews JA, Cho DS (2000) Tiger technology: the creation of a semiconductor industry in East Asia. Cambridge University Press, Cambridge, UK
Mattsson L-G (1998) Dynamics of overlapping networks and strategic actions by the international firm. In: Chandler AC Hagström P, Sölvell O (eds) The dynamic firm. Oxford University Press, Oxford
Metcalfe JS (1992) Varie:1456tructure and change: An evolutionary perspective on the competitive process. Revue d'Economie Industrielle 59
Metcalfe JS (1994) Competition, Fisher's principle and increasing returns in the selection process. Journal of Evolutionary Economics 4: 327–346
Metcalfe JS (1998a) Evolutionary economics and creative destruction: graz Schumpeter lectures, Vol 1. Routledge, London
Metcalfe JS (1998b) Evolutionary concepts in relation to evolutionary economics. CRIC Working Paper #4. Centre for Research in Innovation and Competition, Manchester
Moran P, Ghoshal S (1999) Markets firms, and the process of economic development. Academy of Management Review 24 (3): 390–412
Montgomery CA (ed) (1995) Resource-based and evolutionary theories of the firm: towards a synthesis. Kluwer, Boston, MA
Nelson RR (1989) Capitalism as an engine of progress. In: Carlsson B (ed) Industrial dynamics. Kluwer, Boston Dordrecht
Nelson RR (1994) The role of firm difference in an evolutionary theory of technical advance. In: Magnusson L (ed) Evolutionary and neo-Schumpeterian approaches to economics. Kluwer, Boston Dordrecht
Nelson RR, Winter SG (1973) Toward an evolutionary theory of economic capabilities. American Economic Review (Papers and Proceedings) LXIII (2): 440–449
Nelson RR Winter SG (1982) An evolutionary theory of economic change. The Belknap Press of Harvard University Press, Cambridge, MA
Nuti F, Cainelli G (1996) Changing directions in Italy's manufacturing industrial districts: The case of the Emilian footwear districts of Fusignano San Mauro Pascoli. Journal of Industry Studies 3 (2): 105–118
Penrose E (1959/1995) The theory of the growth of the firm, 3rd edn (with new foreword by the author). Wiley, New York
Peteraf MA (1993) The cornerstones of competitive advantage: a resource-based view. Strategic Management Journal 14: 179–188
Pitelis CN, Wahl MW (1996) Edith Penrose: Pioneer of stakeholder theory. Long Range Planning 31 (2): 252–261
Porter ME (1980) Competitive strategy: techniques for analyzing industries and competitors. The Free Press, New York
Porter ME (1985) Competitive advantage: creating and sustaining superior performance. The Free Press, New York
Porter ME (1998) Clusters and the new economics of competition. Harvard Business Review Nov–Dec: 77–90
Prahalad CK Hamel G (1990) The core competence of the corporation. Harvard Business Review May–Jun: 79–91

Prendergast R (1992) Increasing returns and competitive equilibrium – the content and development of Marshall's theory. Cambridge Journal of Economics 16: 447–462
Richardson GB (1972) The organisation of industry. The Economic Journal 82 883–896
Richardson GB (1975) Adam Smith on competition and increasing returns. In: Skinner AS, Wilson T (eds) Essays on Adam Smith. Clarendon Press, Oxford, UK
Richardson GB (1996) Competition, innovation and increasing returns. DRUID Working Paper #96–10, Danish Research Unit in Industrial Dynamics, Copenhagen Business School (reproduced as paper #12 in Richardson, 1998)
Richardson GB (1998) The economics of imperfect knowledge: collected papers of GB Richardson. Edward Elgar, Cheltenham, UK
Rothbard MN (ed) (1977) Capital, interest, and rent: essays in the theory of distribution by Frank A Fetter. Sheed, Andrews McMeel, Kansas City
Rumelt RP (1984) Towards a strategic theory of the firm. In: Lamb RB (ed) Competitive strategic management. Prentice Hall, Upper Saddle River, NJ
Saxenian AL (1994/1996) Regional advantage: culture and competition in Silicon Valley and Route 128. Harvard University Press, Cambridge, MA
Schmitz H (1999) Collective efficiency and increasing returns. Cambridge Journal of Economics 23 (4): 465–483
Schumpeter JA (1912/1934/1996) The theory of economic development (with introduction by John E Elliott). Transaction Publishers, New Brunswick, NJ
Schumpeter JA (1942/1976) Capitalism, socialism and democracy (with Introduction by Tom Bottomore). Harper Collins, New York
Steiner TL (1996) A reexamination of the relationships between ownership structure, firm diversification, and Tobin's q. Quarterly Journal of Business and Economics 35 (4): 39–48
Stigler G (1951) The division of labor is limited by the extent of the market. Journal of Political Economy 59: 185–193
Swedberg R (1991) Schumpeter's early work. Journal of Evolutionary Economics 2: 65–82
Teece DJ (1986) Profiting from technological innovation: Implications for integration, collaboration, licensing and public policy. Research Policy 15: 285–305
Teece DJ (1992) Competition, cooperation, and innovation: Organizational arrangements for regimes of rapid technological progress. Journal of Economic Behavior & Organization 18: 1–25
Teece DJ, Pisano G, Shuen A (1997) Dynamic capabilities and strategic management. Strategic Management Journal 18 (7): 509–533
Tokumaru N (1999) The self-sustaining mechanism in the industrial cluster. Paper presented at the 1999 EAEPE Conference, Charles University, Prague
Vromen JJ (1995) Economic Evolution: An Enquiry into the Foundations of New Institutional Economics. Routledge, London New York
Wernerfelt B (1984) A resource-based view of the firm. Strategic Management Journal 5: 171–180
Winter SG (1995) Four Rs of profitability: Rents, resources, routines and replication. In: Montgomery CA (ed) Resource-based and evolutionary theories of the firm: towards a synthesis. Kluwer, Boston, MA
Witt U (1992) Evolutionary concepts in economics. Eastern Economic Journal 18: 405–419
Young AA (1928) Increasing returns and economic progress. The Economic Journal 38: 527–542

Transferring exploration and production activities within the UK's upstream oil and gas industry: a capabilities perspective*

John H. Finch

Department of Economics, University of Aberdeen, Edward Wright Building, Dunbar Street, Aberdeen, AB24 3QY, Scotland (e-mail: j.h.finch@abdn.ac.uk)

Abstract. Following Richardson (1972), capabilities comprise tacit, personal, subjective and context-specific knowledge that may be shared in practice only with difficulty across small, task oriented groups within firms or other types of organisation, and are expressed in the form of activities. The definition has been influential, and its focus on tacit knowledge has, arguably, encouraged research activities in the form of studies adopting experimental and simulation techniques, while providing less impetus for complementary empirical inquiry. This paper presents an empirical inquiry into an aspect of the development of capabilities in the UK's upstream oil and gas industry promoted by the changing organisation of activities across oil companies and contracting and supply companies. The main argument is that researchers can gain partial and subjective access to capabilities – distinct from activities – because individuals involved in the industry articulate and codify understandings of capabilities through practical theorising and commercial experimenting. Such articulation and codification plays an important role in the development of capabilities in industrial contexts.

Key words: Upstream oil and gas industry – Capabilities – Supply chains – Industry business cycle – Codified and tacit knowledge

JEL Classification: D21, F23, L23, L71

---

* The research on which this paper is based was supported by a grant from the University of Aberdeen's Faculty of Social Sciences and Law Research Committee. Versions of this paper have been presented at the annual conference of the Network of Industrial Economists, Queens' College, Cambridge, March 2000, and at the Eighth Conference of the International Schumpeter Society, University of Manchester, June 2000. I am grateful to Nicola Dinnie, Paul Hallwood, Alex Kemp, Brian Loasby, Jim Love, Fiona Macmillan, John Mathews, Graeme Simpson, Linda Stephen and Campbell Watkins for their criticisms of previous drafts. The usual disclaimer applies.

1 Introduction

This paper presents an analysis of the capabilities approach through undertaking an empirical inquiry into the transfer of activities in the UK's upstream oil and gas industry (Richardson, 1972; Loasby and Foss, 1998). The empirical inquiry focuses on a period in the industry's development, the mid to late 1990s, when analysis of which companies undertake which activities across supply chains and networks has been much discussed within companies and has been the subject of initiatives by industry-level organisations. Such appraisals elevated what had been considered an operational matter of procurement to a strategic matter of the possible transfer of exploration and production activities to different positions in supply chains and networks. Interviews with personnel of oil companies (henceforth, operating companies[1]) and major contracting and supply companies (henceforth, services companies) suggest that reviews of the organisation of production sought improvements in the cost-efficiency in which existing activities were organised, and in the ways in which the activities themselves were undertaken. It is argued here that these objectives draw upon two distinct though overlapping visions of the industry, which are difficult to reconcile in order that some overall target or objective may be articulated.

An important aspect of the capabilities approach is its focus on how knowledge associated with, or implicated in, a firm may be co-ordinated, and at the same time develop in the face of fundamental uncertainty and its personnel's procedural rationality (Langlois, 1992; Loasby, 1998a, pp. 193–196, 1999, pp. 34–37). The organisational problems facing decision-makers in the UK's upstream oil and gas industry may, following the capabilities approach, be interpreted as: (1) seeking optimal organisational solutions with given knowledge of exploration and production tasks currently undertaken; and (2) seeking a distribution or allocation of knowledge across a supply chain that seems more likely to foster novel solutions to existing exploration and production problems, and also to cope with different exploration and production problems that could occur over forecasting (or corporate planning) investment time frames. Forecasting time frames in the industry for particular hydrocarbon fields or assets can be up to 25 years, including exploration, production and decommissioning activities (Newendorp, 1996; Simpson et al., 1999).

Applying the capabilities approach raises important issues of research method. Few studies have investigated an industry with explicit reference to the capabilities approach (Cohen et al., 1996, pp. 680–682). The problems of research method may be considered under three headings: (1) how to identify capabilities in an industry context, which, following Richardson (1972, p. 888), underlie more easily observed industry activities; (2) how to gain access to the interpretive frames of those involved in the industry, such that inferences may be made about the organisation and development of capabilities; and (3) how to combine the case study with prior

[1] An operating company conducts its activities on a concession, where a concession is a licence, lease or other permit for exploration and/or production in an area, and is usually in the form of a government lease (Oil and Gas Handbook, 1996). Yergin (1993) provides a detailed history of the development of the major operating companies. An important aspect of the organisation of the industry, worldwide, is that it is unusual for services companies to develop into operating companies.

theoretical understandings so that (in a loose sense) novel knowledge claims may be made of more general use than the UK's upstream oil and gas industry, which could not have been made without undertaking the empirical inquiry. The three headings may be combined under the concern to undertake empirical research that is informed by prior theoretical understanding, but which is not so determined or constrained by this understanding that the potential for making novel insights is foregone.

2 The capabilities approach and its application to industry studies

Richardson (1972) describes an industry as "carrying out an indefinitely large number of *activities*" related to a range of business functions located within firms, and "carried out by organisations with appropriate *capabilities*, or, in other words, appropriate knowledge, experience, and skills" (*ibid.*, p. 888, *original emphases*).[2] Penrose (1959) had already written in terms of managerial stewardship of growing firms in which managers connected their perceptions of market opportunities with their perceptions of resources available in the firm (or close at hand) in drawing out productive services. Richardson's emphasis on industry co-ordination leads him to focus on the managerial problem of combining capabilities that are complementary, but similar or dis-similar, in order that activities may be undertaken. Richardson infers that capabilities will be developed within different organisations where dis-similar, but that the complementarity requires co-ordinated through close and on-going working relations.

Activities seem unproblematic to identify, but the capabilities required for undertaking activities are expected to be in the form of "knowledge, experience, and skills," which are problematic to gain access to. Activities of the upstream oil and gas industry include seismic analysis, exploration drilling, well design, infrastructure design, manufacture and installation (CAPEX), production management (OPEX), and all aspects of supply chain management (a longer list of activities is given in Table 3). Activities, though, are not proxies of capabilities. Similar capabilities may have different realised or potential uses. While observation may focus on the organisation or co-ordination of activities in an industrial context, the capabilities approach prepares researchers for explanations that draw on cognitive theories,

[2] Further, Loasby explains capabilities as "a theory which seeks to explain what any particular firm does by what its decision-makers believe they know, what they believe they have learnt, and what they believe they can do now; that seeks to explain the evolution of a firm's activities by the evolution of its knowledge and skills, and in part of the unintended consequences of its actions; and that pays particular attention to the firm's internal arrangements and patterns of behaviour and also its connections with other firms ..." (1998a, p. 163). Others have used "competencies" to mean something similar. Foss (1996a, p.1) argues that: "by competence, we understand a typically idiosyncratic knowledge capital that allows its holder to perform activities – in particular, to solve problems – in certain ways, and typically do this more efficiently than others." "Capabilities" is preferred in this paper. Some authors are not as clear in distinguishing competencies from activities. Patel and Pavitt (1997) use patent information as a proxy measure of competence in an aggregate study of 400 large firms. While they admit that patents "measure codified knowledge, whereas a high proportion of firm-specific competencies is non-codified (i.e. tacit) knowledge," they argue that codified and tacit knowledge are complementary, such that differences in patent data will proxy differences in underlying competencies (*ibid.*, p. 143).

which have been substantiated by experimental research in cognition. Capabilities have though been drawn into a wider theoretical milieu, where discussion has extended and developed to include rules, routines, procedural rationality, fundamental uncertainty, memory and deliberative decision-making. A problem faced in translating the theoretical notion of capabilities into a practical definition for the purposes of guiding empirical inquiry is identifying a clear role for capabilities among these other terms.

Three contributions provide a starting point. (1) Richardson (1953) draws on Ryle's (1949) distinction of declarative or theoretical know-that, and personal and mainly tacit know-how, which may be seen as different types of knowledge, both implicated in experience and skills.[3] (2) Nelson and Winter (1982) develop a notion of extra-personal routines that provide some measure of organisational stability over time, and connect individual skills (which, confusingly, also comprise routines), both of which are discussed by Cohen et al. (1996). (3) Recent research into the implications and meaning of articulating or codifying personal knowledge has added another dimension to theoretical understanding of the nature of capabilities (Ancori, et al., 2000; Boisot, 1995; Cowan, et al., 2000; Malerba and Orsenigo, 2000).

Richardson (1953, pp. 136–138) makes a categorical distinction between objective (codified and articulated) and subjective (personal and tacit) estimates of uncertainty in the context of procedures involved in decision-making. Loasby (1998b, 1998c) argues that capabilities are within the province of direct and indirect know-how (in personal and tacit form), and are categorically distinct from direct and indirect know-that (declarative or theoretical knowledge in codified or articulated form). One implication may be that capabilities cannot be communicated easily and at low cost between individuals, either in the form of know-how, or through attempts at articulating and codifying it in the form of know-that. An empirical inquiry into the transfer of activities – focusing on the capabilities required for undertaking activities – is directed at individuals, but with little expectation of being able to interpret and understand any aspect of an individual's experience with regard to an industrial activity.[4]

This implication is though drawn simplistically. Loasby's reference to direct and indirect know-how involves "gaining control of other capabilities or ... obtaining access to them" (Loasby, 1998b, p. 149). This implies that communication invoking something like the theoretical notion of capabilities is possible, and is expected to occur regularly, either within or between firms. Any combination of cognitive processes that combine know-that and know-how in social and economic interaction promises some, albeit partial and context-dependent, access to capabilities.

[3] Polanyi (1966) describes tacit knowledge as knowing more than we can tell, and Rosenberg develops this in the context of science and technology through "knowledge of techniques, methods and designs that work in certain ways and with certain consequences, even when one cannot explain why" (Rosenberg, 1982; Senker, 1995). It is notable that Polanyi and Rosenberg establish some of the nature of tacit knowledge in terms of individuals reflecting upon their awareness of their activities and trying to articulate and communicate this form of knowledge, but experiencing difficulties in so doing.

[4] This is an age-old problem in the social sciences. Koppl (2000) places Machlup's criticism of behavioural theories of the firm into the context of interpretivism in social sciences, drawing attention to Schutz's notion of ideal types.

The possibility of capabilities including an important extra-personal dimension requires that researchers come to terms with routines as well as capabilities. Different and incomplete accounts have been written about the relationship between capabilities and routines. Nelson and Winter (1982) give routines a critical role of providing stable connections over time between individuals and positions or roles within organisations, and serving as memory store, organisational truce, and target for replication. This is similar to capabilities: both have the effect of reducing the cognitive demands on individuals within a firm. Routines, in Nelson and Winter's terms, seem to be working from the level of the firm (and this can be extended to a firm's external organisation) to smaller sub-groups and individuals involved in the firm's activities. Capabilities seem to be working from individuals and smaller sub-groups, seeking broader forms of connection with the firm's internal and external organisation. It may then be helpful to see routines and capabilities as overlapping but not achieving synonymity. Loasby (1998b, p. 149) argues that: "Divided capabilities typically need to be used in clusters, or in closely-ordered sequences, if the improvements in each new sub-skill which follow this division are to be guided in compatible directions." A firm, in its internal or external organisation, or some combination, may provide such guidance. This distinction between routine and capability is not absolute, but implies that researchers should examine groups within firms.

This inference receives (at best) ambiguous support in Cohen et al.'s (1996) investigation of routines, which conflates routine and capability: "A routine is an executable *capability* for repeated performance in some *context* that has been *learned* by an organization in response to *selective pressures*" (p. 683, *original emphases*). The definition implies that there are non-executable types of capability, that capabilities are connected with repetition, and that part of the context is the organisation per se, rather than some sub-group within (or between) organisations. The definition seems distant from individuals in firms, whose conscious reflections on experiences of capabilities – where articulated and codified – may be helpful for researchers. Cohen et al. (1996, p. 667) recognise another factor, "grain size of routine," which is of some use in distinguishing capability from routine, and for establishing roles for individuals that do not reduce to organisation-wide routines. Organisational routines of small grain size are "broken up" by free-willed individuals who have the capacity to reason, appraise, decide, act and be aware of these activities while they are on-going and also in monitoring and appraising their effects. Reasoning, appraising, deciding and acting have been explained in terms of routines (Holland et al., 1986; Rizzello, 1999). But these seem to be different types of routines, subjective and interpretive, compared with the organisational type referred to in Cohen et al.'s definition. Hence, for the purposes of empirical inquiry, it may be helpful to see capabilities connecting with small-grained organisational routines involving sub-groups that are oriented around particular activities, in the context of wider and larger-grained routines.

Locating capabilities within the province of know-how makes empirical inquiry within the capabilities approach difficult, and may explain the rise of experimental and simulation research undertaken in connection with cognitive hypotheses. While the distinction between know-how and know-that is clear, based on different cognitive activities and memory processes, the possibility of interactions over time between the two types of knowledge is a matter of contention. Ancori et al., (2000), Cowan et al. (2000) and Malerba and Orsenigo (2000) all address the issue of articulating and codifying knowledge, with Cowan et al. going so far as to examine the problem as if those involved consider the costs and benefits of codifying knowledge that would otherwise remain tacit. Instances of attempted articulation and codification of know-how are expected to be partial articulations and codifications, occurring, for example, in moments of individual or organisation-wide reflection and conscious monitoring and reviewing of activities, or in negotiating within or between organisations in establishing the nature of a planned activity and the terms under which it is to be undertaken. These may invoke routines selectively, but are not of themselves routines and may instead be termed commercial experimenting or practical theorising. Instances of knowledge articulation and codification do not translate capabilities from the realm of know-how into know-that, but do offer the hope for researchers in coming to understandings of industrial organisation beyond observing activities.

This section's review has highlighted the possibility of gaining partial access to capabilities. Further, any application of epistemological principles based only on observation and measurement seems inappropriate. Rather, an interpretive endeavour is envisaged in which analysis includes transcripts of open-ended interviews, and in which interviewees are encouraged to provide accounts of, and reflect upon, experiences – in this case – of transferring activities within and between firms in the industry. Overlapping accounts, say of personnel from an operating company and a services company of the same, or similar episodes, are advantageous in providing points of comparison. Grounded theory research procedures are explained in Glaser and Strauss (1967), and developed within economics in Finch (forthcoming).[5] They are not pursued here in the manner set out by Glaser and Strauss. Glaser and Strauss establish an approach that is intended for theorising in situations where prior theoretical principles (or pre-understanding) are judged to be either irrelevant, or harmful to the endeavour of establishing novel knowledge claims. While practical theorising is still presented in Section 4, the role of existing theories has been established in guiding the research. Grounded theory is though still useful in that it allows scope for interviewees to take on a role of articulating practical theorising in giving what are expected to be partial explanations.

3 The organisation of activities in the UK's upstream oil and gas industry

This section sets out the UK industry's recent development, which has been dominated by concerns that, as a location, it is becoming characterised by exploration and production maturity, and higher costs of exploration and production.[6] A useful

[5] The questionnaires are available from the author.

[6] Exploration and production costs are widely held to be greater in the UK industry than in other locations. Wood Mackenzie (1999, p. 1) report that: "Columbia, UK and Norway, appear to be the highest cost areas although the success of ... [industry] initiatives adopted in the UK since the early 1990s has narrowed the gap between lower unit cost countries. On a more positive note, the UK and

starting point is Hallwood's research into the nature of vertical disintegration in the industry, which covers similar operating and services companies to those included in this research project (Hallwood, 1990a, b, 1991, 1992). Hallwood acquired primary information by postal questionnaire and face-to-face interviews during 1984 and 1985 (just prior to the industry recession of 1986).

The industry recession of 1986 is significant for this research. Interviewees in the largest operating companies pointed to the recession as an important period in which the extent of activities undertaken within operating companies, and relations with services companies for activities undertaken beyond operators' internal organisations, became the subject of strategic analysis by senior managers. The main motivation during the late 1980s for such analysis was reducing exploration and production costs (Cibin and Grant, 1996). Following Hallwood, the pre-1986 period is characterised by less variation in the organisation of activities, with a standard, industry-wide approach to mediating relations between operating and services companies. Operating companies issued contracts of fairly short duration following an invited tender/sealed bid auction system (Hallwood, 1990a, pp. 96–103). US-headquartered companies that had established UK affiliates during the early stages of the UK industry's development in the late 1960s and early 1970s have since been joined by UK headquartered companies (ibid., pp. 151–168), although some these have since been taken-over by, or merged with, the US head-quartered services companies.

Hallwood (1991, p. 277) describes his research as a case study of the transaction cost economics paradigm.[7] This focus distinguishes it from the capabilities approach in which the development of capabilities is considered alongside their internal and external organisation. Notwithstanding sections on the organisation of research and development between operating and services companies (Hallwood, 1990a, pp. 51–52, 139–140), much of Hallwood's analysis concerns make-buy decisions relating to exploration and production activities. As a consequence, care is required in interpreting Hallwood's important empirical finding that "the oil companies [operators] provide inhouse less than 10 percent of the volume of inputs required" (Hallwood, 1990a, pp. 29, 36–45, 1990b, p. 576).[8] The estimate provides a sound basis for the research reported in this paper, but greater attention is paid here to explaining variations in this pattern across different operating companies, and to how operating and services companies may go about transferring activities when adjustments to their boundaries within the supply chain are undertaken.

Norway appear to have the largest remaining commercial reserves." Wood Mackenzie estimates that 1999 unit operating costs were \$2.50 barrel of oil equivalent, compare with \$1.50 in Indonesia and \$1.10 in Australia (ibid.). Further, average costs for finding and producing a barrel of oil in the North Sea in 1999 were at least \$13, compared with \$9 in the Gulf of Mexico and \$4 in Malaysia (Oil and Gas Industry Task Force, 1999, p. 2). The UK, in common with other North Sea jurisdictions, can offer the multi-national operating companies advantages in political and fiscal stability, reducing the overall risk and uncertainty of investments, albeit with higher exploration and production costs.

[7] Hallwood developed this aspect of his research into an analysis of transaction cost economics and multi-national enterprises (Hallwood, 1994, 1997; Love, 1995, 1997).

[8] This reinforces points made by Dietrich (1993, p. 179), Kay (1998, p. 289), Kogut and Zander (1995, p. 424, 1995, pp. 510–11), Langlois (1988, 1998, p. 193), Langlois and Robertson (1995, pp. 33–34), and Loasby (1998a, pp. 174–75).

Operators have sought to reduce exploration and production costs by adopting different organisational forms, and encouraging different working practices (Barlow, 2000). A prominent trend has been towards organising both exploration and production activities into project groups, alliances or joint ventures involving at least one operating company and groups of services companies. These terms cover a wide range of organisational forms, which have themselves developed with experience within each company, and at the industry level. The upstream oil and gas industry has a reputation for documenting its procedures. Handbooks are used internally as an attempt to articulate and codify working practices; something which is particularly important given health and safety requirements in the industry. This also extends to contracts between operators and services companies, especially in the period leading up to the 1986 recession (Aase, 1998).[9]

One explanation of the increased use of both formal and ad hoc organisational structures combining operating and services companies is that it was necessary to get individuals from different companies working together closely in order that each could access others' capabilities when directed at a defined set of activities. These could not be articulated and codified in internal handbooks or detailed contracts between separate companies. Indeed, explicit or routine application of such standard operating procedures could be harmful. This was even more so where operating companies sought a different approach to undertaking tasks, or where services companies had accumulated different types of exploration and production experience through working with other operating companies, and in other industries, such as construction, or where novel technical problems were encountered during early phases of exploration activities. Such arrangements were, especially in the early examples of the late 1980s and early 1990s, organisational and commercial experiments for all concerned.

As argued towards the end of Section Two, grounded theory procedures offer some scope for establishing a practical explanation of the transfer of activities in the industry. Fieldwork was conducted during 1998 and 1999. Seven operating companies were visited, along with eight first tier services companies (this covers all major first tier companies as defined by contracts or procurement managers among operating companies), three drilling companies (covering about one-third of such companies active in the UK industry), one specialist services company and three industry organisations. More than one interview was carried out in two operating companies, three services companies and one drilling company. Additional information was drawn from annual reports of companies, sales brochures, company magazines and press reports. Transcripts of semi-structured research interviews are an important source of primary information drawn upon in this and the following section. Interviewees were identified from existing contacts, industry publications, industry conference and seminar participant lists, and by interviewee recommendation.

---

[9] Handbooks serve another function of establishing operating practises in circumstances of personnel changes on operating facilities, either as normal personnel turnover, or when services companies replace one another as part of re-tendering of work, or when an operating company sells assets in production to another operator.

Tables 1, 2, 3 and 4 collate and summarise information from the different primary sources. Table 1 shows that the seven operating companies included in this study account for 44 percent of UKCS gas production and 48 percent of UKCS oil production over June 1998 to May 1999. Within this, two large operators are included, alongside four medium-sized operators and one operator with a small UKCS presence. Operators One and Two have been innovative in promoting different approaches to alliances since the late 1980s, with Operator Four developing similarly innovative approaches since the mid-1990s. Table 2 draws data from three years of annual reports of the services companies included in the study. Nominal total revenues fell from 1998 to 1999 in all but one services company, with Services Companies Two, Four, Five, Six and Seven dominating the sector. Data on proportions of revenues from the upstream oil and gas industry indicate considerable variation in the extent to which services companies are diversified across industrial sectors, with Services Companies Two, Five and Six being major transnational services companies that have developed through periods of acquisition. The proportions of total revenues from the UKCS indicate typically that services companies are involved in exploration and production activities across different provinces.

Table 3 augments the information in Table 2 by setting out the range of exploration and production activities of each services company. It demonstrates that activities themselves are identified fairly easily, with the list drawn from industry publications. The range of activities reflects corporate strategies of acquisitions and these have been contentious in the industry. The pattern is clearest in the case of Services Company Five. Its strategy has been to provide a wide range of sub-surface and production services, along with facilities management and decommissioning. Other large services companies have tended to focus on particular areas of industry activities, seen most clearly in the cases of Services Companies Nine, Ten and Eleven. Service Company Three is unusual in supplying a limited range of operations support, and Service Company Twelve is an equipment producer.

Table 4 is drawn from interviews rather than annual reports and provides insights into the strategies being pursued among services companies in acquiring upstream activities. Two interrelated trends stand out. First, some services companies have broadened the range of activities that they can be involved in through acquisitions within the sector. Second, others have acquired project management (co-ordination and administration) capabilities partly through employing former operating company employees. Operators One and Two (from Table 1) have led both trends by introducing what have become known as integrated services contracts for operations maintenance and support, and in slightly different form for drilling services. Further, alliance arrangements were introduced for major capital projects in the development of oil and gas fields during the early and mid-1990s.[10]

[10] Hallwood (1990a, 1990b) argues that the invited tender/sealed bid auction system is efficient in economising on transaction costs in terms of measuring and metering costs. Hallwood sees another interpretation of transaction costs, in terms of rent appropriation, as less important because work being tendered was codified fairly easily. The modifications made by many operating companies to the invited tender/sealed bid auction method during the 1990s seems to involve elements of measur ing and metering, and of rent appropriation, and, in addition, competence development.

Table 1. Operating companies included in the study

Operator (with 1999 Wood Mackenzie UK reserves value ranking)	UKCS gas production (mmcfd) (June 1998 – May 1999)	UKCS oil production ('000 barrels per day) (June 1998 – May 1999)	Share of total UKCS gas Production	Share of total UKCS oil Production	Number of UKCS operatorships[a]	Value of UKCS reserves $m, discounted to January 1999
One (1)	1569	558	18.2	21.9	190	8289.3
Two (3)	742	269	8.7	12.1	111	5263.1
Three (5)	373	92	4.3	3.8	39	2201.6
Four (6)	470	6	5.5	2.8	65	2037.5
Five (8)	269	109	3.1	4.5	88	1772.1
Six (10)	306	63	3.6	2.6	70	1265.4
Seven (No ranking given)	27	11	0.3	0.4	10	No valuation given

Sources: Wood Mackenzie, UK Upstream Reports, March 1998, and March and July 1999, Wood Mackenzie, Edinburgh.
Notes: [a] Operatorships denote that an operator has day-to-day responsibility for a particular oil or gas field (asset). One company on behalf of a consortium of operators operates many fields. It is usually the largest shareholder who has operating responsibility. Operatorships denote anything from 100 to 20 percent ownership.

Table 2. Financial details of services companies included in study

Companies	Total revenues ($, millions, nominal)	Total assets ($, millions, nominal)	Proportion revenues from oil and gas	Proportion of total revenues from UK
One	1999 $2,190.0	$1,122.6	100%	Not reported
	1998 $2,608.7	$1,298.5	100%	17%
	1997 $1,829.8	$893.4	100%	7%
Two	1999 $5,015.9	$1,802.5	Not reported	55%
	1998 $5,621.0	$2,000.2	19%	55%
	1997 $5,154.5	$1,968.0	23%	72%
Three	1999 $788.0	$4,177.0	100%	Not reported
	1998 $761.8	$3,409.3	100%	32%
	1997 $539.4	$1,677.9	100%	23%
Four	1999 $4,546.7	$7,039.8	100%	9%
	1998 $5,820.6	$7,632.9	100%	9%
	1997 $4,957.9	$6,897.1	100%	8%
Five	1999 $14,898.0	$10,728.0	47%	12%
	1998 $17,353.0	$11,066.0	52%	13%
	1997 $16,272.0	$10,704.0	52%	14%
Six	1999 $9,387.3	$5,648.8	26%	29%
	1998 $10,547.3	$7,286.6	26%	31%
	1997 $9,665.3	$7,717.4	26%	28%
Seven	1999 $8,395.0	$15,081.0	70%	10%[a]
	1998 $10,725.0	$16,078.0	73%	Not reported
	1997 $10,625.1	$13,186.0	72%	Not reported
Eight	1999 $972.0	$414.4	68%	Not reported
	1998 $976.6	$375.1	72%	50%
	1997 $839.4	$375.2	72%	Not reported
Nine	1999 $791.0	$2,264.5	100%	26%[b], 11%[c]
	1998 $1,162.2	$1,971.6	100%	23%[b], 15%[c]
	1997 $1,067.1	$1,421.9	100%	13%[c]
Ten	1999 $614.2	$1,563.4	100%	41%[d]
	1998 $811.4	$1,453.7	100%	38%[d]
	1997 $689.0	$1,161.5	100%	36%[d]
Eleven	1999 $648.0	$6,140.0	100%	19%
	1998 $1,091.0	$1,473.0	100%	32%
	1997 $891.0	$1,051.0	100%	35%
Twelve	1999 $1,031.2	$720.6	15%	33.2%
	1998 $1,034.7	$473.5	23%	33.0%
	1997 $1,055.2	$407.3	21%	Not reported

Sources: Annual Reports of each services company included in the study; and Financial Times (1998) *Energy Yearbook. Oil and Gas Service Companies* 1999, Financial Times Energy, London.

Notes: [a] Interviewee estimate, UK data not reported. [b] Refers to proportion of long lived assets located in the UKCS. [c] Refers to revenue from the whole North Sea rather than the UKCS. [d] Refers to revenue earned from the whole North Sea rather than the UKCS.

Table 3. Exploration and production activities undertaken by services companies included in the study

Upstream activities	One	Two	Three	Four	Five	Six	Seven	Eight	Nine	Ten	Eleven	Twelve
Construction		√			√	√						
Contracting		√		√	√			√				
Decommis-sioning	(√)				√							
Downhole Systems				√	√		(√)					
Drill Bits				√	√							
Drilling and Drillships	√			√				√	√	√	√	
Environmental Services		√		√								
Fabrication	√	√			√	√						
Facilities Management			√									
Field Development					√		(√)					
Geophysical Services				√			√					
Inspection and Survey					√							
Installation and Positioning	√				√							
Maintenance	√	√		√	√	√		√				√
Pipelaying		√			√							
Pipeline Maintenance				√	√	√		√				
Plant Manufacture		√			√							
Platform Construction		√			√	√						
Platform Modules					√	√						
Process Engineering		√		√	√							
Production Facilities	√	√			√			√				

Table 3 (continued)

Upstream activities	One	Two	Three	Four	Five	Six	Seven	Eight	Nine	Ten	Eleven	Twelve
Pumps				√	√	√		√				√
Seismic Processing				√			√					
Software Systems				√	√		√					
Speciality Chemicals					√							
Subsea Systems				√	√	√		√				
Well Services				√	√		√	√	(√)	(√)		

Sources: Financial Times (1998) *Energy Yearbook. Oil and Gas Service Companies 1999*, Financial Times Energy, London; supplemented in parenthesis by personal interview notes 1998 and 1999, and company reports.

Both initiatives have influenced how services companies have arranged their activities, and have encouraged Service Company Five (in particular) to undertake project and supply chain management across a broad range of services within its own organisational boundaries. Conventional practices are though still in evidence among some of the medium-sized operators, including Operators Three and Seven, and there are signs that Operator One is reverting to something like conventional contracts for operations maintenance and support and for drilling services.

The overall picture is one of operating and services companies coming to terms with a proportion of exploration and production knowledge becoming established in the wider industry among services companies. Services companies have acquired a wide range of activities through corporate acquisition and merger in "their sector" of the industry. This has led senior personnel in some operating companies to see services companies as sources of knowledge, mainly as capabilities, which have developed in that sector of the industry, but have not been integrated into the wider organisation of production. Detailed specifications of activities, written for the invited tender/sealed auction method of mediating between operating and services companies, and the fairly short duration of subsequent contracts, were recognised as being inappropriate for integrating knowledge developed by services companies.

4 An industry-level model of the transfer of activities in the UK's upstream oil and gas industry

This section draws upon fieldwork resources and the theoretical capabilities perspective in order to pursue an industry-level explanation of how personnel within the UK's upstream oil and gas industry went about appraising and undertaking

Table 4. Activity acquisition among services companies

Services company	Date of entry into UK market	Corporate status	Summary of oil and gas activities as of 1999	Trends in activity acquisition since mid-1980s	Competitors in UKCS[a]
One	1970	UK division of transnational services company	Engineering, procurement, construction, (ISCs), acquisition of engineering company decommissioning	Operator Two's integrated services contracts maintenance, operations support and and consultancy, joint venture in decommissioning	Two, Six, Five b, Three, Eight
Two	1975	Division of international engineering company	Design, engineering, hook-up and commissioning, project management in green and brown fields, module fabrication, offshore construction, and health and safety systems	Off-shore design and engineering acquired through personnel recruitment, influenced by Operator Two's contracts in early 1990s and more widespread ISCs in late 1990s	6 competitors for larger projects and ISCs, especially Five b and Six
Three	1972	Acquired by larger services group in late 1990s	Operating and maintaining offshore fixed structures and floating production systems, design, engineering and construction, a main contractor since 1994	Alliance with medium-sized operator, taken-over by larger services company	Up to 15 competitors including Five b, Six, One, Eight, Two
Four	Solutions group for engineering activities	Division of transnational services company	Project management to enable selling of engineering capability, subsurface development, integrating products and services, supplying equipment for single wells, projects, exploration wells, and potentially full field development	New contracting environment with Operator Two in 1994, merged with another services company, working in alliances with other services companies	Five a, Seven, and smaller service companies that overlap rather than include a similar range of services, including One, Two, Six and Eight

Table 4 (continued)

Services company	Date of entry into UK market	Corporate status	Summary of oil and gas activities as of 1999	Trends in activity acquisition since mid-1980s	Competitors in UKCS[a]
Five, division a	Long established in US industry	Transnational services company	Relationship management, dealing with diverse range of client expectations, commercial skills, information gathering and management	Operator One-led approaches to alliances and joint ventures, co-location, cultural alignment, win-win ethos, where clients looking for one-stop shop	Four and Seven for comparable ranges of solutions services
Five, division b	early 1970s	Division of transnational services company	From hook-up and commissioning to maintenance and its management, operations and operations support, engineering, construction	Health and Safety requirements of subsurface shut down and safety valves, Operator Two introducing ISCs, general alliances impetus and confidence to broaden skills, employing former operator staff	Two, Eight, Six, One, and sometimes Three
Five, division c	early 1970s	Division of transnational services company sine mid-1990s	Adding value to energy assets, engineering design, procurement, health and safety management, design engineering, fabrication, sub-sea installation, flow lines, and flexible risers	In capital projects alliances, working with other contractors for up to two or three years. Partnerships in operations support and refurbishment, some operators want ISCs others want grouping, corporate acquisitions	8 to 10 in all including Six, Eight, and One

Table 4 (continued)

Services company	Date of entry into UK market	Corporate status	Summary of oil and gas activities as of 1999	trends in activity acquisition since mid-1980s	Competitors in UKCS[a]
Six	late 1960s	UKCS division of transnational corporation	Managing people and services, hard issues of well-head control systems, fabrication, procurement, hook-up, commissioning operations, supplying kit, processes and system	Acquired engineering and construction services company 1995, more early-stage development work undertaken	10 in capital projects, including Five b and Two; 6 in operations support
Seven	Late 1960s	Transnational services company	Project management, seismic, field diagnostics, well development, planning integration and design on non-core to operator activities, logging tools, gaps in reservoir prediction, finding oil and developing fields	Since 1986, consequence of outsourcing as procurement-driven, with core competencies as residual activities, one stop shops for large operating companies, problem solving for independent operating companies, advances threatened by current trend to commodity supply	Five a and Four, then smaller services companies that offer some of the range offered by Seven, such as Two
Eight	mid-1970s		Operations and maintenance, engineering, modifications, brown field engineering, pipeline engineering, project management, design, procurement, and sub-sea installation	1970s low-tech work in logistics, plant hire, offshore engineering; 1980s engineering and design, fire detection and suppression, rig repair; 1990s international diversification and more outsourcing, being client driven, mergers and acquisitions, international diversification	Two, Five b, Six, less so Three and One

Table 4 (continued)

Services company	Date of entry into UK market	Corporate status	Summary of oil and gas activities as of 1999	Trends in activity acquisition since mid-1980s	Competitors in UKCS[a]
Nine	1963	2 UK groups, part of US company	Drilling, renting rigs and crews, project management and ISC, risk management, field economics, reservoir engineering, well design (since 1995)	Operator One's outsourcing policy, its loss of competence and neglect of marginal fields, innovative approaches of smaller operators	8 in rig supply including Ten, 10 for ISCs, including Six, Two, and Five b
Ten	1965	Majority of shares owned by an oil company	From rig supply and drilling contractor in 1989 to well engineering, including cementing, directional drilling and procurement	From Operator Two's 1990 initiative which implied either play and develop or stay renting rigs	Four, Nine, and about 10 others
Eleven	1971	Many mergers & acquisitions, now part of recently merged drilling company	Supplying semi-submersible rigs and drilling services, project management, co-ordinating others in mud, cement and bit purchase, planning, design and operating	Develop internally, with some ex-operator employees in mid-1990s, has large engineering department globally, North Sea sets technical challenges	10 competitors across 3 sectors in drilling: platform contracts, platform jack ups, and semi-submersibles, mentioned Ten
Twelve	mid-1970s	Division of engineering company	TVM contracts on pumping and rotating equipment – often for parent company, project management, reducing maintenance costs	Working on Operator One's pipeline and other sectors, 1990 supplied through Service Company Eight, adapting within longer contracts with operators and prime contractors	Eight, Five b, and Two for its upstream oil and gas activities

Sources: Personal Interviews 1998 and 1999. Interviews held with reference to particular operating divisions and with senior managers of parent groups.
Notes: [a] Recorded in the order mentioned during the interviews.

transfers of activities. The main themes developed from fieldwork resources are: power battle in the supply chain, knowledge differentiation and integration in the organisation of production, and industry business cycle.

4.1 Power battle in the supply chain

The activity of supply chain management has received critical attention within the industry during the 1990s (CRINE Network and Ernst and Young, 1999). The term is useful in capturing different organisational arrangements deployed in transferring and developing activities. The main theme, power battle in the supply chain, has been formulated through a process of coding and recoding interview transcripts and other primary information. It is common to a number of issues that emerged during semi-structured interviews, and is supported in other primary documentation. These issues included:

- An emphasis on rent-seeking contracting rather than value-creating activities.
- Concentration among number of operating and services companies, consequences of this for companies in the respective sectors of the industry.
- Operators seeking commodities rather than specialised and bespoke solutions from services companies.
- Recognition of the impact of strategic decisions affecting upstream activities on corporate share prices.
- Procurement and human resource motives of instigating supply chain management initiatives.
- Analysis in terms of risk rather than uncertainty.
- Designing instruments of control and incentives within well-defined areas of delegation.
- Strong influence of industry initiatives.

Power battle in the supply chain is characterised by decision-making in which fundamental exploration and production knowledge (as know-how and as know-that) is taken as given by senior personnel of operating and services companies, albeit often incomplete. This allows decision-makers to focus on arranging exploration and production activities in order to increase margins by reducing costs. The impacts of the issues listed above on the main theme, power battle in the supply chain, are difficult to discuss separately. Rather, they are well-connected dimensions of the main theme.

The organisation of activities between operating and services companies is influenced strongly by decisions made following the power battle in the supply chain. Performance incentives are often arranged around risk and reward sharing contracts that articulate a share of any profits or losses compared with a pre-contract analysis of normal costs and normal performance, in turn calculated from historical records and benchmarking exercises. The emphasis on risk is often made operational by drawing upon historical performance data and expert opinion to formulate probability distributions across a range of feasible performance outcomes specific to particular contracts (Simpson et al., 1999). Standardisation of processes and of technologies enhances such a way of thinking, and standard well designs, drilling solutions, contracts, and assessments of components suppliers and of services suppliers have been subjects of recent industry initiatives. These have been strong in the sense of being articulated and codified, and publicised widely, illustrating processes of providing approximations or versions of know-how in the form of know-that.

4.2 Differentiation and integration in the organisation of production

Differentiation and integration are tendencies both of specialising knowledge into more and more groupings, each with a successively narrower scope, of connecting areas of specialist knowledge through some general principles, and of facilitating on-going processes of knowledge development.[11] These tendencies can be discerned at the same time as those drawn upon in forming the main theme power battle in the supply chain. They differ in the critical respect of decision-makers not taking exploration and production knowledge as more or less given. Instead, differentiation and integration refers to longer-term processes of knowledge development, which will also be shaped by decision-makers' thinking about power battle in the supply chain arrangements.

The issues that emerged from semi-structured interviews and other primary information, and that can be grouped under the main theme of differentiation and integration, include:

- Deliberately seeking new knowledge through product or service innovation.
- Recognition that such decisions are characterised by uncertainty rather than risk.
- Recognition that the industry is dynamic such that decisions made through static analysis entails simplifications and assumptions.
- There are irreducible limits to managerial control.
- Different approaches to coordination within and across different companies may range from centralised to decentralised.
- Technical considerations alongside commercial considerations.
- Industry initiatives are weak.
- Motivations including seeking and capturing value.

Within this second main theme, decision-makers still enjoy discretion, but it is recognised that processes and outcomes are less clearly related to conscious strategic considerations compared with designing contracts and other incentives under power battle in the supply chain thinking. The focus within this theme is on processes of the growth and diffusion of knowledge, which are partly autonomous. Of course, new knowledge and innovation can yield rents. In order that these may be

[11] "Differentiation and integration" were used by Marshall (1920) to describe a biological-like organic development of companies and groups of companies through increased specialisation of knowledge by subdivision and an increased pressure to co-ordinate, and hence benefit from, growing and specialised knowledge through establishing connecting principles. A similar approach is developed in management research by Thompson (1967) in relation to technology and organisational struc ture, and by Mintzberg (1991) in relation to organisational development.

prolonged as quasi-rents, the activity in which any innovation is embedded may be transferred into power-battle thinking, especially once the manifestations and consequences of it become better understood (which implies an interpretation in the form of know-that).

The two themes, power battle in the supply chain and differentiation and integration, imply some contradictions even among those decision-makers interviewed. Power battle in the supply chain emphasises organising existing activities, and hence existing capabilities directed at these activities, to expected maximum commercial benefit to a particular company. Seeking new technical solutions, for example, through commercial experimenting as well as normal learning-by-doing, implies developing technical knowledge, which may undermine the aim of maximum commercial benefit for a particular configuration of capabilities and activities. The contradiction was most clearly reflected in relations between operating and services companies. There were secondary divisions between operating companies where some companies, such as Operators Three and Six, maintained conventional relations with services companies. They did not embark upon forming alliances so had no need to develop the type of risk and reward sharing approach where longer-term contracting relations were manipulated for maximum commercial benefit. The contradiction had a third, less well pronounced, reflection within companies and even within individuals. Some seemingly closely related issues were framed in static terms and others in dynamic terms, for example with facilities operating (OPEX) work in the first instance and capital project (CAPEX) work in the second instance, or supplier relations in static terms, and personnel development and organisational learning in dynamic terms.[12]

4.3 Industry business cycle

The third main theme, industry business cycle, is practically beyond the influence of the UKCS activities of operating and services companies.[13] Initially, the research project took 1990 as a reference point during interviews. The date proved useful as marking the beginning of a distinctive phase in organising some industry activities, especially among services companies. For the larger operating companies, changes in organising exploration and production activities can be traced to the 1986 fall

[12] This may be a practical manifestation of framing effects (Tversky and Kahneman, 1981). Operators and services companies establish a set of working relations as alliance members for capital projects and field development, and then try to carry forward such relations into operational phases.

[13] Business cycles have usually been interpreted as seemingly autonomous transitions of upswing, boom and recession, associated with the macro-economy. Schumpeter (1939) argued that there are three business cycles operating simultaneously: short cycles are of around 40 months, business cycles of around ten years and long waves of anything between 50 and 70 years. The long waves are associated with revolutionary inventions and their diffusions. Business cycles exhibited regular and periodic crises, but these are brought on by significant phases of innovation, adaptation and diffusions to do with the dominant technology of the long wave, and short cycles were more of a statistical regularity, lacking clear explanation, which may have been adaptations. Schumpeter's characterisation is appropriate in the case of the UK upstream oil and gas industry because it highlights distinct but interrelated tendencies of maturity and cycle.

in oil prices. The research project hence covers a complete industry business cycle between the low oil prices of 1986 and 1999.

The theme industry business cycle emerged from a series of smaller issues that were discussed during research interviews. Hence:

- Different opportunities arise during the business cycle and affect the balance of power in the industry.
- The business cycle affects the balance of attitudes between short and long termism.
- Business cycles are exogenous regulators of activity levels with in the UKCS industry.
- Decisions have consequences in terms of problem-seeking or problem avoidance, with further consequences for whether decision may be analysed in terms of risk or uncertainty.
- Decisions made with respect to the UKCS industry have little effect on the business cycle, which is most strongly influenced by the level of world oil prices.
- Decision-makers tend to become reactive.

Industry business cycle differs from the other two main themes in that it is much closer to ideas formulated by interviewees. Interviewees spoke about business cycles, which typically were approximated by drilling rig rates as well as world oil prices, and on their effects on activity levels with the UKCS. Interviewees from some services companies saw the business cycle as a predictable and inevitable process of adjusting revenues over time.

4.4 Reversibility and discretion

The main themes are recombined and organised for purposes of analysis and explanation with reference to two analytical categories – reversibility and discretion – that allow different types of connections between the main themes. Reversibility is used in the framework in the sense of experiment or hypothesis. Two of the main themes, power battle in the supply chain and industry business cycle, are reversible in that *if* each could be isolated and other factors held constant, any initial changes pertaining to the themes may be negated by equal and opposite changes, leaving the industry's set of capabilities unchanged. A business cycle could transform from recession to boom and back to recession with the two points of recession requiring identical sets of capabilities. Exploration and production activities may be arranged between operating and services companies through the invited tender/sealed bid auction method and then through alliances. A return to the previous regime would leave the set of industry capabilities unchanged. Only the distribution of activities and revenues would change.

Differentiation and integration in the organisation of production is irreversible because learning and knowledge creation are captured within this main theme. Following Section 2, capabilities comprise skills and knowledge that are tacit to some extent, such that their performance and communication is also bound up in a complex combination of tacit know-how and declarative know-that. Activities may be performed differently over time without agents in performance or review being fully aware of changes, and will be affected further as personnel change positions and develop new roles within companies, move between companies and leave the industry altogether. The development of capabilities has also been affected, as alliances have been formed, changed and broken-up. Incremental or radical innovations are likely to alter the set of capabilities in the industry in a way that makes any recovery of previous states difficult and also of little practical value.

Where reversibility is a categorical (yes-no) term, discretion is a matter of degree. The main theme industry business cycle is beyond the discretion of decision-makers in the UKCS upstream industry. Discretion varies across the other two main themes and is most prominent in the theme power battle in the supply chain where operating and services companies negotiate terms concerning which companies under which commercial terms undertake exploration and production activities. Two or more companies involved in sets of vertical relationships may possess the capabilities to undertake particular activities within its own boundaries, or at least within its boundaries to a greater extent. Descriptions given during interviews include "making fields attractive to clients," "developing relations with operators," and "formulating value propositions." Supply chain management is though becoming a contested area between operating and services companies, especially regarding changes introduced by Operator One and, to some extent, Operator Two (which did not implement such a radical, devolved approach in the early 1990s). Differentiation and integration in the organisation of production also highlights discretion for decision-makers, especially in choosing to configure devolved knowledge within organisations, and search for new solutions, either of an adaptive or more radical nature (March, 1991, 1999). The development of capabilities over time is though partly an autonomous process, which casts managerial discretion in a different role than that envisaged under the power battle in the supply chain theme.

4.5 A framework for explaining the transfer and development of activities

The framework is illustrated in Figure 1. It is analytical because the main themes can be connected in different ways by drawing upon reversibility and discretion, and upon different states of the industry business cycle, in order to provide different explanations of the transfer and development of capabilities in the UKCS upstream oil and gas industry. In Figure 1, the main theme industry business cycle is related indirectly to the development of competencies because it is, in the abstract, reversible and exogenous from the perspective of decision-makers in the UK industry. Hence, industry business cycle has the role of catalyst through its effects on the other two main themes, through changing the balance of power in the supply chain between operating and services companies, and in changing the supply of technical challenges faced within the industry.

Power battle in the supply chain is hypothetically or experimentally reversible and is also connected indirectly to the development of capabilities. The issues from which the theme emerged are subject to a great deal of conscious reflection and

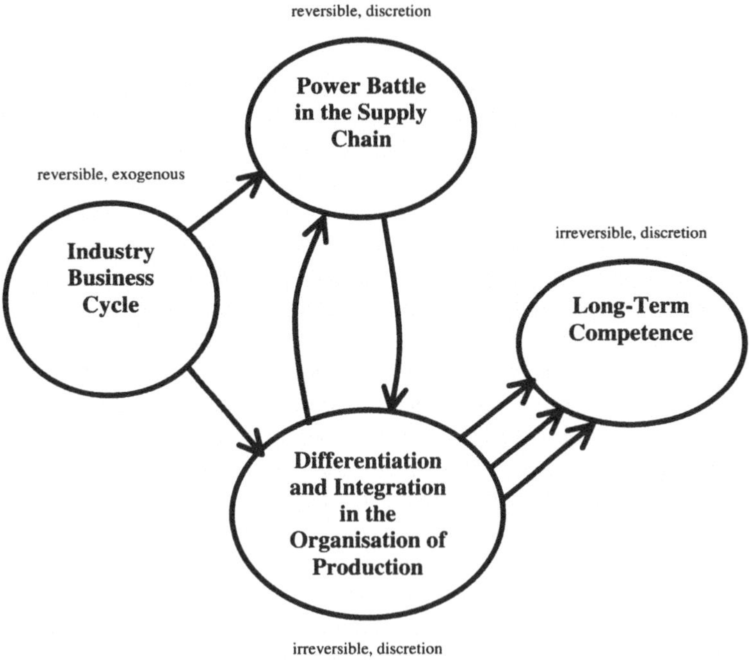

Fig. 1. Conceptual framework for developing conmpetencies in the upstream oil and gas industry

discretion by those involved in the industry. Decisions and outcomes of negotiations within this theme have an indirect effect on the development of competencies by resetting the parameters that shape the processes of differentiation and integration. Attempts have been made within alliances to use contracts and incentives in order to provide working conditions conducive to developing innovative solutions, especially in capital project work. In other words, power battle in the supply chain thinking has been drawn upon to encourage innovations, which is subject to issues formulated under differentiation and integration of knowledge. This has had mixed success as novel problems require innovative solutions in designing capital equipment and in developing working relations across different companies. In other words, processes and outcomes are characterised by uncertainty rather than risk.

Differentiation and integration in the organisation of production has the role in Figure 1 of collecting together exogenous and endogenous influences on the development of capabilities. Differentiation and integration are irreversible and not easily amenable to the intentional analyses of operating and services company personnel as if articulated as declarative knowledge (know-that), irrespective of the amount of power such companies may enjoy within the industry at that time. The diffusion of knowledge has an intentional aspect to it, in the form of instructions that are articulated in contracts. Some diffusion is though autonomous from the perspective of any individual involved in the industry. For example, most services companies have worked over long periods with different operating companies through various types of relationship, and become network nodes around activities fostering the diffusion of know-how and know-that.[14]

The irreversible nature of differentiation and integration, shown in Figure 1, connects with the experimentally or hypothetically reversible themes of power battle in the supply chain and industry business cycle, giving the whole framework an irreversible character. Starting points cannot *in practice* be recovered.[15] The industry business cycle regulates the supply of technical problems into the system, affecting capital project expenditure in particular, by regulating the rate at which development proposals made within operating companies are sanctioned. This may seem a rather reactive model of industry development, but reflects the path of development in the UKCS upstream oil and gas industry during the 1990s.

Outcomes of negotiations about which companies undertake which activities under which types of commercial relationship, established within the main theme power battle in the supply chain, guide and shape the processes of differentiation and integration. But processes of differentiation and integration can also shape the negotiations undertaken under the main theme power battle in the supply chain. Service Company Five, and to a lesser extent the other larger service companies, have acquired exploration and production activities slowly. This has been taken a step further by services companies in undertaking additional acquisitions in order to internalise part of the supply chain management task and offer operating companies larger, integrated packages of services which has reduced the number of alternative sources of supply for operating companies.

Differentiation and integration may be seen as opposing tendencies, both of which can cause coordination problems within companies and within the overall industry if one dominates the other. Differentiation has occurred within most operating companies during the 1990s as managerial discretion has been devolved to asset managers who are responsible for clusters of exploration and production activities in the UK industry. Additional outsourcing, for example through alliance relationships, has encouraged further decentralisation as tasks that were described in some detail in contracts have to some extent been replaced by devolving discretion to services companies concerning how still well-defined outcomes are to be achieved. Differentiation though complicates sharing of knowledge, both within and between companies operating in the more differentiated industry. Some large operating companies have responded to problems caused by differentiation and decentralisation by integrating some business support functions, such as procurement. This is partly for power battle in the supply chain reasons, of creating dependencies among the smaller number of larger services companies on that operating company's business.

[14] Interviewees in operating companies noted, consistently, disappointment with potential gains from the network node role of services companies, in having noticeable impacts on overall operating performance.

[15] Teece and Pisano explain such irreversibility or path dependency in terms of history mattering: "[A] firm's previous investments and its repertoire of routines (its 'history') constrains its future behavior." This is mainly because learning within the firm, which is to some extent experimental, is a rolling process of "trial, feedback and evaluation," which, to be meaningful, requires that some other parts of a firm's activities remain at least stable (Teece and Pisano, 1998, p. 203).

5 Concluding remarks

Three concluding remarks may be made concerning the contributions that can be made to the capabilities approach through undertaking empirical inquiry. First, the definition of capabilities requires critical reflection if it is to guide empirical inquiry. Capabilities have become conflated with routines, possibly under the influence of research that had adopted computer simulation techniques based on behavioural rules and evolutionary conjectures. This has diverted attention from the creative capacities of individuals who are undertaking activities in industrial contexts, or has aimed to explain these in terms of complex cognitive models. Empirical inquiry though depends on individuals involved in industrial activities being able to provide verbal accounts of activities, and the ability of individuals to do so indicates that conscious reflection about capabilities is normal, albeit occasional. The approach applied in this paper involves capabilities mediating between routines of small and large grain-size, and recognising the possibility of partial articulation and codification of capabilities in the form of mediations between know-how and know-that.

Second, and related, those involved in empirical inquiry require some set of procedures by which individual accounts may be compared, but which do not go to the opposite extreme of computer simulation modelling by over-emphasising individuals' explanations. Open-ended interviews provide a way in which interviewees can adopt the role of practical theorist, and also provide accounts of commercial experimenting undertaken within their organisations.

Third, the framework presented in Section Four raises further questions for researchers about how individuals within organisational and industrial contexts interpret and manipulate different types of knowledge that appear contradictory or incommensurable. Transferring activities is a commercial experiment undertaken with some set of expectations that are, typically, articulated and codified as part of appraisal, decision-making and review. There is a strong emphasis on short-term (power-battle) motives in planning the organisation of activities. There is also wide-spread recognition and articulation of practical theories about which firms and which types of organisational relations may be best suited to gaining access to capabilities within the industry to undertake existing activities differently, and to propose novel technical solutions to exploration and early-stage development problems that have yet to be experienced fully. These are different types of knowledge, framing the same issue in different and incommensurable ways, with recourse to interpretations of information that are held with different levels of confidence.

References

Aase K (1998) Handbooks as tools for organizational learning. A case study. Journal of Engineering and Technology Management 15: 201–228

Ancori B, Beruth A, Cohendet P (2000) The economics of knowledge. The debate about codification of tacit knowledge. Industrial and Corporate Change 9: 239–253

Bank of Scotland and Bank West (1996) Oil and gas handbook, 4th edn. Bank of Scotland and Bank West, Edinburgh

Barlow J (2000) Innovation and learning in complex offshore construction projects. Research Policy 29: 973–989

Boisot MH (1995) Information Space. A framework for learning in organizations, institutions and culture. Routledge, London New York

Cibin R, Grant RM (1996) Restructuring among the world's leading oil companies, 1980–92. British Journal of Management 7: 283–307

Cohen MD, Burhart R, Dosi G, Egidi M, Marengo L, Warglein M, Winter S (1996) Routines and other recurring action patterns of organizations. Contemporary research ideas. Industrial and Corporate Change 5: 653–698

Cowan R, David PA, Foray D (2000) The explicit economics of knowledge codification and tacitness. Industrial and Corporate Change 9: 211–253

CRINE Network and Ernst and Young (1999) Supply chain management in the UK oil and gas sector. CRINE Network, London

Dietrich M (1993) Transaction costs ... and revenues. In: Pitelis C (ed) Transaction costs, markets and hierarchies, pp 166–186. Blackwell, Oxford

Finch JH (forthcoming) The role of grounded theory in developing economic theory. Journal of Economic Methodology

Foss NJ (1996a) Introduction. The emerging competence perspective. In: Foss NJ, Knudsen C (eds) Towards a competence theory of the firm, pp 1–12. Routledge, London New York

Foss NJ (1996b) Knowledge-based approaches to the theory of the firm. Some critical comments. Organization Science 7: 470–476

Glaser BG, Strauss AL (1967) The discovery of grounded theory. Strategies for qualitative research. Aldine, Chicago

Hallwood CP (1990a) Transaction costs and trade between multinational corporations. A study of offshore oil production. Unwin Hyman, Boston

Hallwood CP (1990b) Measuring cost and the organization of exchange in the oil gathering business. Journal of Institutional and Theoretical Economics 146: 576–593

Hallwood CP (1991) On choosing organizational arrangements. The example of offshore oil gathering. Scottish Journal of Political Economy 38: 227–241

Hallwood CP (1992) Perceptions of market efficiency, transaction costs, and vertical disintegration in offshore oil gathering. Journal of Economic Studies 19: 36–49

Hallwood CP (1994) An observation on the transaction cost theory of the multinational firm. Journal of Institutional and Theoretical Economics 150: 351–361

Hallwood CP (1997) The transaction cost theory of the multinational firm – a reply. Journal of Institutional and Theoretical Economics 153: 682–687

Holland JH, Holyoak KJ, Nisbett, RE, Thagard PR (1986) Induction. Processes of inference, learning and discovery. MIT Press, Cambridge, MA London

Kay, NM (1998) Clusters and collaboration. The firm, joint ventures, alliances and clubs. In: Foss NJ, Loasby BJ (eds) Economic organization, capabilities and co-ordination. Essays in Honour of GB Richardson, pp 222–242. Routledge, London New York

Kogut B, Zander U (1995) Knowledge, market failure and the multinational enterprise. A reply. Journal of International Business Studies 25: 417–426

Kogut B, Zander U (1996) What do firms do? Coordination, identity, and learning. Organization Science 7: 502–518

Koppl R (2000) Fritz Machlup and behavioralism. Industrial and Corporate Change 9: 595–622

Langlois RN (1988) Economic change and the boundaries of the firm. Journal of Institutional and Theoretical Economics 144: 635–657

Langlois RN (1992) Transaction cost economics in real time. Industrial and Corporate Change 1: 99–127

Langlois RN (1998) Capabilities and the theory of the firm. In: Foss NJ, Loasby BJ (eds) Economic organization, capabilities and co-ordination essays in honour of GB Richardson. pp 183–203. Routledge, London New York

Langlois RN, Robertson PL (1995) Firms, markets and economic change. a dynamic theory of business institutions. Routledge, London New York

Loasby BJ (1998a) The concept of capabilities. In: Foss NJ, Loasby BJ (eds) Economic organization, capabilities and co-ordination. Essays in honour of GB Richardson, pp 163–182, Routledge, London New York
Loasby BJ (1998b) The organisation of capabilities. Journal of Economic Behavior and Organization 35: 139–160
Loasby BJ (1998c) On the definition and organisation of capabilities. Revue Internationale de Systémique 12: 13–26
Loasby BJ (1999) The significance of Penrose's theory for the development of economics. Contributions to Political Economy 18: 31–45
Love JH (1995) Knowledge, market failure and the multinational enterprise. A theoretical note. Journal of International Business Studies 26: 399–407
Love JH (1997) The transaction cost theory of the multinational firm – note. Journal of Institutional and Theoretical Economics 153: 674–681
March JG (1991) Exploration and exploitation in organizational learning. Organizational Science 2: 71–87
March JG (1999) The pursuit of organizational intelligence. Blackwell, Oxford
Malerba F, Orsenigo L (2000) Knowledge, innovation activities and industrial evolution. Industrial and Corporate Change 9: 289–314
Marshall A (1920) Principles of economics, 8th edn. Macmillan, London
Mintzberg, H (1991) The effective organization. Forces and forms. Sloan Management Review, Winter
Newendorp PD (1996) Decision analysis for petroleum exploration. Planning Press, Aurora, CO
Oil and Gas Industry Task Force (1999) A template for change. A report to government and industry. HMSO and Department of Trade and Industry, Aberdeen
Patel P, Pavitt K (1997) The technological competencies of the world's largest firms. Complex and path-dependent, but not much variety. Research Policy 26: 141–156
Penrose EET (1959) The theory of the growth of the firm. Blackwell, Oxford
Polanyi M (1966) The tacit dimension. Routledge and Kegan Paul, London
Richardson GB (1953) Imperfect knowledge and economic efficiency. Oxford Economic Papers 5: 136–156
Richardson GB (1972) The organisation of industry. Economic Journal 82: 883–896
Rizzello S (1999) The economics of the mind. Edward Elgar, Cheltenham
Rosenberg N (1982) Inside the black box. Technology and economics. Cambridge University Press, Cambridge
Ryle G (1949) The concept of the mind. Hutchinson, London
Schumpeter JA (1939) Business cycles, 2 Vols. McGraw-Hill, New York London
Senker J (1995) Tacit knowledge and models of innovation. Industrial and Corporate Change. 4: 425–447
Simpson GS, Finch JH, Lamb FE (1999) Risk, uncertainty and relations between "strategic" and "within-strategy" decision-making in the upstream oil and gas industry. Discussion Paper 99–01, Department of Economics, University of Aberdeen
Teece DJ, Pisano G (1998) The dynamic capabilities of firms. In: Dosi G, Teece DJ, Chytry J (eds) Technology, organizations and competitiveness, pp 193–212. Oxford University Press, Oxford
Thompson JD (1967) Organizations in action. Social science bases of administrative theory. McGraw-Hill, New York
Tversky A, Kahneman D (1981) The framing of decisions and the psychology of choice. Science 211: 453–458
Wood Mackenzie (1999) UK Upstream Report. Number 313. Wood Mackenzie, Edinburgh
Yergin D (1993) The prize. The epic quest for oil, money and power. Pocket Books, London

Uncertainty, institutional structure and the entrepreneurial process

Keith Jakee and Heath Spong*

Royal Melbourne Institute of Technology, School of Economics and Finance, 239 Bourke St., Melbourne, Australia 3000 (e-mail: {Keith.Jakee; Heath.Spong}@rmit.edu.au)

Abstract. While there exist numerous theories of entrepreneurship, we aim to construct an account that is thoroughly process-oriented and is thus consistent with non-teleological evolutionary foundations. To accomplish this, we combine theories of structural uncertainty with recent work in the theory of social institutions. From such a perspective, creatively thinking and acting entrepreneurial individuals can account for endogenous social change through their effect on institutions. Our approach helps to clarify many of the inconsistencies that arise in the existing entrepreneurial literature and we are able to clarify issues of entrepreneurial failure, self-employment versus entrepreneurship, entrepreneurs versus managers, and incentive for entrepreneurs in formal versus informal institutional settings.

Key words: Entrepreneurship – Radical Uncertainty – Non-Teleological Processes – Institutions and Institutional Change – Socio-Economic Evolution – Creativity

JEL Classification: L2, O1, O3, M13

---

* We would like to thank Rodney Beard, Bob Miller, Pavel Pelikan, Mark Picton, David Prychitko, S.M. Riordan, Mario Rizzo, Annette Weier, and Xiaokai Yang for their suggestions. We appreciate the comments and insights from numerous individuals who attended the presentation of this paper at the 2000 *International Joseph Schumpeter Conference*. Jakee gratefully acknowledges research support from the *Australian Research Council* and Spong gratefully acknowledges support from the *Austrian Economics Program* at New York University. Any shortcomings are, of course, our own.

1 Introduction

Economics has developed elaborate models of market structure, cost functions and individual optimisation, yet we still do not have a generally accepted theory of entrepreneurial behaviour and its subsequent effects. This lacuna is surprising in light of the recent attention paid to entrepreneurs by the media and policymakers. While students of the entrepreneurial process have been few, the most well known account is likely to be Schumpeter's, which is considered the first to explicitly discuss the role of the entrepreneur (1934 [1911]). The eminent Austrian economist Israel Kirzner is one of the only scholars in the profession to devote most of his career to examining entrepreneurship in great detail; his seminal work of 1973 stands out in an otherwise neglected field. Recently, writers such as Lachmann (1986), Buchanan and Vanberg (1991), Choi (1993) and Harper (1996) have each made theoretical contributions to the entrepreneurial literature.

The entrepreneurial role is often modelled from a psychological perspective. Kirzner's "alertness" (1973, 1985), Shackle's "enterpriser" (1979), and Lachmann's "entrepreneur" (1986) are all such examples. In contrast, this paper returns to the Schumpeterian position, wherein innovative entrepreneurship is portrayed as a distinct *type of activity* (see, for example, 1934: 81-84). We will explore a variant of the entrepreneurial role, which differentiates between the activities of entrepreneurs and those of non-entrepreneurs. Some implications of this distinction will also be explored later in the paper.

A fundamental issue that has surfaced over the past two decades should be of interest to scholars of evolutionary processes: most theories of entrepreneurship are teleological in their methodology. For example, Kirzner was criticized as early as the mid-1970s for his reliance on equilibrium to model the entrepreneur (see, e.g., Lachmann, 1976). Buchanan and Vanberg (1991) have most clearly outlined the problems of mooring entrepreneurial theories to teleological foundations, especially for evolutionary processes. They argue that viewing the market as naturally converging towards equilibrium implies that the process is, in one way or another, pre-determined. They emphasise that an equilibrium framework is fundamentally incompatible with an evolutionary, process-oriented account of the market's development. While many authors are now aware of these issues and have attempted to free their approaches from equilibrium theorising, we would argue that none has adequately accomplished the task.

As a result of this shortcoming, our description of entrepreneurship as an activity will be placed in the context of Shackle's (1961, 1972) and Buchanan's (1969) work in true uncertainty. Relying on uncertainty allows us to emphasise choice, imagination, and creativity – elements that are lacking in standard neoclassical analyses. Our use of the uncertainty framework also maintains a thoroughly non-teleological, process-oriented approach to entrepreneurship.

However, we also endeavour to acknowledge stability in socio-economic activity. We recognise that radical uncertainty is mitigated by the institutional structures defined by Hayek, Lachmann, Buchanan, North and others. Within this framework, we develop a rudimentary suggestion by Lachmann (1971) that entrepreneurs alter institutional structures. We believe that this modified approach, with its numerous implications, improves upon existing entrepreneurial theory. While we reconcile some differences in the established literature, we do so by diverging from each of these other approaches.

Section two of the paper briefly reviews some background on entrepreneurial theory, focusing particularly on the issue of teleology. Section three presents our own description of entrepreneurial action and the entrepreneurial process. We begin by constructing a detailed framework of structural uncertainty and by emphasising the importance of imagination and creative choice in the socio-economic process. This is followed by a review of the role institutions play in enabling individuals to plan and coordinate their activities in complex socio-economic processes. We then define the entrepreneur and her impact on the market process. In section four, we examine some of the key implications of this approach. We discuss its application to individuals, business practices, and social evolution.

2 Some foundations

In his account of the market process, Lachmann (1986) emphasises Schumpeter's depiction of entrepreneurial innovators, whose imagination is the driving force of economic development. Lachmann argues: "[W]e are entitled to regard Schumpeter as a predecessor of Professor Shackle, who in recent years has done more than any other thinker to insist on the significance of the human imagination as a source of expectations" (Lachmann, 1986: 109). On the other hand, he also contends that Schumpeter relies heavily on a general equilibrium framework (1986: 109) and that in the years since Schumpeter's monumental contribution, neoclassical theorists have come to rely overwhelmingly on equilibrium-oriented approaches to economic phenomena (1986: 112–114). One shortcoming of this practice is that it presents economic activity as a deterministic process: real imagination and individual choice cannot exist in such a framework. Lachmann concludes this work with an appeal to the profession to retreat from the determinism inherent in the standard equilibrium view and to re-introduce genuine imagination and choice into market process theories.

In a similar vein, Buchanan and Vanberg (1991) are critical of the Austrian "middle ground", particularly its leading proponent, Kirzner.[1] They show that Kirzner's attempt to reconcile a framework of uncertainty with a variant of neo-classical equilibrium is problematic. Kirzner claims he is working within a process-oriented framework of uncertainty, yet his theory also incorporates the concepts of *equilibrium* (or "equilibrating tendencies"), *error-correction* within markets, and the

[1] An account of the Austrian "middle ground," and the debate with its subjectivist critics, is given in Kirzner (1992: 6–7). Kirzner defines the Austrian "middle" by reference to two polar extremes. He begins with the description of the entrepreneurial role given by Schultz (1975) who claims that the entrepreneurial role is one which restores the market to equilibrium. Schultz constructs demand and supply curves for the entrepreneur's services, thus imputing an economic value on the ability to deal with disequilibria. Kirzner argues that the other extreme is given by Shackle, whose heavy emphasis on the originality and creativity of the individual challenges the entire neoclassical equilibrium construct. Kirzner claims that his own version of the entrepreneurial role lies between these two extremes. The Austrian middle ground retains the neoclassical equilibrium as representative of the systematic "tendencies" of markets, while still recognising, putatively, the importance of individual creativity.

act of *discovery*.[2] Buchanan and Vanberg argue that a true process-oriented, evolutionary theory must be non-teleological; the system that it models must not be characterised as converging toward any equilibrium, or as having "equilibrating tendencies". Moreover, non-teleological models cannot be reconciled with what is essentially a pre-ordained future that is simply "discovered" by the entrepreneur.[3]

Buchanan and Vanberg's criticisms can be similarly levelled against the work of others who have recently discussed the role of the entrepreneur. While space precludes us from entering into a lengthy discussion of each, we briefly mention a few of the more prominent works. Harper (1996) and particularly Choi (1993) attempt to integrate the notion of real uncertainty in their works, but do so by building on Kirzner's approach. Choi constructs a theory of paradigm change in the face of uncertainty and while he does not explicitly build on an equilibrium framework, the concept of discovery still retains a primary role in his discussion of the market process. Harper combines a Popperian "growth of knowledge" theory with Austrian fundamentals in constructing a theory of learning within the competitive market. Like Choi, Harper does not use an equilibrium framework, however, he still implies that entrepreneurial opportunities have an objective existence that is discovered by entrepreneurs.

In his well-known work, North (1990) attempts to introduce an entrepreneurial role into his theory of institutions and institutional change. While provocative, it is difficult to argue that North provides an especially coherent description of the role of the entrepreneur herself, or of the interaction between entrepreneurial activity and institutional processes. In our view, North's theory of institutional change is, like the other theories discussed, deterministic – albeit at the level of institutional change.[4]

3 A modified theory of entrepreneurial activity

We begin our own approach to entrepreneurial activity by explicitly recognising the existence of the structural uncertainty inherent in socio-economic processes and the necessity of incorporating this uncertainty within a theory of entrepreneurial behaviour. Following Buchanan and Vanberg's lead, we will argue that entrepreneurial opportunities do not have an objective existence that is simply discovered, but rather are *created* by individuals. On the other hand, this process of creation and true choice must be considered within a framework of stability-producing social institutions.

---

[2] In a related paper, the authors elaborate more fully on a critique of Kirzner's use of "equilibrium", "error-correction" and "discovery" (Jakee and Spong, forthcoming).

[3] "If the market is genuinely perceived as an open-ended, non-determined evolutionary process in which the essential driving force is human choice, any insinuation, however subtle, of a "telos" toward which the process can be predicted to move must be misleading" (Buchanan and Vanberg, 1991: 180).

[4] North's only real elaboration on his view of the entrepreneur is the suggestion that, "Incremental change comes from the perceptions of the entrepreneurs in political and economic organizations that they could do better by altering the existing institutional framework at some margin" (1990: 8). However, he does not provide a clear description of that process, or of the implications for entrepreneurial activity.

3.1 Uncertainty, time and creativity

As many will recall, Frank Knight differentiates between risk and true uncertainty early in his classic *Risk, Uncertainty and Profit* (1921: 19–21).[5] Buchanan (1969), Knight's student, and the school of radical subjectivism led by Shackle (1961, 1972, 1973, 1990a) have probed considerably deeper into the notion of uncertainty, extending it well beyond Knight's discussions. Buchanan and Vanberg (1991) revisited the notion of structural uncertainty in their discussion of entrepreneurship and evolutionary market process theories.

Shackle's exposition of radical uncertainty reminds us that the uncertainty underlying daily life is vastly different from the conventional economic approach to risk and probability. Probability analysis requires the construction of a list of possible outcomes that is both *specific* and *complete*.[6] In a vast array of economic decisions, the individual actor cannot construct such a list; all the *possible* outcomes of an event are unlikely ever to be known. Indeed, within the Shacklian framework, the future is so profoundly uncertain that the full range of socio-economic possibilities is often beyond imagination (e.g., Shackle, 1990a: 54).

In developing his theory of uncertainty, Shackle makes the distinction between two different conceptions of time: the *outside* versus the *inside* view. "Outside" time is, essentially, an *ex post* perspective of events. For example, reading history allows us to examine different events occurring at different points in time. Importantly, the individual reads from a position that is completely detached from each historical event: "[a]ll this long process presents itself to him in one panorama, as a unity, every part of it as real as every other part; he is an outside observer, not himself part of what he describes" (Shackle, 1990b: 15).

The outside conception of time is, of course, useful for conventional economic analysis. From this outside perspective, we can examine events in retrospect, identifying changes over time and apparently logically predicted sequences in events. Neoclassical economic theory has successfully utilised such an approach to make restrictive assumptions and to enable the construction of formal models and statistical analyses. However, as pointed out by Lachmann (1986), the by-product of this perspective is that the market is depicted as a mechanical process, in which activity is completely determined by exogenously existing choice parameters.

The "inside" perspective of time refers to the individual's actual experience of time. Shackle suggests the life experience, from the individual's perspective, is composed of "one solitary moment" after another. This moment can contain thoughts, feelings, imagination and decisions: newly inspired or created concepts that do not retain meaningful relation to that which has gone before are distinct possibilities. The individual, at the solitary moment of inside time, will have in mind some conception (based on memory and experience) of past events, and

[5] Knight also discusses the role of the entrepreneur (1921: 268–82). His description is similar to the one later expounded by Kirzner, where the entrepreneur engages, principally, in *arbitrage*. Knight's entrepreneur is one who orchestrates the production of goods or services and conducts a search for the lowest cost inputs available. The entrepreneur will then lay claim to the residual profit after all wages and rents are removed (1921: 280).

[6] A list of outcomes is "specific" if every possibility has a specific probability attached to it, and "complete" if it encapsulates all possible outcomes (Shackle, 1961: 49).

some anticipation (based on imagination) of what is yet to come.[7] The individual's chosen action is based on this subjective mental construct, and the range of possible activities includes both logically predictable actions as well as purely original ones. While the large proportion of individual actions do appear to indicate a logical response to memory of past events and circumstances, this does not preclude the possibility of individuals acting in ways that bear no logical connection to the past.

Unlike the mechanical *ex post* perspective of outside time, inside time is a purely *ex ante* perspective and redirects attention to the actual process of individual contemplation and action. This subjective description of time has the potential to account for creativity in our version of the entrepreneurial process.[8] Indeed, imagination and creativity are concepts that must be integral in accounting for changes and developments in socio-economic processes, a fact so often neglected by mainstream analysis.

We conclude this sub-section by emphasising the dichotomy between the initial formation of an economic plan, and individual action. Within the framework of radical uncertainty, the individual's plan of action is formed as he continuously interprets his environment from the perspective of inside time. Thus, the individual's interpretation of the socio-economic environment is an input into the formation of his plans. However, until the individual *acts*, a plan remains nothing more than imaginative figment. Thus, the possibility for radically uncertain long-run processes, so emphatically emphasised by Shackle, is unleashed only once the economic agent acts.

3.2 Rules, institutions and plan coordination

Despite our emphasis on the uncertainty of socio-economic outcomes and the importance of individual creativity in the formulation of economic plans and decisions, casual observation suggests that a broad range of social interaction can be described as "stable", or persistent.[9] This stability, which guides a large degree of economic behaviour and promotes the construction of individual plans (economic

---

[7] While not discussing the concept of "inside" time *per se*, Buchanan (1969) provides a detailed account of the subjectivity inherent in individual decision-making, as well as the profound implications for economic theory.

[8] Shackle begins to tie the concepts of "enterprise" and the "solitary moment" together in his 1973 work, *An Economic Querist*, but he does not develop a coherent and comprehensive theory of entrepreneurship based on these concepts. Furthermore, he argues that "enterprise" is descriptive of each and every decision, by each and every decision-maker. Thus, in our view, Shackle's later articulation of enterprise is much too broad to be of practical use. O'Driscoll and Rizzo have also illustrated the value of using a more dynamic conception of time in economic analysis (1996). They present limitations of what they call the "Newtonian" conception of time (akin to Shackle's "outside" perception of time), while also examining some of the more technical aspects of "real" time (a similar conception to Shackle's "inside" perception of time). However, the real time discussed by O'Driscoll and Rizzo is less concerned with the dichotomy between the individual agent and the outside observer. Instead, their depiction of "real" time allows one agent to observe the actions of another *at the same time*.

[9] Rizzo (2000) contends that while Shackle argues that uncertainty is bounded, he still does not adequately account for the extent of observed stability. A similar criticism could possibly be levelled at Buchanan and Vanberg.

or otherwise), is made possible through the existence of individual rules and social institutions.

The role of social institutions as guides to economic behaviour is explored by a number of prominent scholars, including more recently, Vanberg (1994).[10] He acknowledges that many rules are adopted by society to reduce uncertainty and are therefore constraining devices, e.g., traffic laws. However, he emphasises that rules are often followed purely by choice, as no coercive threat for non-compliance exists. He defines this as "genuine rule following" behaviour, which is essential for understanding how structural uncertainty can be reduced (although, importantly, not eliminated).

Social institutions can be formal or informal. Formal institutions exist in the political and judicial realm, as well as in governing the general economy. North (1990: 47) outlines a hierarchy of formal institutions, ranging from the most general, such as constitutions, to the particular, including statutes, common laws, specific bylaws and individual contracts. And while informal institutions are beyond the legal framework, this does not lessen their importance: as Hayek (1973), Buchanan (1975), and North (1990) each suggest, informal institutions, or social norms account for much of the observed socio-economic stability in society.

Dramatic changes in the institutional structure typically occur only over a substantial period of time (see, e.g., Lachmann, 1971: 91). Persistence of institutional structures thus enables the individual to form reasonably dependable expectations for much of his life experience: individuals – even from the stark inside moment of time – often do correctly anticipate how others will behave and what the effects are likely to be as they pursue their own, individually specific, plan.[11] The resulting stability makes it possible for individual agents to contemplate intertemporal transactions and to enter into long-run contracts with some amount of confidence as to outcomes, as well as providing some avenue for recovery in the event of default. Stable institutions within the business world reduce transaction costs associated with long-run uncertainty. The persistence of dependable legal systems in most developed economies permits the implementation of sophisticated plans and contracts involving enormous financial resources over relatively long periods of time.

Hence the potential to mitigate profound structural uncertainty is possible through rules and social institutions. Individuals who wish to benefit from reduced uncertainty will frequently find it in their own interest to follow established customs, as Vanberg has pointed out. Importantly, the expectation that others will also follow established laws and conventions allows the individual to rely on much broader social outcomes, such as generalised social stability; in socially stable environments, individuals are likely to have greater confidence that their personal safety

[10] Other comprehensive accounts include Hayek's discussion of social order (1967, 1973), Lachmann's description of institutional structure (1971), Buchanan's discussion of moral order (1975), Brennan and Buchanan's examination of the "reason of rules" (1985), and more recently North's discussion of institutional stability (1990: 83). Also, a more general theory of rule following behaviour has been proposed by Vanberg (1994).

[11] An account of the coordination of inter-agent economic plans that results from the institutional stability of the economy is represented by O'Driscoll and Rizzo's *plan coordination* (1996: 88). We would argue that the relative persistence of institutional structure is what allows this plan coordination to occur.

and property are secure, which allows them to devote more energy and resources to productive outlets.

While generalised social "order" or stability has public good characteristics, and therefore may give rise to problems of insufficient individual contributions, it will often be in the individual's self interest to follow the rules: activity that is congruent with existing social institutions is most likely to elicit the expected results and reactions from others. Activity that is familiar and "acceptable" will elicit approval and understanding from others, rather than objection and confusion, as Choi argues (1993: 55–68). Of course social approbation is a critical social device that has the effect of inducing individual contributions to public goods, as many social theorists are now widely recognising. [12]

The overwhelming degree of individual rule-following behaviour serves as an input into reinforcing existing social institutions, both in the mind of the economic agent, as well as in those who are observing the action and its subsequent effects. On the other hand, action that deviates from accepted thinking typically provokes hostility. Institutions therefore reduce our uncertainty, not only regarding the long-term likely result, but also regarding the immediate perception others have of our actions.[13]

In concluding this subsection, we wish to emphasise some key differences between our description of socio-economic stability, and the teleological conception of equilibrium. Depending upon the particular context, the equilibrium concept necessarily implies any one or more of the following: optimality, determinism, and convergence. The reader will note that the socio-economic state we have described has no role for these characteristics. The claim that institutional structure allows individuals to form plans of action is not intended to suggest that the market process entails the coordination of plans at some optimal level. We do not deny that individuals attempt to improve their subjective position in the market, perhaps by identifying a new source of good or service. Furthermore, while we discuss some persistence in the basic institutional structure, no single point can be said to be any more (or less) final than another. Instead, the institutional environment is continuously changing and evolving. Lastly, as there is continuous marginal adjustment and development in institutional structure, the plans of individuals also need to be fluid and constantly changing. In sum, our depiction of the entrepreneurial process is not deterministic or converging toward any point or position.

[12] Moreover, the reader who follows the burgeoning psychobiological literature will recognise a recurring theme: humans appear to be "hard wired" to behave socially. As a result, it is claimed that we are physiologically sensitive to the approval and disapproval of our peers, whoever they may be.

[13] An "economic" example of this type of phenomena can be taken from the finance literature where it is recognised that asset traders are often prepared to follow the trading strategies of other market participants without investigating the relevant market information. A good introduction to "noise trading", and the subsequent tendency of asset market traders to "jump on the bandwagon", is given by Shleifer and Summers (1990).

3.3 Entrepreneurial activity and its impact

As we explained, individuals tend to act in a manner that is consistent with established norms and institutions. However, there remains the ever-present potential for imaginative and creative insight on the part of individuals. The individual who implements (either intentionally or inadvertently) a novel approach that is beyond the scope of existing institutions, which guide or constrain the behaviour of most other agents, is behaving *entrepreneurially* in our framework. For the reasons previously outlined, such individuals will themselves bear increased uncertainty, at least in the short-term, as a result of their innovative methods, techniques, products, or services.

Our definition of entrepreneurial action is straightforward yet powerful. It puts Schumpeter's "creative destruction" in a modern *institutional* context: creative *individuals* can potentially upset institutional structures. These structures can be "local" to a specific firm or region. Or, they can be extremely general as in the case of cultural norms and even generally accepted mental constructs (or "models") of, for example, the physical world.

It is, moreover, not necessary that the individual consciously break the institutional structures, or even intend to be "entrepreneurial". The effect of being defined as an entrepreneur may be purely a by-product of an individual's own, subjective intentions. The point to be emphasised is that creative individuals have the capacity to potentially change the structures that bind many others.

Indeed, it is likely that even when individuals do intend to alter institutional structures that they will necessarily assume a large degree of socio-economic stability, if their economic plan is to have any chance whatsoever of success. The increased uncertainty faced by breaking an institution cannot be extended to every aspect of the entrepreneur's activity: there will be other institutions and norms that the entrepreneurial agent specifically relies upon as she implements her plan. Thus, not only must the great majority of individuals be able to imagine the future within certain bounds in order to construct long-term plans, but our institution-breaking entrepreneurs must too.[14]

Our definition naturally raises the question of what motivates entrepreneurs to act beyond the scope of the existing institutional norms and to face greater uncertainty in results. While most other theorists contend that the entrepreneur operates under the expectation of obtaining greater levels of profit (Kirzner, 1973;

[14] Several commenters on our paper have wondered how, if all rule-breaking activity is "entrepreneurial", we are able to distinguish between meaningful entrepreneurial actions and those of the "village idiot", which, for all intents and purposes, have little connection to reality and are unlikely to ever have any influence on the surrounding community. We would argue that institutions provide stability by reducing the uncertainty of potential outcomes, and the more an activity diverges from these institutional structures, the more uncertain the result of such action will be. Thus, activity that breaks with a host of well-recognised institutions simultaneously is less likely to be considered a meaningful plan of action. While we do not have space to pursue this issue here, we would argue that as action moves further from the institutional guidelines, the chance of it being "successful", in any conventional sense, decreases dramatically. Therefore, such activity can be classified as foolish or reckless in this framework, and can be distinguished, to some extent, from entrepreneurial activity. We wish to thank Mario Rizzo for helping us work through this issue.

and Choi, 1993), from our perspective this is not necessarily the case. The greater degree of uncertainty inherent in operating outside the accepted institutions, and the well-recognised fact that most entrepreneurial endeavours do not meet the individual's (usually optimistic) expectations, mean that an individual's pursuit of monetary gains may indeed be more successful by following proven institutional methods; indeed, the variance of the pay offs should be lower. While we would not suggest that entrepreneurial individuals ignore the prospect of large profits, we would posit that non-pecuniary objectives are an important motivational force for entrepreneurial activity. Some entrepreneurial individuals surely derive greater satisfaction from the act of pure creation than they do from following conventions. This assertion, we might add, corresponds with the considerable empirical literature on the psychological and behavioural analyses of entrepreneurs.[15]

We now turn to the potential effects entrepreneurial acts will have on the socio-economic system. As the reader should now suspect, the impact on institutional structures by the entrepreneur will be profoundly uncertain. At one extreme, an entrepreneurial activity may cease in very short order, having had no noticeable effect on the market process. In fact, the swift termination of such activity may serve to reinforce the established configuration of institutions and the associated stability, as the individual is rebuked for having broken the established rules.

At the other extreme, the impact of entrepreneurial activity is potentially dramatic. As other market participants witness the viability of some new type of activity, they may also revise their own pattern of behaviour to realise the benefits of the new practice. As market participants increasingly adopt the new techniques, formerly trusted processes and guidelines may become obsolete. In this way, entrepreneurial activity will alter the institutional structure, although the replacement of existing institutions with new ones will, as we have already discussed, typically not happen instantaneously. Of course the issue of how new techniques and processes are taken up, or adopted links our story of entrepreneurial activity to a well-developed literature in innovation and technology adoption and diffusion.[16] On the other hand, we want to clearly differentiate the act of entrepreneurial creation from the issue of how – *or even whether* – such creation is taken up by others. Most other accounts of entrepreneurship collapse these two very distinct phases of the process.

---

[15] See Ronen's (1983) suggestion, which, on the basis of interviews with highly successful entrepreneurs, contends that they often have a non-monetary motive that sets them apart from standard manager "types":

"There is restlessness, tension in the entrepreneur; by contrast, what he perceives in the usual professional manager is preference for financial security within a structured environment. The restlessness of the entrepreneur spells outright boredom with routine – the urge is to innovate and forge ahead in new endeavours. Some of those questioned indicated that four to five years was the longest they could conceivably stay put with a given product or in one line of business. They were irresistibly drawn to the untried, the unknown – a new venture, a new company" (1983: 143–144).

[16] A seminal theoretical discussion is presented by Rosenberg (1976), who examines the factors that affect the process of diffusion and adoption of technological innovation. Some empirical evidence and theoretic testing of Rosenberg's theory is provided by Weiss (1994). An update of Rosenberg's theoretical discussion and a case study is provided in David's (1990) well-known article.

Innumerable possibilities lie within these two extremes. For example, successful entrepreneurial activity may see the internal rules changed within a single firm, or it may alter the institutional structure in a single industry without having any impact on any other sector. But in each scenario, until such time as the institutional structure settles once more, the degree of uncertainty surrounding the entrepreneurial event or practice will have increased.

Now, consider the case where the institutional structure changes independently of entrepreneurial activity: this could be via truly exogenous "shocks", like war, earthquakes and storms, or it might be via socially-induced shocks, like the adoption of "free trade" principles in a formally protectionist economy. Established institutional guidelines are often made redundant as a result of such externally changing physical or socio-economic events. The absence of relevant guiding institutions will result in certain individuals, firms and industries bearing increased uncertainty until a new set of institutional guidelines emerges.

In these circumstances there may be some individuals who act creatively in the face of institutional disarray. This, too, is entrepreneurial action in our view and if certain modified practices survive the environmental challenges, they may indeed become the basis of imitation, and eventually new institutional forms. A new stability will, in time, be established.

It should be apparent that our approach has much in common with Schumpeter's *process* of creative destruction. Yet, we can also potentially reconcile such a perspective with the principal theme of Kirzner's work, namely, that the entrepreneur elicits *increased stability* in the economic system. But in our version, the stability-enhancing effects of entrepreneurship are a special case, occurring only when institutional change *precedes* entrepreneurial action; and, of course, these can only be assessed *ex post*.

The contrasting perspectives of both of these great antecedents are incorporated in our approach – though within a very different context from the equilibrium foundation that each of them, in his own way, is reliant upon.

4 Implications

In the foregoing analysis we have attempted to elaborate and extend upon the non-teleological foundations advocated by Buchanan and Vanberg and pioneered by Lachmann and Shackle. In what follows, we will address several implications that follow from such a perspective.

4.1 Implications for the individual

Our theory emphasises that it is the very creativity inherent in any individual that accounts for the actions and visions of entrepreneurial behaviour. The entrepreneur is one who, at the individual level, bears the uncertainty that comes with breaking established institutions. Our approach re-focuses attention on both the uncertainty of an unfolding process through which entrepreneurial individuals pursue their creative impulses, and on the key role played by existing institutions and prejudices.

By explicitly emphasising the subjective thought process of the individual, we provide a theoretical representation of the imagination and creativity that would have sparked well-known entrepreneurial agents such as Henry Ford, or Richard Branson as they constructed their plans of action.

Our perspective should remind the analyst of the difficulty in viewing the socio-economic system as waiting to be pushed to some equilibrium by an entrepreneurial actor who "discovers" what we will find out, *ex post*, to have been lying just below the surface. We may learn that the assembly line has changed production techniques, or that the personal computer has changed workaday habits in offices around the world. However, simply assuming that these developments were pre-ordained and were just waiting to be discovered does not provide an accurate account of the entrepreneurial process. Evaluating entrepreneurial activities from the more typical *ex post* perspective can ignore the profound difficulty of imagining and then implementing a plan that will potentially confront the institutions that bind the behaviour of most other individuals; in other words, such a perspective ignores *creativity*. Entrepreneurial activity is much more than picking up dollar bills lying on the street that others cannot "see" – as Kirzner, Choi and others seem to suggest.

Our use of Shackle's inside time also keeps us from confusing the *effects* of agents' activities (which may indeed be by-products), with their planned objectives. It is important to stress the error of confusing *ex post* observations, which we can scrutinise with the advantage of historical perspective, with the actual conditions and institutional constraints that an individual will find herself in through the process of entrepreneurial decision making. The ability to formulate a plan of activity, and then actively pursue this plan, should not be conflated with the ability to somehow visualise the ultimate effects of such action.

This point is of paramount importance for our version of entrepreneurship because it underscores the fact that *ex ante*, or from the inside perspective of time, neither the individual agent nor the social analyst can predict exactly where or when entrepreneurial actions will set off an unravelling of institutional structure. Neither will we be able to predict where and when the system will stabilise as new norms and institutions begin to coalesce.[17] This is just as true in the case of externally-induced entrepreneurship. Even our most celebrated entrepreneurial protagonists could not have predicted the momentous consequences of their activities. It does not challenge the genius of figures such as Bill Gates or Steve Jobs to suggest, for example, that they could not have foreseen the very broad class of activities associated with modern personal computing power at the moment Gates secured early software contracts with IBM, or as Jobs began producing easy-to-use computers with mass appeal.

Our emphasis on entrepreneurship as an activity does not restrict our analysis to any specific type of individual. Entrepreneurial activity, as we define it, is behaviour that takes place within a specific socio-economic context, and as such, can

[17] This view is consistent with Lachmann's point regarding change in institutional structures, in which he emphasises the impact entrepreneurial innovation has had on the transformation of our world during the last two centuries: "Not merely were they accompanied by considerable shifts in data other than those pertaining to technical knowledge. It took in most cases several decades before their effects had percolated all sectors of the economic system" (Lachmann, 1986: 110).

be undertaken by any willing-and-able individual. Such a view has some useful implications for the way that we analyse business activity in general, as we discuss in the following sub-section.

In addition, our theory can be useful in analysing behaviour not traditionally considered market activity. For example, legal, political, social, scientific, and even artistic pursuits are each comprised of institutional structures that determine the general class of "acceptable" practices. Agents that break from these institutional guidelines threaten the established stability of their respective field and challenge the expectations of others. Figures such as Galileo, Columbus, and even the Beatles are, from our perspective, entrepreneurs. The effects their creative actions have had on their respective spheres of activity, as well as on the greater socio-economic environment are indisputable.

Our approach yields additional implications that follow from our emphasis on the institutional setting. Whether the institutional barriers to entrepreneurial activity are *formal* or *informal* is expected to affect the incentive structure for entrepreneurial action. Consider a potential entrepreneur who is faced with an established industry norm, or with informal guidelines set down by the executives of the firm. The prospect of breaking such informal rules will raise the level of uncertainty: the individual will herself confront greater uncertainty as those around her are surprised by her entrepreneurial actions, while the long-term industry – as well as broader social effects – can hardly be predicted. But uncertainty may also occur at some intermediate level. For example, will the individual be denounced, or esteemed? Will new production techniques take hold? Will the firm make higher or lower profits?

Thus, where informal institutions guide market interactions, the "costs" to the individual of acting entrepreneurially are essentially comprised of psychic disutility arising from the potential displeasure of others and the increased variance of intermediate effects. Presumably, whether an agent undertakes entrepreneurial activity depends on the very personal calculus of the perceived net benefits; and we re-emphasise that the benefits can be pecuniary or non-pecuniary.

On the other hand, laws and formal regulatory structures bring with them formal and explicit sanctions for non-compliance. The consequences of operating outside formal institutions have legal implications and hence the potential costs to an individual of breaking society's formal strictures can be considerably higher than in the case of breaking informal ones. Thus, to the extent that formal sanctions replace informal ones, we would expect that the entrepreneur's threshold of perceived net benefits to be increased: her likelihood of breaking with conventions will be more attenuated.

4.2 Implications for business activity and the firm

Consistent with descriptions of the entrepreneurial role provided by Schumpeter and Kirzner, our theory can accommodate behaviour that is congruent with the notion that some agents follow standard maximisation guidelines while others do not; this allows us to distinguish between activities that are *managerial* and those that are *entrepreneurial*. We define a "manager" as one who relies upon recognised methods to achieve some maximising (or "satisficing") outcome within the given constraints. But it is entrepreneurial activity that gives rise to new methods outside of the conventional framework, and which can potentially upset the established maximisation problem. In other words, entrepreneurial activity dispenses with the notion that the best firms can do is maximise subject to the *given* constraints. Indeed, our entrepreneurial actors have the potential to change the constraints.[18]

Our distinction thus accommodates the notion that entrepreneurial activity can exist *within firms*. An entrepreneurial individual may build and utilise a firm to reduce particular elements of uncertainty, or in the Coasean sense of reducing the transaction costs of contracting for various services; services that will serve as inputs into the entrepreneurial activity. However, even within the protection of the firm, an individual (or the firm) undertaking entrepreneurial activity will bear greater uncertainty in the results of her (its) actions.

Such an approach reveals our disagreement with many students of business and many empirical approaches to the study of entrepreneurialism. We insist that – at the theoretical level at least – we must distinguish between *real entrepreneurial activity* and *self-employed* activities. A simple shorthand method has apparently arisen in much of the existing literature suggesting that the *self-employment* status of individuals is synonymous with "entrepreneurialism" (see, e.g., Bates, 1990; or Evans and Leighton, 1989). In our view, not all self-employed individuals would qualify as entrepreneurial. We have defined entrepreneurial activity as the creation and utilisation of original methods and insights that break with conventional institutions. Indeed, while many successfully self-employed people are doing exactly this, many others are operating well within the confines of recognised norms or conventions. The typical fish and chip shop, small grocer, or even business consultant undoubtedly carries out its affairs within established institutional confines.[19]

We believe our modified definition of entrepreneurial activity better fits the stylised facts than the commonplace definition that all new start-ups are en-

[18] The point is that the "profit maximiser", or "satisficer" does not, by definition, have control over a myriad of parameters when executing the maximisation/satisficing decision. Readers familiar with the "Constitutional Political Economy" literature pioneered by Buchanan and others, will note the echo of Buchanan's insistence that we clearly delineate between "choice within constraints" and "choice among constraints" (1990). Also, in this regard, we can revisit one of the shortcomings of North's account of the entrepreneur. He asserts that entrepreneurs consciously change the existing institutional framework because of the (presumably personal) benefits that they will derive from an altered institutional structure. His assertion implies that entrepreneurs can (correctly) anticipate how they will fare under some new set of institutional rules, *once all the implications and counter-effects resulting from the entrepreneurial action have worked themselves through the system*. In North's version, it seems that opportunities objectively exist within the (sub-optimal) institutional framework and entrepreneurs presumably take advantage of these (1990: 73, 83). Such an approach is essentially what Buchanan and Vanberg argued vigorously against in their criticism of Kirzner: North simply elevates the preferred equilibrium target to a higher level (i.e., the institutional level) than Kirzner did (see, for example, our footnote number three).

[19] Witt (1999) discusses the role of cognitive leadership by the entrepreneur within the confines of the firm. While we agree with his discussion of this entrepreneurial leadership and the premise that entrepreneurial activity can indeed exist within firms, he appears to suggest that all firm start-ups involve some degree of entrepreneurial imagination (1999: 108). Within our framework, firms may well be started with the aim of operating *within* the confines of established institutional boundaries, and are therefore not entrepreneurial.

trepreneurial, while large corporations are hopelessly non-entrepreneurial.[20] While we appreciate the necessity of simplifying complex theoretical constructs for empirical studies, this particular simplification has the potential, we believe, to misguide those trying to understand entrepreneurial activities and the associated processes, not least of whom would include policymakers.

4.3 Implications for socio-economic evolution

The implications of our thoroughgoing process-oriented approach on the development and evolution of society are numerous. In creating new methods, products, and even industries, entrepreneurial activity plays a key role in influencing the unknown directions that economies will take. Thus, in our theory, entrepreneurial activity based on the pure originality and creativity of the individual is one of the key loci of *truly endogenous social change*. Even in an environment in which all resources and quantities are stable and exogenous parameters are constant, socio-economic activity will always have the potential to develop in new, unexpected directions.

Yet, while we have identified entrepreneurial activity as one locus of endogenous change, we wish to carefully point out that entrepreneurial activity is not synonymous with business "success". Even though we emphasise the potential for the actions of entrepreneurs to have a hugely unexpected and far-reaching positive impact, our focus on processes and our emphasis upon entrepreneurship as a type of *activity* (as opposed to an outcome) force us to acknowledge that *many, if not most, entrepreneurial efforts are not successful* – at least not initially. Entrepreneurship scholars in the management sciences have long recognised this facet of the entrepreneurial process (e.g., Aldrich and Auster, 1986). Thus, our description of the entrepreneurial process cannot concentrate solely on the so-called winners in entrepreneurial undertakings.

In contrast, virtually all versions of entrepreneurial theory retain a focus, explicitly or implicitly, on the beneficial outcomes of entrepreneurial efforts: Schumpeter's, Kirzner's, and Choi's are all such examples.[21] In our view, theories that focus on specific outcomes, like entrepreneurial success, or on specific individuals such as Henry Ford, fail to recognise the uncertainty that exists in socio-economic activity. Even though the vast proportion of entrepreneurial attempts do not succeed, each attempt has the *potential* to play an important role in the evolutionary

---

[20] In making this distinction between entrepreneurial activity and self-employment we are able to extend a discussion of Audretsch and Thurik (1998) who consider two types of entrepreneurship. The first type is what they refer to as "refugee" or "shopkeeper" entrepreneurship. This is essentially entrepreneurship in response to unemployment. It has no great bearing on the aggregate economy and "... tends to generate marginal firms with a low likelihood of survival and lower wages or at best stable but marginal firms"(1998: 9). In our framework, we refer to start up activity that is within the confines of established institutions simply as self-employment. Their second type of entrepreneurship, the "Schumpeterian" type, creates new opportunities and higher wages. This latter description is compatible in many ways to what we would classify as genuine entrepreneurial activity.

[21] Among economists, Harper (1996) seems to stand alone in providing a theoretical representation of entrepreneurial failure.

market process. These "failures" need to be considered just as carefully as the well-known successes, if we are to account for the *actual unfolding* of entrepreneurial processes.[22]

Another aspect of the socio-economic processes set off by entrepreneurial action has to do with the earlier discussed dichotomy between formal and informal institutions. Not only will the formality of institutions be expected to influence the incentive structure facing the individual considering entrepreneurial activity, but in addition, the specific mix of institutional types will also impact upon the magnitude of entrepreneurial activities in general. The formal institutions of an economy shape its evolution in a manner that is often different from the informal ones.

We might consider limited liability laws that allow individuals to disassociate their personal financial prospects from the uncertainty that they face as entrepreneurs.[23] Clearly, such rules are likely to encourage greater amounts of uncertainty taking, assuming the smooth functioning of other institutional features like accounting standards and legal recourse in cases of negligence or malfeasance. To the extent that society wants to promote greater uncertainty taking, the legal enshrinement of such provisions is one means by which to accomplish it.

On the other hand, we would expect that formal institutions are not as easily challenged as informal institutions by would-be entrepreneurs. The more formal the rules and the more onerous the regulatory environment that codifies the existing or established institutions, the higher we expect the costs associated with entrepreneurial activity to be. Thus, while limited liability laws may result in increased entrepreneurial activity, formal institutions that regulate, standardise, or control sectors of the economy may suppress entrepreneurial action and uncertainty taking.[24]

In those situations where the relationship between some specific business activity and the legal framework is unclear (or is made unclear), we can even potentially reverse the causation: some entrepreneurial activity can be expected to have an impact upon formal institutions. As entrepreneurial activity is creative, it will not always be clear how this new activity fits within the pre-existing legal framework. Lawmakers cannot see the future any better than others and can therefore not be ex-

---

[22] Pelikan (1993) discusses one of the principal advantages that market systems have over non-market, or centrally planned ones: the rules of the market system not only allow for, but encourage a large number of "trials and errors". From an evolutionary perspective, more trial and error (or experimentation) is likely to lead to greater "adaptability" to environmental conditions than a system that does not allow much experimentation.

[23] We want to thank Mark Picton for raising this possibility.

[24] The discussion of formal versus informal institutions forces us to contemplate the role of policy and the "rules of the game". As the Constitutional Political Economy literature emphasises, rules are paramount in encouraging or discouraging a vast array of economic and social activity. This proposition is supported by recent empirical work done by Davis and Henrekson (1999) and Henrekson and Johansson (1998). While these two papers do not clearly distinguish between many of the issues raised in the current paper, (such as entrepreneurial versus managerial activity, or entrepreneurship versus self-employment), they distinctly investigate the claim that formal "rules of the game" affect the success of small firms, the capital intensity of firms, individual and family ownership of firms, and the size and distribution of firms. They find that, compared with most other OECD countries, Sweden has imposed a myriad of industrial regulations and tax policies that disadvantage firms that are not already well established (i.e., small firms and small firm start up, family businesses, non-capital intensive firms, etc.).

pected to frame laws that readily anticipate all future eventualities. Indeed, it should come as no surprise that lawmakers appear to be playing "catch up" in the domain of intellectual property law as a result of the rapid and unforeseen developments in information and technology.[25]

A critical implication of our theoretical position on entrepreneurial activity is that individual actors may *knowingly* break formal institutions: at least some criminal activity must be considered in any generalised theory of entrepreneurship. The long list of business agents who have broken financial laws in their pursuit of higher monetary gains is a testament to the existence of such activity. Indeed, criminal activity breaks the recognised institutional framework and represents a potential threat to the stability provided by existing institutional structures.[26]

Some readers may be surprised by our preparedness to include such destructive activity within the confines of our theory.[27] Yet, the refusal to consider these possibilities highlights what we view as the ongoing theoretical bias in most entrepreneurial approaches (see also Jakee and Spong, 2002). In most other versions, entrepreneurs are viewed as heroes and their (always successful) activities (always) lead to better production techniques, better products, and even a better society. Surely this is not only a lopsided view of the entrepreneurial process, but it smacks of an *ex post* whiggish view of events. While not denying there is a place for normative judgements about the entrepreneur-induced evolutionary process, for our current purposes we are attempting to retain a steadfast adherence to non-teleological theory: some acts of some entrepreneurs will clearly not have beneficial results for society.

On the other hand, deliberate law breaking is not always aimed at securing personal monetary gain. Entrepreneurial activity may be undertaken at the political or social level, when for example social activists break laws in protest: consider Gandhi's peaceful disobedience protests against English occupation or Nelson Mandela's continual opposition to the policy of apartheid. Political and social activity such as this should not be ignored or precluded from a more general entrepreneurial theory of uncertainty, institutions and socio-economic change.

5 Conclusion

Our discussion of the market process and the role of the entrepreneur is explicitly non-teleological. We emphasised both the profound uncertainty of the future, and

[25] Consider the much-publicised case of "Napster" where the extant legal framework had clearly not been constructed to deal with such entrepreneurial activity.

[26] In addition to entrepreneurial activity that violates formal and informal rules, another potentially fruitful area of entrepreneurial analysis is possible: the large zone of human action that would lie outside strict institutional structures. This "grey area" might include an array of actions that are not associated with any particular sanction, either formal or informal. Entrepreneurs might well consciously exploit these behavioural grey zones in order to evade explicit social sanctions. We envision such entrepreneurial activities invoking confusion and even consternation on the part of observers, but technically, no rule (formal or informal) will have been broken. We suspect that a considerable amount of entrepreneurial activity occurs in this institutional grey area. We would like to thank Rodney Beard for helping us clarify this point.

[27] See also North (1990: 77) who suggests that piracy was an entrepreneurial activity.

the role of the individual as the source of original ideas. However, economic activity is, to a considerable degree, guided and constrained by social institutions that mitigate profound uncertainty. We defined entrepreneurship as creative individual activity that breaks from these prevailing institutions. Such activity has the potential to account for endogenous social change through its effect on institutions.

We want to re-emphasise that our approach shares much in common with others. For example, Schumpeter's "creative destruction" view of entrepreneurial innovation as new combinations of methods and or resources (Schumpeter, 1934: 74–75) can be easily placed within our framework of uncertainty and institutional structure: new combinations of economic resources and methods represent activity that is not guided by the established institutions and norms; and such activity has the potential to severely disrupt industrial sectors and even whole economies. Kirzner's view that entrepreneurial activity is essentially equilibrating – or "stability enhancing", from our perspective – is also incorporated. Indeed, both Schumpeter's and Kirzner's approaches become special cases of our more general approach.

On the other hand, we also diverge from the other authors on entrepreneurship, at one point or another. We have discussed at length our attempt to break with the equilibrium approaches of many authors, including Schumpeter and Kirzner. But another important point of divergence regards the concept of arbitrage. Kirzner (1973) and to some extent Knight seem to base their definition of entrepreneurship, fundamentally, on the act of arbitrage.[28] From our perspective, simple arbitrage carried out by traders and "middle-men" is *not* classed as entrepreneurial behaviour for reasons which should, by now, be clear.

In our departure from some of the well-known theories we believe that we have potentially improved our understanding of the entrepreneurial process. Firstly, our account of the entrepreneurial process is explicitly built upon the tension that exists between institutional stability and the profound uncertainty of daily life. Second, our construction provides a clear distinction between managerial and entrepreneurial behaviour, unlike many authors, including North. Yet, this distinction also allows us to reconcile our theory of entrepreneurial behaviour with standard maximising theories of economics. Furthermore, we find it misleading to define entrepreneurial activity only by *ex post* successful endeavours (i.e., that of the Bill Gates and Henry Ford variety) as most others seem to do: entrepreneurs often – even usually – fail.

We also clearly distinguish between self-employment and entrepreneurship, and our approach suggests that entrepreneurial incentives will differ between formal and informal institutions. Indeed, we highlight the importance of the "rules of the socio-economic game" in the encouragement or discouragement of entrepreneurial activity. A number of comparisons come to mind, such as the greater emphasis on "orderliness" in some cultures' socio-economic structures compared to others. On cursory inspection, it would appear that a *relative* lack of cumbersome formal institutions and barriers is concomitant with considerably more entrepreneurial vitality.

As for future directions and work, our version of entrepreneurial activity must prove to be of some practical use should it pass its own test of survival. In related

[28] Jakee and Spong (forthcoming) includes a lengthy discussion on the place of arbitrage within Kirzner's theory of entrepreneurship.

work, we intend to re-examine several well-known industrial developments from the perspective set out here. We are hoping to gain additional insights into the role played by the entrepreneurial process in these developments. One key implication of our approach as it pertains to empirical work involves looking much more carefully into history for the so-called "losers" who are all-too-often forgotten in the rush to anoint entrepreneurial "winners". As we have suggested, it is the entrepreneurial *process* that matters and through that process there will be many less well-remembered and less successful "entrepreneurs". A careful looking back in this manner should also help us to delineate why some individuals did turn out to be more successful than others. Such an approach, we believe, might considerably alter views on entrepreneurial policy as well.

Other issues, about which we can now only speculate, also come to mind. Is there something successful entrepreneurs have in common, compared with those who are not? Or, might it be largely the institutional environment that determines success? How, for example, does *culture*, which we might describe as the myriad of informal institutions, affect entrepreneurial activity? Are there robust empirical connections between societies with more "entrepreneurially"-oriented norms and their economic development?

In sum, we have attempted to develop a far-reaching evolutionary approach to entrepreneurial theory that is grounded in subjectivist individual action. By emphasising individual creativity in the face of institutional stability, our approach is applicable not just to "enterprise" and "profit making", the focus of most other entrepreneurial theories. Rather, our insights are equally applicable to entrepreneurial activity in the realm of politics, as well as general intellectual, scientific, and even artistic processes. Indeed, it is some of these presumed "non-economic" areas where entrepreneurial activity has had some of the most profound effects on social – *and economic* – structures.

References

Aldrich H, Auster E (1986) Even dwarfs started small: liabilities of age and size and their strategic implications. In: Staw BM, Cummings (eds.) Research in Organizational Behavior Vol. 8. JAI Press, Greenwich, CT: pp 165–198

Audretsch D, Thurik R (1998) The knowledge society, entrepreneurship and unemployment. Discussion Paper 1170. Centre for Economic Policy Research. London

Bates T (1990) Entrepreneur human capital inputs and small business longevity. Review of Economics and Statistics Vol. 72(4): pp 551–559

Brennan G, Buchanan J (1985) The reason of rules. Cambridge University Press, Cambridge

Buchanan J (1969) Cost and choice: an inquiry into economic theory. Markham Publishing Company, Chicago

Buchanan J (1975) The limits of liberty: between anarchy and Leviathan. University of Chicago Press, Chicago

Buchanan J (1990) The domain of constitutional economics. Constitutional Political Economy Vol. 1 (1): pp 1–18

Buchanan J, Vanberg V (1991) The market as a creative process. Economics and Philosophy Vol. 7: pp 167–186

Choi YB (1993) Paradigms and conventions: uncertainty, decision making, and entrepreneurship. University of Michigan Press, Ann Arbor

David P (1990) The dynamo and the computer: an historical perspective on the modern productivity paradox. American Economic Review Vol. 80: pp 355–361
Davis S, Henrekson M (1999) Explaining national differences in the size and industry distribution of employment. Small Business Economics Vol. 12: pp 59–83
Evans D, Leighton L (1989) Some empirical aspects of entrepreneurship. American Economic Review Vol.79(3): pp 519–535
Hayek F (1967) Studies in philosophy, politics and economics. University of Chicago Press, Chicago
Hayek F (1973) Law, legislation and liberty: rules and order Vol. 1. University of Chicago Press, Chicago
Harper D (1996) Entrepreneurship and the market process: an inquiry into the growth of knowledge. Routledge Publishing, New York
Henrekson M, Johansson D (1998) Institutional effects on the evolution and size distribution of firms. Working Paper No. 497, 1998. The Research Institute of Industrial Economics, Stockholm
Jakee K, Spong H (forthcoming) Praxeology, entrepreneurship, and the market process: A Review of Kirzner's Contribution. Journal of History of Economic Thought
Jakee K, Spong H (2002) The normative bias in entrepreneurial theory. Workung paper presented at the Australian Conference of Economists, Adelaide University, Australia (October)
Knight F (1921) Risk, uncertainty and profit. Houghton Mifflin Company, New York
Kirzner I (1973) Competition and entrepreneurship. University of Chicago Press, Chicago
Kirzner I (1985) Discovery and the capitalist process. University of Chicago Press, Chicago
Kirzner I (1992) The meaning of market process: essays in the development of modern austrian economics. Routledge, New York
Lachmann L (1971) The legacy of max weber. Glendessary Press, Berkeley, California
Lachmann L (1976) On the central concept of austrian economics: market process. In: The Foundation of Modern Austrian Economics. The Institute of Human Studies, California
Lachmann L (1986) The market as an economic process. Basil Blackwell, New York and Oxford
North D (1990) Institutions, institutional change, and economic performance. Cambridge University Press, New York
O'Driscoll G, Rizzo M (1996 [1985]) The economics of time and ignorance. Routledge, New York
Pelikan P (1993) Ownership of firms and efficiency: the competence argument. Constitutional Political Economy Vol.4(3): pp 349–392
Rizzo M (2000) Real time and relative indeterminacy in economic theory. In: Baert P (ed.) Time in Contemporary Intellectual Thought. North-Holland Press, Amsterdam
Ronen (1983) Some insights into the entrepreneurial process. In: Ronen J (ed.) Entrepreneurship. D.C. Heath and Company, Massachusetts
Rosenberg N (1976) Factors affecting the diffusion of Technology. In: Perspectives on Technology. Cambridge University Press, Cambridge: pp 189–212
Schumpeter J (1934 [1911]) The theory of economic development – an inquiry into profits, capital, credit, interest, and the business cycle. Oxford University Press, London
Shackle G (1961) Decision order and time – in human affairs. Cambridge University Press, Cambridge
Shackle G (1972) Epistemics and economics – a critique of economic doctrines. Cambridge University Press, Cambridge
Shackle G (1973) An economic querist. Cambridge University Press, Cambridge
Shackle G (1979) Imagination and the nature of choice. Edinburgh Press, Edinburgh
Shackle G (1990a) The expectational dynamics of the individual. In Ford, J.L (ed.) Time, Expectations and Uncertainty in Economics – Selected Essays of G. L. S. Shackle. Edward Elgar Publishing, England
Shackle G (1990b) Time and thought. In: Ford JL (ed.) Time, Expectations and Uncertainty in Economics – Selected Essays of Shackle GLS. Edward Elgar Publishing, England
Shultz T (1975) The value of the ability to deal with disequilibria. Journal of Economic Literature. Vol. 5: pp 224–232
Shleifer A, Summers L (1990) The noise trader approach to finance. Journal of Economic Perspectives Vol. 4(2): pp. 19–33
Vanberg V (1994) Rules and choice in economics. Routledge, London and New York
Weiss A (1994) The effects of expectations on technology adoption: some empirical evidence. Journal of Industrial Economics Vol. 42(4): pp 341–360
Witt U (1999) Do entrepreneurs need firms? a contribution to a missing chapter in austrian economics. Review of Austrian Economics Vol. 11: pp 99–109

Weber, Schumpeter and Knight on entrepreneurship and economic development

Maria T. Brouwer

University of Amsterdam, Department of Economics and Econometrics, Roetersstraat 11, 1018 WB Amsterdam, The Netherlands (e-mail: mariab@fee.uva.nl)

Abstract. This paper interprets the discussion on entrepreneurship and economic development that started off with Weber's papers on the Protestant Ethic. Weber sought the reason for the relatively rapid growth of the Occident in the rational, Calvinist attitude to life. Calvinism – in his view – exactly suited a society of free labourers, who were not tied to master and soil by extra-economic considerations as in tribal and feudal societies. Schumpeter gave an alternative explanation, emphasizing the importance of innovation and entrepreneurship. Knight, who stressed neither rationality nor innovation but uncertainty and perceptiveness as the sole source of progress and profits, followed up German language writing on this subject. Only the investor who can detect hitherto hidden qualities in people can gain. The paper demonstrates how these three authors influenced each other. The debate between these three authors has raised many issues of governance and organization that feature contemporary thinking.

Key words: Entrepreneurship – History of economic thought – Organizational behavior – Bureaucracy – Uncertainty

JEL Classification: B25, D23 , D73, D81, K11

1 Introduction

Theories of entrepreneurship span a long period, at least from Cantillon's time up to the present. Many scholars have contributed to the literature on the subject. All wanted to unravel the way in which entrepreneurial initiative contributed to economic development (Hébert and Link, 1982). This paper analyzes the contributions of three leading social scientists: Max Weber, Joseph Alois Schumpeter and Frank Hyneman.Knight. All three authors wrote their theories during the first two decades of the 20th century and were well aware of each other's writings. This paper will indicate how they reacted to one another. Their work constitutes a fascinating debate on the motives and effects of entrepreneurship, their theories explaining why new ventures emerge and how they are financed. Weber held the view that Calvinist parsimony would finance investment, whereas both Schumpeter and Knight considered external finance the main source of entrepreneurial investment.

Entrepreneurship had only been a sideshow in economic theory after World War II. Neo-classical economics had read the entrepreneur out of the economic model, leaving no room for enterprise and initiative but only for passive calculation (Baumol, 1968). Entrepreneurship escapes neo-classical modeling by definition due to its relationship to novelty and change. The lack of interest in entrepreneurship in the second half of the past century can be attributed to the widespread idea that entrepreneurship would become more and more obsolete as capitalism developed. Weber contributed to this idea by emphasizing that economic life would get ever more rational. Large bureaucracies would take over as the predominant organizational form of capitalism. Entrepreneurship would fade out as a consequence. Schumpeter also became convinced that large firms would become the main vehicles for innovation and economic progress. He predicted that market societies would evolve from competitive to trustified capitalism, which with time would give way to socialism. Schumpeter also contended that the demise of capitalism would be hastened by an increase of rationality in all realms of life. People would no longer tolerate the irrational elements of capitalism such as the incidence of business cycles and income inequalities. Hence, he considered entrepreneurship in the sense of new firm formation a relic of the past. Knight did not have such a strong opinion on the matter, but he expected that large diversified companies would become predominant due to their uncertainty reducing capacities.

The view that large firms would become increasingly important was supported by the facts in the developed world during the larger part of the twentieth century. Firms did get larger on average, but this trend has been reversed since 1973. The increased importance of small firms and start-ups is indicated by empirical research, which indicates that the share of large firms in employment has decreased since the seventies in Europe, Japan and Northern America (OECD, 2000). This course of events indicates a deviation from a long-term trend of increasing dominance of large firms[1]. Political and economic events of the turn of the 21st century thus seem to mark a break from the evolutionary path that these authors foresaw. We want to find out what elements of the theories of Weber, Schumpeter and Knight are most relevant for contemporaneous entrepreneurship.

2 Life and work of Weber, Schumpeter and Knight

The German social scientist Max Weber (1864–1920) exerted a large influence on American social science. Weber studied law in Berlin, but held chairs in economics in Freiburg and Heidelberg before he became seriously ill and was kept from work-

[1] The alleged superior innovative and job creating capacities of new and small firms have spawned a large theoretical and empirical literature (Acs and Audretsch, 1988).

ing for about four years. After that time he lived as a private scholar in a state of semi-invalidism in Heidelberg. He only accepted a regular appointment at the university of Munich one year before his sudden death in 1920 [2].

Joseph Schumpeter (1883–1950) was born in Moravia, which now belongs to the Czech Republic but at that time to the Austro-Hungarian empire. As a young boy he moved to Vienna with his mother after the sudden death of his father. He was a brilliant student of law and economics at the famous University of Vienna, where prominent economists such as Boehm-Bawerk and von Wieser taught. Schumpeter was an economic prodigy, who wrote a history of economic thought at the age of 25 (Schumpeter, 1908). After taking stock of existing theory, he wrote his own treatise, *Theorie der Wirtschaftlichen Entwicklung*, which was first published in 1911. Schumpeter contended that all economic theories up to that date only applied to a stationary economy and could not explain change. His theory diverges from that of his teacher Boehm-Bawerk on several crucial points, and it may be due to this divergence of opinions that Schumpeter's career had a slow start. He got a professorship at the University of Cnernowitz, located at the borders of the empire and was later appointed in Graz, in which provincial intellectual climate he did not feel well at ease. A brief career in politics followed in the aftermath of World War I, when he became minister of finance in the Renner Coalition cabinet in 1919. But he was forced to resign after one year due to heavy political struggles on the issue of nationalization. Schumpeter became a banker, but this career was also unfortunate, when the Biedermeier bank, of which he was president failed in 1925. He became a professor of economics in Bonn in the same year and moved to Harvard in 1932, where he stayed until his death in 1950. He wrote two major works while at Harvard: *Business Cycles* (1939) and *Capitalism, Socialism and Democracy* (1942).

Frank Hyneman Knight (1885–1972), a farmer's son from Illinois, studied chemistry, German drama, and philosophy at the universities of Iowa and Cornell. He also studied with Max Weber in Heidelberg. He completed his doctoral dissertation in economics in 1916 at Cornell, published in 1921 as *Risk, Uncertainty and Profit,* while he was a professor of economics at the University of Iowa. Knight moved to the University of Chicago in 1928 and published several works on ethics and economic reform, of which *The Ethics of Competition* (1935) and *Freedom and Economic Reform* (1947) are most well known [3].

This paper starts with Weber's articles on Protestantism and the Rise of Capitalism, which were first published in 1904 and 1905 in the *Archiv fuer Sozialwissenschaft und Sozialpolitik*. Schumpeter's *Theory of Economic Development*, which was first published in 1911, can be considered a refutation of Weber's theory. He stated that not the Puritan ethic, but innovation could explain economic development. Weber's books on social and economic organization and on economic history, which were published posthumously, can be interpreted as a response to Schumpeter's (implicit) criticism. Weber sketched a historical sequence of business organizations from antiquity till the early twentieth century in *General Economic*

[2] These biographical details are from Talcott Parson's introduction to Weber's *The Theory of Social and Economic Organization* (Oxford University Press, 1947).

[3] These biographical details are from J. Buchanan's foreword to Frank H Knight (1982).

History (GEH). He pointed out that most organizational forms were designed to support the status quo and thus unable to cope with change.

Knight translated Weber's book on economic history, which was published in German in 1923. Weber's book was compiled out of lecture notes of a class he had taught a year before his death. Knight was an ardent admirer of Weber, to whom he referred as *the most outstanding name in German social thought since Schmoller* in his preface to *General Economic History* (GEH). Weber had more admirers among American academics. Talcott Parsons, a leading sociologist, who worked at Harvard, translated Weber's essays on the protestant ethic in English [4]. The book was first published as *The Protestant Ethic and the Spirit of Capitalism* (PE) in 1930. Parsons also translated part II of Weber's *Wirtschaft und Gesellschaft* which had been published posthumously in 1921 and came out in English as *Social and Economic Organization* (SEO) in 1947. Apparently, prominent American scholars took great efforts to get acquainted with German scientific publications, tantamount to the fact that the German-speaking world was considered foremost in the social sciences at that time.

Knight contributed to the debate by making uncertainty instead of rationality or innovation the central feature of his theory of entrepreneurship. In this way he solved some of the puzzles which had emerged out of the (implicit) debate between Weber and Schumpeter.

3 Weber on Calvinism and economic development

Weber sought the reason for the relatively rapid growth of western capitalism in the specific attitude to life of the Calvinist Puritan. In *The Protestant Ethic and the Spirit of Capitalism* (PE), his most well known book, he associated the economic rise of Holland, England and the American colonies with the presence of Puritan and Calvinist religious groups in those regions, such as the 'Gereformeerden', the Mennonites, the Methodists and the Baptists. These Calvinist groups distinguished themselves from Catholics and Lutherans by their specific concept of salvation, which cannot be attained through the church, but will only fall upon the predestined 'elect'. The concept of predestination did not lead to fatalism, as Weber explained. *The true believer held it to be an absolute duty to consider oneself chosen and to combat all doubt as temptations of the devil* (PE, 111). Lack of self-confidence could be seen as a sign of insufficient faith. People were thus largely self-elected, although a community of believers needed to endorse their beliefs. Calvinism favored rationality in business matters, because material success acted as 'proof' of being one of the chosen (PE, 114). Weber calls the Calvinist attitude to life a rationalization of the world, because magic had been banned as a means of salvation (PE, 117). Calvinism also honored the acquisitive motive, whereas other Christian denominations had often denounced riches. Confession and good works had been replaced by duty and hard work and by abstinence of worldly pleasures. Calvinism

[4] The Protestant Ethic was the translation of the first part of Weber's *Gesammelte Aufsatze zur Religionssoziologie* (1920) and contained the revised version of the articles that were published in the *Archiv fuer Sozialwissenschaften und Sozialpolitik* in 1904 and 1905.

- in Weber's view- perfectly fitted a society of free laborers, who were no longer tied to master and soil by extra-economic considerations as in tribal and feudal societies. Such a situation existed in 17th century England, after the enclosure movement had driven many peasants from the land. The Puritans, in Weber's view, followed the example given by the Catholic monks, who also had applied rational methods to economic activities. Weber considered rationality the outstanding characteristic of both (industrial) capitalism and Calvinism. He drew a sharp line between adventurous and rational capitalism (PE, 20), locating speculative voyages for land and booty the results of which could not be calculated in advance in the first category.

Weber thus portrays the Puritans as a group that was driven by religious zeal to apply rationality to the pursuit of economic activities. This differs sharply from the idea of the Austrian economists of his days, such as Menger and Boehm-Bawerk, who saw the equation of (marginal) utility to revenue as the main motivation of rational economic man. Such a scheme of things was insufficient to explain saving and entrepreneurship in Weber's view, because economic man would only put in as many hours as were required to meet his daily needs. The Puritan, however, would be forward-looking. He would save to obtain wealth, which was considered a sign of godly approval. There was no room for feelings of class resentment in the Puritan world, because the unequal distribution of goods of this world was seen as divinely ordained in a pursuit of secret ends unknown to men (PE, 177). However, the wealthy could never rest in comfort, because their wealth would make it more difficult for them to lead the life of the righteous. But Puritanism had its hard, judgmental side. The consciousness of divine grace by the elect and holy was accompanied by utter hatred and contempt towards people who were considered sinners. Since the dividing line between saints and sinners was never known, sectarian divestitures were common among Puritans in order to keep a 'pure' church (PE, 122). Moreover, the formation of a new sect gave its followers the opportunity to escape from ecclesiastical regimentation of life as this had happened in the Calvinist State churches, which amounted almost to an inquisition (PE, 152). Such despotism would enforce external conformity but would weaken the motives for rational conduct, according to Weber (PE, 152). Calvinism thus had its authoritarian features, which is also apparent from the iron collectivist way Calvin had organized his church in Geneva[5]. However, protestant religion could not prohibit the foundation of new sects due to its lack of central control.

The late 15th and 16th century constituted a period of religious revolts. Many cities in southern Germany and northern Italy had gained autonomy after the German Empire had disintegrated around 1100 (McNeill, 541). The cities used their newly won freedom to promote trade and commerce. The Reformation also emerged in this relatively liberal era. Religious sects such as the Huguenots and 'Wederdopers' emerged that wanted to establish their ideal communities. Many towns and villages were converted, but commercial cities stuck to secular laws and government. The protestants clashed fiercely with rulers who based their claims to authority on the inmutable truths as professed by the Mother Church. Spain and its Inquisition championed the Counter Reformation. The Jews and Moors were evicted

[5] Tawney in his foreword to the english translation of the Protestant Ethic.

from Spain to create a mono-religious country. The same happened to the French Huguenots, who were persecuted by the unified French State under Henry IV. The Italian cities were brought under Spanish rule in 1498.

The Dutch Freedom war against Spain (1568–1648) was also inspired by religious rebellion, but did not result in religious tolerance right away. The Reformers pillaged the Dutch Catholic churches and monasteries in 1566 at the beginning of their insurrection. The Dutch Calvinists succeeded in grabbing state power and constituted a theocracy for a limited period of time. However, some forces counteracted the movement towards a Calvinist state. Most important was the loose federal character of the United Provinces, which lacked strong state power. The cities could, therefore, take their own stance in economic and judicial affairs. The city lawyers and magistrates were opposed to harsh persecution of the non-reformers. Moreover, the cities needed extra hands and, therefore, welcomed immigrants from regions where the Calvinists were persecuted such as France and Belgium, but also from other areas. Jews, who fled from Spain and Portugal after being extradited, could take shelter in Dutch cities, predominantly in Amsterdam. Catholics and adherents of other religions were tolerated after 1630, which meant that they could practice their religion, if it was done non-conspicuously. However, only members of the Reformed Church could fulfill official positions in the 16th and 17th centuries in the Netherlands (Schama, 1987). Some Dutch practices seem to fit Weber's portrayal of the Calvinist. Weber describes the repugnance of idleness and begging as a trait of the Puritan for whom work is a holy duty. This attitude is reflected in the harsh policies pursued by the 17th century Dutch cities towards vagabonds, beggars and idle people, who were either put to work in the towns' workhouses or deported (Schama). This contrasted with Catholic (and Buddhist) attitudes in which begging was seen as a respected way of living.

A Calvinist state was never founded in England, because the Puritan sect members were outnumbered by the Anglicans. The English Puritans contested the authority of the state in religious and economic matters and were considered a threat to the English nation. Their fight against state monopolies and military conscription are cases in point. Weber considers this a main reason for the early emergence of a professional army in England. As a consequence they were kept from land leases and official occupations.

Weber's theory of the Puritan ethic was not explicitly directed towards entrepreneurship. The Puritan ethic did apply equally to the businessman, the professional and the laborer (PE, 177). However, many Puritans took up entrepreneurship in England, since they constituted a religious minority to which other routes of social advance were closed. The same had applied to other religious minorities in the civilized world: Jews and Christians were the merchants and moneylenders of the Ottoman Empire. But those occupations were not highly esteemed in Oriental civilizations that placed them at the bottom of the social pyramid. A career in the state bureaucracy or army constituted the only road to social prestige in those countries. England and the Low Countries, just like the northern Italian cities of the 13th and 14th century, had less absolutist forms of government and gave more scope to commerce and manufacturing.

4 Schumpeter's theory of economic development

Schumpeter contributed regularly to the *Archiv fuer Sozialwissenschaft und Sozialpolitik*, which was edited by Weber in the 1920s[6]. It can, therefore, be assumed that he was well aware of Weber's work. Schumpeter sketched a model of a dynamic economy in his *Theorie der Wirtschaftlichen Entwicklung* (WE) in 1920. His theory can be considered a refutation of Weber's hypothesis on the importance of the Calvinist attitude for economic development. Schumpeter defined the entrepreneur as the founder of a new firm and as an innovator, who breaks up established routines and opposes the old way of doing things. Schumpeter's entrepreneur only undertakes those ventures which turn out to be successful (WE, 177). The entrepreneur's special leadership qualities enable him to see the right way to act. Others will follow in his wake. There are some overtones of the Nietzschean leader in Schumpeter's description of the entrepreneur.

In order to introduce his innovations, the entrepreneur needs to withstand the opposition of the environment, which is usually hostile to deviating behaviour and novelty. *All deviating behaviour of a member of a community meets with disapproval from the other members* (WE, 118). Moreover, the individual is also restrained from doing something new, due to the psychic and physical efforts it requires to leave familiar paths (WE, 120); it is the difference between swimming with the current and against the current (WE, 121). Schumpeter's entrepreneur, however, takes a delight in this opposition (WE, 132). Schumpeter's entrepreneur is a creative non-conformist and not a religious dissident. He is not shunned by society, but warmly welcomed by the banking community that grants him credits to finance his attack on established positions. Bankers are supposed to be gifted with perfect foresight, for they can discern the best and brightest entrepreneurs without difficulty.

The entrepreneur breaks up the 'circular flow', actively steering the economy away from old paths and opening up possibilities hitherto unknown. The circular flow describes a stationary economy, in which economic processes are repeated period after period without change. Prices and quantities do not vary in the circular flow and can be completely deduced from the data. The interest rate is equal to zero, and net investments are absent. Schumpeter used the concept of the circular flow as a point of reference to indicate the changes that are caused by the introduction of innovations. His concept of the circular flow had many predecessors as is indicated by the various references to the concept of a stationary state in his *History of Economic Analysis* (HEA). The first reference is to Plato's utopian vision of a Perfect State, described in his *Republic*. Born out of dissatisfaction with the changes that Athens went through in his lifetime Plato outlined the preconditions for a stationary Utopia such as a stationary population, constant wealth, division of labor according to ability and limited freedom of speech (HEA, 55-6). The authoritarian aspects of Plato's Perfect State are absent in Schumpeter's description of a circular flow, which is not directed by command, but by perfect competition and established routines. Schumpeter unravels the way in which productivity increases are produced

[6] Schumpeter published 27 times in the *Archiv* during the period 1910–1920, when Weber was one of the editors of the journal. Massimo M. Augello (1990).

by the bunchwise appearance of innovative firms, which set off (cyclical) waves of investments and disinvestments.

Entrepreneurs found new firms to introduce innovations, because established firms are reluctant to change their routines. Moreover, established businesses will postpone innovation until their old assets have become obsolete. New firms are not impeded by former investments and will, therefore, speed up economic progress by introducing innovations at a date before incumbent firms would. Schumpeter pointed out that innovation could inflict losses on incumbent firms, which he labeled *creative destruction*.

Schumpeter explains the way in which the innovative method of production obtains a premium caused by its inherent superior efficiency. The differential with previous methods accrues to the entrepreneur until it vanishes due to competition by imitators. Schumpeter argued that this premium constitutes the source of all profits and of interest payments. Consequently, all former interest theories, including Boehm-Bawerk's were incorrect. Neither waiting nor the lengthening of the production period constitutes the source of interest and profits, but rather the source is innovation. Capital deepening means that labor is replaced by capital, but it does not need to entail overall productivity increases. This can only occur, if fewer production factors are used to produce a certain product. Schumpeter sketched a theory, in which the interest rate equals the profitability of the marginal entrepreneur (WE, 383). This presumes that all entrepreneurs can be ranked unambiguously according to their profitability. Hence, innovative profits are split up into interest payments to bankers and a residual that accrues to entrepreneurs. Entrepreneurial income reflects entrepreneurial quality. The best entrepreneur receives the highest incomes, whereas the marginal entrepreneur's profits just suffice to pay his banker.

Schumpeter agrees with Weber on the non-hedonic nature of the entrepreneur. The entrepreneur does not resemble 'economic man', who weighs (marginal) costs and benefits and stops working at the moment when the costs of the extra effort (fatigue) exceed the extra satisfaction (WE, 126). Such behaviour, according to Schumpeter is characteristic of the circular flow. The entrepreneur – by contrast – is prepared to work countlessly more hours in order to achieve his goal. But, Schumpeter's entrepreneur is not a Puritan. He does not abstain from worldly pleasures, but participates fully in politics and culture. He is not motivated by a belief to belong to the 'elect', since the banking community has already chosen him. Schumpeter's entrepreneur is motivated by the joy of creating and by the pleasure success brings (WE, 141). Schumpeter's entrepreneur – in contrast to Weber – is not motivated by rewards beyond his lifetime, but by improving his social position (and that of his family) in this world. Leaders of former times had based their leadership mainly on military and bureaucratic expedience and less on commercial qualities. But innovative qualities can occur everywhere and are thus not system-specific, in Schumpeter's view. The clan-leader could also lead his people into new territories, as could the feudal knight. However, not all economic and social systems are equally well equipped to innovation; he mentions India and China as two countries without much innovation to prove his point that economic development is not obvious and automatic (WE, 113). He also remarks that it is harder for individuals to break up established routines in primitive societies without bothering others due to their communal way of living as is exemplified by the man-house (WE, 119).

Schumpeter considered entrepreneurial behaviour non-rational, because it did not fit in with the model of 'economic man' designed by the marginalist School. Weber's Puritan and Schumpeter's entrepreneur are forward-looking, whereas rational economic man is supposed to live only by the day. But, such long-term thinking could be called rational by our modern standards. People will be inclined to forgo immediate consumption, if interest rates are sufficiently low. Weber's concept of rationality was not always perfectly clear. Schumpeter pointed this out in a seminar, which he gave at Harvard in 1940; among the attendants were Parsons, Sweezy and Leontief[7]. Schumpeter distinguished between formal or objective and substantive or subjective rationality in this seminar. Formal rationality applies, if costs and benefits can be calculated accurately; a means-end relationship. Subjective rationality refers to the achievement of absolute values irrespective of costs. Salvation fits in with the latter rationality concept.

However, the two concepts coincide in Weber's portrait of the Calvinist, who believes he is saved if he succeeds in making ends meet. Weber combined both strands of rationality as he depicted capitalist enterprise as calculable. Schumpeter rejected Weber's portrait of the scrooge capitalist and replaced it by the well-mannered gentleman, who wanted to build his own estate. But calculability also featured largely in Schumpeter's work. The absence of uncertainty could explain why financiers are only meagerly rewarded by base rate interest payments, whereas the entrepreneur obtains the rest of the innovation premium in accordance with his capabilities. Hence, innovative investment is considered devoid of any risk. Schumpeter, however, did mention risk in the first German edition of *Theory of Economic Development,* where he stated that foreseeable risk can be reduced to costs (WE, 49). Interest rates will be elevated by a certain percentage, if some of the new ventures are expected to fail (WE, 387). The percentage of failures could be calculated based on experience. Moreover, losses would mainly occur at the old firms, which were unable to adapt to new economic conditions in time (WE, 493). Schumpeter's ideas on this matter lie at the foundation of Knight's theory of profit, as will be demonstrated below.

5 Knight on investment and entrepreneurship

Frank Knight, in his seminal contribution to economics *Risk, Uncertainty and Profit* remedied Schumpeter's disregard of uncertainty and in so doing laid the foundation for modern finance and organization theory. He was one of the founding fathers of the famous Chicago School in economics. Knight was acquainted with German economics texts. It seems, therefore, plausible to assume that he was familiar with Schumpeter's *Theorie der Wirtschaftlichen Entwicklung*, when he wrote his thesis *Theory of Profits and Uncertainty in* 1916, which was published as *Risk, Uncertainty and Profit* (RUP) in 1921.

[7] The article was found in the Harvard Archives and was first published *in Zeitschrift fuer die Gesamte Staatswissenschaft* in 1984.

Knight agreed with Schumpeter on the matter of capital deepening. He stated that the length of life of capital goods is a matter of choice and can never be a source of profits (Knight, 1939). Knight also agreed with Schumpeter on the point, that profits and interest can only exist in a progressive society, although interest could be paid for consumption loans (RUP, 328). Innovation is the source of profits and can only occur when investment is used to create new resources. Knight differed somewhat from Schumpeter on the point of entrepreneurial motivation. He considered the desire to excel, to win at a game, the biggest and most fascinating game yet invented, not excepting even statecraft and war most important (RUP, 360).

But his theory differs from Schumpeter's on an essential point. Schumpeter did not deal with selection problems, as was mentioned above. Bankers would always pick the right entrepreneurs. It might have been Schumpeter's (and Weber's) familiarity with the credit mobilier type of banking and the disastrous consequences of the failures of these banks that made them emphasize calculability. Knight, however, contended that only uncertainty could explain profits (and losses). Not all ventures will become successes; some will fail. However, which ventures will succeed and which fail cannot be predicted in advance. He also remarked that the profits of change come largely in the form of readjustments of capital values. Hence, where Schumpeter assumed that bankers do not make mistakes in selecting entrepreneurs, Knight made errors or uncertainty the basis of his theory of entrepreneurial profits (and losses). The crucial type of decision in all organized activity, according to Knight, involves the selection of men to make decisions. Any other sort of decision-making or exercise of judgment is automatically reduced to a routine function (RUP, 297).

Uncertainty needs to be sharply distinguished from risk in Knight's view. Risk is calculable a priori and can, therefore, be treated as a cost. Experience can teach us what percentage of bottles is going to burst in a champagne factory. These damages can be included in our cost calculations (RUP, 213). Other types of risk, such as the incidence of fires, can be insured. *Uncertainty, in contrast, is uninsurable, because it depends on the exercise of human judgment in the making of decisions by men and although these estimates tend to fall into groups within which fluctuations cancel out and hence to approach constancy and measurability; this happens only after the fact* (RUP, 251). The major difference between risk and uncertainty thus consists of the possibility of making *ex ante* calculations of the incidence of an event. That can be done for fires, but not with respect to the outcomes of investment projects that can only be calculated after everything is said and done. Knight borrowed from Schumpeter the idea that entrepreneurs are not self-selected but chosen by investors, who – in contrast to Schumpeter's portrayal – are not infallible but subject to error. But the vision of the investor is central to his theory and not that of the innovator.

Knight's theory portrays investment as a discovery process. Many new ventures will be launched, but only a few will survive and prosper. Such a sketch of events fits actual developments. Many new businesses were launched in the 1990s, of which a few obtained astronomically high valuations. This happened before they actually made profits, which supports Knight's thesis that not actual profits but profit expectations are the decisive element in investment. Knight's theory thus foreshadowed things to come. The development of financial markets (particularly in the US) has made it possible to measure success by readjustments of capital values before actual profits have been reaped. We can also conclude that investment under conditions of uncertainty requires other methods of finance than debt. (High) chances of failure require a risk premium on interest, which curtails investment. Equity capital is therefore much more suited for the task of financing uncertain ventures than debt.

Knight pointed out that only the investor with above average skills of perception would earn (excess) profits, whereas investors, whose perceptiveness was below average, would lose money on their ventures. Perceptiveness refers both to project choice and to the timing of investment. Knight contended that the average rate of return on investment does not need to exceed the riskless rate of return in the long run (RUP, 284). Hence, both Knight and Schumpeter argued that investors as a group do not need to earn excess profits. But, the interest rate equals marginal profitability in Schumpeter's view and average profitability in Knight's. Knight's investor/entrepreneurs need to be compensated for losses caused by uncertainty. Schumpeter's entrepreneur was entitled to the residual, which indicated his capacity at innovation. But entrepreneurs are not entitled to the residual in Knight's view. Their incomes should take the form of salaries.

It is obvious that uncertainty in investment, just as in sports, only exists, if more than one company/team vies for the same prize and if the result cannot be predicted with any accuracy. The match would be superfluous, if the 'best' team could be indicated before the event. The same applies to the business world. Only one innovative firm needs to be launched, if the best innovator can be discerned ex ante. The best employee will also be the best entrepreneur, if standards of excellence apply equally to entrepreneurs and employees. As a consequence, all investors would flock towards the designated winner, which would drive up the price of that person to the point at which no profits are left for the investor. Moreover, this innovator has no incentive to start his own company but will earn exactly as much as an employee (Brouwer, 2000). I hold the view that Knight has corrected Schumpeter's theory on an essential point. Profits (and losses) can only appear, if uncertainty is present. In fact, all profits would vanish, if the winning person or company could be indicated ex ante. If everybody had known beforehand that Microsoft would become the most successful company of the 1990s investors would have rushed to provide funding and the list of prospective employees would have been infinite. As a consequence Bill Gates could have obtained all the money he wanted at riskfree rates of interest and could have paid his employees just standard wage rates. But, Gates was funded by venture capital firms, which took equity shares in his firm. The same applies to himself and his co-founders, who were also paid (partly) by equity shares. Most founders of 'high tech' start-ups receive modest salaries, but get equity shares in compensation. Microsoft became an enormous success and everybody was handsomely rewarded for his wise choices. However, a host of start-ups, which turned out to be failures and whose investments were largely lost counterbalance the Microsoft success story.

Uncertainty can explain many features of innovation, such as the relatively small size at which innovative ventures are launched. The rationale for this is that losses are limited in the case of failure. It can also explain why many start-ups are short-lived and why some grow very rapidly. Only firms that are expected to become successful will receive several rounds of finance so that they can expand. Moreover, uncertainty is also a big motivating force. This applies to sports, but also to economics. No firm would wage investment in innovation, if it knew that it would lose out to an objectively better competitor. But this competitor in turn would have no incentive to disrupt existing routines, since this could entail creative destruction. Both Weber and Schumpeter believed that innovation could proceed unhindered in socialism. This becomes plausible when we realize that both authors did not incorporate uncertainty into their theories.

6 Weber's history of organizations

Weber's *General Economic History* (GEH) and *Social and Economic Organization* (SEO) can be considered a response to Schumpeter's thesis, that not the Protestant ethic but innovation causes economic development. These two books also attempt to refute Schumpeter's claim that innovation is not system-specific. Weber gives a powerful sketch of economic history from ancient times until 1920 in his *General Economic History*. He analyses several consecutive historical forms of economic organization, such as the clan, ancient bureaucracy, the manor, the guilds and the business enterprise. These organizations can be embedded in larger political organizations, such as an empire, state or city.

Most primary organizational forms were directed towards protection of its members against outside aggression. This applies to the clan, which Weber saw as the primal organizational unit and as an agency of blood revenge and the prosecution of feuds (SEO, 128). The clans did not possess written laws; clan members were totally dependent on their leaders for their lives and livelihood. The clan and later the sedentary agricultural village were composed of several families. The power of the clans could only be broken, if a political system was established, which eroded the absolute power of the chieftain. All early civilizations, in his view, such as the Egyptian, Greek and Roman civilizations, had loosened the grip of clan and family on people – usually by force – and had established more rational forms of organization. (Weber mentioned ancient Egypt as an example of a huge bureaucracy with a very extensive division of labor). Officers in a bureaucracy are chosen on the basis of some kind of proven expertise. This contrasts with the selection of officers (such as the medicine man) in a tribal society, which relies largely on magic.

Feudalism constitutes another organizational form, which is not (completely) based on kinship and blood ties and thus embodies some rational aspects in Weber's view. In feudalism, the tribes are subjugated to an overlord, who in turn has been appointed by the emperor or prince. Feudalism emerged gradually through either the contracting out of taxation benefices to administrators (such as to the mandarins in China) or through the granting of fiefs to militaries (the feudal lords of the occident and the Ottoman empire). The power of the clans was completely broken in the occident, but not in the orient according to Weber (GEH, 45). Life on the countryside in occidental feudalism revolved around the estate or manor. Peasants were either obliged to pay fixed or variable fees (such as in sharecropping) to their landlords and/or were subjected to compulsory labor on the estate. A landed aristocracy replaced the clan leader in its task of defending the life and good of the people. But, the manor or *oikos* was not organized along rational lines. Holdings were usually of sub-optimal size. Income was largely determined by privileges and obligations, which were passed on from generation to generation.

Part of the medieval populations lived in cities, which had their own organizations. The inhabitants of the towns in Western Europe usually did not start out as free men, but fell under the jurisdiction of a feudal lord. It was only when they became engaged in trade that they reached the status of mercator and of free citizen. The citizens of the medieval towns were relieved from feudal obligations but had to pay taxes to the emperor or king. The lords of the manor were opposed to town privileges. The prince and the emperor, however, were willing to grant privileges to the towns in order to increase tax incomes.

The guild organization of craftsmen dates back to antiquity. Ancient guilds were mainly based on clan lines. The guilds as associations of free craftsmen did not develop outside Western Europe (GEH, 137). Instead, every individual belonged to a clan or caste in the Orient. Guilds never existed in China, because clan organization was dominant. Guilds did not develop in India either, since the division of labor was organized along lines of castes, which can be seen as ritualistic guilds (GEH, 137).

Craftsmen and merchants populated the towns, whereas peasants, who were attached to the land by feudal obligations and rights, populated the countryside. The establishment of the towns introduced trade, because the town people needed to be fed by agricultural surpluses, which were exchanged for handicraft and imported products on local markets or by peddlers. Towns were much more numerous in occidental than in eastern Europe. Weber remarks that the power of the towns diminished that of the estates. *With the decrease of the frequency of towns on the map there is an increase in the frequency of estates* (GEH, 89).

Its military basis made occidental feudalism rather unstable. The power of a king or prince was never absolute, as he had to rely on the military capabilities of local lords to sustain his power. The knights could also use their power at arms to fight each other or the emperor. The inherently instability seems to have contributed to the development of towns and therewith to commerce and enterprise in western Europe. The emperor or king could curb the power of the aristocracy and expand its tax revenues by granting royal privileges to cities in exchange for a guarantee of their corporate liberties and protection against the feudal lords (McNeill, 542). However, some cities developed their own military capability and supplanted the feudal lords. Towns bought up the surrounding countryside and established their own states. Italian towns such as Florence and Venice are cases in point. Free guilds became the dominant organizational form for craft workers in Western Europe after craftsmen had thrown off the feudal shackles. This happened first in the 'free' city-states of southern Germany and northern Italy.

7 Weber on the emergence of capitalist enterprise

Weber indicated that several organizational forms were not up to the transition to a market economy. The manorial system in agriculture, which had existed in Europe for many centuries, was bound to shatter, when faced with market forces. The newly developed bourgeois interests of the towns promoted the weakening of the manor to expand their market opportunities. The manorial system curbed the purchasing power of the rural population due to the compulsory services and payments it demanded from the tenants. The manorial system also prevented the creation of a free labor market because it attached the peasants to the soil (GEH, 94).

The guilds promoted production for the market, which gradually supplanted the division of labor within the manor. Where the guilds were not victorious or did not arise at all, house industry and tribal industry persisted (GEH, 147). This applies to eastern European feudalism. Russian peasants could not break loose from the soil because of eternal rights and obligations to the estate and to the communal mir organization.

The western guild organization was of a more rational nature than the manor. Craft guilds promoted professional competence. People could only become a master and open up their own shop after a long period of training. More importantly, admission to the western guilds was usually based on skills and not on family ties, although minorities were often excluded from the guilds. The guilds were first regulated by the town lords, who demanded both taxes and military services from guild members, but all these prerogatives were later acquired by the guilds (GEH, 148).

After having won these battles the guilds used their newly won freedom to establish an effective monopoly in their trade. Guild regulation protected its members from competition. It was impossible to set up a shop without the guild's consent. Moreover, the number of masters and therewith of shops was strictly regulated in order to restrain output. Equality prevailed within the guild organization. No master could improve his position beyond that of another. Hence, no shop could ever increase its market share. The guilds' initial boost to progress petered out, when the adverse effects of their regulations became dominant. Moreover, guild members could own no capital of their own, and were therefore restricted to non-capital intensive production methods. The few capital goods in existence such as grain and wood mills were often owned by the town or by a cooperative.

New organizations appeared in the 16th and 17th century, which broke the power of the guilds. Journeymen, who were not allowed to become masters, set up their own businesses. They could do that outside the jurisdiction of the towns in the countryside. The putting-out system resulted from the attempt to escape guild regulation. Rural labor was amply available in England due to the enclosure movement. Production could expand beyond local demand to supply export markets (GEH, 94). But guild regulation persisted in many trades especially in Germany. The German princes more or less operated on behalf of the guilds and the guilds remained strong within Germany for a long time. The remnants of the guild system can still be found in some German trades, in which the *handwerkergesetze* apply.

The putting-out form of enterprise could thus only flourish outside the guild organizations. The same applies to the trading companies of the 16th and 17th centuries, which were organized in a corporate form. The Dutch East India Company (VOC) is a prominent example of such a joint stock company, which raised capital by issuing shares to participating merchants. The VOC existed from 1602 till 1790 and employed several thousand people at its hey-day. It was a multi-plant company consisting of six local chambers (Amsterdam, Rotterdam, Delft, Middelburg, Enkhuizen and Hoorn). It exploited shipyards and warehouses apart from organizing maritime expeditions. The VOC and other trading companies had many traits in common with the modern corporation. However, they needed to be chartered by the state, because there was no free incorporation in Holland in those days. As a consequence the VOC was considered to represent the national interest and could rely on military support to fight rival maritime nations and pirates. The English situation differed from the Dutch since 'the Statute of Monopolies' was adopted in the early 17th century, which ended the Crown's prerogative to grant monopoly rights. Monopolies could now be challenged in courts and disappeared rapidly from the English scene. Incorporation of voluntary groups spread rapidly in 17th century Elizabethan England as is shown by the popularity of the joint stock company (North and Thomas, 154).

We may conclude that the guilds were initially conducive to progress, but arrested progress later on. The guilds came to eschew novelty and wanted to maintain existing privileges and therewith the status quo. They began to show traits of the caste organizations such as in India, which forbade all use of new production methods. Guild organization gradually gave way to capitalist organizations such as the joint stock company; but this process was initially limited to England.

Other Western European countries upheld monopoly privileges much longer and only allowed free incorporation at a much later date (Thomas and North, ch. 10).

Weber wanted to demonstrate in *General Economic History* that institutional change is not immanent to all types of societies. Only some western countries succeeded in establishing free enterprise. Weber emphasized that capitalist business enterprise could only emerge after production factors had become mobile and property rights had been established. Labor should not be bound to master and soil by hereditary and, therefore, irrational traditional ties. Such migration dates back to ancient times, when people moved from the hills to the riverbed valleys of Mesopotamia and Egypt. Consequently, the clan chieftain and also the father as the head of the family were robbed of (part of) their possessions and privileges. The same happened when Greek and later Italian cities rose to prominence. Contractual labor relations gradually replaced serfdom, royal privileges and communal land ownership.

Apart from labour, land also needed to become freely negotiable before capitalism could take off. Private property in land emerged during the 17th and 18th centuries in Western Europe, but the process knew several different forms and the dissolution of the feudal system therefore resulted in the different agricultural systems of today. *In part the peasantry was freed from the land and the land from the peasantry as in England; in part the peasantry were freed from the proprietors as in France; in part the system is a mixture as in the rest of Europe, the east inclining more to the english conditions* (GEH, 108).

France became a country of small and medium sized agricultural firms, when the estates were broken up and distributed among the peasants after the French revolution (GEH, 99). The absence of primogeniture – in contrast to England – also contributed to the emergence of small lot sizes in France. The landlords were also expropriated in south and western Germany. The change was the same as in France except that it took place slowly and according to a more legal process (GEH, 100). *With the dissolution of the manors and of the remains of the earlier agrarian communism through consolidation, separation etc., private property in land has been completely established* (GEH, 111). The occident thus turned into a world which was dominated by a division of labor based on free labor and private property in land.

An additional condition, which needed to be met before capitalism could flourish - in Weber's view - is the establishment of a rational state. This means a state which is based on rational law and expert officialdom (GEH, 339). He contrasted this with states whose officials were trained in ancient literature or religious texts as was common in the Orient. Many feudal empires such as in China were foremost concerned with preserving their centralized power structure. Their local representatives were not selected by military or technical criteria but by their expedience in Chinese literature. Such vassals had to be completely loyal to the emperor, because another recruit could easily replace them. The mandarins were also prevented from building a local power base among the population, because they were continually transferred from one province to another and did not speak the local dialect.

Rational law in Weber's view was based on Roman law, which revived in great parts of the occident, but not in England, during the late Middle Ages and Renaissance. Weber considers its calculability the main advantage of Roman formal-legalistic law, which contrasts to materialistic, theocratic or absolutist law (GEH, 340-1). Roman law also differed from other law systems by its emphasis on civil matters, which greatly facilitated trade. Roman law provided reasoned and consistent principles of jurisprudence, just like English common law. Most empires had only developed criminal law apart from the famous Hammurabi Code of ancient Mesopotamia that also addressed commercial matters.

8 Authority and change

Weber gave an overview of institutional history, but failed to explain why these institutions had come and gone. His analysis of charismatic authority in *Social and Economic Organization* can be interpreted as an attempt to fill this void. Here, he distinguished three forms of authority; legal-rational, traditional and charismatic. Some leaders derive their authority from tradition, such as the clan leader and the lord of the manor. Bureaucracy, by contrast, is based on rational-legal authority. We can summarize Weber's historical view by stating that the western world had changed from traditional towards formal-legal societies. However, he failed to explain why such change occurred only in some societies and was absent in others. Parsons discusses this point in his introduction to the American edition of *Wirtschaft und Gesellschaft* (1947). Traditional authority is opposed to change by definition. Legal-rational authority spurs efficiency, but it is doubtful whether it can break up existing routines. Both traditional and rational-legal authority seem, therefore, suited to an established social system of a routine character (SEO, 64). Charismatic authority, however, wants to lead people away from existing routines and beliefs and is, therefore, an agent of change. The charismatic leader - as described by Weber - is always in some sense a revolutionary and a dissident. He promises people a better future. But, a charismatic leader has to deliver. The accuracy of his vision needs to be proven by events, or people will turn their backs on him.

Schumpeter's innovative entrepreneur, who leads the production process into new grids, has many things in common with Weber's charismatic leader. They both oppose established authority and lead their flock in new directions. Weber's analysis of charismatic authority can, therefore, be considered an attempt to blend the Calvinist and the innovator. But Weber situates charisma mainly in the realm of religion and the military and not in the business world. A founder of a new religious sect and his disciples and a military band are the types of organization he had in mind. Charismatic authority loosens all former ties to family and clan and forges completely new loyalties. The people who have 'recognized' the charismatic leader are completely devoted to him in a non-rational way. There is no such thing as promotion or dismissal in the early stages of new religious movements or military bands; there is no hierarchy either. Hence the charismatic organization is exactly the opposite of the rational-legal bureaucracy. The charismatic organization does not demand formal qualifications. Everybody who recognizes the greatness of the charismatic leader can follow him and become a member of his community. The charismatic leader expects much of his disciples. They are expected to achieve extra-ordinary things. Rewards are not formalized but consist of free gifts or booty, which are distributed on a communal basis. However, this charismatic organization cannot endure – in Weber's view – and will in time turn into an organization based on either traditional or legal-rational authority.

Some elements of charismatic organizations can be found in entrepreneurial start-ups. The (innovative) entrepreneur is often mission-oriented. He believes in his own capabilities and those of his employees. No entrepreneur would ever start his business, if he were not convinced of its ultimate success. Most of the time he needs to convince at least a few others to join him in his venture and to defer income until the moment of success has arrived. He also needs to convince financiers and ultimately customers of the superiority of his product or process. Many start-ups are organized along non-bureaucratic lines and the founders are prepared to put in many extra hours. The lack of hierarchy and formal roles and the vicissitudes of income as determined by stock options are also characteristic of start-up firms. This applies most forcefully to the founders of the organization, but also to employees, as the example of Microsoft demonstrates. More than thousand of its (early) hires became millionaires thanks to stock options.

Weber's hypothesis of the charismatic leader is certainly powerful, because it can explain why people are prepared to leave their old ways of life in order to start anew. His theory of the charismatic leader attempts to explain the *rise* of protestant sects, instead of analyzing their characteristics. However, heretics are of all times, but their rise was often arrested by repression. Examples on the persecution of deviants or dissidents abound. We can think of ancient Rome, which persecuted the Christians, as well as of the Russia of Catharine the Great, which sent its religious dissidents off to Siberia. It was only when non-absolutist forms of government arose that both freedom of religion and of business incorporation emerged. Hence, freedom of religious incorporation went hand in hand with its counterpart in business. We could, therefore, conclude that charismatic leadership could only occur when certain conditions were met

9 Schumpeter's economic development revisited

Schumpeter wrote a second edition of his *Theorie der Wirtschaftlichen Entwicklung* of about two thirds the size of the original version. The new edition was published in 1926 and became the basis of the English translation, which appeared as *The Theory of Economic Development* (ED) in 1934.

The second edition – apart from its smaller size – differs from the first on a number of points. Firstly, all passages criticizing Boehm-Bawerk and other economists had been eliminated. Schumpeter presented his work as a piece of economic analysis and himself as an economist. The professionalization of the social sciences had advanced to a stage which seemed to require a strict demarcation between economics, sociology and history. Schumpeter thus complied with his critics, who had claimed that Schumpeter had not produced a historical, evolutionary theory of economic change. Schumpeter returned to history in his *Business Cycles* (1939) and *History of Economic Analysis* (1954). Another criticism involved the emphasis Schumpeter had put on the role of the individual in economic development. *One of the most annoying misunderstandings that arose out of the first edition of this book was that this theory of development neglects all historical factors of change, except one namely the individuality of entrepreneurs (ED,* 60).

Schumpeter is not prepared to give up his original position that the entrepreneur is the dynamic element in capitalism. He now puts a greater emphasis on the intuitive capacities of the entrepreneur. *He has the capacity to see things in a way, which afterwards proves to be true, although it cannot be proven at the time.* Schumpeter even adds*; thorough preparatory work, and special knowledge, breadth of intellectual understanding, talent for logical analysis may under certain circumstances be sources of failure* (ED, 85).

Schumpeter made some superficial changes in his description of entrepreneurial motives, but he stuck to his former statement that entrepreneurial motivation is unlike that of economic man, who is balancing probable result against disutility of effort. The entrepreneur is still motivated by the dream to found a private kingdom, which most closely resembles the position of medieval lordship. New motives involve the will to conquer, the impulse to fight, to prove oneself superior to others, to succeed for the sake of success. Pecuniary gain is therefore not the main motive for entrepreneurial action, but it is a very accurate expression of success, especially of relative success (ED, 93-4). We may conclude that Schumpeter rephrased his entrepreneurial motives somewhat to get them more in line with Knight's description of what moves the entrepreneur. Knight emphasized sportsmanship to explain entrepreneurship, whereas Schumpeter had stressed creativity in his first edition. They now both share the view that entrepreneurship can best be understood as an element of the game spirit and the ambition to win. But, Schumpeter rejects Knight's definition of the entrepreneur as the capitalist and risk-taker. Schumpeter argues that the usual shareholder has no impact on company policy and hence cannot fulfill entrepreneurial functions (ED, 75).

Schumpeter also strongly repudiates the Marshallian definition of entrepreneurship, which treats the entrepreneur as a manager. He draws a sharp line between managers and entrepreneurs. *The manager chooses from the most advantageous among the methods which have been empirically tested and become familiar at a certain point in time, whereas the entrepreneur looks for the best method possible at the times* (ED, 83). He also firmly rejects Weber's definition of the entrepreneur as the person with a calling. *The carrying out of new combinations can no more be a vocation than the making and execution of strategic decisions* (ED, 77).

The elements of charismatic leadership that could be discerned in the first edition have now faded out. Schumpeter still describes the entrepreneur as a leader, yet the heroism with which the entrepreneur was adorned in the first edition is largely gone. He now calls the entrepreneur a leader against his will, whose only followers are the imitating firms that rob him of his profits. The entrepreneur only needs to convince or impress one man, that is the banker who is to finance him (ED, 89). Schumpeter contended that the days of the heroic entrepreneur were almost over. The entrepreneur still needed to confront the opposition from the environment, but this opposition has largely succumbed in modern times. *The importance of the entrepreneur must diminish just as the importance of the military commander has already diminished* (ED, 86). He attributed this to the increased calculability and rationality of the capitalist world. Schumpeter had first depicted the entrepreneur as the man of action who could lead society into new ways of doing things. This figure was replaced by the socially much less apt entrepreneur, who only cared about business and left politics to the politicians. This line of reasoning made Schumpeter predict that innovation could be delegated to experts without affecting performance. Schumpeter foresaw that innovation would become a matter of routine, executed by employees in R&D laboratories. He did not like this course of events, because the world would lose much of its splendor due to these changes. But he thought that increasing rationalization would make it inevitable.

Hence, Schumpeter held the view that increased rationality spells the end of capitalism. The origins of Schumpeter's bleak outlook on the viability of capitalism, which he has exposed magnificently in *Capitalism, Socialism and Democracy,* can thus already be found in the second edition of *Theorie der Wirtschaftlichen Entwicklung*.

Schumpeter, however, came to retrace his steps and emphasized competition and liberalism as crucial conditions for economic development in his posthumously published *History of Economic Analysis* (Brouwer, 1996). Schumpeter describes in HEA how competitive capitalism could only unfold unfettered after liberalism had become the main political movement. Liberalism - in Schumpeter's words - entailed the subjugation of politics to commerce and reigned only from the end of the 18th century till about 1900 (HEA, 761). We could, therefore, conclude that Schumpeter narrowed his theory even further at the end of his life. His theory now only applied to 19th century western and especially English capitalism, where the state had largely withdrawn from the economic arena and the bourgeoisie had become the leading class

10 Rationality of capitalist enterprise

Weber defined capitalism as rational calculation applied to the pursuit of economic gain. Rational capitalist enterprise exists -in Weber's view- when rational methods such as bookkeeping and capital accounting are used (GEH, 275). He distinguished rational from adventurous capitalism. The latter is represented by the overseas trading companies of the 16th and 17th century whose ventures were subject to high uncertainty. This is attested by the fact that many VOC ships never returned from their voyages. Weber dates the beginning of rational capitalism back to 17th and 18th century mercantilist states. This period was characterized by monopolies granted by the state. Weber calls mercantilism the first rational economic policy, because this policy was primarily directed towards economic welfare instead of reassuring peace among rival groups and classes. Mercantilism meant that the state was run as a single firm. The objective of the state was to generate as much tax income as possible through national monopolies. Mercantilism thus favored managed instead of free trade.

Schumpeter tracked the origins of capitalism back to the medieval city-states that largely depended on commerce. *Capitalist institutions, such as big business; stock and commodity speculation and 'high finance' had already established themselves firmly at the end of the 15th century, and had entailed the ascent of the bourgeoisie* (HEA, 78). Schumpeter contended that capitalist progress was arrested by the rise of the absolutist nation states that ascended from the 15th century onwards. *The rising bourgeoisie had to submit for centuries to come to the rule of a warrior class of feudal origins that milked the bourgeoisie to fight their endless series of wars (*HEA, 144). It was only in the 19th century that capitalism regained its bloom under liberal governments, especially in England.

Weber considered the ability to calculate profits in advance as the most distinctive feature of (rational) capitalism to mark it off from its (irrational) predecessors. Capitalist enterprise would take the form of large bureaucracies that would reward people according to their abilities. However, to restrict rationality to capitalism seems somewhat misplaced, as Schumpeter remarked. The lord of the manor also behaved rationally within the feudal setting. The same applied to courtiers, who lived at the French royal court in Versailles, or aspirant Chinese mandarins, who took great efforts to become proficient in reciting Chinese verse. All these people used the possibilities open to them to improve their social position; a clear means-end relationship (Schumpeter, 1991, 325).

Knight launched a more devastating criticism on Weber's identification of capitalism with rationalism. He explained that all ex ante calculations could only be educated guesses of prospective returns. So, it is not so much rationality, but perception that makes a capitalist entrepreneur successful. Perception can be considered subjective rationality, which can only be borne out by the facts after the investment has been made. Subjective rationality resembles Weber's description of the charismatic leader, who is convinced that he is right and is eager to prove it. We might, therefore, argue that (objective) rationality is not the distinctive factor of capitalism. Investment can only be subjectively rational in competitive capitalist societies. Moreover, certain political and institutional conditions are required before competitive capitalism could occur. Absolutist rule needed to disappear to make room for both religious and economic freedom and experimentation.

Modern authors on western history also hold the view that absolutism and its inherent monopolization of economic life impeded western economic ascent. North and Thomas explain the economic bloom of England and the Low Countries after 1500 by the absence of absolutism in those countries. A delicate balance was struck between central and local powers, in which neither got the upper hand. In England the power of the crown was curbed by parliament, which obtained substantial power under the Tudors (North and Thomas, 147). A federal form of government, which derived from a league of autonomous medieval towns, ran the seven unified Dutch provinces (McNeill, 581). However, these were the exceptions. Absolute monarchs ruled in France, Spain Scandinavia, Poland and Hungary. Local taxes were abandoned and supplanted by national taxes such as the French taille and the Spanish alcabala. Absolute rulers created vast bureaucracies whose functionaries were loyal to the crown. Guild rules and local monopolies were upheld. Trade, free incorporation and labor mobility were impeded. These measures prevented the emergence of a merchant class in countries that were governed along absolutist lines, which greatly hampered their economic progress (North and Thomas, ch. 10).

11 Conclusions

This essay traced the linkages between the theories of Weber, Schumpeter and Knight on the issue of entrepreneurship and economic development. Weber occasioned one of the great debates in modern intellectual history with his publications on the importance of the Calvinist ethic for economic development. He emphasized the rational i.c. anti-magical features of Protestantism and especially Calvinism. Schumpeter contested Weber's view. He put the innovative entrepreneur center stage. Knight differed from both Weber and Schumpeter in his analysis of investment under uncertainty. He cast the perceptive capitalist in the leading role.

Comparing the three authors, we can note that Weber did not pay attention to either financial matters or uncertainty. Savings funded investment and uncertainty was antithetical to his conception of capitalism as a rational system. Weber sketched the course of western economic history as a process of increasing rationalization, which resulted in modern capitalism. Schumpeter contested Weber's view of capitalism as a predominantly rational system, but adopted some of Weber's points in his later work. But rational capitalism, in his view, would make entrepreneurship obsolete and would blur the lines between capitalism and socialism.

Schumpeter attributed a central role to the financial sector, but largely neglected uncertainty in his work. His emphasis on credit as one of the main institutions of capitalism derives from his idea that capitalism allows people from all social classes to introduce innovations. It was Knight, who explained how developed financial markets could absorb uncertainty. Knight's analysis stresses the game-like and intuitive and, therefore, largely non-rational aspects of entrepreneurship. The entrepreneur can perceive qualities in people, which remain hidden to the average observer. This gives the investor/entrepreneur some prophetic features. Only the more than averagely perceptive investor can make a profit, if capital markets are well organized. But capital and IPO markets (Initial Public Offerings) have only recently and in only a few countries achieved the degree of sophistication required for the smooth operation of investment under uncertainty. Many aspects of the 'new economy' such as the vital role of equity markets and of uncertainty can already be found in embryonic form in Knight's work.

References

Acs Z, Audretsch D (1988) Innovation in large and small firms; an empirical analysis. American Economic Review 78: 678–690

Augello MM (1990) Joseph Alois Schumpeter: a reference guide. Springer, Berlin Heidelberg New York

Baumol WJ (1968) Entrepreneurship in economic theory. American Economic Review 58: 64–71

Brouwer M (1991) Schumpeterian puzzles, technological competition and economic evolution. Michigan University Press, Harvester Wheatsheaf

Brouwer M (1996) Economic evolution and transformation of the economic order. In: Helmstaedter E Perlman M (eds) Behavioral norms, technological progress and economic dynamics, pp 355–370. Michigan University Press, Ann Arbor

Brouwer M (2000) Entrepreneurship and uncertainty: innovation and competition among the many. Small Business Economics, an International Journal 15 (2): 149–160

Hébert RF, Link AL (1982) The Entrepreneur. Praeger Publishers, New York

Knight FH (1961) Risk uncertainty and profit. (First edition 1921). Kelley

Knight FH (1939) Ethics and economic reform; the ethics of liberalism. Economica 6: 1–29

Knight F. H. (1982) Freedom and economic reform: essays in economics and social Philosophy (with a foreword by J. Buchanan). (First edition: 1947) Harper & Row Liberty Press, Indianapolis

Loring AR (1991) Opening doors: the life and work of Joseph Schumpeter, vol 1. Transaction Publishers, Europe

McNeill WH (1963) The rise of the West, a history of the human community. University of Chicago Press, Chicago

North DC, Thomas RP (1973) The rise of the western world: a new economic history. Cambridge University Press, Cambridge, MA

OECD (June 2000) Employment outlook: the partial Renaissance of self-employment, pp 155–193

Schama S (1987) The embarrassment of riches: an interpretation of Dutch culture in the golden age. Knopf, New York

Schumpeter JA (1908) Das Wesen und Hauptinhalt der Theoretischen Nationaloekonomie. Duncker & Humblot, München Leipzig

Schumpeter JA (1912) Theorie der Wirtschaftlichen Entwicklung, 1. ed. Duncker & Humblot, Leipzig

Schumpeter JA (1978) The theory of economic development: an inquiry into profits, capital, credit, interest and the business cycle, 2. ed. (translated by Redvers Opie). Oxford University Press, London New York (first published in 1934)

Schumpeter JA (1943) Capitalism, socialism and democracy. Allen & Unwin, New York London (first published in the USA in 1942 by Harper and Brothers, New York)

Schumpeter JA (1962) Eugen von Boehm-Bawerk. In: Ten great economists, pp 143–190. Unwin Unversity Books, London (first published in 1951 Oxford University Press; with a foreword by Elizabeth Boody Schumpeter)

Schumpeter JA (1981) History of economic analysis (with an introduction by Elizabeth Boody Schumpeter). Allen & Unwin, New York London (first published in 1954 by Oxford University Press, New York, and Allen & Unwin, London)

Schumpeter JA (1991) The meaning of rationality in the social sciences. In: Swedberg R (ed) The economics and sociology ofcapitalism. The Lowell Lectures. Princeton University Press, Princeton [First published in: (1984) Zeitschrift für die gesamte Staatswissenschaft 4: 577–593, edited and introduced by Stolper WF, Richter R, from a seminar given at Harvard in 1940)

Weber M (1965) The protestant ethic and the spirit of capitalism (with a foreword by Tawney RH), 7th impression (first published in 1930). Unwin University Books, London

Weber M (1995) General economic history (translated by Knight FH). Transaction Publishers, New Brunswick London (sixth printing, first published in 1927)

Weber M (1947) The theory of social and economic organization (translated by Henderson AM, Parsons T), with a foreword by Talcott Parsons. Oxford University Press, Oxford (reprinted by Macmillan as a Free Press paperback)

Connecting principles, new combinations, and routines: reflections inspired by Schumpeter and Smith

B. J. Loasby

Department of Economics, University of Stirling, Stirling FK9 4LA, Scotland
(e-mail: b.j.loasby@stir.ac.uk)

Abstract. Schumpeter's distinction between allocative theory and development 'from within' provides Walras and Schumpeter with non-competing domains: Adam Smith is a major contributor to the latter. The domains have distinctive methods: proofs of rationality and equilibrium, which assume a fully-connected system, contrast with evolutionary processes of knowledge creation, in which connections are problematic, selective and changing – exemplified by Smith's *History of Astronomy*. Human cognition relies on routines and (less restrictive) rules which, as Schumpeter emphasised, provide a necessary baseline for innovation; they are also important elements in innovation processes, linking management and enterprise – and allocation and development.

Key words: Innovation – Connections – Cognition – Routines and Rules

JEL Classification: B41, M13, O12, O31

1 Joseph Schumpeter and Adam Smith

In Joseph Schumpeter's *History of Economic Analysis* (1954) Adam Smith is not allotted a leading role. This is entirely appropriate, because the focus of economic analysis has been price theory, gradually refined into a formal and intendedly complete model of the allocation of economic resources; and at the time that Schumpeter was producing his treatise the culmination of this development was to be found in the work of Walras, even though it was recognised that Walras's analytical system did not quite realise his vision. In this perspective Smith is indeed not a dominant figure. Despite the claims of some economists who have attempted to rescue his reputation as a theorist, it is seriously misleading to interpret Smith as a 'precursor', even unintentionally, of general equilibrium in a Walrasian sense of rationally co-ordinated activities, and even worse to interpret him as a precursor of post-Walrasian theory or of other models which rely on closed-system rationality to deduce equilibria. Nor indeed did Smith consider theory, formal or informal, to be sufficient to explain the working of an economic system. As we shall shortly explain, he had very different ideas about both the content and the method of economics.

However, we should notice that Schumpeter also considered economic analysis, as defined above, to be only part of economics; and from his other writing we may infer that he did not believe it to be the most important part. In his *Theory of Economic Development* he makes a sharp distinction between an analytical focus on the co-ordination of a set of economic activities that is assumed to be well-defined – which is still the central tradition today – and the explanation of economic growth 'from within' (Schumpeter, 1934, p. 63). Economic analysis provides models of the 'circular flow of economic life as conditioned by given circumstances' (Schumpeter, 1934, p. 3), and although since Schumpeter's time these circumstances have been extended to include the specification of a very wide range of locations, future dates and possible states of the world, they are no less 'given' (though it is not at all clear by whom) than in the models with which he was familiar. However, '[d]evelopment in our sense is a distinct phenomenon, entirely foreign to what may be observed in the circular flow or in the tendency towards equilibrium' (Schumpeter, 1934, p. 64). Co-ordination, Schumpeter asserted, is not the result of rational choice or efficient contracts, but is achieved by allowing 'things [to] have time to hammer logic into men' (Schumpeter, 1934, p. 80); thus although conduct eventually appears to be 'prompt and rational', it is simply the manifestation of evolved routines. Walrasian theory may therefore be used for the contingent prediction of behavioural regularities, but it does not explain how that behaviour is generated – and certainly cannot explain how it is changed. A comprehensive attempt at such an explanation, embedded in the ideas of the Scottish enlightenment, is to be found in the corpus of Smith's work; Skinner (1996) provides a guide.

That evolution, not rationality, provides the correct explanation for economic order was the theme of a famous argument by Alchian (1950), which was notably criticised by Penrose (1952) for excluding human purpose. However, Schumpeter did not bother with this issue, preferring to use the dependence of co-ordinated behaviour on time rather than rational choice to explain why economic development is necessarily cyclical. He invokes the limitations of human cognition to justify his claim that 'stability is indispensable for the economic conduct of individuals' (Schumpeter, 1934, p. 40) – a theme to which we shall return; but the major innovations which he identified as the prime movers of economic development undermine established routines and thus destroy the established basis of co-ordination. The creation of a new basis is no easy task, and no Walrasian mechanisms are at hand – indeed, as we shall see, no such mechanisms are possible within a Walrasian system; therefore a period of extensive innovation is necessarily followed by recession. In Schumpeter's own vision of the economic system, the theory of business cycles and the theory of growth are inseparable; it is illegitimate to add cycles to a trend which is independently modelled, for without the cycle there can be no growth to produce a trend. This proposition was developed at length in his study of business cycles (Schumpeter, 1939). That the analysis of business cycles cannot properly be treated as a discrete topic has recently been reasserted, but proposals to assimilate this analysis to models of macroeconomic equilibria or of equilibrium growth are clean contrary to Schumpeter's position.

Schumpeter's admiration for Walras, who is very firmly located on one side of his dichotomy between 'economic analysis' and the theory of economic development 'from within', therefore leaves ample room for Schumpeter himself on the other side, which many people, especially members of the Schumpeter Society, might think more important. However, in a history of the theory of economic growth, Smith is not so easily left in a subordinate place. Perhaps Schumpeter himself realised this in recognising the quality of Smith's *Essays on Philosophical Subjects* (1980), and especially his *History of Astronomy* (Schumpeter, 1954, p. 182). Smith's account of 'the principles which lead and direct philosophical enquiries' is essentially a theory of development from within, motivated by the human desire for some means of understanding phenomena which cannot presently be explained; Smith exemplifies his theory by analysing a succession of theoretical visions of cosmological systems, not dissimilar from the visions that Schumpeter invokes both in his *History of Economic Analysis* and in his emphasis on major innovations. In contrast to the deductive theory of co-ordination based on a closed data set, for both Smith and Schumpeter the growth of ideas is not a logical procedure; it is a product of the imagination. It therefore belongs in a theoretical system in which radical or structural uncertainty is not suppressed by fictional probability distributions, but is taken seriously.

In the course of his exposition, Smith also observes that growing specialisation within the scientific community leads to greater attention to the detail of theoretical systems, which increases the likelihood that inadequacies will be perceived within those systems; and these perceptions, working on the motivation to avoid the psychological discomfort of the inexplicable rather than the search for gain (Smith, 1980, p. 61), provide the incentive for the creation of new patterns which appear to be locally appropriate. The result is an increasing and self-sustaining diversity of increasingly specialised systems of thought. Scientific knowledge grows by division, and this growth is accelerated by the specialised studies that result from a finer division. In the *Wealth of Nations* Smith transformed the division of labour from an extension of his psychological theory of the growth of knowledge into the foundational principle of his own theory of economic development, with the emphasis now on productive knowledge (Smith, 1976b). Smith differs from Schumpeter in emphasising the importance of incremental change, as Marshall was to do (drawing explicitly on Darwin as well as Smith); but we should not forget Smith's (1976b, p. 22) reference to the innovative role of 'philosophers or men of speculation' who 'combine together the powers of the most distant and dissimilar objects' – in other words, those who, in Schumpeter's (1934, p. 66) phrase, make 'new combinations'.

2 Two conceptions of economics

It would be a mistake to treat Smith's interest in economic development, and its relationship with the growth of knowledge, as a side-issue in the history, or the future, of economic thought. In another paper (Loasby 2002), I have argued for the presence of two substantively different traditions in economics. What might be conveniently labelled the cartesian tradition emphasises the search for proven knowledge, where proof is a logical operation performed on knowledge which is beyond doubt; it is easy to see how this tradition encourages the conception of choice as optimisation, where the optimum is deduced by applying a criterion function to a possibility set which is precisely – and correctly – defined, and to the representation of any stable situation or stable process as an equilibrium, the existence of which is deduced from the initial specification of the system. (Proofs of stability and uniqueness are also usually thought to be desirable, although multiple equilibria may be welcome to policy-minded economists if they suggest policy options). The progress of knowledge in economics is then deemed to consist in improving the internal coherence of models, increasing the range of circumstances that are included in the specification, and finding methods, compatible with this methodological imperative, of modelling phenomena or relationships that have hitherto proved recalcitrant. These directions may conflict – as an obvious example game-theoretic models ignore the requirements for proofs of a general equilibrium; but the formal properties of closed-system logic provide a common link across these varieties of formal theorising.

David Hume ([1739–1740] 1978) rejected this emphasis on formal proof as a route to empirical knowledge; indeed he rejected the possibility of proving the truth of any general empirical proposition, either by deduction or induction. This we might call Hume's Impossibility Theorem. Certainty was not achievable; and axiomatic reasoning was insufficient as a method of scientific enquiry. Instead of a fruitless search for a route to infallible truth, Hume therefore proposed an inquiry into the processes by which people come to develop particular empirical propositions, and to believe that these propositions are true. It was precisely such an enquiry that Smith ([1795] 1980) undertook in his *History of Astronomy*, and the terms of his enquiry led him to see Newton's theory – astonishingly for his time – not as demonstrable truth but as the product of Newton's imagination, and consequently liable eventually to be superseded, as earlier cosmological theories had been. He drew particular attention to the appeal of Newton's theoretical system to the imagination of others, and not least to its aesthetic quality, which endowed it with great rhetorical power, a theme to which we will return in a later section. This exploration of Hume's agenda subsequently led Smith to emphasise the psychological influences on human decision making in his *Theory of Moral Sentiments* (Smith, [1759] 1976a), with due recognition of human fallibility both there and in the *Wealth of Nations*.

Alfred Marshall studied mathematics at Cambridge when Euclidean geometry was widely believed to have demonstrated the power of axiomatic reasoning to demonstrate empirical truth, only to experience the clearest refutation of this belief by the creation of non-Euclidean geometry (Butler, 1991). This refutation was later reinforced by the perception that Cournot's axiomatic reasoning, which led to the conclusion that falling costs necessarily implied monopoly, was falsified by Marshall's own statistical and personal observations of the contemporary British economy. (For a fuller account of Marshall's intellectual development, see Groenwegen, 1995.) Though still impelled towards the development of theory, he became cautious about its application, and extremely cautious about its refinement, because of the likelihood that errors in and exclusions from the specification of any model would be carried through to its conclusions, thus creating knowledge which might be dangerously false. (How right he was is abundantly illustrated in the record of economists' policy recommendations.) He also gave the development of knowledge a central role in economic systems, although this role was underdeveloped in his own work, and even proposed treating organisation as a factor of production, because of its importance in the generation and use of knowledge. Allyn Young (1928) presented a forceful summation of the Smith-Marshall theme of the development of knowledge in economic systems, and Penrose (1959) virtually reinvented it (Loasby, 1999). Evolutionary economics belongs in this tradition, and the accounts of economic development, based on the growth of knowledge, that are offered by Smith and Schumpeter rely on similar elements, similarly structured.

3 Connections

To understand these common conceptual bases, and the contrast with formal modelling, I shall draw on Jason Potts' (2000) analysis. As one might expect of an analytical method that was borrowed from nineteenth-century physics, formal economic models typically assume an integral space, in which every element is directly connected to every other element. Just as in Newton's universe every object with mass influences every other such object, so in the full general equilibrium system all preferences are directly connected to all goods and all production possibilities, even when these preferences, goods and production possibilities are distributed across space, time, and conceptual states of the world; and in conventional game-theoretic models all players know the rules of the game, are able to compute the full consequences of any action, and know that all other players have the same knowledge and capabilities. (In evolutionary game theory, agents know very little, but the analyst knows everything.) All actions are therefore simultaneously determined, even when they take place in different locations and at different times, and even though some of them never occur at all in many states of the world; the genius of the Arrow-Debreu model is to resolve the problems of distance, time, and uncertainty by abolishing them. It is no surprise that completeness implies efficiency – in relation to the problem specification, and that incompleteness, in the form, for example, of inadequately-specified contracts, is a threat to efficiency. This efficiency, though incorporating all relevant information about future dates, is assessed at a single point of time.

Now there are opportunity costs in economic theorising, as in all choices. The assumption of integral space which makes general equilibrium analysis possible excludes the prospect of analysing a path to equilibrium. As Richardson (1960, p. 57) warned us forty years ago, '[a] general profit opportunity, which is both known to everyone, and equally capable of being exploited by everyone, is, in an important sense, a profit opportunity for no one in particular'. A careful reading of Richardson's argument will show that he was sensitive to the deep foundations of this difficulty; as he was later to comment, he 'knew enough economics and enough physics to know the difference' (Richardson, 1998, p. xii). In a fully-connected world, the possible ramifications of any decision to change a local configuration cannot be calculated; because any or every element is liable to change in response to any initiative, nothing can be relied on. Rational choice by any individual is therefore possible in such a world only if all other agents are already committed to their choices. Thus the standard practice of explaining rational choice by an economic agent on the assumption that everyone else is already in equilibrium is no innocuous simplification; it is the only assumption on which such analysis is valid. As Alan Coddington (1975, p. 154) pointed out, there can be at most one omniscient person; presumably that person should be appointed as central planner.

For a fully-connected system, the only alternative to omniscient central planning is pre-reconciled choice, which is isomorphic to it. That, of course, is why simultaneity is essential to proofs of equilibria, and this is what the fixed-point theorem provides. Both planning and the pre-reconciliation of choice must occur outside the system which is being analysed. If markets are to be invoked as a possible means of reaching a fixed point, their operations must be shunted off into an unanalysable realm in which neither production nor the physical exchange of commodities is permitted, because any non-equilibrium action changes the initial conditions on which the calculation of that equilibrium depends, as Walras came to realise in replacing conjectural production by the exchange of tickets (Walker, 1997). Simultaneity of reasoning before play starts is also necessary – though it is not always sufficient – for standard game-theoretic analysis, typically in the form of Nash equilibria, which as a complete and coherent set of best responses to best actions is by definition fully-connected. Sequences must be precisely, if contingently, co-ordinated in advance, and based on knowledge which is already complete. It follows, as a trivial consequence, that this analytical system excludes any notion of development 'from within'. Schumpeter was quite right.

A conceptual switch from integral to non-integral space is a switch from fully-connected to partially-connected systems. In both conceptions, connections matter, but in the former connections are ubiquitous and therefore require no explicit attention; in the latter, however, they are problematic and therefore an essential component of analysis. The human mind is a partly-connected system; though the connections may be very extensive, they are incomplete, and can be only a very small proportion of the connections that exist in the environment – especially when that environment includes many other minds. Formal organisations and social groups may also be quite intensively connected, but their connections will also be meagre in relation to their environment. Many externalities will necessarily be ignored, and also many internalities, or detailed structures which may in some circumstances be extremely important. The gap between the set of mental or organisational connections and the phenomena with which individuals or organisations are trying to cope is filled with uncertainty and the possibility of error – but also with the possibility of discovery or the creation of novelty.

All interesting problems lie in this gap. A wide range of heterodox economic inquiries postulate, though not always explicitly, incomplete connections, which provide their various foci of interest. For example, Kirzner's (1973) entrepreneur is alert to profit opportunities that are created by the failure of others to make connections between markets, Keynesian unemployment is made possible by the absence of any credible connection between future demand and present investment, and Simon explains how problems are decomposed to keep connections manageable and decision premises are established to simplify and co-ordinate decision making. (Even property-right explanations of firms rely on a very specific exception, in the form of information asymmetry, to the completeness of connections, which is then remedied by an adjustment which is presumed to leave untouched the standard analysis of production and exchange.)

In the broadest sense all these theorists investigate the relationship between structure, conduct, and performance (which indeed is a research strategy followed in different ways in many disciplines), but with a much deeper concern for structure than was common in old-style industrial organisation theory, in which the notion of 'market structure' was rarely extended to include any reference to the institutions which guided the operations of each particular market, and did not, as Coase ([1972] 1988, p. 57) commented, provide any explanation of the ways in which economic activities were divided up between firms. What heterodox theorists also have in common is that outcomes all turn on the kind, quality and distribution of knowledge, which is itself a structure of connections; indeed, as Smith saw, knowledge is created by the invention of connecting principles, and we may nowadays think of these connections as having a physical representation within the brain. The growth of knowledge, and the possibility of innovation, depend on the incompleteness of present connections. To put the same point in the language of George Shackle, it is the pervasiveness of uncertainty that provides scope for imagination and makes possible the joy of discovery.

4 Cognition and routines

The basic facts of human cognition are that our brains have the capacity to establish an extremely large number of possible networks of connections, but can actually establish and maintain only a small fraction of this potential. As a consequence, the total number of connections, and therefore the total amount of knowledge and skills, within a community can be greatly increased if the members of that community act in ways which lead them to make different connections. That is why Smith was right to make the division of labour an accelerator of scientific knowledge and the foundation principle of economic growth, and why Marshall was right to emphasise the role of organisation, of various kinds, in the development and use of knowledge. Differentiation between individual entrepreneurs or between firms (sometimes labelled as Schumpeter I and Schumpeter II) then provides the basis for distinctive innovations and thus the foundation of the positive case for competition as a generator of novelty – and, as Schumpeter (1943) declared, against perfect competition. An important corollary, which is not considered in this paper, is that they also provide the basis for distinctive pathologies, and thus the foundation of the defensive case for competition as a protection against systematic error.

Why, then, in Schumpeter's account, are routines essential to entrepreneurial processes? The routines that matter in his theory are those of the non-entrepreneurs, because these sustain the circular flow of economic activity and thus provide the assurance of predictable prices and quantities which allow the entrepreneur to calculate that the resources needed for innovation will be available at a price that will ensure a profit, given the prices and characteristics of the products which are to be displaced (Schumpeter, 1934, p. 141). If the innovation is in the institutions of marketing or organisational structure and practice, then the established routines which the entrepreneur intends to challenge permit a reliable evaluation of the advantages of the new system. Why it is important that prices should be sufficient statistics is that, because of cognitive limitations (or missing connections) the entrepreneur cannot hope to understand the workings of the economic system that generates these prices, and relies on their stability to give him confidence in his calculations. His own cognitive powers will be fully employed in creating the productive, marketing, and administrative systems (Chandler, 1990) that are needed to realise his vision.

However, the effect of a major innovation is to disrupt these routines, by introducing novel connections to existing businesses, through competition from directions or of a kind that had never been contemplated within those businesses; understanding what has happened, let alone finding a new pattern of new routines appropriate to the changed circumstances, is a formidable challenge to most people (Schumpeter, 1934, pp. 81–86). Thus a period of recession is inevitable while non-entrepreneurs gradually work out how to behave in a world they had never imagined. However, it is not only those who depend on routines for their daily activities who are affected. The flow of entrepreneurial ideas is continuous (though the volume may vary); but the introduction of these ideas depends on the calculation of their advantage, which requires a stable environment. Not only do entrepreneurs not bear financial risks (Schumpeter, 1934, p. 137), which are left to the capitalists (though an entrepreneur may choose to be a capitalist as well, or a capitalist might conceive an entrepreneurial vision, and invest his own funds in it); they do not take risks in launching their projects. The practice of entrepreneurship is therefore suspended until recession has run its course, and a new set of routines has been established (Schumpeter, 1934, pp. 235–236). Thus Schumpeterian innovation might be seen as a process which replaces a very large set of interconnected routines with another set; this is a complex and time-consuming process, unlikely to be facilitated, in Schumpeter's opinion, by government intervention. (How we might decide whether the new set is better, and by a sufficient margin to compensate for the disruption, is a question I will make no attempt to answer in this paper.) We may here see a justification for taking routines as the unit of analysis; but we should remember that in Schumpeter's theory the essential importance of this stabilisation of routines is that it makes possible the next major innovation.

Now there is a valid underlying logic to this insistence on the prevalence of routines in the decision environment, which derives from the confrontation between complexity and human cognitive limitations. Reasoned assessment – to say nothing of rational choice – is impossible unless one can take a great deal for granted;

and the larger the number of changes that are incorporated in the project that is being assessed the smaller the number of changes that can be seriously considered outside it. Businessmen's appeals for a stable environment may be overdone, and sometimes exceed what any government or any social system can provide, short of the suppression of all change; but without substantial assumptions of environmental stability, purposeful action is impossible.

Nevertheless, the contrast between entrepreneurial creativity and external routines, though rhetorically effective, and justifiable in those terms, is overdone; for there are typically routines within the entrepreneurial project also. This is concealed in Schumpeter's (1934) original presentation because of his desire to emphasise – and justifiably emphasise – that major innovations cannot possibly be produced by any identifiable procedure from what already exists (this is Knight's (1921) definition of entrepreneurship, which relies on incompletable connections) and, in particular, that the economic analysis of co-ordination, in which all connections are already in place, has no direct relevance to the explanation of innovation, which depends on the creation of new connections. Schumpeter's original identification of the typical innovator as an outsider epitomises this thesis with striking rhetorical power. However, if we ask what is connected, the answer very often is a collection of existing routines. This can be inferred both from Schumpeter's list of types of innovation, and from his discussion of what does not constitute innovation, notably invention.

As many readers will be aware, the theme of innovative discontinuity was also the core of Thomas Kuhn's (1962, 1970) theory of scientific revolution, and here too discontinuity was placed in a relationship of complementary antagonism to the routines of normal science. There are important differences, notably in Kuhn's identification of an accumulation of failures of routine, or 'normal science', as the incitement to search; but the process of paradigm creation is no less mysterious in Kuhn's theory than in Schumpeter's. Now Kuhn's explanation of the development of science through a succession of paradigms, linked by incremental changes which eventually produce an unacceptable accumulation of anomalies, is at first sight very similar to Smith's; but Smith, unlike Kuhn, sets out to explain the transition, which he does by examining the ways in which the successful creators of new paradigms frame their search and the elements which are carried over from one system to another, typically being modified, and their significance sometimes being transformed, in the process – for example, by the redefinition of a 'planet' and the abrupt dissociation of 'earth' from 'stationarity'. Instead of 'paradigm', Smith refers to 'connecting principles', which he insists are principles of cognition rather than 'the real chains which Nature makes use of to bind together her several operations' (Smith, [1795] 1980, p. 105), and here is the clue, for principles are less constricting than routines; they supply categories for classification, premises for argument, rules of procedure and criteria for choice, while leaving scope for imaginative conjectures. Neither the content nor the success of these procedures can be predicted – knowledge cannot be attained before its time – but they may often be readily understood in retrospect.

5 Routines and rules

The difficulty of agreeing on a definition of routines is clearly displayed in the discussion reported by Cohen et al. (1996), not least in describing their topic as '[r]outines and other recurring action patterns of organizations'. Among the questions discussed is whether the term 'routine' should be applied to a regularly recurring sequence of actions, typically in response to a regularly recurring cue, or to single repeated actions, whether or not such actions are regularly combined with a particular set of other actions in a standard way. My own view is that both conceptions are needed for different purposes. The second is particularly useful in thinking about the innovation process, because the rearrangement of elementary routines may be guided by rules. I have noted a tendency for authors to slide between 'rules' and 'routines', applying both, for example, to institutions; but, although Marshall's principle of continuity applies here as to most other attempts to define mutually exclusive categories, it is nevertheless useful to think of routines as fully-specified programmes (which may allow for contingent actions as long as the contingencies and the actions are precisely defined) and rules as devices for limiting the range of actions, either by exclusion ('we are not in that business') or by endorsement ('expectations should always be assumed to be rational'). In any situation requiring a decision, they indicate the sort of things that should not be considered or the sort of things that should. Thus they simplify the ex-ante selection of a problem-definition and the ex-post selection from the options that emerge from this problem definition, or in other words they supply the procedural framework for procedural rationality. In any complex decision process, many of the detailed steps may be committed to routines, tacit or explicitly formal; selection among routines is guided by rules.

In his *Theory of Moral Sentiments*, Smith ([1759] 1976a) sets out to explain how a particular category of rules influences human behaviour, not only when interacting with others but in private; and the prevalent rule among the more orthodox new institutional economists that institutions should be explained solely as devices for co-ordinating interactions seems to me an unnecessary obstacle to understanding. As Keynes (1937) pointed out, when we don't know what to do we often look to others for some sort of rule which might work for us, and in general borrowing rules from other people is a highly efficient (though sometimes disastrous) means of economising on the scarce resource of cognition by exploiting the division of labour. In a recent article, Ekkehard Schlicht (2000) has reminded us of the importance of aesthetic criteria in choosing both our actions and our theories, and Smith was certainly aware of this throughout his work. To cite only his *History of Astronomy*, we may note Copernicus's wish to replace the disorder into which the Ptolemaic system had degenerated with 'a new system, that these, the noblest works of nature, might no longer appear devoid of that harmony and proportion which discover themselves in her meanest productions' (Smith, [1795] 1980, p. 71), Kepler's choice of the ellipse because it was the most perfect of shapes after the circle, which clearly would not fit the evidence (Smith, [1795] 1980, pp. 86–87), and the aesthetic appeal of Newton's unification of celestial and terrestrial phenomena. In his *Lectures on Rhetoric*, Smith recommended the 'Newtonian method' of deriving all explanations from a central principle as the best way to give an account of a system, because it 'is vastly more ingenious and engaging', even while observing that Descartes' use of this method had gained widespread acceptance for a theoretical system that 'does not perhaps contain a word of truth' (Smith, 1983, p. 146).

To make sense of innovation processes, it will often be helpful to switch the balance of attention from the contrast between routines and origination to the intermediate range of rules, premises, and criteria, while simultaneously taking a very generous view of the factors which may appear in these rules, premises, and criteria. One consequence of doing so is to reduce the apparent contrast between the inexplicable visions of the outsider in Schumpeter's earlier work and his later account of the large organisation which apparently delivers innovation to order by following what, at a casual glance, look very like routines (Schumpeter, 1943); for though research departments cannot meet innovative targets simply by following routines, nevertheless research sequences and research assessments may make extensive use of routines, guided by rules, premises and criteria which develop as members of each research organisation find their own ways of construing their experiences. Popper's logic of scientific discovery is that new ideas, though conjectural, have deducable implications, and that these implications may be testable; indeed, that ideas should be testable (though not necessarily by experiment) is a fundamental principle of scientific research.

Much of the work of a research department dedicated to commercial success may be understood as following a set of procedures, often incorporating mini-routines, that are inspired by this logic, which has very wide applicability. However the task of delivering 'what is required' is not as straightforward as Schumpeter (1943, p. 132) seems to suggest. Many attempts to achieve prescribed targets fail completely; in other cases these procedures may generate novel problems which may inspire fresh ideas, and so the end-product of innovative processes does not often correspond at all precisely to the original intentions. Indeed, although research departments may make great efforts to anticipate the conditions of use, they are unlikely to be completely successful, because of the different ways of thinking which result from the division of labour, and so it is likely to be a mistake, for both practitioners and analysts, to exclude diffusion from the process of innovation.

In a recent paper, Paul Nightingale (2000) investigates the contemporary process of pharmacological research. Although many research targets are very obvious in terms of the desired effects, it is not possible to derive from these effects a specification of the product that will achieve that target; thus the standard rational choice model, which relies on the logic of consequences, cannot be used. It is necessary instead to rely on what James March has called the logic of appropriateness, the basic principle of which was stated by Frank Knight (1921, p. 206): 'It is clear that, in order to live intelligently in our world . . we must use the principle that things similar in some respects will behave similarly in certain other respects even when they are very different in still other respects'. This principle applies even though we do not know which of the similarities are decisive for behaviour or why the differences don't matter. There is an obvious corollary: we also assume that things different in some respects will behave differently in certain other respects even though they are similar in still other respects, without knowing which differences are decisive and why the similarities are irrelevant. Of course, these rules sometimes break down, and certain phenomena, or certain actions, are transferred from one category to the other as a result of experience. Nightingale investigates the effect on pharmaceutical research routines of a major change of rules for the selection of candidate molecules. The rules are not very prescriptive – except, crucially, by comparison with the vast numbers of molecules which might be investigated, and a long sequence of testing is necessary in order to reject almost all of them. Thus the rule is not a routine; but the testing has become highly routinised, and the sequence of tests is guided by rules which are derived from scientific understanding.

6 Enterprise and management

Though Schumpeter suggested that innovation could be managed, he maintained his distinction between innovation and the ongoing management of a business. Smith, however, included the innovative role of 'philosophers' among the consequences of the division of labour, alongside the improvements produced by workmen and specialist machine makers, and Marshall (1920) identified the principal cause of progress as the 'tendency to variation' that resulted from experiments which were based on the particular experiences and temperaments of ordinary business people. That this progress included changes in organisation is made explicit in his definition of increasing return (Marshall, 1920, p. 318). Marshall wished to maintain a close connection between his account of the daily operation of an economy which generated prices and quantities and his explanation of economic progress, and consequently made extensive though hesitant use of equilibrium theorising. However, in developing Smith's and Marshall's ideas Young (1928) contrasted increasing return as a process of continuous change with the standard concept of equilibrium; and Penrose (1959, p. 10) also sharply distinguished her theory of growth from the equilibrium models of price theory, and linked it with Schumpeter's vision, though observing that '[t]he Schumpeterian "entrepreneur", though more colourful and identifiable, is too dramatic a person for our purposes' (Penrose, 1959, p. 36n).

Penrosian firms develop capabilities, including both productive and managerial routines and rules for applying them, in the course of their business; but, unlike the inputs of standard production functions, these capabilities are not automatically connected to a prescribed set of productive services. Thus new connecting principles may be created by entrepreneurial conjecture to generate new productive opportunities; and like larger-scale Schumpeterian visions, these productive opportunities are images in the entrepreneur's mind (Penrose, 1959, p. 5). They rest on developed internal routines (and, no doubt, on external routines, though Penrose does not investigate the firm's connection with its environment); but their exploitation necessarily requires much non-routine activity, which crowds out other possibilities (Penrose takes the boundedness of cognition for granted), and so, as in Schumpeter's theory, but at the level of the firm instead of the economy, the resumption of innovation requires the establishment of a new set of routines. Stability is a pre-condition of innovation.

7 An open question

We began with Schumpeter's dichotomy between theories of co-ordination and theories of innovation, and ended with Penrose's endorsement of this dichotomy. But neither Smith nor Marshall thought of economic theory in this way; for both of them, what had to be co-ordinated was a process of continuous change. Theoretical development since the time of Smith left Marshall much more conscious of the theoretical difficulties in this concept; and Marshall's own attempts at compromise were not so much rejected as incomprehensible according to the rules of formal theorising. But whereas in Schumpeter's scheme periods of innovation and of co-ordination may be treated as temporally distinct, in Penrose's theory it is precisely the activity of resource allocation that generates change. If resource allocation, the focus of formal theory, is part of the process of economic development, is there any place for concepts of equilibrium and optimisation? Is there any sense in trying to appraise efficiency at a point of time? We know what Schumpeter (1943, p. 83) thought.

> A system – any system, economic or other – that at *every* given point of time fully utilizes its possibilities to the best advantage may yet in the long run be inferior to a system that does so at *no* given point of time, because the latter's failure to do so may be a condition for the level or speed of long-run performance.

How we should appraise long-run performance against a background of knowledge which can never be complete but which may be augmented by a trial and error process of seeking together (which is the etymological meaning of 'competition') is left as an exercise for the reader.

References

Alchian AA (1950) Uncertainty, evolution and economic theory. Journal of Political Economy, 58: 211–221

Butler RW (1991) The historical context of the early Marshallian work. Quaderni di Storia dell'Economia Politica, 9, 2-3: 269–288

Chandler AD (1990) Scale and scope. Belknap Press, Cambridge, MA

Coase, RH ([1972] 1988) Industrial organization: a proposal for research. In Fuchs VR (ed) Policy issues and research opportunities in industrial organization. National Bureau of Economic Research, New York: 59-73 Reprinted in Coase RH The firm, the market, and the law. University of Chicago Press, Chicago: 57–74. (Page reference is to this reprint)

Coddington A (1975) Creaking semaphore and beyond. British Journal for the Philosophy of Science, 26: 151–163

Cohen MD, Burkhart R, Dosi G, Egidi M, Marengo L, Warglien M, Winter S (1996) Routines and other recurring action patterns of organisations: current research issues. Industrial and Corporate Change, 5: 653–698

Groenewegen PD (1995) A soaring eagle: Alfred Marshall 1842–1924. Edward Elgar, Aldershot, UK and Brookfield, US

Hume D ([1739-40] 1978) A treatise of human nature, ed L.A. Selby-Bigge, (2nd edn.) ed P. H. Nidditch. Clarendon Press, Oxford

Keynes JM (1937) The general theory of employment. Quarterly Journal of Economics, 51: 209–223

Kirzner IM (1973) Competition and entrepreneurship. University of Chicago Press, Chicago

Knight FH (1921) Risk, uncertainty and profit. Houghton Mifflin, Boston. Reprinted University of Chicago Press, Chicago 1971
Kuhn TS (1962, 1970) The structure of scientific revolutions (1st and 2nd edn.) University of Chicago Press, Chicago
Loasby BJ (1999) Edith Penrose's place in the filiation of economic ideas. Économies et Sociétés-Cahiers de l'ISMÉA, XXXIII, 8, Série Oeconomica: 103–221
Loasby BJ (2002) Content and method: an epistemic perspective on some historical episodes. European Journal of the History of Economic Thought, 9 (1)
Marshall A (1920) Principles of economics, 8th (edn.) Macmillan, London
Nightingale P (2000) Economies of scale in experimentation: knowledge and technology in pharmaceutical R & D. Industrial and Corporate Change, 9: 315–359
Penrose ET (1952) Biological analogies in the theory of the firm. American Economic Review, 42: 804-819
Penrose ET (1959) The theory of the growth of the firm. Basil Blackwell, Oxford. (3rd edn.) Oxford University Press, Oxford 1995
Potts J (2000) The new evolutionary microeconomics: complexity, competence, and adaptive behaviour. Edward Elgar, Cheltenham and Northampton, MA
Richardson GB (1960) Information and investment. Oxford University Press, Oxford. (2nd edn.) 1990
Richardson GB (1998) The economics of imperfect knowledge. Edward Elgar, Cheltenham and Northampton, MA
Schlicht E (2000) Aestheticism in the theory of custom. Journal des Economistes et des Etudes Humaines 10 (1): 33–51
Schumpeter JA (1934) The theory of economic development. Harvard University Press, Cambridge, MA
Schumpeter JA (1939) Business cycles: a theoretical, historical and statistical analysis of the capitalist process. McGraw-Hill, New York and London
Schumpeter JA (1943) Capitalism, socialism and democracy. Allen and Unwin, London
Schumpeter JA (1954) History of economic analysis. Allen and Unwin, London
Skinner AS (1996) A system of social science: papers relating to Adam Smith. Clarendon Press, Oxford
Smith A ([1759] 1976a) The theory of moral sentiments, ed. Raphael DD and Macfie AL. Oxford University Press, Oxford
Smith A ([1776] 1976b) An inquiry into the nature and causes of the wealth of nations, ed. Campbell RH, Skinner AS and Todd WB.: Oxford University Press, Oxford
Smith A ([1795] 1980) Essays on philosophical subjects, ed. Wightman WPD. Oxford University Press, Oxford
Smith A (1983) Lectures on rhetoric and belles lettres, ed. Bryce JC. Oxford University Press, Oxford
Walker D (1997) Walras's market models. Cambridge University Press, Cambridge
Young AA (1928) Increasing returns and economic progress. Economic Journal, 38: 527–542

Innovating routines in the business firm: what matters, what's staying the same, and what's changing?*

Keith Pavitt

SPRU: Science and Technology Policy Research, Mantell Building, University of Sussex, Brighton, BN1 9RF, UK (e-mail: K.Pavitt@sussex.ac.uk; K.Pavitt@britishlibrary.net/)

Abstract. One challenge in evolutionary economics is to give greater operational content to the notion of "innovating routines" inside the firm. Historical and contemporary evidence suggests that such routines always have to deal with increasing specialisation in knowledge production, increasing depth in knowledge sources and complexity in physical artefacts, and with the continuous matching of specific corporate competencies and organisational practices to the nature of the market opportunities offered by specific technologies. As a consequence, some innovating routines have always been important, such as those dealing with co-ordination and integration within the firm, and with reducing uncertainty through learning. Others are becoming more so, such as those co-ordinating technological resources external to the firm, coping with systems and simulations, and adapting organisational practices to the requirements of radically changing technological opportunities.

Key words: Management – Innovation – Routines

JEL Classification: D20, L20, M13, O31, O32

1 Introduction

The purpose of this paper is to give greater empirical content to the notion, first elaborated by Nelson and Winter in 1982, of "innovating routines". They defined routines in general as the regular and predictable behavioural patterns within firms

*This paper is based on one of the keynote speeches at the meeting of the Schumpeter Society in Manchester on July 1, 2000.

who are coping with a world of complexity and continuous change that precludes decisions and behaviour that maximise anything of importance. Whilst acknowledging the considerable uncertainties surrounding innovation, Nelson and Winter agreed with Schumpeter that " ....organizations have well-defined routines for the support and direction of their innovative efforts." (p. 134). Since then, considerable brainpower has been mobilised to dissect the notion of routines, and to compare it with other useful abstractions like "skills", "operating procedures", "capabilities", "competencies", and "distinctive corporate advantage". But we still have only a very hazy idea of what innovating routines are in practice. How would practising managers respond when asked "what are your innovating routines?" How would observant scholars recognise an innovating routine, when (and if) they visit a business firm? These important questions remain largely unanswered.

At the same time, business historians, management specialists, sociologists, economists and other scholars have accumulated evidence on what happens inside the innovating firm, and some have developed explanatory models for parts of the innovation process, and for distinguishing success from failure.[1] This rich but scattered body of knowledge enables us to define three key features of innovation processes that firm-specific innovating routines must cope with:[2]

- *Specialisation* in the production of scientific and technological knowledge
- *Complexity* in the linkages between such scientific and technological knowledge, on the one hand, and operating artefacts, on the other
- *Co-evolution* of specific technologies with specific organisational practices that transform them into useful artefacts

For each of the three features defined above, I shall identify associated tasks that innovating routines must carry out. I shall also show that some of these tasks are invariant and others are evolving. Innovating routines are made up of technological competencies and related organisational practices. They are embodied in organisations rather than individuals. I shall concentrate here on the functions (or desired results) of the tasks that innovating routines carry out, rather than on the detail of their implementation.

2 Increasing specialisation in knowledge production

2.1 What's the same: internal co-ordination, increasingly multitechnology products and diversification

As Adam Smith rightly predicted, specialisation in the production of knowledge has turned out to be just as efficient as in the production of goods and services

[1] One of the earliest was the distinction, made by Burns and Stalker (1961), between "mechanistic" and "organic" forms of organisation, with the latter better able to deal with changing and uncertain environments.

[2] I have benefited greatly in this process of identification from earlier work with my colleagues Ed Steinmueller (Pavitt and Steinmueller, 2001), Stephano Brusoni and Andrea Prencipe (Brusoni et al., 2001), and P. Nightingale (Nightingale, 2000), and from reading Coombs and Metcalfe (2000).

(Pavitt, 1998; Pavitt and Steinmueller, 2001). Three interrelated dimensions of such specialisation can be distinguished: – between disciplines in universities, between functions in firms, and between institutions in countries. Specialisation requires co-ordination, but to differing degrees. Empirical studies show that some processes of co-ordination can and should be consciously managed and reasonably tightly coupled: like the links within the firm between R & D and other functions, and between corporate professionals in various scientific and technical disciplines. Others require a softer touch: like the establishment between universities and business of partly non-market based networks between like-minded researchers.[3] These findings confirm the notion that the more costly and less uncertain processes of technological selection should be more tightly constrained than the less costly and more uncertain processes of search.

Increasing specialisation in disciplines has also meant that an increasing range of fields of knowledge are being deployed in business firms to solve technical problems and to reach technical targets. Products incorporate an increasing number of fields of knowledge, and so do the firms that make them. One consequence is that technologies (e.g. computing) should not be confused with closely related artefacts (e.g. computers), since the former is used in many more products than the latter (Granstrand et al., 1997). Another consequence has been that new combinations of fields of knowledge[4] have enabled firms based in rich fields of science and technology to diversify by creating and entering cognitively related product markets. Yet another consequence has been the difficulties facing managers in large firms in decomposing their organisation, in order to match unique fields of technology with unique classes of product. Increasingly multi-technology products in diversifying firms have made the pure M-form of organisation impossible to sustain. Some form of central corporate laboratory or technical competence persistently proves necessary in order to help mix and match changing technologies with changing products.

2.2 What's changing: external co-ordination – the 'third face' of corporate R & D

An important and relatively recent manifestation of the increasing specialisation in knowledge production has been the growth of so-called "strategic alliances" between large firms, designed to exchange knowledge in rich technological fields like ICT and biotechnology. One explanation for this growth has been that – as with production – firms have been "outsourcing" R & D in order to reduce technology-related costs, and to concentrate on their core technological competencies. But the evidence shows the contrary: both corporate R & D costs and their dispersion across fields have continued to increase in large firms.

A more plausible explanation is the continuously increasing number of technological fields that firms must monitor and master. These have increased not only with

[3] See, for example, Pisano, 1991.

[4] Although Schumpeter spoke extensively of "new combinations", he was not thinking of knowledge (Tunzelmann, 1995, p. 76–78), to which Adam Smith gave more central importance in the innovation process (Pavitt, 1998).

the division of labour in the production of knowledge, but also with the division of labour in the production of goods and services. The latter has led to increasing vertical disintegration (great "roundaboutness") in the production system. Changes and improvements in the production system can be handled satisfactorily through purely market mechanisms, when the links between the various production stages can be made modular, namely, when they have a standardised physical interface between each stage, which allows improvements to be made within each stage. Firms are then able to outsource current production of components and sub-systems, in order to benefit from cost advantages arising from competition, as well as from changes and improvements still possible within established modular configurations.

However, complete modularity is not sustainable, for two reasons (Richardson, 1972; Brusoni et al., 2001). First, increasing product complexity increases the probabilities of unforeseen systemic interactions amongst components and subsystems. Second, rapid technical change and increases in performance in one part of complex systems can create both bottlenecks and opportunities in other parts. Under such conditions, firms at the centre of complex or fast-changing supply systems – be they physical supply systems as in the automobile industry, or knowledge supply systems as in pharmaceuticals – must also have the means to co-ordinate change in these systems, especially when they are designing and implementing major changes.[5] Empirical studies show that such firms maintain in-house a systems integration capability: first, in order to monitor and stimulate improvements by suppliers within the modular constraints of established systems; and second, in order to integrate major changes periodically into new and improved systems. With increasingly systemic complexity (of which more in Sect. 3 below), it is likely that a growing share of corporate technological activities are being devoted to these activities. In addition to innovation and imitation, corporate R & D now therefore has a third face (Cohen and Levinthal, 1987): co-ordinating change and improvement in increasingly complex external product and knowledge networks.

2.3 What's changing: "useful" university research – opportunities not incentives

Another (perhaps related) recent change has been in the role of university-based research in underpinning corporate innovative activities. Here it is particularly important to distinguish rhetoric from reality. The current rhetoric – mainly from government sources – is that university research has been too "blue sky" and "academic", and should now be focussed more closely towards application. The current reality – based on statistical studies and surveys – shows that the research appreciated most by corporate practitioners is publicly funded, undertaken in high prestige research universities, and published in high quality academic journals (Mansfield, 1995; Narin et al., 1997). It also shows that the training of high quality researchers is generally more appreciated by practitioners than direct inputs of knowledge to corporate practice (Salter and Martin, 2001).

However, there is scattered evidence that university research is beginning to make more direct inputs into corporate technological activities, but not because

[5] For theoretical explorations of this phenomenon, see Frenken et al. (1990), Marengo (2000).

of government-led policies and exhortations to be more "useful". For example, a recent study shows that licensing income from research in major US universities has increased dramatically in the past 20 or more years (Mowery et al., 2001). The rise began in biotechnology before the Bayh-Dole Act, suggesting that the increase has been opportunity- rather than incentive-driven. There have also been related increases in university-based spin-off firms, not only in biotechnology, but also in fields benefiting from improvements in large-scale data processing like speech recognition and computational chemistry (Mahdi and Pavitt, 1997; Koumpis and Pavitt 2000).

One possible explanation for this apparently greater involvement is that, whilst large firms previously explored the opportunities emerging from fast-moving fields in their central laboratories in direct contact with university departments, they are now doing so increasingly through "loose-coupled" arrangements to explore potential fields of application through university spin-offs. This may be linked to another change, namely the increase in the range of technological opportunities emerging from university-based research in certain fields, because applied experiments have become much cheaper, following improvements in fundamental understanding, in measurement, and in computation and simulation (see Sect. 3.3 below). Given the high uncertainties involved, companies prefer to stimulate developments through loosely coupled arrangements rather than close integration (Pisano, 1991).

3 Increasing complexity

3.1 What's the same: coping with uncertainty – "management" vs. learning

Schumpeter was right about the growing predictability of corporate *inputs* into innovative activities, but wrong about *outputs*. Specialised R & D and related activities have certainly become institutionalised and predictable sources of of discoveries, inventions, innovations and improvements. But technologists and managers are still not able to make accurate predictions about the emergence and acceptability of major new products, about the technical performance of newly designed artefacts, about the costs or time to develop them, or about the size of market for specific innovations. This is because the world of innovation is complex, in that it involves many variables, the properties and interactions of which are understood only very imperfectly. As a consequence, we are not able to explain fully and predict accurately either the technical performance of major innovations, or their acceptability to potential users (or even who the potential users are).

Corporate management therefore continues to have difficulties in deciding how to deal with innovative activities, which have some of the elements of conventional investments activities, but which are also uncertain and require continuous feed back from experience and experiment to learning. The broad differences between search and selection activities has been recognised for a long time in practice with the distinction between corporate and divisional R & D activities, and in theory with the distinction between "knowledge building" and "strategic positioning" on the one hand, and "business investment " on the other (Mitchell and Hamilton, 1988).

However, as the recent history of corporate R & D shows, maintaining balance and linkages between the two is not an easy task. There are swings in fashions and management practices with periodic attempts – often following "failures" or examples of "waste" – to "manage" R & D more effectively through "better" techniques forecasting, more "rigorous" methods of personnel management, and of project selection and control. These are often subverted by practice and by evidence of missed opportunities. Professional judgement and experience then displaces management technique as the main basis for decisions, and the co-ordination of learning across organisational and professional boundaries becomes paramount. This style of innovation management is likely to be more successful, but more difficult to achieve, since it depends on the person-embodied skills and informal networks, rather than on codified techniques and procedures. So failure can start off the whole cycle again.

3.2 What's the same: increasing scientific understanding and increasing technical complexity

One interesting paradox in the development of new technology is that, despite the massive increase in scientific understanding in the past 200 years, technological practice stills runs ahead of what science can predict. Most expenditures in business firms are still on the development and testing of specific artefacts rather than on the development of underlying theory. The sheer combinatorial complexity of useful artefacts precludes accurate predictions of practice based purely on theory (Girin, 2000). This is why technology advances through the practices of scientists and engineers that Constant (2000) has recently described as recursive, involving "alternate phases of selection and of corroboration by use. ...... The result is strongly corroborated foundational knowledge: knowledge that is implicated in an immense number and variety of designs embodied in an even larger population of devices, artefacts, and practices, that is used recursively to produce new knowledge." (p. 221).

However, there are grounds for thinking that technical complexity cannot run too far ahead of scientific understanding. The feedback loops in both directions between improvements in scientific understanding and improvements in technical performance have been well documented by historians. More specifically, increases in technical complexity and in associated increases in combinatorial complexity will by themselves increase the risks and costs of search and selection. One factor that can reduce them is improved scientific understanding of cause-effect relations, very often emerging from advances in the technologies of measurement and manipulation of the increasingly small. This has been the case in the past decades in molecular biology and materials, both of which have opened major new opportunities for technical change.

3.3 What's different: ICT in complexity and understanding

Improvements in ICT are increasing both complexity and understanding. First, advances in network and digital technologies are opening major new possibilities of products and services with much greater systemic complexity, involving not just the spheres of production and exchange, but also those of distribution and consumption. Second, the major advances in large-scale computing and simulation technology are reducing considerably the costs of exploring alternative technical configurations. This is another major method of reducing the costs and risks associated with increasing technical complexity.

Nightingale (2000) has shown recently that experimental techniques in the pharmaceutical industry have in the past ten years seen major changes resulting from all the mechanisms described above: first a shift towards more fundamental science, for example, linking biochemical mechanisms to the expression of genes; second, using simulations and data banks to conduct virtual experiments complementary to real ones; third, using high throughput screening techniques. His findings tend to corroborate the recent suggestion by Perkins (2000) that improved search strategies involve both "code" (i.e. theories) and "construction" (i.e. prototypes). However, his (and others') use of "Fitness Landscapes" does not grapple with most of the central features of corporate activities for technological search and selection that can be found today in sectors as different as aircraft and pharmaceuticals: namely, the design, testing and re-design of hierarchical and interdependent systems, subsystems and components, based on bodies of increasingly specialised knowledge and practice, that are continuously improving, and at different speeds.

4 The co-evolution of technology and organisation

4.1 What's the same: matching specific technologies and specific organisational practices

It is a commonplace today to argue that technologies and organisational practices co-evolve. It is less common to expose oneself to accusations of "technological determinism" by arguing that, on the whole, corporate organisational practices adapt, in order to exploit emerging technological opportunities. The case can be made on historical grounds. For example, Chandler (1977) has shown that the rise in the USA at the end of the nineteenth century of the large, multi-unit firm, and of the co-ordinating function of professional middle managers, depended critically on the development of the railroads, coal, the telegraph and continuous flow production. Similarly, the later development of the multi-divisional firm in part reflected the major opportunities for product diversification in the chemical industry opened up by breakthroughs in synthetic organic chemistry. Technical advances normally precede organisational advances, because of their firmer knowledge base and the lower costs of experimentation. This does not mean that technology imposes one organisational "best way": variety in the characteristics of technologies, their continuous change and uncertain applications lead to variety in organisational practices. But it also does not mean that "anything goes" in organisation. For example, a firm practising conventional cost-benefit analysis and strict cost controls with all its investment decisions will not prosper in the long term in a competitive market governed by the exploitation of a rich, varied and rapidly advancing body of technological knowledge[6].

Based on the empirical literature, the first two columns in Table 1 identify the key features of technologies that must be matched with corporate organisational practices. The richness of the technological opportunities and the scale of technical experiments will determine the appropriate share of resources devoted to technological search, as well as the degree of centralisation and fluidity in organisation structures. Supporting skills and networks will define the specific competencies to be accumulated, professional networks to be joined and key functional interfaces across which learning must take place within the firm. And the strategic position of the firm will determine which technologies are supported as part of its distinctive core advantage, or as necessary background technologies.

Differences amongst technologies are therefore reflected in differences in organisation practices. Thus, both pharmaceutical and consumer electronics firms devote substantial resources to technological search, but the former tends to have centralised and formal procedures for launching new products, whilst the latter is more likely to be decentralised and informal. Similarly, both pharmaceutical and automobile companies have centralised decision structures, but the former will stress interfaces between corporate R & D and public research in bio-medical fields, whilst the latter will stress links between R & D and production.

4.2 What's the same: widespread adoption of revolutionary technologies

The past 200 years has seen periodic step-jumps in technological understanding and performance in specific fields, that have reduced considerably the costs of key economic inputs, and have therefore been both widely adopted and the catalysts for major structural changes in the economy. These include steam power, electricity, motorization, synthetic materials and radio communications (Freeman and Louçã, 2001). The contemporary example is of course the massive and continuing reductions in the costs storing, manipulating and transmitting information brought about by improvements in ICT (Information and Communications Technologies).

Each wave of radically new technologies has been associated with the emergence of firms that have mastered the new technologies, and that have led in the development and commercialisation of related products, processes and services. In the current jargon of corporate strategy, these firms have developed *core competencies* in the new technologies, which have become a distinctive and sustainable competitive advantage. They must be distinguished from the far more numerous firms who adopt and integrate the new technologies with their current activities. For these firms, in-house competencies in the new technologies are *background*: in other words, necessary for the adoption of advances made outside the firm. Paradoxically, the very fact that the new technologies allow step-jump reductions in the costs of a key input simultaneously make their adoption both a competitive imperative, and an unlikely source of their own distinctive and sustainable competitive

[6] See, for example, the history of the UK General Electric Company under Arnold Weinstock (Aris, 1998).

Table 1. Matching corporate technology and organisational practices

Corporate Technology→	Matching Organisation Practices→	Dangers in Radical Technological Change
Inherent Characteristics		
1. Richness of opportunities	1a. Resources for exploring options	1a. Greater opportunities not matched by resources for exploring options
	1b. Matching Technologies with Product Divisions	1b. Matching opportunities missed
2. Costs of specific technical experiments	2. Degree of centralisation in decision-making	2. Reduced cost of experiments not matched by decentralisation
Supporting Skills and Networks		
1. Specific sources of external knowledge	1. Participation in specific professional knowledge networks.	1. Difficulties in recognising and joining new knowledge networks
2. Specific families of products, production methods, supply chains, customers and distribution channels	2a. Accumulated knowledge of specific customers' demands, distribution channels, production methods, supply chains.	2a. Difficulties in recognising and responding to new customers' demands, distribution channels, production methods, supply chains.
	2b. Focus on learning across key functional interfaces	2b. Difficulties in recognising new key functional interfaces
		3. Scepticism and resistance from potentially obsolescent professional and functional groups
Strategic Position in the Firm		Inability to tell core from background:
1. Core technologies (Central to sustained competitive advantage. Difficult to imitate)	1. Strong commitment of technical resources to maintain state of the art	1. Excessive outsourcing of core technologies;
OR	OR	OR
2. Background technologies (Necessary for use of outside technologies in supply chain and pervasive applications)	2. Commitment of technical resources sufficient to monitor and assimilate technologies developed outside the firm	2. Inability to sustain competitiveness with background technologies

advantage. For example, in the past many factories had no choice but to adopt coal and steam – and later electricity – as a source of power, given their cost and other advantages. The same is true today for many ICT-based management practices. In neither case were – or are – these revolutionary advances by themselves a source of sustainable competitive advantage for the adopting firms. This means that much of the emphasis by writers on corporate strategy – like Barney (1991) and Porter (1996) – on the importance of establishing a distinctive and sustainable advantage does not, and cannot, apply to the major transformations now inevitably happening in many companies through the adoption of ICT. Their framework helps understand *CISCO* (a major US supplier of equipment for the Internet), but doesn't help much with *TESCO* (a major UK supermarket chain).

4.3 What's different: creative destruction in organisational (not technological) practices

Ever since Schumpeter associated the advent of revolutionary technologies with "waves of creative destruction", there has been debate about the relative role of incumbent large firms and new entrants in exploiting them. Over the past 20 years, most of the analytical writing has been stacked against incumbents, although recent empirical studies can point to evidence in favour of both (Methe et al., 1996). Over time, the weight of the arguments against has shifted. Earlier studies emphasised the difficulties facing incumbents in mastering new fields of technological knowledge (Cooper and Schendel, 1976; Tushman and Anderson, 1986; Utterback, 1993). More recently, there has been a shift towards emphasising the difficulties in changing and matching organisational practices to the opportunities opened up by revolutionary technological changes, because of changes in product architectures (Henderson and Clark, 1990), of resistances from groups with established competencies (Leonard-Barton, 1995), and of the unexpected emergence of new markets (Christensen, 1997; Levinthal, 1998).

It is becoming clearer that, contrary to a widely held belief, the nature and directions of radical new technological opportunities are easily recognised by the technically qualified: for example, miniaturisation, compression and digitalisation today in ICT. The technological consequences of these trends can be explored in corporate R & D laboratories: thus, a growing number of large firms in a growing number of industries is now technically active in ICT (Granstrand et al., 1997). The difficult, costly and uncertain task is that of combining radically new technical competencies with existing technical competencies and organisational practices, many of which may be threatened or need to be changed. Experimentation and diversity is therefore necessary, not in exploring the directions of radical technological changes, but their implications for products, markets and organisational practices.

The third column of Table 1 tries to identify some of the reasons why such experiments may fail in incumbent firms. Some are a consequence of the need to modify organisational practices, and some of the inevitable uncertainties in the early stages of radically new technologies. The likelihood that established firms will fail increases with the number of organisational practices that need to be changed. Here the comparison between two recent industry studies is instructive. Klepper and Simons (2000) have shown that firms already established in making radios were the most successful later in the newly developing colour TV market. On the other hand, Holbrook and his colleagues (2000) have shown that none of the firms established in designing and making thermionic valves were successful in establishing themselves in semiconductors. With the benefit of hindsight, we can see that success in semiconductors required more changes in both technological competencies and organisational routines amongst incumbents than success in colour TV. The valve firms required the new competencies and networks in quantum physics, a much stronger interface between product design and very demanding manufacturing technology, and the ability to deal with new sorts of customers (computer makers and the military, in addition to consumer electronics firms). For the radio firms, the shift to colour TV required basically the same technological competencies, augmented by screen-technologies. Otherwise, the customers and distribution channels remained unchanged, as did the key networks and linkages both inside and outside the firm.

5 Conclusions

This paper has argued that innovating routines in business have to deal with three fundamental features of technical change in modern societies: increasing specialisation in knowledge production, the tendency for technological practice to run ahead (but not too far ahead) of scientific theory, and the continuous matching of the specific features of changing technologies and specific organisational practice that transform technology into useful artefacts

Some of the tasks of innovating routines remain the same: co-ordination and integration of internal knowledge sources and functions, technology-based product diversification, learning by analysing and doing. Other tasks are changing as a consequence of increasing specialisation in knowledge production: co-ordination and integration of external as well as internal knowledge sources, dealing with the dangers of creative destruction in corporate organisational – rather than technological – practices. And yet others are changing – at least in part – because of the effects of rapid improvements in performance in ICT: increasing technical complexity, reduced costs of technical experiments and more direct links in university research to practice.

The paper has two obvious limitations, and (hopefully) one achievement. First, it has not discussed in detail how the various tasks of innovating routines should be implemented. There is a rich research agenda here on how the execution of these tasks can be influenced by incentives, transaction costs, information flows, mobility of personnel, training and research, and a variety of other factors.

Second, the paper has been satisfied with identifying three underlying trends in innovative activities, and with compiling an extensive check-list of tasks that corporate routines must try to accomplish. It has not developed anything like a formal model or theory. This is because its author takes the same position as the scholars from a variety of disciplines who contributed to a recently published book about technological change, and concluded that "its categories and their interactions are too imprecise and contextual to be represented realistically by a computable algorithm. ... ... A mathematical simulation can never be true to life. It . . . is a metaphor in its representation of a real system with complex unquantifiable structural relationships between its elements" (Ziman, ed. 2000, p. 312). In such circumstances, it is probably better to keep things simple, at least to start with.

However, the paper has made at least some progress in its initially stated purpose, namely, to give more operational content to the notion of innovating routines. As academics, we can ask questions like "What are your key external knowledge sources, and how do you access them?" and expect to get meaningful answers. And as advisors, we can use the various headings in the checklist, not to tell corporate managers what to do, but to identify where they should exercise their judgement, and learn from subsequent experience. After all, this is what they are paid to do.

References

Aris S (1998) Arnold Weinstock and the making of GEC. Arum

Barney J (1991) Firm resources and sustained competitive advantage, Journal of Management 17: 99–120

Brusoni S, Prencipe A, Pavitt K (2001) Knowledge specialisation, organizational coupling and the boundaries of the firm: why do firms know more than they make?. Administrative Science Quarterly 46:597–621

Burns T, Stalker G (1961) The management of innovation. London: Tavistock (republished in 1994 by Oxford UP)

Chandler A (1977) The visible hand. The Managerial Revolution in American Business. Cambridge, MA: Belknap

Christensen C (1997) The innovator's dilemma. Boston: Harvard University Press

Cohen WM, Levinthal DA (1987) Innovation and learning: the two faces of R & D, Economic Journal 99: 569–596

Constant E (2000) Recursive practice and the evolution of technological knowledge. In: Ziman (ed.) Op. Cit.

Coombs R, Metcalfe S (2000) Organizing for innovation: co-ordinating distributed innovation capabilities. In: Foss, Mahnke (eds) Op. Cit.

Cooper A, Schendel D (1976) Strategic responses to technological threats. Business Horizons (February): 61–69

Foss N, Mahnke V (eds) (2000) Competence, governance and entrepreneurship: Advances in Economic Strategy Research Oxford University Press, Oxford

Frenken K, Marengo L, Valente M (1999) Interdependencies, Nearly Decomposability and Adaptation. In: Brenner T (ed.) Computational Techniques for Modelling Learning in Economics Kluwer Academics

Freeman C, Louçã F (2001) As time goes by: from the industrial revolutions to the information revolution. Oxford University Press, Oxford

Girin J (2000) Management et Complexite: comment importer en gestion un concept polysemique? Centre de Recherche sur La Gestion, Ecole Polytechnique, Paris

Granstrand O, Patel P, Pavitt K (1997) Multi-technology corporations: why they have distributed rather than distinctive core competencies. California Management Review 39: 8–25

Henderson R, Clark K (1990) Architectural Innovation: the Reconfiguration of Existing Product technologies and the Failure of Established Firms. Administrative Science Quarterly 35: 9–30

Holbrook D, Cohen W, Hounshell D, Klepper S (2000) The nature, sources, and consequences of firm differences in the early history of the semiconductor industry. Strategic Management Journal 21: 1017–1041

Klepper S, Simons K (2000) Dominance by birthright: entry of prior radio producers and competitive ramifications in the U.S. television receiver industry. Strategic Management Journal 21: 997–1016

Koumpis K, Pavitt K (2000) Corporate activities in speech recognition and natural language: another new-science-based technology. International Journal of Innovation Management 3: 335–366

Leonard-Barton D (1995), Wellsprings of knowledge. Harvard Business School Press, Boston, MA

Levinthal D (1998) The slow pace of rapid technological change: gradualism and punctuation in technological change. Industrial and Corporation Change 7: 217–247

Mahdi S, Pavitt K (1997) Key national factors in the emergence of computational chemistry firms. International Journal of Innovation Management 1: 355–386

Mansfield E (1995) Academic research underlying industrial innovations: sources, characteristics and financing. Review of Economics and Statistics 77: 55–65

Marengo L (2000) Decentralisation and market mechanisms in collective problem-solving. Dept. of Economics, Un. of Trento, Italy

Methe D, Swaminathan A, Mitchell W (1996) The underemphasized role of established firms as sources of major innovations. Industrial and Corporate Change 5: 1181–1203

Mitchell G, Hamilton W (1988), Managing R & D as a strategic option. Research Technology Management 31: 15–22

Mowery D, Nelson R, Sampat B, Ziedonis A (2001) The growth of patenting and licensing by US universities: an assessment of the Bayh-Dole act of 1980. Research Policy 30: 99–119

Narin F, Hamilton K, Olivastro D (1997) The increasing linkage between U.S. technology and public science. Research Policy 26: 317–330
Nelson R, Winter S (1982) An evolutionary theory of economic change. Cambridge MA: Harvard University Press
Nightingale P (2000) Economies of scale in experimentation: knowledge and technology in pharmaceutical R & D. Industrial and Corporate Change 9: 315–359
Pavitt K (1998) Technologies, products and organisation in the innovating firm: what Adam Smith tells us and Joseph Schumpeter doesn't. Industrial and Corporate Change 7: 433–51
Pavitt K, Steinmueller W (2001) Technology in corporate strategy: change, continuity and the information revolution. In: Pettigrew A, Thomas H, Whittington R (eds) Handbook of Strategy and Management Sage
Perkins D (2000) The evolution of adaptive form. In: Ziman (ed.) Op. Cit.
Pisano G (1991) The governance of innovation: vertical integration and collaborative relationships in the biotechnology industry. Research Policy 20: 237–249
Porter M (1996) What is strategy? Harvard Business Review November/December: 61–78
Richardson G (1972) The organisation of industry. Economic Journal 82: 883–896
Salter AJ, Martin BR (2001) The economic benefits of publicly funded basic research: a critical review. Research Policy 30: 509–532
von Tunzelmann GN (1995) Technology and industrial progress: the foundations of economic growth. Aldershot: Edward Edgar
Tushman M, Anderson P (1986) Technological discontinuities and organisational environments. Administrative Science Quarterly 31: 439–465
Utterback JM (1993) Mastering the dynamics of innovation. Boston MA, Harvard Business School Press
Ziman J (ed.) (2000) Technological innovation as an evolutionary process. Cambridge UP, Cambridge

The emergence of a growth industry: a comparative analysis of the German, Dutch and Swedish wind turbine industries*

Anna Bergek and Staffan Jacobsson

Department of Industrial Dynamics, Chalmers University of Technology, 412 96 Göteborg, Sweden (e-mail: annbe@eki.liu.se, stja@mot.chalmers.se)

Abstract. The objective of this paper is to compare the evolution of the wind turbine industry in Germany, the Netherlands and Sweden. Four factors stand out in explaining the relative success of the German industry: (1) creation of variety in an early phase, (2) establishment of legitimacy of wind energy, (3) the employment of advanced market creation policies in a later phase and (4) the use of industrial policy to favour the domestic industry. Implications for policy include fostering legitimacy for the new technology and creating powerful, predictable and persistent economic incentives.

Key words: Wind energy – Innovation system – Industrial dynamics – Policy – Functional analysis

1 Introduction

In recent years, the 'innovation system' perspective has obtained increased legitimacy as a way of analysing industrial development. For such a system to support the growth of an industry, a number of functions have to be served within it, e.g. the supply of resources. We suggest that we can evaluate the performance of an innovation system by assessing its 'functionality', i.e. how well these functions are

* Financial support from the Swedish National Energy Administration is gratefully acknowledged. We would also like to express our deepest gratitude to all the people in Germany, the Netherlands and Sweden who have contributed to the empirical content of this paper by allowing us to interview them. We are also grateful to Bo Carlsson and Michael Durstewitz for useful comments on an earlier draft.

served. The analytical objective of the paper is to develop a framework that enables us to make such an assessment in different phases in the evolution of an industry.

We apply this framework to a cross-country comparative analysis of the evolution of the wind turbine industry in Germany, the Netherlands and Sweden over a period of about twenty years.

The wind turbine industry is a non-high tech growth industry (Jacobsson and Johnson, 2000), in which the knowledge base is mechanical and electrical engineering mixed with software and aerodynamics. Since its inception, it has been dominated by Danish firms, which currently supply about 44 percent of the world sales (BTM Consult, 2000).

Other countries have also tried to develop a wind turbine industry, but with varying success. *Sweden* developed very large turbines in the early 1980s, but a domestic industry never quite materialised in spite of a substantial government R&D programme. Today, there are a few Swedish firms at the tail of the global industry. In *the Netherlands*, a range of firms entered the industry in the 1970s. At the end of the 1980s, the Dutch industry was relatively advanced, but today there is only one Dutch-owned firm left, which accounted for less than 1 percent of the world sales in 1999 (BTM Consult, 2000). *Germany* shared the Swedish emphasis on large wind turbines in the early 1980s, but in the mid-1980s a set of firms supplying smaller turbines emerged. These firms now constitute the nucleus of the German industry, which grew phenomenally in the 1990s and is now the second largest industry in the world. The four largest German firms accounted for approximately 27 percent of the world sales in 1999.[1] The empirical objective of this paper is to analyse the evolution of the wind turbine industry in these countries and to explain their relative success and failure in terms of the functionality of their respective innovation systems.

The paper is structured as follows. In Section 2, we present our analytical framework. Section 3 contains a brief description of the technology and of the market for wind turbines. In Section 4, we describe the development of the wind turbine industries in the three countries by mapping the functional pattern of their respective innovation system. In Section 5 the three cases are compared and some policy implications are discussed.

2 Analytical framework

As is argued in an expanding literature on innovation systems, the innovation and diffusion process is both an individual and collective act. The determinants of industrial development and growth are not only found within individual firms; firms are embedded in innovation systems that aid and constrain the individual actors within them.

The innovation system approaches share an understanding of a set of basic functions that are necessary for an innovation system to work (Johnson, 1998). Earlier,

[1] These firms and their market shares are Enercon (12 percent), Nordex (7.8 per cent), Tacke (5 percent) and Dewind (1.5 percent) (BTM Consult, 2000). Tacke is now owned by a US corporation, but develops and produces its turbines in Germany. Nordex was originally Danish, but is now owned by a German firm and develops its turbines in Germany.

we have suggested that a technology or product specific innovation system may be described and analysed in terms of its 'functional pattern', i.e. in terms of how these functions are served (e.g. Johnson and Jacobsson, 2000). The pattern stems from the character of, and interaction between, the components of an innovation system, i.e. actors, networks and institutions (Carlsson and Stankiewicz, 1995), which may be specific to one innovation system or 'shared' by a number of different systems.

The first, and maybe most obvious, function of an innovation system is to *create 'new' knowledge*.[2] This function may also be viewed as an overall goal of a system since an innovation system may be defined in terms of knowledge generation, diffusion and utilisation (Carlsson and Stankiewicz, 1995).

A second function is to *guide the direction of the search process* of the suppliers of technology and customers, i.e. to influence the direction in which actors deploy their resources. This function includes providing recognition of a growth potential (e.g. in terms of identifying technological opportunities), which is closely connected to the legitimacy that a new technology has in the eyes of various actors. The function also includes guidance with respect to both technological choice (i.e. the choice of specific design configurations) and market choice. Individual actors may be guided by inducement mechanisms such as the identification of problems of a technical nature, changing factor prices, the formation of standards or regulation and relationships to competent customers, or by various policy interventions. This is, of course, a particularly important function in the process of forming a new industry.

A third function is to *supply resources*, i.e. capital, competence and other resources. Capital is partly needed to distribute risks and may, sometimes, come with competence, for instance in the form of venture capital. Competence refers to a whole range of competencies, including technological competencies.

A fourth function is to facilitate the creation of *positive external economies* in the form of an exchange of information, knowledge and visions. Indeed, this function lies at the heart of the systemic approach to innovation and involves the formation of networks and meeting places and, perhaps, changes in culture.

A fifth function is to *facilitate the formation of markets*. Markets do not necessarily emerge in a spontaneous fashion, but may need to be created. Firms need to make investments of various types in order both to identify and to reach new customers. Governments may need to improve social acceptance by legitimising the new technology or removing legislative obstacles.

The functions are not, of course, independent of one another, and a change in one function may, thus, lead to changes in other functions. For instance, the creation of an initial market may act as an inducement mechanism for new entrants that bring new resources to the industry. The linkages between functions may also be circular, which may set in motion a virtuous circle. For instance, the resources brought into the industry by a new entrant may be used to develop the market further.

The framework provides us with a tool for analysing the dynamics of an innovation system. In addition to studying evolutionary processes in terms of changes

---

[2] This and the following paragraphs constitute a synthesis of a large body of literature and are largely based on Johnson (1998). For detailed references see Johnson and Jacobsson (1999).

in entries and exits, network formation, institutional adaptation etc., attention can be paid to the way in which the functional pattern of an innovation system evolves and what drives its evolution.

The framework also allows us to evaluate an innovation system in terms of the way it supports the development of an industry. Since all of the functions need to be served for a new industry to evolve and perform well, we suggest that a particular innovation system may be evaluated in terms of its 'functionality', i.e. in terms of *how well* the functions are served within that system.

What 'well served' means is to be expected to differ depending on what particular stage of evolution an industry is in (Utterback and Afuah, 1998). Several cyclical models of product/industry development have been developed in order to capture regularities in the evolution process (e.g. Utterback and Suarez, 1993; Tushman, Anderson and O'Reilly, 1997). These models are based on the idea that most (if not all) products or industries go through identifiable phases which differ in terms of the character of technical change, the patterns of entry/exit and the rate of market growth. The number of phases differs between models, but it is usually possible to differentiate two main phases.

The first is one of experimentation with frequent entries and exits, many different competing technological alternatives and a small market (Nelson, 1994; Utterback and Suarez, 1993). The outcome of the competition is highly uncertain both in terms of which alternative(s) will be the winner (Nelson, 1994) and in terms of industry leadership. Innovators compete as much against market scepticism as against rivals (Utterback, 1994).

The second is characterised by market growth (Utterback, 1994), fewer new entrants (Utterback and Suarez, 1993) and, possibly, a shake-out of firms. In some of the literature, the transition from the first phase to the second is driven by the emergence of a 'dominant design'. The selection of that design results in a change in the nature of technical change from radical product innovation to process innovation and incremental product innovation (Utterback, 1994).

We have chosen not emphasise the concept of dominant design for two reasons. First, a shift between the phases may occur in the absence of a clear dominant design. A dominant design might, indeed, occur as a result of a large-scale diffusion rather than cause it. Second, and more important, even if a dominant design does emerge, the 'radicalness' of technical change does not necessarily have to decrease. Technological discontinuities may very well occur within the frames of a dominant design (Hidefjäll, 1997) and, hence, a technology-driven turbulence may continue to exist even after the selection of the dominant design (Ehrnberg and Jacobsson, 1997, Tushman, Anderson and O'Reilly, 1997).

In view of the different characteristics of the two phases, the functions clearly have different roles to play in the two phases. Whereas the discussion below is not exhaustive, we point to some main differences between the phases in terms of these roles.

In the first phase, the key words are experimentation and the generation of variety. A necessary condition for this to occur is that the direction of search of new or established firms is guided towards the new product. Due to the need for legitimacy for the new technology, it may be especially important that the entry of respectable actors is stimulated. The creation of variety is, however, central and the system needs to ensure the creation of new knowledge within different technological approaches. This may involve the entry of firms from different backgrounds and the provision of special incentives for experimentation, e.g. in the form of resources.

The system must also facilitate the creation of external economies, e.g. via problem-solving networks in the form of user-supplier links. A necessary condition for these links to form and for new firms to enter is, of course, that markets are open to new sources of supply (Nelson, 1994) or that new (niche) markets are identified and stimulated.

In the second phase, the key words are diffusion and firm expansion. The system needs to support a shift towards cost reduction. This is, in part, achieved by exploiting economies of scale. The system must, thus, identify and facilitate the formation of mass markets. At the same time, it needs to prevail in its support of a variety of actors and technologies. Continued legitimacy contributes to actor variety and firm growth, since it guides the direction of search of firms and attracts private capital that supply firms with resources. Such resources are necessary for firm expansion and technology development, including development required to handle technological discontinuities within the frame of a dominant design.

In short, the 'functionality' of an innovation system may be assessed in terms of how it supports firm entry, variety and the formation of niche markets in the first phase, and market expansion and the supply of resources to exploit that market in the second phase. In order to make such an assessment, we need to analyse the dynamics of the innovation system by mapping the evolution of the functional pattern. This will be done in Section 4. Before that we will outline some key features of two phases in the wind turbine industry in the next section.

3 Salient features of the two phases in the wind turbine industry

In the wind turbine industry, we can identify two phases that correspond to the ones described above. The first is characterised by substantial technological variety (and uncertainty), underdevelopment of the market and entry of many firms. The second is characterised by a considerable turbulence, driven by rapid growth in the market and an up-scaling of the turbines (corresponding to a set of minor technological discontinuities), as well as by many exits but also some new entrants, including some larger firms. In this, as well as in the subsequent section, we will therefore distinguish between a phase of experimentation (roughly 1975–1989) and one of turbulence and growth (roughly 1990–1999).

3.1 The phase of experimentation

In the 1970s and early 1980s, there was a large number of fundamentally different designs: horizontal- and vertical-axis turbines,[3] turbines of varying sizes (5 kW to

[3] The difference between horizontal-axis and vertical-axis turbines lies in the orientation of the axis of rotation – horizontal with respect to the ground (and roughly parallel to the wind stream) vs. vertical with respect to the ground (and roughly perpendicular to the wind stream).

3 MW) and turbines with different numbers of blades (one to four). Firms with a broad range of backgrounds (shipbuilding, gearboxes, agriculture machinery, aerospace, etc.) experimented with a variety of approaches, bringing their specific competencies to the industry.

The market developed quite slowly during the 1980s; even in the peak year of the Californian 'boom'[4] in the mid-1980s, only 420 MW[5] were installed (see Fig. 3.1).

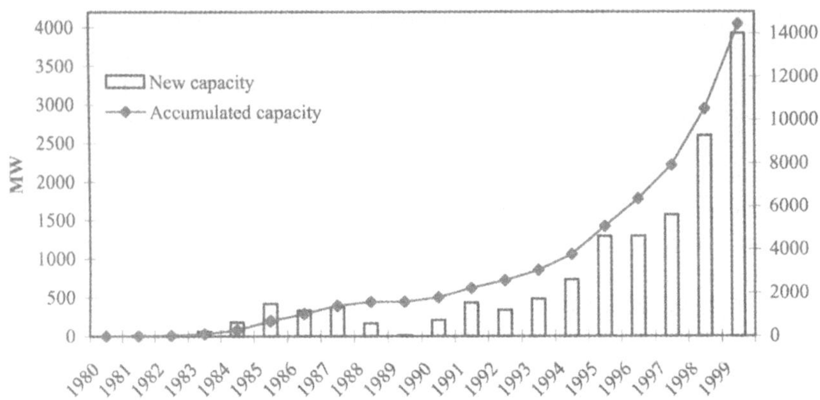

Fig. 3.1. Global wind turbine installations 1980–1999 (Sources: 1980–1990 Kåberger (1997); 1991–1993 European Commission (1997), Volume 5, Table 2.2; 1994–1999 BTM Consult (2000), Table 2-1)

In the late 1970s and early 1980s, large firms such as MBB and SAAB entered the industry on the basis of government contracts to develop MW turbines. Most of these firms left in the 1980s or early 1990s, mainly because of the non-viability of MW turbines at that time. In parallel, many smaller firms entered, either as entrepreneurial start-ups or as a result of the diversification by mechanical engineering firms. These firms focused on smaller turbines and came from, e.g., Denmark, Holland and Germany. In Denmark as many as 26 firms had sold more than three turbines in the 1980s (Karnoe, 1991, Appendix 2, Table 2.2). In Germany, 13 firms were active in the 1980s[6] (elaboration on Durstewitz (2000)). In the Netherlands,

[4] This 'boom' was driven by tax incentives and involved the setting up of many thousands of wind turbines.

[5] MW stands for megawatt, which is a unit for power. Wind turbines are usually classified by their rated power, i.e. the maximum power that the turbines may produce.

[6] In this case, we included all German firms that had sold a turbine on the German market.

15–20 firms entered the field in the late 1970s and early 1980s (elaboration on Verbong (2000)).[7]

3.2 The phase of turbulence and growth

The horizontal-axis three-bladed design was selected from among the many alternatives.[8] A technology driven competition continued, however. First, three alternative horizontal designs competed throughout the 1990s.[9] Second, the commercial turbines greatly increased in size. This was especially apparent in Germany, where the average size of newly installed turbines increased from perhaps 50 kW in the mid-1980s to roughly 185 kW in 1992 (BTM Consult, 1999, Table 2-3) and over 900 kW in 1999 (BWE, 2000).[10]

In the mid-1990s, a very rapid market growth began. Indeed, the average annual market growth in terms of installed capacity was 38 percent during the 1990s (see Fig. 3.1). The leading markets were Germany (with a growth of 4,336 MW in 1992–1999), followed by Spain (1,796 MW) and the USA (1,348 MW) (European Commission, 1997, Volume 5, Table 2.2; BTM Consult, 1999, Table 2.2; BTM Consult, 2000, Table 2-6).

A process of concentration and growth in the size of the firms took place. Some firms grew organically, whereas other firms grew by mergers and/or acquisitions.[11] The rapid growth in the market also led to more entries, primarily in Germany and Spain. Among these were both entrepreneurial firms and large firms that entered the industry by acquiring small, established firms.

---

[7] In 1989, the German and the Dutch industry supplied in the order of 10 MW each (elaboration on Durstewitz, 2000; IEA, 1997a), which may be compared with the Danish industry, which supplied about 130 MW (Hantsch, 1998). NB: The supply and installation of turbines does not always take place in the same year.

[8] The horizontal axis design has historically been equipped with one, two or three blades. However, the one-bladed design was never successful and from around 1990, the two-bladed design lost ground for aesthetic reasons, leaving the three-bladed design as the dominant one. However, as late as around 1990 the vertical axis design was seen as one of four competing designs (Karnoe, 1991).

[9] The first design is the 'Danish stall' design, which combines stall control, constant rotation speed and an asynchronous generator. Stall control is a way of power control in which the aerodynamic design of the blades causes the lift to decrease and the drag to increase above a certain wind speed, thus limiting the power output of the turbine. The stall-control is now giving way to pitch control, which constitutes the second design. In pitch control, the blades are rotated in their longitudinal axis above a certain wind speed, which limits the power output of the turbine to its rated value. The third design involves the use of pitch control, variable rotation speed and a synchronous generator. In addition, there are semi-variable designs and a couple of other power-control principles, e.g. the 'active stall' control.

[10] There is reason to believe that the further up-scaling will lead to a preference for the third design approach as the pitch and variable speed features make it easier to handle large aerodynamic forces and to monitor and control the turbines (van Kuik, 2000). The German industry, which is leading the up-scaling process, is now shifting towards this design principle (Müller, 1999; Hansen, 1999).

[11] The German market leader Enercon and the Danish firm Vestas are examples of organic growth, whereas the Danish firm NEG Micon is an example of growth by mergers and acquisitions.

4 Development of a wind power industry in Germany, the Netherlands and Sweden

In this section, we will map the functional pattern of German, Dutch and Swedish innovations systems in the two phases.

4.1 The phase of experimentation

4.1.1 The German case

In the German case, the key function in this phase was 'Guide the direction of search' (see Fig. 4.1). This function was initially influenced by an R&D policy, which via the function 'Supply resources' induced a search in many directions. Even though the projects aiming at developing MW turbines received much international attention,[12] this R&D programme was large enough to finance most projects applied for and flexible enough to finance most types of projects (Windheim, 2000a). In the period 1977–1991, about 46 R&D projects were granted to as many as 19 industrial firms and a range of academic organisations for the development or testing of small (e.g. 10 kW) to medium sized (e.g. 200–400 kW) turbines (elaboration on Windheim (2000b)).[13] Both horizontal- and vertical-axis turbines received support, as did turbines with different numbers of blades. These experiments stimulated the creation of variety through an influence on the function 'Create new knowledge'.

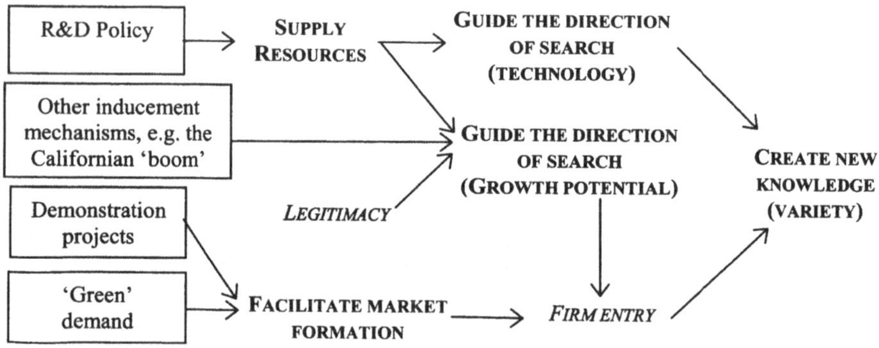

Fig. 4.1. The German innovation system in the phase of experimentation

[12] The most prominent was the Growian machine, developed by MAN, erected in 1982 and dismantled in 1987 (Gipe, 1995).

[13] The numbers exclude funding given for the purpose of demonstrating wind turbines. In addition, there was support for projects that could benefit all sizes of turbines.

Some of this new knowledge was also exploited commercially by German suppliers, beginning in 1984 when MAN sold a 20 kW turbine. Another thirteen German firms sold turbines in the 1980s, and eleven of these firms still existed in 1989 (elaboration on Durstewitz (2000)).[14]

A condition for the firm participation in turbine development and production, within and outside the R&D programme, was an early legitimacy to wind turbines. The legitimacy was due partly to a political consensus on the benefits of wind power,[15] and induced firms to begin a search towards wind turbines.

Given the legitimacy, firms were induced to enter for a number of reasons. Of course, the resources provided by the R&D programme made the area seem attractive. Moreover, in several cases the firms' existing markets were in recession at the same time as the Californian boom, and the expansion of the Danish wind turbine industry sent clear signals about the attractiveness of the wind turbine market (Tacke, 2000; Schult and Bargel, 2000). The new entrants were also induced by emerging local niche markets, supported by the function 'Facilitate market formation', which was served by two mechanisms. First, the green movement in Germany was strong, which led to the emergence of a 'green' demand from some utilities. As there were as many as 800 different utilities in Germany, there was ample room for diverse opinions with regards to technology choice (Reeker, 1999).[16] There was also an early niche of environmentally concerned farmers (Schult and Bargel, 2000; Tacke, 2000). Second, the federal R&D policy subsidised investment in wind turbines in a number of demonstration programmes (Hemmelskamp, 1998). At least fourteen German suppliers of turbines received funding for 124 turbines in the period 1983–1991 (elaboration on Windheim (2000b)).[17]

Table 4.1. The German market for wind turbines 1982–89

	Number of new turbines	Accumulated number of turbines	New power capacity (MW)	Accumulated power capacity (MW)
1982	1	1	0.02	0.02
1983	1	2	0.06	0.08
1984	4	6	0.10	0.18
1985	12	18	0.24	0.42
1986	15	33	0.52	0.94
1987	44	77	1.94	2.88
1988	61	138	4.99	7.87
1989	87	225	11.8	19.67

14 We approximate entry with the year of the first sales and exit as the year of the last sales.

15 This consensus was particularly clear after the Chernobyl accident (Molly, 1999).

16 Several people we interviewed claimed that the Green movement reached the Universities of Technology (Fachhochschule) and that some of the engineers who graduated began to develop wind turbines, both as suppliers and customers.

17 According to Hemmelskamp (1998), 214 turbines were supported.

However, the domestic market still remained weak throughout the experimental phase. For instance, in 1986 and 1987, when seven new German firms entered the industry, only 15 and 44 turbines were sold respectively (elaboration on Durstewitz (2000)) and the total installed power was just 19 MW by the end of 1989 (see Table 4.1).

Together with the R&D policy, the large number of entries nevertheless contributed to the broad range of experiments undertaken and the consequent accumulation of knowledge and competence. Indeed, the diversity in experiments undertaken is the main characteristic of this early phase in Germany.

4.1.2 The Dutch case

In the Dutch case, the first phase consists of two, quite different, sub-phases partly running in parallel.

The outcome of the first of these sub-phases was similar to that of the first phase in Germany, although the functional pattern of the innovation system was somewhat different (see Fig. 4.2).

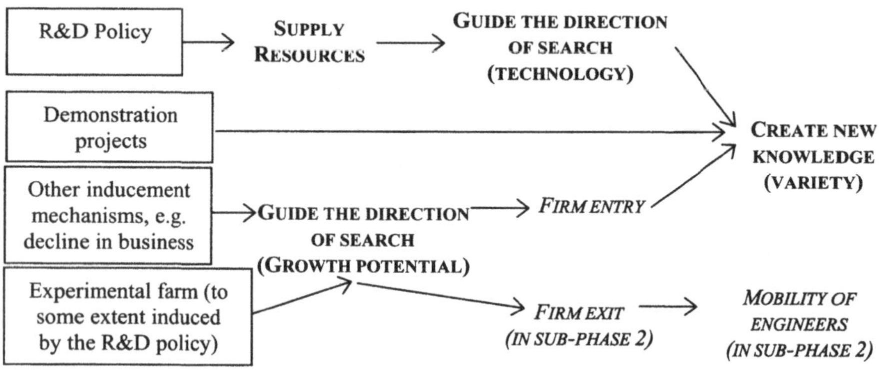

Fig. 4.2. The Dutch innovation system in the first sub-phase of the phase of experimentation

In the late 1970s and early 1980s, a multitude of firms entered and experimented with various designs, thus influencing the function 'Create new knowledge' in such a way that variety was created. Most of the firms were driven by the decline in their original business (van Holten, 2000). The first commercial turbines were erected in 1980, and in the mid-1980s, there were 15–20 firms developing or producing a large variety of turbines (mostly 10–80 kW).

The function 'Create new knowledge' was also influenced by two government-financed wind energy programmes. Within these programmes, the development of a large variety of turbine types was supported. Basic research and development of larger turbines was conducted at the Dutch energy research institute (ECN) and the technical universities, sometimes in co-operation with the larger industrial firms. In addition, the development activities of the emerging wind turbine industry were funded; all types of turbines could receive support (Janssen and Westra, 2000; Versteegh, 2000), with roughly half of the investment cost. This was, of course, very important, especially for the smaller firms (Versteegh, 2000).[18] Thus, the function 'Guide the direction of search' was influenced, via the function 'Supply resources', so that variety was created and sustained. There were also some other mechanisms serving the function 'Create new knowledge'. Firms could get free help with testing their turbines at a test field run by ECN (Janssen and Westra, 2000; van Holten, 2000) and there were also some small demonstration projects, in which new prototypes and turbines in new applications were supported, e.g. by fiscal incentives ('t Hooft, 2000).

On the market side, the influence on the function 'Facilitate market formation' appears to have been much weaker than in Germany; the green demand was not so strong and the interest from the utilities was weak. However, in 1982 some electricity producers decided to build an experimental wind farm with 300 kW turbines (Kuipers, 2000a). Half of the planned cost of 50 MNLG was provided by the wind energy programme ('t Hooft, 2000). However, although the stated objectives were achieved,[19] there were problems with the turbines; when they were put into operation, many components failed (IEA, 1991) and in the longer run, the maintenance cost turned out to be too high to keep the turbines in operation (Kuipers, 2000b).[20] The primary impact on the industry structure was that two large firms left the wind turbine industry.[21]

When the efforts of the larger firms faded, some of their R&D people moved to the smaller firms (van Holten, 2000; van Kuik, 2000). The mobility in the labour market also increased as people from research institutes and universities went to the industry (Versteegh, 2000). This mobility later proved to be important for the choice of technology in these firms (see below).

In the second sub-phase, the situation changed substantially (see Fig. 4.3). The political interest in wind was revived by an energy price crisis in 1984 ('t Hooft, 2000) and an official goal of 1,000 MW by the year 2000 was set in 1985. As the demand was still weak, one of the primary government goals was to influence the function 'Facilitate market formation'. For this purpose, another programme was introduced in which an investment subsidy was awarded to utilities and independent customers (Carlman, 1990; 't Hooft, 2000). This resulted in a small market

[18] The second programme even had as an explicit goal to involve the Dutch industry in the development and production of wind turbines (van Holten, 2000; Verbong, 1999).

[19] One of the objectives of the experimental farm was to develop further wind turbine technology and to optimise the application of wind turbines (Kuipers, 1986). Other objectives were to study the aerodynamic effects of putting many turbines close together (Kuipers, 2000a) and to study the social and environmental aspects of the farm (Kuipers, 2000b).

[20] As one of the objectives was to develop further wind turbine technology, a new prototype was developed and installed in the farm. However, this meant that there was no room for gradual improvements and this may have contributed to the turbine problems, although they were probably also caused by the supplier's lack of experience.

[21] One was Stork FDO that lost its enthusiasm for wind turbines partly as a result of competing for the contract without getting it (Verbruggen, 1999). The other was the supplier, Polenko/Holec, that after the turbine problems decided that wind turbines was not a core business (van Kuik, 1999).

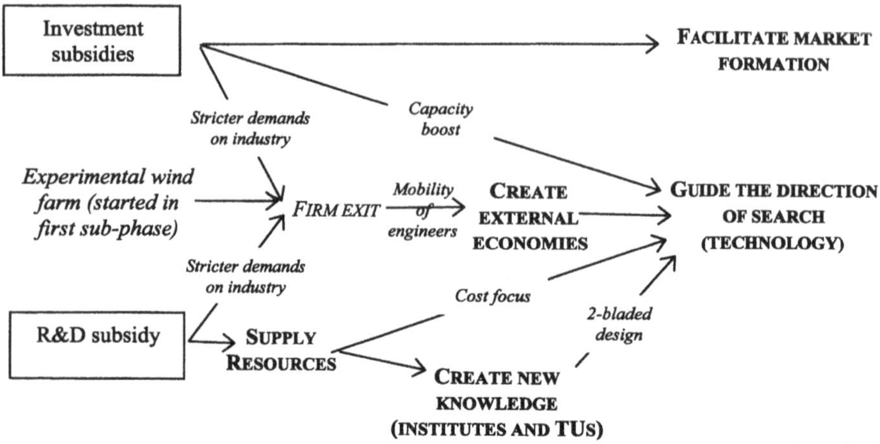

Fig. 4.3. The Dutch innovation system in the second part of the first phase

expansion (see Table 4.2). The majority of the turbines were Dutch, even though it was not required by the programme ('t Hooft, 2000).

Table 4.2. Installed capacity in the Netherlands 1980–1989

	New power capacity (MW)	Accumulated power capacity (MW)
1980		0.02[a]
1981	0.38	0.4[a]
1982	1.1	1.5[a]
1983	1.7	3.2[a]
1984	0.5	3.7[a]
1985	1.8	5.5[a]
1986	1.5	7[b]
1987	9	16[b]
1988	6	22[b]
1989	11	33[b]

Sources: Elaboration on:
[a] *Kamp (1999)*
[b] *IEA (1997a)*

The programme also influenced the function 'Guide the direction of search' in three ways with respect to technology choice. First, the design of the investment subsidy made some firms 'boost' the generator of the turbines in order to maximise the subsidy for their customers (Gipe, 1995; 't Hooft, 2000).[22] Some of the Dutch turbines that were developed were, therefore, not cost competitive internationally. Second, the programme's focus on finding a breakthrough in cost-effectiveness was

[22] The customer received an amount of money per kW installed (up to a maximum amount). The customer, thus, got more support for a turbine with a large generator than for a turbine with a smaller generator.

one of the reasons why some of the Dutch firms developed two-bladed turbines.[23] The programme contained an R&D subsidy for which only firms that aimed at developing cheaper turbines were eligible.[24] When the firms sought for cost-effective designs, they were directed towards the two-bladed design by the researchers in engineering firms, institutes and technical universities ('t Hooft, 2000).[25] In part, this was probably made possible by the mobility of engineers (see above), which influenced the function 'Create External Economies', increasing the receiver competence of the firms.[26] Third, the focus on cost-effectiveness also drove firms to develop relatively large turbines since these were considered to be better in this sense ('t Hooft, 2000).[27]

The programme also involved stricter demands on the firms; instead of supporting basically all firms, project proposals were evaluated and ranked more systematically ('t Hooft, 2000) and firms were, thus, more obviously made to compete for funding. The selection pressure increased and many firms left the industry (Verbong, 1999). In 1988, only seven Dutch firms had certified turbines, which was a prerequisite for obtaining building permits and investment subsidies, and these firms were also the only firms that had received funding for turbine developments (Hack and de Bruijne, 1988).[28] The experiences in relation to the experimental wind farm (see above) further induced firm exit. At the end of the phase, the industry consisted of one firm in the category below 100 kW and four firms in the category 200–500 kW[29] (de Bruijne, 1990).[30]

In summary, the result of the second sub-phase was a growing domestic market, the selection of a number of firms and the choice of a, by international standards, quite unusual dominant design. Thus, at a time when the German industry was still in a phase of variety creation, the Dutch industry had taken its first step into a phase of market growth and selection, both in terms of technology choice and commercial success.

---

[23] For example, Newinco introduced its first two-bladed design in 1989 (Versteegh, 2000). Another firm, Lagerwey, actually had two-bladed turbines from the very beginning.

[24] The firms had to state how they could contribute to the aims of the programme (the installation of 150 MW, a cost-effective wind turbine and a self-supporting industry in the period 1986–1990) in order to receive the 70 percent subsidy for their development costs (Hack and de Bruijne, 1988).

[25] From the very beginning, the Dutch researchers focused their research on the two-bladed design (van Holten, 2000; van Kuik, 2000; Versteegh, 2000). It was, and still is, claimed to be less expensive than the three-bladed.

[26] Several of our interviewees have stated that the competence gap between, e.g., ECN and the firms had been much too large for firms to adopt the research results earlier.

[27] The development of larger turbines was probably also a 'natural' choice for the firms that saw utilities as their main customers.

[28] A seventh firm, Stork, had certification for its NEWECS turbines (Hack and de Bruijne, 1988), but at this time it had already left the wind turbine industry and was concentrating on providing engineering consultant services.

[29] However, one of these firms seems to have been active on paper only.

[30] There were probably also some manufacturers of small ($<$ 50 kW) turbines.

4.1.3 The Swedish case

In Sweden, an R&D policy provided substantial influence on the function 'Supply resources' until the mid-1980s. It began in 1975, when Saab received funding for the design of a 60 kW experimental turbine (DFE, 1979). In 1977, a more substantial R&D programme for wind energy was initiated, 105 million SEK over a period of three years (Carlman, 1990).[31] Until 1979, Sweden spent more government money on wind energy R&D than either Germany or the Netherlands, and the Dutch accumulated R&D funding did not reach the Swedish level until 1985 (IEA, 1997b).

However, the supply of resources influenced the function 'Guide the direction of search' in such a way that the function 'Create new knowledge' was restricted to very large turbines. The aim of the programme was to develop 2–3 MW turbines and there was no support for small or medium-sized turbines. Two full-scale MW turbines were erected in 1982 and 1983 respectively, one by a shipyard and one by a mechanical engineering firm (which later became part of Kvaerner Turbin) (Göransson, 1998). Much due to these turbines, Sweden was considered to be one of the leading countries in wind energy at this time (Carlman, 1990).

Apart from the government funding, which induced a few large firms to enter the wind turbine industry, there was hardly any positive influence on the function 'Guide the direction of search' (in terms of growth potential). The local market was close to non-existent since there was hardly any positive influence on the function 'Facilitate market formation', such as demonstration programmes or 'green' demand.[32] The lack of demand was further aggravated by the fact that several new nuclear power plants were taken into operation after the referendum, which led to an expansion in the supply of electricity and low prices. More importantly, neither the market growth in the 1980s in California and Denmark, nor recessions in other areas led to a search into wind turbines, presumably due to the lack of legitimacy of the technology.

This lack of legitimacy had its roots in the Swedish nuclear power issue, which had been discussed since the early 1970s and which led to a referendum in 1980 after the Harrisburg accident. It was decided that the Swedish nuclear époque was to end in 2010, but the issue has still not been settled. The energy-intensive industry, the capital goods industry and the two dominant utilities formed a powerful alliance to stop the threat of nuclear power being dismantled.[33] In the other camp, the anti-nuclear power movement referred to the results of the referendum and demanded the dismantling of the first nuclear power station.

The Social Democrats in power had considerable problems to balance the demands of the two camps, which led to an uncertainty with regards to policy and an

[31] The political force behind this programme, as well as its 1981 follow-up, was the Centre party. The Centre party, which is a non-socialist party, has always been the main political force in favour of renewables, whereas the other parties, with the exception of the Communist party, have had either a cool or a very hostile attitude. For more details, see Carlman (1990).

[32] As late as in 1992, there were only 39 turbines in Sweden (BTM, 1998, Table 2-5).

[33] These actors had successfully worked together earlier to develop a good infrastructure to provide industry and consumers with cheap electricity (Kaiser, 1992).

associated lack of predictability of the conditions in the energy field (Göransson, 1998). Over time, a 'nuclear power trauma' emerged, which reduced all energy issues to one: the issue of whether or not to dismantle the Swedish nuclear power plants. In the very heated debate, renewable energy technology was seen only as a substitute for nuclear power.[34] Consequently, an interest in, for example, wind power was automatically assumed to involve an anti-nuclear stance and, thus, a 'betrayal' of Swedish industry, which enjoyed the benefits of cheap nuclear power. Thus, it was not surprising that few industrial actors wanted to be associated with wind power and, obviously, it did not gain any legitimacy either in the eyes of the capital goods industry or among potential industrial users.

In 1985, the new Social Democratic government drastically reduced the level of ambition in the wind energy programme (Carlman, 1990). The few existing firms were severely constrained, as they needed subsidies to sell their turbines (as did all other large turbines in this period).[35]

In the parliament, there were, however, several demands made for government financed or subsidised demonstration plants, both in 1985 by the Centre party and in 1986 by an expert group (Carlman, 1990). The responses to the latter proposal were typical for the 'trauma': The federation of industries was critical; the farmer's association wanted demonstration plants for small turbines and an environmental group wanted an expansion of both small and large turbines (Carlman, 1990). This group also suggested the implementation of measures to ensure easy grid connection for the turbines. In 1989, the Centre party argued that the utilities should be obliged to accept electricity from wind turbines and that there should be a guaranteed minimum price (Carlman, 1990).[36] None of these suggestions were, implemented and the market for wind turbines developed very slowly. In 1990, there were fewer than 30 commercial turbines (Carlman, 1990) and a total stock of 4.4 MW (Elforsk, 1996).

The Swedish industry was still almost non-existent, even though a small mechanical engineering firm entered the industry and Kvaerner took up its work on large turbines again when approached by the German firm MBB (Göransson, 1998).[37] The former firm supplied three 250 kW turbines, which were the first medium-scale turbines ever built by a Swedish firm.[38]

Hence, in a phase when Germany and the Netherlands developed a lot of knowledge and a set of industrial firms with experience in building a few hundred turbines, Sweden's main strength lay in designing very large turbines for which there was no market at that time.

---

[34] This also influenced the function 'Guide the direction of search'. Since renewable energy technologies were measured against the yardstick of a nuclear power plant by the utilities, the subsequent technology choice in renewables was biased in favour of large sizes, the only technology which could have an impact on the power balance in the short and medium term (Johnson and Jacobsson, 1999).

[35] For example, Kvaerner's preparations for a series production of 2 MW turbines were never realised.

[36] As we will see, these were key features of the German Electricity Feed-in Law to come a year later.

[37] A 3 MW turbine was developed partly with support from the Swedish wind energy programme and Vattenfall and was erected in 1992 (Göransson, 1998).

[38] The entry of this firm, Zephyr, was induced by the municipal utility, which had an ambition to develop a 'green profile' (Svensson, 1998).

4.2 The phase of turbulence and growth

4.2.1 The German case

In the second phase, the German case was characterised by virtuous circles, in which the functions influenced each other in a self-reinforcing process (see Fig. 4.4). These circles were initially induced by measures affecting the price of wind electricity, which influenced the function 'Facilitate market formation' and led to a rapid market expansion.

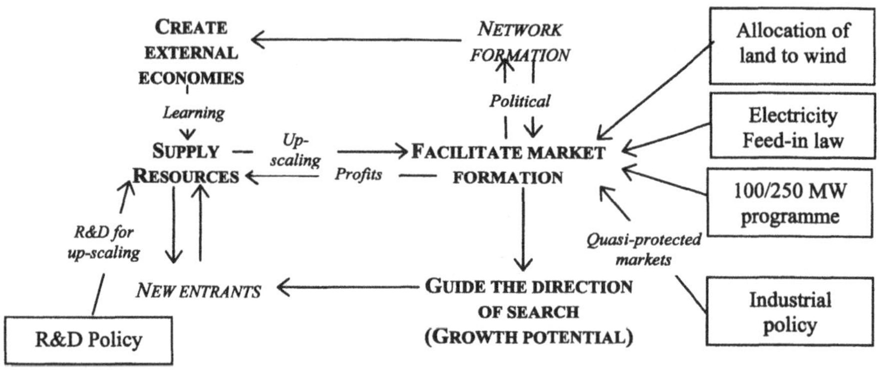

Fig. 4.4. The German innovation system in the phase of turbulence and growth

The first measure was a federal combined market stimulation and scientific programme, which was initiated in 1989. This programme initially aimed at installing 100 MW of wind power – a huge figure compared to the stock of 19 MW in 1989 – and was later expanded to 250 MW. The programme mainly involved a guaranteed payment per kWh electricity produced.[39] The bulk of the sales within the programme took place 1990–1995 and the programme accounted for most of the close to 60 MW that were sold in the years 1990–1992 (ISET, 1999b, Table 3).

The second measure, the Electricity Feed-in Law (EFL), came into force in 1991.[40] It required utilities to accept electricity delivered to the grid by independent wind turbines and to pay 90 percent of the average consumer electricity price. The payment stipulated by the EFL was put on top of the 100/250 MW programme subsidy as well as of various state programmes (DEWI, 1998), which resulted in very high payments.

The powerful combination of market stimulation measures resulted in an 'unimaginable'[41] market expansion from about 12 MW in 1989 to close on 490

[39] In addition, private operators, e.g. farmers, had the possibility to obtain an investment subsidy (Durstewitz, 2000a).

[40] The Law was backed by all parties in the parliament (Ahmels, 1999).

[41] This was the word used by a central person in the evolution of the German wind turbine industry and market.

MW in 1995 (BWE, 2000). Since the payment was based on a law and not a temporary programme, the income generated from wind turbines was both high and predictable, which greatly reduced the risks associated with investment. Farmers, private individuals and firms had a clear economic incentive to invest in wind turbines and, as a consequence, private capital was mobilised on a large scale.

Not surprisingly, some of the economic benefits spilled over to the suppliers in the form of high prices, which through the function 'Supply resources' induced product development, in particular in terms of the up-scaling of turbines (Molly, 1999). One of the reasons for the rapid up-scaling during the 1990s was the allocation of land to wind turbines stipulated by the federal government in 1997; if the states did not designate areas for the erection of wind turbines, operators would be free to erect them anywhere. As the land available for wind turbines became more restricted, the demand for larger turbines increased.[42]

The function 'Supply resources' was also served by the federal R&D programme, which continued to co-finance the industry's development work. For example, some of the early entrants received funding to design medium-sized and large turbines (500 kW, 750 kW and 1 MW), which helped them to up-scale their turbines further. Resources were also provided by some of the new entrants, which were induced by the market expansion (Molly, 1999).[43]

Aided by industrial policies at federal and state levels, German suppliers managed to capture a significant part of the expanding market. An umbrella of implicit and explicit federal and state policies created a temporary quasi-protected market and German firms were able to increase their supply to the German market from 62 turbines (9 MW) in 1989 to 719 turbines (325 MW) in 1995 (elaboration on Durstewitz, 2000). Thus, these policies may be said to have influenced the function 'Facilitate market formation' to the advantage of the German suppliers.

First, an industrial policy element seems to have been present in the 100/250 MW programme. Projects were selected so that there was a wide range of experiments in terms of different applications (state and operator) and types of turbines. Due to this selection process and the large number of applications – as many as 8,000 applications were received and only 1,500 granted (Windheim, 2000a) – there were ample opportunities to manipulate the selection of projects so as to favour German industry (Hoppe-Kilpper, 2000; Molly, 1999; Ahmels, 1999).[44] Moreover, since a ceiling of 40 turbines was set on the sales of each turbine category (DEWI,

---

[42] Moreover, new firms emerged, specialising in erecting and managing wind parks, primarily built with larger turbines, which further stimulated the fast market growth.

[43] As many as 16 German firms started to sell turbines on the German market in the period 1990–1993, although most of them stopped soon thereafter (elaboration on Durstewitz (2000)). Later, two large corporations, Enron and Balke-Dürr, acquired the firms Tacke and Nordex, providing the capital they needed to participate in the race towards larger turbines.

[44] The programme was especially significant for the present German market leader, Enercon. In 1990–1992, between 40 and 50 percent of Enercon's turbines were sold within the programme (elaboration on Durstewitz (2000)). Interestingly, the high share was reached before Enercon developed the E40 model, which used a unique technology and therefore could be assumed to constitute a special 'type'. In total, Enercon sold 325 turbines within the 250 MW programme, which represented about 45 per cent of the total sales by German firms. Other firms which benefited greatly were Tacke, Husumer, Seewind and Ventis (which later spun off the fast growing Dewind).

1998), [45] small firms were able to benefit from the programme (the ceiling was, of course, only relevant to the large and dominant firms). This also worked in favour of the German firms since they were relatively small, especially in comparison with the Danish market leaders. Yet, the dominant Danish industry was not locked out of the market entirely since it was neither possible, due to EU regulations, nor seen as desirable as the benefits of competition would be reduced (Hoppe-Kilpper, 2000).[46] Eventually, the Danish industry received about 35 percent of the projects, the Dutch (Lagerwey) about 7 percent and the German industry the remaining 57 percent (more than 700 turbines) (elaboration on ISET (1999b), Table 11).[47]

Second, at the state level there were explicit or implicit policies to foster a local turbine industry. For instance, at the end of the 1980s Tacke benefitted when Nordrhein-Westfalen created a programme where one of Tacke's turbines was the only one eligible for a 50 percent investment support (Tacke, 2000).[48] Other firms had close local user-supplier relationships. For example, Enercon sold its first units of an early turbine to local utilities (Reeker, 1999) and Husumer experienced that there were strong local biases in the choice of supplier (Schult and Bargel, 2000).

The growing strength of the suppliers and users allowed for the formation of two types of powerful networks. First, it led to the emergence of learning networks via the function 'Create external economies'. These developed primarily between wind turbine suppliers and local component suppliers due to the need to adapt the turbine components to the particular needs of each turbine producer.[49] The benefits of learning also spilled over to new entrants, since they influenced the function 'Supply resources'; subsequent entrants could rely on a complete infrastructure.[50] Indeed, some new entrants were able to work as design firms, i.e. without the production of components or in-house assembly (Mayer and Delabar, 2000), which made it possible for them to minimise capital investments.[51] Second, political networks were formed between competitors with a common interest in influencing the institutional framework to the benefit of the whole industry. These networks were manifested in an active industry association that enrolled both turbine suppliers and

[45] The ceiling probably became higher as the programme changed from 100 to 250 MW.

[46] The former CEO of Vestas, Finn Hansen recalled however that the Danish firms found the German market difficult to penetrate but never understood why (Hansen, 1999).

[47] This is based on information on the firms that had installed more than 40 wind turbines by 1998. Thus, the tail of German firms, which sold fewer units, has been left out. Therefore, the share of German firms is slightly underestimated. As earlier, we have treated Nordex as a German firm.

[48] This program was seen as a second start for the firm, which later grew to be the second largest in Germany. The CEO considered this programme only next to their earlier California participation in importance to the development of the firm.

[49] Relationships to customers seem to be of less importance, which is not surprising since most of the customers have been farmers. Some larger customers are, however of importance today for feed-back. (Hansen, 1999).

[50] Today, the supplier industry is well developed in Germany and several of the interviewees emphasised the importance of having access to this industry locally.

[51] Other parts of the infrastructure were set up as well, for example the German wind energy institute (DEWI), which is an organisation that bridges industry with customers and government (Molly, 1999), and wind turbine test sites, which have been of great importance for the turbine manufactures (Windheim, 2000a).

turbine owners,[52] and proved to be of great importance during the ensuing battle over the feed-in law.

At the time of the design of the EFL, neither the opponents nor the proponents of wind power could have imagined the scale of the diffusion that ensued (Ahmels, 1999). However, in the mid-1990s the rapid diffusion of wind turbines led to a response from the larger utilities. Intense discussion and lobbying followed, which reintroduced substantial uncertainty and the market stagnated. In 1997, a select committee with 15 members of parliament was responsible for investigating whether or not the law should be amended. In the end, the wind turbine lobby won the political battle, although it was a close call; in the select committee, the proponents of a continued law won the vote by eight to seven (Ahmels, 1999). This was largely a result of the rapid diffusion of wind turbines in the first half of the 1990s and the associated growth of the German industry, which made it possible to add economic arguments to environmental ones in favour of wind energy. Additionally, the utilities did not get support from any of the political parties (Molly, 1999). Indeed, as one CDU member of the industry said, "In this matter we collaborate with both the Greens and the Communists". Nor did the German federation of industries (VDMA) choose to support the power companies when they opposed the EFL (Tacke, 2000).

In conclusion, the market formation set in motion a set of virtuous circles, which resulted in the German industry narrowing the gap to the Danes considerably. Today, there are at least nine German firms active in the industry.[53] Yet, this evolution would hardly have been possible without the phase of extensive experimentation in the 1980s, which led to the emergence of a German wind turbine industry strong enough to respond to the effects of the function 'Facilitate market formation'.

4.2.2 The Dutch case

In the Dutch case, the virtuous circles of market growth, increased industry resources and growing political strength did not appear for two main reasons: The domestic market did not develop as expected and the Dutch industry failed to exploit the growing German market (see Fig. 4.5).

In 1990, the total installed capacity was less than 50 MW and the Dutch goal of 1,000 MW in 2000 was still far away. The government tried to influence the function 'Facilitate market formation' through continued investment subsidies (Verbong, 1999; Wolsink, 1996), but it did not have the intended effect. One of the reasons for this was the problem of finding sites for the turbines. The population density is

[52] In addition, it collaborated with other industry associations in the renewable field in an umbrella organisation which had 8,000 members in 1999 (Ahmels, 1999).

[53] Enercon, Tacke (bought by Enron and recently by GE), Dewind, Husumer (recently bought by Jacobs), Jacobs Energy, Fuhrländer, WTN, Frisia and Seewind; these firms sold at least one turbine on the German market in 1999. There may also be other firms operating, selling very small turbines or selling only abroad.

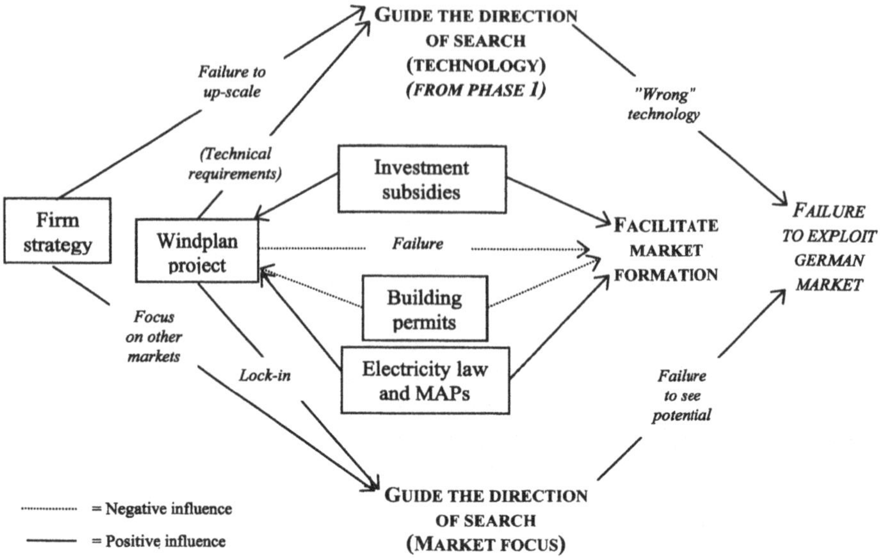

Fig. 4.5. The Dutch innovation system in the second phase

very high and the building permit procedure was slow and time-consuming (Gipe, 1995; ‘t Hooft, 2000).[54]

However, at the end of the 1980s a new electricity law created a demand for wind energy. The law separated electricity production from electricity distribution (Verbong, 1999) and the electricity distributors could not produce or import electricity with the exception of electricity produced by renewable energy technology (Kip, 1999).[55] Moreover, the distributors were allowed to finance their investment in renewable energy technology via a new electricity tax (Wolsink, 1996).

Since the distributors believed that co-ordinating their investments in wind energy would result in better and cheaper turbines, they formed the Windplan foundation aiming at installing 250 MW over a five-year period (Kuipers, 2000a). Compared to the market size at this time, the promised 50 MW per year during five years was a lot and the hopes on the Windplan project were high.[56] The Dutch firm Newinco/Nedwind started to deliver turbines, but the project was quite suddenly abandoned in 1993 (Verbong, 1999). The primary reason was that the electricity distributors started to question the benefit of joint procurement; they believed that they could get lower prices and better guarantees if they bought the turbines themselves (Kuipers, 2000a). Another reason was the problem to obtain building permits

[54] Obtaining building permits for wind turbines involve changing the zone plans of the municipalities and applying for a building permit. Both these decisions may be appealed at numerous levels and in total, a normal project takes over five years to complete. (Janssen and Westra, 2000; ‘t Hooft, 2000).

[55] The electricity distributors saw wind power as an opportunity to strengthen their bargaining power in relation to the large electricity producers and improve their environmental image (Verbong, 1999).

[56] This was also a lot more than the German 100 MW programme, which was initiated at about the same time (at that time it was not yet clear that it would be extended to 250 MW).

described above (Janssen and Westra, 2000; Kuipers, 2000a; Wolsink, 1996), which had the same effect on the Windplan members as on other customers.

The central authorities did, however, make an attempt to solve the building permit issue. They made an agreement with some of the provincial authorities about how to distribute the 1,000 MW (Kuipers, 1991). However, neither central nor provincial authorities had any real decisive power over building and environmental permits – these were obtained from the local authorities, and they were not included in the agreement (Janssen and Westra, 2000; Kip, 1999) and had no real interest in complying with the agreement.[57] Thus, the building permit issue continued to block the function 'Facilitate market formation'.

In spite of the problems with the Windplan project and the building permits, the Dutch investment subsidy was withdrawn in 1996 (Novem, 1999; 't Hooft, 2000).[58] A couple of other measures were implemented to support wind power: tax schemes designed to support environmental investments in general (Novem, 1999) and, later, a new electricity law, which specified the share of the utilities' electricity sales that had to come from renewable energy sources.[59] The result of these measures was not, however, very impressive. After 1995, the market stayed at a stable level of 40-50 MW per year and in 1999 the total installed capacity was still nowhere near the goal of 1,000 MW.

One reason why more powerful measures were not used to facilitate the market formation seems to have been that wind power was not really an important political issue (Janssen and Westra, 2000). For example, it is in principle possible for the government or parliament to impose directives for land usage on the local authorities in cases of 'national interest', but wind power has not been considered sufficiently important to get that kind of support ('t Hooft, 2000).

The second reason for the lack of virtuous circles in the Dutch case was that the Dutch industry failed to exploit the German market, which started to expand a few years prior to the failure of the Windplan project.[60] There were two primary causes of this failure.

First, some of the Dutch turbines turned out not to be in demand on the German market due to an 'inappropriate' choice of technology. During the second phase, a change to only three-bladed turbines was dictated by the German market ('t Hooft, 2000; van Kuik, 2000) and there was not yet any demand for large turbines. However, as mentioned earlier, most Dutch turbines were two-bladed and quite large. In addition, some claim that although the requirements of the Windplan project were technically feasible, they were not commercially feasible with respect

57 There were few benefits and many disadvantages associated with wind energy for local communities (van Kuik, 2000). In addition, the central government was fairly arrogant towards them, which made them unenthusiastic (Kip, 1999).

58 Ironically enough, this caused the market to 'boom' in 1995, much due to the perceived uncertain future for wind power in the Netherlands (IEA, 1998).

59 It also included a 'green labelling' system where every kWh of electricity produced from renewable energy sources got a label that may be bought and sold (van Zanten, 1999).

60 The exception was Lagerwey, which entered the German market and was fairly successful initially, largely because it had a niche market (in the farmers) and a good reputation. However, when the rapid up-scaling begun, Lagerwey's managers made the strategic mistake of ignoring it (Boursma, 1999) and Lagerwey quickly lost its market position.

to other markets.[61] Thus, the requirements may have influenced the function 'Guide the direction of search' in terms of technology choice even further away from the demands of the international (including the German) market.

Second, and probably more important, the function 'Guide the direction of search' in terms of market focus steered the firms away from the German market so that they failed to see and to react upon its potential. The function was influenced in part by the strategic choices made by the firms and in part by the Windplan project. Both Nedwind and Windmaster put their 'export bets' partly (if not mostly) on the North American and Indian market. They also invested a lot of time and effort in the Windplan project answering the tender and developing new turbines and had, therefore, less management time and resources to develop other markets.[62] Indeed, the promise of the Windplan project probably made it seem unnecessary to look for opportunities elsewhere. Even though over 90 percent of the first 75 MW to be built were reserved for Dutch firms,[63] a large number of foreign firms answered the tender (Kuipers, 2000a). This clearly indicates that the Dutch market was perceived as very interesting for the future.[64] In this perspective, the choice of the Dutch firms to concentrate on their home market is not so difficult to understand.

Without access to a booming market and the associated economic benefits, the Dutch firms had neither the resources to develop their technology fast enough to keep up with the German suppliers nor the political strength to influence the vital building permit issue. Thus, most of the industry stagnated and failed. In 1998, Windmaster went bankrupt and was acquired by Lagerwey. Nedwind also got into trouble and was acquired by NEG Micon. Lagerwey is now the only large Dutch wind turbine firm left.

4.2.3 The Swedish case

In Sweden, there continued to be R&D support for the development of Swedish turbines, which was now carried out mainly in three firms: Kvaerner Turbin, Zephyr and Nordic Windpower.[65] However, the firms had to co-finance the projects to a

---

[61] There are some contradictory statements on this issue – some claim that the turbines developed were not that different from what was already commercially available, whereas others claim that the turbines were very different from other turbines and that they were too advanced to be in demand in other markets.

[62] Of course, when Newinco/Nedwind was chosen as the only supplier (Kuipers, 2000a) it was quite natural for it to concentrate its efforts even more on this project.

[63] In order to receive investment subsidies within the frames of the wind energy programme, Windplan was required by the Ministry of Economic Affairs to buy a large part of the turbines from Dutch producers (Kuipers, 2000a).

[64] One of the interviewees even stated that "at that time, no firm could afford to not be part of the project".

[65] Nordic Wind Power was founded in 1990. Its turbines (a 400 kW turbine and a 1 MW turbine) have an unusual design; the turbine is significantly lighter than other turbines and is, therefore, expected to be cheaper.

larger extent than before (Göransson, 1998) and the technology had to be 'new' in order for the firms to receive support (Svensson, 1998).[66]

It was also difficult for the firms to find venture capital or other partners that could co-finance technology and market development.[67] One reason for the difficulties was that the deregulation of the electricity market changed the status of Vattenfall so that it had to reduce its role as a development partner (Averstad, 1998), a role that had been of importance to several Swedish wind turbine firms. Most importantly, however, potential industrial partners were not interested, which was clearly associated with the 'nuclear power trauma'. As one CEO explains: "There was mental resistance to wind power".

Thus, there was a lack of resources, which was troublesome for all firms; the smaller firms had very weak resource bases and wind turbines were not a prioritised part of the larger Kvaerner corporation. The lack of resources severely constrained the firms, as they needed reference installations to gain creditability in the market and to get enough 'staying power' to challenge the Danish suppliers with their early mover advantages.

The function 'Facilitate market formation' was influenced somewhat by policy measures. A market expansion programme in the form of an investment subsidy was started in 1991 (Averstad, 1998). It was supplemented by an environmental bonus in 1994, and from 1996 utilities were required to buy the wind power produced by independent producers (Averstad, 1998). Although these inducement mechanisms were much weaker than in Germany and Denmark, the market began to expand. The diffusion could have been much faster, though, had there been fewer problems in securing building permits (Grahn, 1998).

However, in sharp contrast to the German and Dutch cases, the phase of experimentation did not lead to the development of a strong Swedish industry with response capacity. In addition, there were no mechanisms that favoured Swedish suppliers.[68] Without a quasi-protected local market, development partners and legitimacy, no virtuous circles were set in motion and the market was handed over almost completely to Danish suppliers.

The remaining industry now consists mainly of two firms: Nordic Wind Power and ScanWind, which is a Swedish-Norwegian spin-off of Kvaerner's wind turbine activities.[69] Although both these firms have some experience in large turbines, they

[66] For example, the firm Zephyr had to continue to develop its second turbine by expanding the wing by 2 meters in order to fulfil the conditions for receiving subsidies for technical development work. As a result, the turbine failed (Svensson, 1998).

[67] For example, it took the founder of a new firm, Nordic Windpower, five years to find a first customer and initial financing. After that, it took eight more years of search before NWP received risk capital from three venture capital firms. (Bergqvist, 1998).

[68] In fact, the Swedish rules for the investment subsidies worked more in favour of the Danish suppliers. The turbines had to be certified in order to receive investment subsidies, and since there was no certification authority in Sweden, it was decided that turbines that were certified in Denmark would automatically be certified in Sweden as well (Svensson, 1998).

[69] After some economic problems, Kvaerner decided to discontinue its wind turbine activities and some of its employees then founded a new firm (Energimagasinet, 2000). Zephyr left the industry in 1998, due to problems in keeping up with the international trends in size of the turbines, which in turn was much due to lack of resources.

are up against Danish and German firms that are now mass-producing 1.5 MW turbines and that are in the process of designing much larger turbines (5 MW). The risk is therefore obvious that Sweden will end up without any firm in this growth industry.[70]

5 Conclusions: Why does performance differ and what implications can we draw for policy?

The objectives of this paper were to develop an analytical framework and to use this framework to explain the performance of the wind turbine industry in three countries.

The analytical framework is based on an innovation system approach in which the system is analysed in terms of its 'functional pattern'. With this framework, we can scrutinise, for instance, the direction of search of various actors, the subsequent type and variety of knowledge generated and the resources supplied to exploit that variety. We can analyse in both static and dynamic terms, i.e. how these functional patterns evolve. In Section 4, we outlined such patterns in the three countries studied.

Looking at the patterns through the 'filter' of life cycle models allowed us also to assess the 'functionality' of the innovation systems i.e. *how well* the functions were performed. We argued in Section 2 that the meaning of functionality would be expected to differ between phase in the evolution of an industry. In an early phase, functionality may be assessed by analysing how the innovation system supports firm entry, the formation of niche markets and the creation of variety, whereas in a later phase, the emphasis is shifted to mass market formation and resource supply to exploit that market.

We suggest that it is useful to analyse the functionality of the German, Dutch and Swedish innovation systems under four headings: (1) variety creation in the first phase, (2) legitimacy of wind energy, (3) market formation in the second phase and (4) the use of industrial policy.

(i) *Variety*

In an early period of an industry's evolution, as that of wind turbines in the 1980s, technological uncertainty is high and industry needs to place its bets widely in terms of experimenting with a variety of designs.

This was done in Germany as well as in Holland through several mechanisms. In both countries, R&D policy encouraged a broad range of technical experiments and some of the resulting turbines were exploited commercially on the market. In Germany, but not in the Netherlands, the Californian and Danish 'booms' and the formation of niche markets were clear inducement mechanisms for firm entry. In addition, some German and Dutch firms responded to the decline in their original

[70] In 2000, ABB announced the launching of a new wind turbine, Windformer. The experience, resources and legitimacy of such a large actor could very well have given the Swedish wind turbine industry a new chance, but ABB exited the industry after less than a year.

business by entering the wind turbine industry. So, at the end of the 1980s, a large number of actors, firms and universities had developed and tried out a range of different designs. Many failed, but this is a necessary ingredient in the formation of a new industry.

In Sweden, the picture was very different. Policy guided the firms in one direction only – MW sized turbines. A couple of these turbines were erected, but apart from that there was hardly any local demand. Nor were any Swedish firms stimulated to enter by the Californian/Danish experience. Only one firm supplying smaller turbines entered at the end of the decade.

Whereas both Germany and the Netherlands managed to create variety both in terms of the knowledge generated and in terms of the actors exploiting it, the Swedish knowledge was, thus, limited to larger turbines and to mainly one firm, Kvaerner. The functionality of the German and Dutch innovation systems was, thus, far superior to that of the Swedish in this respect.

(ii) *Legitimacy*

A key feature in the process of generating variety in Germany and the Netherlands was the early legitimacy of wind turbines. Already in the 1980s, there was a political consensus that wind turbines should be supported and it was legitimate for private capital to exploit wind turbine technology. The legitimacy meant that firms responded to various stimuli, e.g. the Californian 'boom', R&D programmes etc., by diversifying into wind turbines or by starting a new firm. Without these entrants, the variety in terms of knowledge generated would, of course, have been much less.

In Sweden, wind turbines lacked legitimacy. This meant that Swedish firms responded differently to the very same stimuli that made some German firms move into the wind turbine industry. The Danish and Californian experiences simply passed by most of Swedish industry. Due to the 'nuclear trauma' and the associated lack of legitimacy, the few individuals and firms who saw a future in wind turbines faced severe limitations in terms of access to resources (capital), partners, markets and government support (apart from R&D funding to MW turbines). This meant that the very considerable competence built up by government R&D programmes, some of which could have been exploited for smaller turbines, came to little use. Thus, legitimacy is a key concept in an explanation of why the German and Dutch innovation systems had a superior functionality in terms of variety creation in the first phase.

(iii) *Market formation*

The legitimacy of wind turbine technology in Germany also meant that there was little opposition to creating a market formation programme. The 250 MW programme and the initial formulation of the Electricity Feed-in Law (EFL) met little or no opposition from the proponents of centralised power production. The market expansion that followed on these programmes set in motion virtuous circles where the variety generated in the first phase was exploited.

When the debate over the EFL started in the mid-1990s, it took place in a context where the initial legitimacy appears to have been strengthened, in part by the growing economic importance and political strength of the wind turbine industry. The legitimate nature of wind energy meant that the 'battle' over the EFL was won by the infant German wind turbine industry and, as a result, market formation continued to be the driving force in a set of virtuous circles.

In the Netherlands, the local industry[71] was locked into a local market that did not grow very fast in the 1990s. The Windplan project of the early 1990s kept the Dutch firms largely focused on the local market and the Dutch turbines were not in demand abroad due to an inappropriate choice of technology. The Windplan project failed, in part because the siting issue was not resolved. No virtuous circles were put in motion and the initial variety was not exploited. We interpret the failure to solve the siting issue as a failure to develop further the initially reasonably strong legitimacy. The central government did not take adequate steps to overrule the local governments which controlled the planning process.

In Sweden, although the market developed in the 1990s, virtuous circles for a Swedish wind turbine industry were not started simply because it was too weak to respond to the growing demand. Unlike in Germany and the Netherlands, there was an absence of an initial variety from which winners could be selected.

Thus, in the second phase, the functionality of the German innovation system was superior to that of the Dutch in that a larger market was formed and, through virtuous circles, more resources were supplied to the industry. Underlying the greater functionality of the German innovation system was, however, a greater legitimacy.

(iv) *Industrial policy*

In the first half of the 1990s, the German industry was aided by industrial policies at the federal and state levels that created a 'quasi protected' market and a German market share of more than 50 percent, which is especially remarkable considering the otherwise dominant position of the Danish industry. This was clearly of vital importance to the ability of infant German industry to benefit from the powerful market formation locally.

In the Dutch case, there were also some elements of industrial policy. For example, more than 90 percent of the initial order of 75 MW in the Wind Plan project was awarded to Dutch firms. However, since the market never really materialised, the protectionist element had instead the effect of locking the firms to the local market.

In Sweden, energy policy never really had an industrial policy element. When the market expanded in the 1990s, no efforts were made to foster a local industry and the few firms that had entered the industry had almost no chance to be part of a virtuous circle of market expansion and increase in the supply of resources flowing into the industry.

There are a number of lessons for policy. First, when the technological uncertainty is large, as in the case of wind turbines in the first phase, diversity must be

[71] The exception was Lagerway.

fostered. The Swedish R&D policy only financed MW turbines whereas the German and Dutch stimulated knowledge creation with respect to both small and large turbines. Diversity in design developments may have to continue to be supported for a lengthy period. Take for example the two-bladed turbine. It was, and still is, considered to be better than the three-bladed in terms of economic performance, and at the end of the 1980s almost half of the turbines on the German market were two-bladed. Yet, the two-bladed design was defeated by the three-bladed on the German market only a few years later.

Second, the creation of variety is closely connected to the number of actors within a field since these may bring different types of visions, competencies and complementary assets to the new industry. It is, therefore, central to guide the direction of search of a variety of firms towards the new field. The guidance may, as we have seen in our cases, come in many different forms and may be case-specific. Therefore, we will limit this discussion to what we believe is its most important aspect, i.e. legitimacy.

As evident in the three cases, it is vital that legitimacy is created for the new technology or industry. Without legitimacy, private capital will not flow into the industry and without an industry active on the political arena it will be difficult to remove institutional blocking mechanisms, as in the Dutch case of building permits, or to get institutional inducement mechanisms, as the present German version of the feed-in-law, in place. Thus, even if a number of support systems are implemented, a poor legitimacy will obstruct the evolution of a virtuous circle of resource supply, market development and firm growth. Legitimising a new technology may, therefore, be a key policy objective.

The third lesson is that the exploitation of variety is strongly associated with the creation of powerful, predictable and persistent economic incentives, which was evident in the German case. Powerful incentives create the profitability needed to attract private investors. Predictability reduces the uncertainty for the actors involved, and persistent policies are required since the development of an industry takes time – the German economic incentives have been in place for ten years and are only now beginning to bear fruit.

Fourth, it is not the volume of resources supplied in government policy programmes that matter, but how the funds are used to generate a self-reinforcing process. As we have seen, a large government financed R&D programme was not enough to create a successful Swedish industry. In contrast to the Swedish case, policy agents[72] need to be concerned with all the required functions of an emerging system, and to intervene, if necessary, to support those functions that are relatively poorly served (or not served at all).

How to do this is, however, by no means self-evident. First, each function may be served in several different ways.[73] For example, resources may be supplied through a number of sources (e.g. government programmes, private investors and venture capital firms) to different recipients (e.g. suppliers or buyers) and in a variety of

[72] These include not only government bodies but all types of actors who have an interest in influencing the functionality of a system.

[73] See Rickne (2000) for an analysis of the relations between actors and functions in the case of biomaterials.

forms (e.g. subsidies and loans). Thus, a well functioning system may, presumably, not come about in one way alone, as illustrated in the phase of experimentation where Germany and the Netherlands managed to create variety in quite different ways.

Second, in some cases a number of different mechanisms are needed in order for a function to be served. Take the example of the function 'Facilitate market formation'. The Dutch experience shows there is more to it than relative prices and financial incentives; the early investment subsidies clearly had limited effect due to the problems of obtaining building permits. Likewise, the German 'boom' was not created by financial incentives alone; it took a combination of investment subsidies, legislation, legitimacy and industrial policy to start and to maintain the virtuous circle that eventually made the German market the fastest-growing in the world. This means that policy makers may have to work with a number of mechanisms simultaneously.

Third, due to the systemic character of industrial development, it may be impossible for one function to be served unless other functions are served as well. For example, some knowledge can only be created through a process of learning-by-using, which requires a market. The presence of a market may also be necessary for the direction of search of industrial actors to be guided towards a new technology. Policy makers therefore need to consider and understand the interdependencies between the various functions.

These features make it difficult to know how to influence a particular function as well as to predict the outcome of an intervention. We will illustrate this with market formation programmes in Germany and Holland. At about the same time, around 1989, Germany and the Netherlands designed market formation programmes of similar sizes. Clearly, nobody could have foreseen the formidable success of the German programme in creating a market (and indirectly influencing functions), nor the failure of the Dutch. Had instead the Dutch been successful with their programme (as many Dutch and foreign firms expected them to be) and the German programme failed (something which was entirely conceivable), the Dutch may today had been the ones catching up with the Danes.[74] Indeed, for an observer in the late 1980s, the Dutch industry must have seemed as likely to succeed as the German (if not more). This should make us humble with respect to our ability to control the sequences of events leading to the growth of new industries.

References

Ahmels H-P (1999) Interview with Dr. Hans-Peter Ahmels, BWE, October 8th
Averstad K (1998) Interview with Kenneth Averstad, Vattenfall MiljöEl, October 22nd
Bergqvist B (1998) Interview with Bruno Bergqvist, Nordic Windpower, November 5th
Boursma R (1999) Interview with Remco Boursma, Lagerwey, October 13th
BTM Consult (1998) International Wind Energy Development. World Market Update 1997. Ringkoebing

[74] For instance, who could have foreseen that by the time the EFL was questioned, the wind turbine industry would have the lobbying strength to counteract the resistance? How could one have foreseen that the Dutch government would not succeed to solve the siting problem or that the electricity suppliers would suddenly decide not to go through with the Windplan project?

BTM Consult (1999) International Wind Energy Development. World Market Update 1998. Ringkoebing
BTM Consult (2000) International Wind Energy Development. World Market Update 1999. Ringkoebing
BWE (2000) Statistik Windenergie in Deutschland. URL: http::www.wind-energie.de/statistik/deutschland.html (Acc. 000223)
Carlman I (1990) Blåsningen. Svensk vindkraft 1973–1990. Geografiska Regionstudier Nr 23. Kulturgeografiska Institutionen vid Uppsala Universitet (In Swedish)
Carlsson B, Stankiewicz R (1995) On the nature, function and composition of technological systems. In: Carlsson B (ed.) Technological Systems and Economic Performance: The Case of Factory Automation. Kluwer Academic Publishers. Dordrecht, 1995
de Bruijne R (1990) Wind Energy in the Netherlands. Paper presented at the European Community Wind Energy Conference, September 10–14, Madrid, Spain
DEWI (1998) Wind energy information brochure. German Wind Energy Institute, Wilhemshaven
DFE (1979) Värdering av insatserna inom området vindenergi. DFE-rapport nr 18. Delegationen för energiforskning. Stockholm (In Swedish)
Durstewitz M (2000) 250 MW wind-programme. ISET. Kassel University
Durstewitz M (2000) private communication with Michael Durstewitz
Ehrnberg E, Jacobsson S (1997) Technological discontinuities and incumbent's performance: an analytical framework. In: Edquist C (ed.) Systems of Innovation, Technologies, Institutions and Organizations. Pinter Publishers. London, 1997.
Elforsk (1996) Elforsks driftuppföljning av vindkraft, årsrapport 1995. Elforsk-Nutek (In Swedish)
Energimagasinet (2000) Svensk-norska ScanWind utvecklar 3 MW-vindkraftverk. Energimagasinet vol. 21, no. 5
European Commission (1997) Windenergy – the facts. Directorate-General for Energy.
Gipe P (1995) Wind energy comes of age. Wiley, New York
Grahn P (1998) Interview with Peo Grahn, Marketing Manager Vestasvind Svenska AB, April 24th
Göransson B (1998) Interview with Bengt Göransson, Kvaerner Turbin/Nordanvind, November 16th
Hack RK, de Bruijne R (1988) The development of wind energy in the Netherlands. Paper presented at the European Community Wind Energy Conference, June 6–10, Herning, Denmark
Hansen FM (1999) Interview with Finn M. Hansen, Managing Director of Tacke Windenergie, October 12th
Hantsch S (1998) Wege zum wind. Diplomatarbeit der Univeristät Wien
Hemmelskamp J (1998) Wind energy policy and their impact on innovation – an international comparison. Institute for Prospective Technology Studies, Seville, Spain
Hidefjäll P (1997) The pace of innovation. Patterns of innovation in the cardiac pacemaker industry. Doctoral Thesis. TEMA Technology and Change, University of Linköping, Sweden
Hoppe-Kilpper M (2000) Interview with Martin Hoppe-Kilpper, ISET, February 8th
IEA (1997a) IEA Wind Energy Annual Report 1996
IEA (1997b) IEA Energy Technology R&D Statistics 1974–1995. OECD/IEA
IEA (1998) IEA Wind Energy Annual Report 1997. National Renewable Energy Laboratory. Golden, Colorado, USA
ISET (1999a) Annual Installation Rate in Germany. URL: http://www.iset.uni-kassel.de:888/reisi/owa/www_page.show?p_name=121007&p_lang=eng (Acc. 990608)
ISET (1999b) WMEP – Jahresauswerung 1998. Institut für Solare Energieversorgungstechnik (ISET), Verein an der Universität Gesamthochschule Kassel, Kassel (In German)
Jacobsson S, Johnson A (2000) The diffusion of renewable energy technology: an analytical framework and issues for research. Energy Policy vol. 28, pp. 625–640
Janssen B, Westra C (2000) Interview with Bert Janssen and Chris Westra, ECN, March 6th
Johnson A (1998) Functions in Innovation System Approaches. Mimeo, Department of Industrial Dynamics, Chalmers University of Technology, Sweden
Johnson A, Jacobsson S (1999) Inducement and Blocking Mechanisms in the Development of a New Industry. In: Johnson AM: Renewable Energy Technology: A New Swedish Growth Industry? The Influence of Innovation Systems on Industrial Development. Thesis for the degree of Licentiate Engineering, Chalmers University of Technology, Gothenburg

Johnson A, Jacobsson S (2000) Inducement and blocking mechanisms in the development of a new industry. To appear in: Coombs R, Green K, Walsh V, Richards A (eds.) Technology and the Market: Demand, Users and Innovation. Edward Elgar. Cheltenham and Northhampton, Massachusetts (This is a revised and shortened version of Johnson and Jacobsson, 1999)
Kamp L (1999) Data supplied by Linda Kamp, University of Utrecht, October 6th
Kaiser A (1992) Redirecting Power: Swedish Nuclear Power Policies in Historical Perspective. Annu. Rev. Energy Environ vol. 17, pp. 437–462
Karnoe P (1991) Dansk Vindmölleindustri – en overraskende international succes. Samfunslitteratur, Frederiksberg
Kip W (1999) Interview with Wilhelmina Kip, EnergieNed, October 6th
Kuipers J (1986) Quality assurance during engineering and manufacture of the Sexbierum wind farm project in the Netherlands. Printout of paper intended for the official proceedings of the European Wind Energy Conference in Rome, October 7th-9th, provided by Mr. Kuipers
Kuipers J (1991) Promotion of wind-generated electricity by cooperation of Dutch utilities. Printout of paper intended for Windpower 1991 Conference in Palm Springs (California, USA), provided by Mr. Kuipers
Kuipers J (2000a) Interview with Mr. Joop Kuipers, Windplan project leader, March 7th
Kuipers J (2000b) E-mail communication with Mr. Joop Kuipers, October 10th
Kåberger T (1997) Data supplied by Tomas Kåberger, Department of Physical Resource Theory, Chalmers University of Technology. Gothenburg, Sweden
Köhler N (2000) Billig vindkraft tar upp kampen med oljan. Ny teknik, 2000:23 (In Swedish)
Mayer, Delabar (2000) Interview with Mr. Mayer and Mr. Delabar, Dewind, January 11th
Molly JP (1999) Interview with Molly, JP DEWI, October 7th
Müller D (1999) Interview with Dirk Müller, Nordex Planungs- und Vertriebsgesellschaft, October 11th
Nelson RR (1994) The co-evolution of technology, industrial structure, and supporting institutions. Industrial and Corporate Change vol. 3, no. 1, pp. 47–63
Novem (1999) 100 MW per year. URL: http://www.novem.org/netherl/subjects/wind100mw.htm (Acc. 990712)
Rickne A (2000) New technology based firms and industrial dynamics – evidence from the technology system of biomaterials in Sweden, Ohio and Massachusetts
PhD Thesis Department of Industrial Dynamics, Chalmers University of Technology, Gothenburgh, Sweden
Reeker C (1999) Interview with Carlo Reeker, BWE, October 11th
Schult C, Bargel A (2000) Interview with Christian Schult and Angelo Bargel, Husumer Schiffswerft, January 12th
Svensson L (1998) Interview with Leif Svensson, managing director Zephyr Energy, April 17th
Tacke F (2000) Interview with Franz Tacke, founder and former managing director Tacke Windenergie, March 9th
't Hooft, J (2000b) Interview with Jaap 't Hooft, Novem, March 7th
Tushman ML, Anderson PC, O'Reilly C (1997) Technology cycles, innovation streams, and ambidextrous organizations: organization renewal through innovation streams and strategic change. In: Tushman ML, Anderson PC (eds.) Managing Strategic Innovation and Change: A Collection of Readings. Oxford University Press. New York, 1997
Utterback JM (1994) Mastering the dynamics of innovation: how companies can seize opportunities in the face of technological change. Harvard Business School Press, Boston, Massachusetts. (Chapter 2, pp. 23–55 and chapter 4, pp. 79–102)
Utterback JM, Afuah AN (1998) The dynamic diamond: A technological innovation perspective. Econ. Innov. New Techn. vol. 6, pp. 183–199
Utterback JM, Suarez FF (1993) Innovation, competition, and industry structure. Research Policy, vol. 22, no. 1, pp 1–21
van Holten T (2000) Interview with Professor Theo van Holten, TU Delft, March 8th
van Kuik G (2000) Interview with Professor Gijs van Kuik, TU Delft, March 8th
van Zanten W (1999) Green labels. CADDET Renewable Energy Newsletter issue 1/99. IEA/OECD
Verbong GPJ (1999) Wind power in the Netherlands 1970–1995. Centaurus vol. 41, pp. 137–170
Verbong GPJ (2000) Personal communication with and printout from Gert Verbong, Technical University of Eindhoven, March 10th

Verbruggen T (1999) Interview with Theo Verbruggen, Stork Product Engineering, October 13th

Versteegh C (2000) Interview with Cees Versteegh, Garrad Hassan (earlier Lagerwey, Newinco and WindMaster), March 9th

Windheim R (2000a) Interview with Dr. Rolf Windheim, Jülich Forschungscentrum, March 7th

Windheim R (2000b) Data supplied by Dr. Rolf Windheim, Jülich Forschungscentrum, on projects funded within the federal R&D wind energy progamme

Wolsink M (1996) Dutch wind power policy: Stagnating, implementation of renewables. Energy Policy vol 24, no 12, pp 1079–1088

Intangible investment and human resources

Michael Peneder

Austrian Institute of Economic Research, WIFO P.O. Box 91, 1030 Vienna, Austria
(e-mail: peneder@wifo.ac.at)

Abstract. To make intangibles more 'tangible' for empirical analysis, statistical cluster techniques are applied in the development of two new taxonomies of manufacturing industries. The first focuses on the distinction between exogenous, location dependent comparative cost advantages, such as the relative abundance of capital or labour, and endogenously created firm specific advantages resulting from intangible investments in marketing or innovation. The second taxonomy discriminates between industries according to their employment of skilled labour. Finally, econometric tests are used to investigate the presumed complementarity between intangible investments and human resources.

Key words: Intangible investments – Human resources – Endogenous sunk costs – Market process – Industry structure

JEL Classification: L1, L6, M3, 03

1 Introduction

Our understanding of the competitive process remains fundamentally incomplete, until we acquire at least some basic knowledge of the empirical relationships between the economy and its intangible factors of production. Because intangibles are, by nature, difficult to measure and to value, the lack of reliable, comprehensive and internationally comparable data is a major barrier to their inclusion in empirical analysis. In response to the increasing awareness of the important role played by intangible factors of production, the specific purpose of this paper is to make at least some of the 'intangibles' more 'tangible' to quantitative analysis.

The creation of new taxonomies in this paper simultaneously serves a practical and an analytic purpose. Reflecting the lack of comprehensive data on intangible investments across a number of different countries or economic areas, it enables us to apply the more easily available basic indicators of economic activity such as value added, employment or trade flows in an economically meaningful way. This is the practical side of the exercise. Conversely, our analytic interest arises from the implicit hypothesis that, despite their apparent firm-specific realisations, many intangible determinants of nonprice competition must nevertheless be strongly affected by the particular characteristics of the market.[1] Hence, during the process of creating the new taxonomies we also expect to learn more about the underlying nature and extent of systematic differences between industries, which can be observed with respect to intangible investments and human resources.

The paper is organised as follows: first, several examples illustrate the consequences of the inclusion of intangibles for some major predictions of economic theory. Secondly, a new taxonomy of manufacturing industries, based on typical combinations of factor inputs, is created and documented in detail, revealing many pronounced structural differences related to the intangible factors of production. Thirdly, a complementary taxonomy is developed, which is based upon data on labour skills and reflects the dimension of human resources. In the final section of this paper, the two taxonomies are applied in a test of the presumed complementarity between intangible investments and the employment of skilled labour.

2 Why intangibles matter

During the past decades, several economic disciplines have witnessed a profound reshaping of some of their major theoretic predictions, which at the utmost general level are characterised by the common inclusion of intangible factors of production in the new generation of models. In an attempt to motivate the more technical empirical analysis, this section collects some supportive, albeit highly stylised, examples of this claim.

We begin with the well-known example of *growth theory*. The conventional Solow–Swan model was famous for its prediction of 'conditional convergence' (whereby the steady state rate of growth varies across economies, depending on the savings rate, population growth and the shape of the production function). However, the assumptions of constant returns to scale and diminishing marginal productivity of each input also implied the eventual end of per capita growth, if there were no exogenous improvements in technology. In contrast, starting with Arrow's (1962) model of 'learning by doing', the large body of literature on endogenous growth (Grossman and Helpman, 1991; Aghion and Howitt, 1998) stressed the cumulative nature of knowledge, the incentives for purposeful investment in its creation and spillovers to the rest of the economy. Long-term growth then is determined by the balance of incentives to innovate, on the one hand (raised e.g. by the increasing returns to learning by doing and the prospects for Schumpeterian rewards of temporary monopoly power), and its own propensity to generate external benefits, on the other hand. As a consequence, one of the most significant economic implications of the new growth theory – which results from explicitly including purposeful

[1] The elasticity of demand with respect to investments into advertising (Dorfman and Steiner, 1954; Sutton, 1991) or technological innovations (Sutton, 1998) is such an example for the importance of a structural variable affecting firm behaviour.

investments in the production of new and intangible knowledge (which is difficult to appropriate and therefore a source of spillover) – implies that long-term growth remains feasible even in developed countries with high levels of per capita income.

Similarly, in *industrial organizations,* the notion of 'endogenous sunk costs' (Sutton, 1991, 1998) – which reflects the irreversible nature of intangible investments in such areas as advertising or research – brought about a major leap forward in our understanding of the evolution of market structure. Within that framework, advertising and R&D are understood as being typically sunk investments intended to raise the consumer's willingness to pay for the firm's output. Broadening the traditional concept of the production function, these can be considered productive inputs to the generation of revenue. The distinctive feature is that for the firm, advertising and R&D are the strategic variables of choice. In contrast, the sunk costs involved in physical investments (e.g. the acquisition of new plants at the minimum efficient scale) are determined exogenously by the underlying technology, and consequently are equal across firms. Investigating the effect of a rise in market size on supplier concentration, Sutton shows that exogenous sunk costs are reflected in a general and unbounded tendency of the equilibrium level of concentration to decline with the ratio of market size to setup cost. *Ceteris paribus*, growing market demand produces an increasingly fragmented market structure. However, depending on the responsiveness of demand to advertising and R&D, endogenous sunk costs allow for a competitive escalation of expenditures, raising the equilibrium level of sunk investments in the particular industry. Thus, even in the presence of increasing market size, sunk costs can effectively act as barriers to further entry and may offset the tendency towards fragmentation. In short, the presence of endogenous sunk costs such as intangible investments in advertising and R&D implies that 'under very general conditions a lower bound exists to the equilibrium level of concentration in the industry, no matter how large the market becomes' (Sutton, 1991, p. 11).

Apart from exogenous technological boundaries, intangible investments such as advertising and R&D, as well as user-specific supplier services, more generally define the potential scope for product differentiation and surplus income. Besides industrial organization, this aspect is of special importance in two complementary fields of *international economics*, i.e., trade theory and the theory of multinational enterprises.

On the one hand, product differentiation (and thus intangible investments) is a necessary precondition for the 'escape' of high-wage countries from the traditional prediction of factor price equalisation and the according downward pressure on labour incomes in trade theory. This downward pressure on factor incomes might otherwise be expected from the increasing integration of global markets accelerated by the high mobility of international capital flows. On the other hand, we could also take the theory of multinational enterprises into consideration, by which locationally bounded comparative cost advantages can only explain a (rather small) fraction of total transborder investment flows. On the contrary, the motivation for multinational investment is largely explained by the exploitation of firm-specific assets such as accumulated organizational and technological knowledge, or reputation and the creation of brands (see e.g. Dunning, 1994, or Caves, 1996). Again, it is precisely

Table 1. Why intangibles matter

Stylized examples of major differences in assumptions and economic predictions	... exclusively based on tangible factors	... considering intangible sources
	of production and revenue generation	
Growth theory	Diminishing marginal productivity of factor inputs; convergence of per capita income; zero growth for high income countries;	Allowing non-diminishing marginal productivity; divergence, and sustainable growth (Arrow, 1962, etc.);
Industrial organization	Increasing market size must lead to increasing fragmentation if sunk costs are exogenous;	Lower bounds to market concentration as market size increases in the presence of endogenous sunk costs (Sutton, 1991);
Multinational enterprises	Motivated by locational cost advantages; market access and transport costs;	Multinational investment in order to exploit firm specific assets (e.g., Dunning, 1994);
International trade	Homogenous goods, comparative cost advantages, and factor price equalisation ;	Increasing returns to scale, product differentiation, and differential incomes (e.g., Krugman, 1979);
The nature of competition	Static equilibrium, and allocative cost efficiency	Entrepreneurial discovery and evolutionary change (e.g., Kirzner, 1997; and Metcalfe, 1998);

their intangible, non-commodity-like nature, which makes these assets difficult to trade and therefore largely specific to the firm. As a consequence, such assets are often exploited more effectively through organization within the firm rather than through purely contractual exchange relationships on (factor) markets.

The creation of firm-specific competitive advantages by intangible investments and knowledge-based resources is also strongly linked to Austrian economics (Kirzner, 1997) and evolutionary models of *the competitive process* (Metcalfe, 1998). Both share an emphasis on the necessary diversity of a firm's capabilities and behaviour, driving the dynamics of innovation and selection by differential growth in the marketplace. Competitive advantage therein is often defined in terms of the firm's specific 'dynamic capabilities', i.e. 'the subset of the competencies/capabilities which allow the firm to create new products and processes, and respond to changing market circumstances' (Teece and Pisano, 1998, p. 197). This definition suits our purpose particularly well, since it stresses not only the responsiveness to 'fast moving' external conditions such as technological change and shifting consumer tastes, but also the key role of strategic choices (among them most notably the choice in which competitive assets to invest) as the fundamental criteria of competitive selection. The particular relevance of the latter relates to the presumed pro-active capability of firms to increase perceived quality and thus also to raise the willingness of consumers to pay for their products. The entrepreneur does not take demand as given, 'but rather as something he ought to be able to do something about' (Penrose, 1959, p. 80).

With respect to these pro-active sources of entrepreneurial opportunity, there are two archetypes of intangible investments capable of raising the perceived quality of products: (i) Research and innovation, which enables 'firms to turn aside the process of "creative destruction" and thrive on the novelty which might otherwise have destroyed them' (ibid., p. 115), and (ii) advertising, or marketing more generally, which 'is perhaps of greatest importance for firms whose productive processes are either highly specialised with respect to the kind of product for which they are suitable, or are simple and easily imitated and of a kind where research yields little that provides particular firms with any competitive advantage' (p. 116).

Despite their admittedly short and simple presentation, these examples bring to the surface a broad but seemingly robust interpretative framework of related pairs of opposites, characterised by terms such as exogenous and endogenous, natural and strategic, or location-bound and firm-specific, which emerge in different fields of economic theory. All of them are also related to the distinction between tangible and intangible factors of production (interpreted more broadly in terms of revenue generation). In particular, the intangible nature of some productive resources implies a lower ability to be traded on the markets, rendering them dependent on strategic choices to invest in their generation within the specific firm. Since such strategic choices are sensitive to public policy, this distinction deserves even more attention.

3 Taxonomy I: typical factor input combinations

3.1 References and novel features

This section presents a new taxonomy of manufacturing industries based on typical patterns of factor input combinations. The new approach focuses on the distinction between tangible and intangible factors of production/generation of revenues. We go considerably beyond the popular, manifold 'high-tech' versus 'low-tech' distinctions, which perhaps have found their most comprehensive update in Hatzichronoglou (1997). In contrast, the new taxonomy presented in this chapter is based on two entirely different sources of inspiration. First of all, Schulmeister's (1990) extension of a classification by Legler (1982) must be mentioned in light of its successful combination of the usual high-, medium- and low tech differentiation with a more comprehensive coverage of factor inputs such as capital investment, labour costs, research expenditures and energy consumption. Secondly, investigating the economic impact of endogenous sunk costs at the industry level, Davies and Lyons (1996) introduced and applied an influential taxonomy based on a firm's intangible investments in advertising and R&D.[2] All of these taxonomies rely on traditional cut-off procedures, by which a certain discriminatory edge is defined exogenously by the researcher before the analysis. In choosing not to use more powerful statistical tools for categorising multidimensional data, the underlying structure within the data is more or less presumed, rather than extensively explored. In contrast, it is precisely on such an exploration that this section will focus.

In short, the new taxonomy is characterised by three distinctly novel features: (i) From an analytic perspective, the particular choice of variables reflects exogenously given technology on the one hand, as well as the firm's targeted expenditures on innovation and marketing on the other. It thus combines elements of Schulmeister (1990), and Davies and Lyons (1996). (ii) From a methodological standpoint, the new taxonomy uses statistical cluster techniques to reveal typical patterns, hidden within the data, across a multidimensional set of variables. This is clearly a more powerful tool than the traditional cut-off procedure. (iii) Finally, from a purely practical perspective, the new taxonomy is the first of its kind to target the 3-digit level of EUROSTAT's NACE rev. 1 classification of industries.

3.2 Choice of variables

In any attempt to classify and categorise a number of observations, the most sensitive step is the initial decision concerning the appropriate dimensions against which individual cases should be measured and discriminated. In the present case, the particular choice of variables reflects the different dimensions along which firms can increase their competitive performance in the market place. Following the stylised distinction of two opposing poles in the first section of this chapter, the new taxonomy tracks both (i) comparative cost advantages stemming from exogenous and location dependent factors such as relative endowments of capital and labour; and (ii) firm specific advantages stemming from targeted investment in intangible assets such as advertising and R&D. It is only with regard to the latter that an evolutionary perspective of the market process, stressing Schumpeterian innovation and diversity in firm behaviour, can be credibly established.[3]

For the calculation of mean values for the latest years available in each case, the following variables were chosen: (i) *labour intensity* (average ratio of gross wages and salaries to value added from 1990 to 1995); (ii) *capital intensity* (average ratio of total investments to value added from 1990 to 1994); (iii) *advertising sales ratio* (average ratio of advertising outlays to total sales from 1993 to 1995); (iv) *R&D*

[2] See also Davies, Rondi and Sembenelli (1998).

[3] For obvious reasons, the complex issues of competitive advantage from firm specific organizational knowledge cannot be included in this cross-sectoral setting. If we assume, however, that organizational tasks will be more difficult and complex, the more dynamic (or 'fast moving') the business environment is, we might reasonably expect a high degree of correlation between organizational complexity (and thus the organizational capabilities required) and e.g. the research intensity of production.

sales ratio (average ratio of expenditures on research and development to total sales from 1993 to 1995).

Due to the lack of equally disaggregated data for the European Union across all four dimensions, data refer exclusively to US-manufacturing industries. Besides this technical necessity, the USA is an attractive source of reference first because of its status as one of the economically most advanced nations, of which the general patterns in the combination of factor inputs constitute a good benchmark, and secondly, being a large economy there are lower risks of distortions in the data due to highly particular local patterns of specialisation. However, the exclusive reliance on US data remains a critical feature, as we cannot generally assume consistency between different economic areas with respect to the typical combinations of factor inputs. But this objection actually highlights one of the major advantages of the taxonomic approach, that the latter is not a necessary assumption for international comparisons. It only requires consistency of factor input combinations as far as membership within the broad boundaries of the final classification is concerned. This obviously is a much weaker assumption and generally allows for more robust results. Nevertheless, we must always remain aware of the fact that much heterogeneity within each individual category can still be found.

All four variables were used in their standardised form, thus eliminating the overall effects of differences in the size of the variables and adjusting for differences in their variability. A final point of concern was potential correlations between variables. However, in the current data set, correlations are low or non-existent. The highest negative correlation is in the relationship between advertising and capital investment, where the value of -0.316 implies that no more than 10% of total variation in the variable for advertising outlays can be explained by opposing variation in the variable for capital investment. No significant positive correlation occurs at all.

3.3 Statistical clustering

Cluster analysis produces a classification scheme of individual observations, depending on their relative similarity or nearness to an array of variables. The basic idea is one of dividing a specific data profile into segments by creating maximum homogeneity within and maximum distance between groups of observations. It is important to remember that, despite its mathematical sophistication, cluster analysis represents a *heuristic* method for the exploration and identification of underlying patterns in the data.

The most important advantage of statistical cluster techniques relative to conventional cut-off methods of dividing given data sets into distinct groupings, is their endogenous determination of the edges and boundaries that discriminate between observations, thus supporting the exploration and screening for regularities from within the initial data set. Especially when dealing with multidimensional phenomena, the analytical capacity of cluster analysis by far surpasses that of traditional cut-off procedures.

After selecting the set of variables, as a first step, an optimisation cluster technique, based on the minimisation of within-group dispersion, is used to classify 100

NACE 3-digit manufacturing industries into clusters of maximum homogeneity. In the second step, the resulting 32 clusters of the first partition enter a hierarchical clustering algorithm as observations. The *cosine* of the vectors of the variables was applied as a measure for the similarity in patterns. In the following agglomerative algorithm (*average linkage* method), all observations are initially treated as independent single clusters. In the iterative process, the similarity of all pairs is compared, and those pairs exhibiting maximum similarity are grouped together to form a common cluster. The outcome is a hierarchical structure, beginning with many single observations that finally unite at different levels, until all are unified by one single trunk (for more details see Peneder, 2001).

The resulting *dendrogram* has to be interpreted a follows: the closer to the origin (i.e. the left hand side of Figure 1) observations are clumped together by the vertical lines, the more similar they are in their underlying patterns of input combinations. Groups of observations that are connected only far out (i.e. towards the right hand side of the graph) are the most dissimilar. Figure 1 shows a surprisingly robust and sharply edged structure of about four clusters and some outlying satellites, which emerged as a rather robust pattern for a number of variations for at least two measures of similarity (cosine and correlation of vectors) and a variety of different algorithms (average linkage between groups, complete linkage, etc.).

3.4 The resulting classification

The graphical representation of relative similarity in the dendrogram of Figure 1 provides a surprisingly sharp-edged discrimination of four broad categories, each one characterised by a rather pronounced reliance on one of the four input-dimensions. Since no successful alternative pattern emerged, this outcome constitutes the basic information for defining the final taxonomy.

The first grouping of industries of the dendrogram in Figure 1 ranges from cluster 11 to cluster 4, and is characterised by particularly high expenditures on research and development (we label them ‘technology-driven industries’, TDI); a second grouping ranges from cluster 27 to 32, and is unequivocally linked to particularly high rates of investment in physical capital (‘capital intensive industries’, CI); the span of the third distinct grouping is from clusters 2 to 22, and exhibits high shares of advertising outlays (‘marketing driven industries’, MDI). Finally, the block ranging from cluster 1 to 18 comprises cases with particularly high labour costs (‘labour intensive industries’, LI).

Clusters 16, 20 and 28, which all include quite a great number of individual industries, could in principle be allocated to their respective groupings of research, capital, or labour intensive industries. However, a closer look at their input combinations reveals that these industries are mostly distinguished by their lack of a pronounced reliance on any of the four factor inputs. In consideration of their position on the relative fringes of their respective agglomerations, they have been grouped together in a residual category, labelled ‘mainstream manufacturing’ (MM), representing more or less the input combination of a typical 3-digit manufacturing industry. This also provides a convenient group of comparison for econometric exercises.

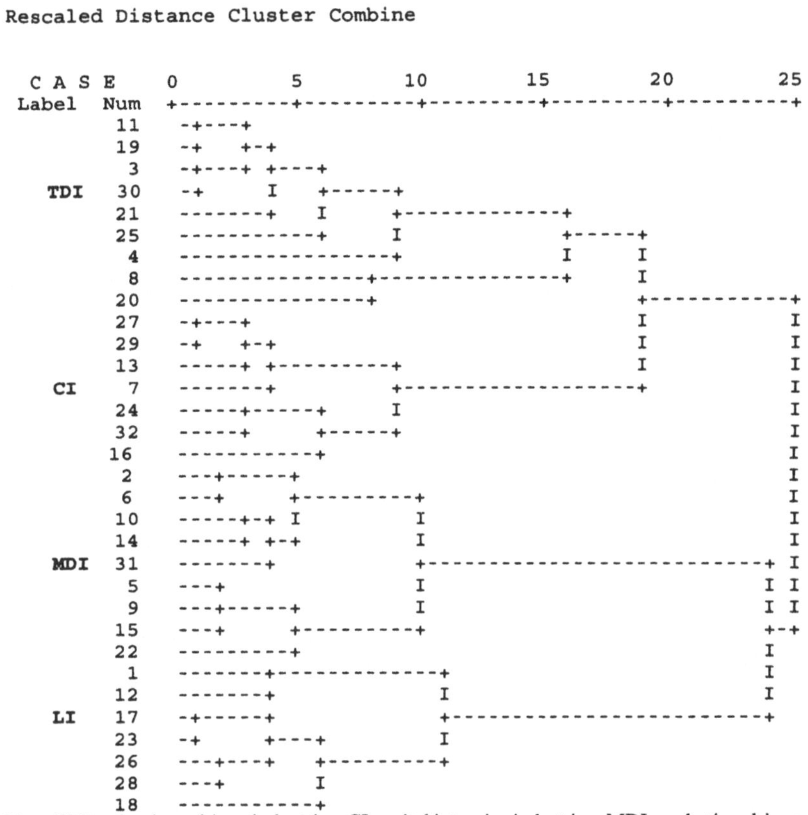

Note: TDI technology driven industries; CI capital intensive industries; MDI marketing driven industries; LI labour intensive industries.

Fig. 1. Dendrogram using average linkage between groups and the cosine of vectors of values

There is one exception to the general robustness of the results with regard to variations in clustering algorithms and measures of similarity: cluster 8, which comprises only one single 3-digit industry – namely agrochemical products. Despite its rather average share of research expenditures in total net turnover, it was allocated to the grouping of research intensive industries, as a result of overall similarities in particular combinations across all four factor inputs. However, when different measures of distance were tried, it was also sometimes more closely associated with capital intensive production. Its ultimate classification within the grouping of technology driven industries rests on reported results from patent analysis across technology fields (Andersen, 1997) and specific industry monographs (Achilladelis, Schwarzkopf and Cines, 1987), which stress its particularly high shares in overall innovative activities.[4]

[4] In Andersen's comparison of accumulated patent stocks across a total of 399 patent classes, agrochemicals even hold the 10th position in a ranking of 'technological size' for 1990. She concludes that these 'have been among the fastest growing technological sectors throughout this century' and 'have never grown as much in accumulated patent stock as in recent times' (Andersen, 1997, p. 38).

Table 2. Shares in total manufacturing: EU15, 1998 in %

Type of industry	Value added	Employment	Exports	Imports
Mainstream				
Manufacturing (*MM*)	25.6	27.2	26.0	16.9
Labour intensive (*LI*)	15.6	21.9	12.6	15.7
Capital intensive (*CI*)	14.6	10.6	14.5	18.2
Marketing driven (*MDI*)	21.2	22.1	12.1	12.3
Technology driven (*TDI*)	23.0	18.1	34.9	37.0

In the end, precisely 100 NACE 3-digit manufacturing industries were categorised under five mutually exclusive groupings (see Appendix):

1. *Mainstream manufacturing*: This residual category was created out of 25 industries, in which input combinations did not show a pronounced reliance on any particular input factor. Summing up all manufacturing industries in the European Union in 1998, this group generated 25.6% of total value added, 27.2% of employment, 26.0% of its exports to non-member countries, but only 16.9% of imports (Table 2). Among others, this group includes the machinery sector, articles of paper, plastics, electronic equipment and motorcycles.
2. *Labour intensive industries*: Comprising 25 individual 3-digit industries, the common share of total employment in the European Union in 1998 amounts to 21.9%. This is contrasted by a rather low share in total value added of about 15.6%. Its shares in the total of (extra-EU) exports and imports are 12.6% and 15.7%, respectively. Typical examples are textiles and clothing, wood processing, construction material and metal processing.
3. *Capital intensive industries*: In this subgroup, only 10.6% of the European Union's total manufacturing employment produces 14.6% of its value added. A share of 14.5% of total exports corresponds to 18.2% of imports. Typical examples are pulp and paper, refined petroleum, basic chemicals and iron and steel.
4. *Marketing driven industries*: This category comprises 24 advertising intensive industries, together accounting for 21.2% of the EU15 total value added and 22.1% of total employment. This is in sharp contrast to the low shares of only 12.1% in total exports and 12.3% of total imports. The archetypal example is the food sector, which is allocated entirely to this category. Other industries within this category produce articles associated with leisure and entertainment such as perfumes, sports goods, musical instruments or games and toys.
5. *Technology driven industries*: The 14 industries within this group are characterised by particularly high expenditures on R&D and account for 23.0% of total value added as well as 18.1% of total employment. Technology driven goods are more highly traded than the products of any other category. Although similar in size to the other categories with regard to value added and employment, their share in total exports and imports amounts to an outstanding 34.9% and 37.0%, respectively. Industries concentrate around three distinct technology fields: (i)

chemicals and biotechnology; (ii) new information and communication technologies, and (iii) vehicles for transport.

Like any broad classification, this taxonomy must be interpreted with care, since industries within the five categories are still very heterogeneous in nature. One particular concern can be raised with regard to the surprisingly neat designation of cases into each of the four chosen dimensions. Certainly, all the industries produce their output using particular combinations of more than one factor. Advertising, in particular, is often modelled as a complement to vertical product differentiation, in order to provide information to consumers about the quality and innovative features of the product. This tendency is especially relevant in the cases of *pharmaceuticals* (cluster 19) and *optical instruments* (cluster 11) under the heading of technology driven industries (TDI), as well as *detergents* (cluster 14), *games and toys* (cluster 15) and *publishing* (cluster 31), which ultimately were labelled marketing driven industries (MDI). Similar combinations can also be found with regard to the other input variables. However, in the final clustering stage, no such pattern of an especially pronounced and characteristic combination of factor inputs, supporting the introduction of an additional category, was observed.

3.5 Characterisation in factor space

This section intends to provide a more precise understanding of the systematic patterns in the typical combinations of tangible versus intangible inputs to production. Simultaneously displaying information about the shape and the dispersion of a distribution, 'boxplots' provide a convenient mode of summarising descriptive statistics (Figs. 2–5). The box itself comprises the middle 50 percent of observations. The line within the box is the median. The lower end of the box signifies the first quartile, while the upper end of the box corresponds to the third quartile. In addition, the lowest and the highest lines outside the box indicate the minimum and maximum values, respectively. Maximum values do not include outliers, which are separately indicated. For each of the four dimensions, the boxplots immediately illustrate one striking characteristic of the new taxonomy: in three of the four input-dimensions, there is one single group which absorbs by far the largest amount of total variation across industries. Besides justifying the rather uni-dimensional shortnames, which invoke a rather convenient interpretation for each grouping, we can also take this as a first and strongly supportive observation regarding the overall accurateness of the classification.

The boxplots also show that *labour intensity* is the one variable with the least clearly cut discrimination, and the one which exhibits the most variation within each grouping. All five categories include individual industries with high shares of labour costs in total value added, comparable to the mean value for the distinct grouping of labour intensive industries. But even in this case, inspection of the other boxplots reveals what constitutes the distinct feature of the group of so-called labour intensive industries, that is the particular combination of high labour costs in combination with the lack of any pronounced reliance on complementary investments either in physical capital or the two types of intangible inputs to production.

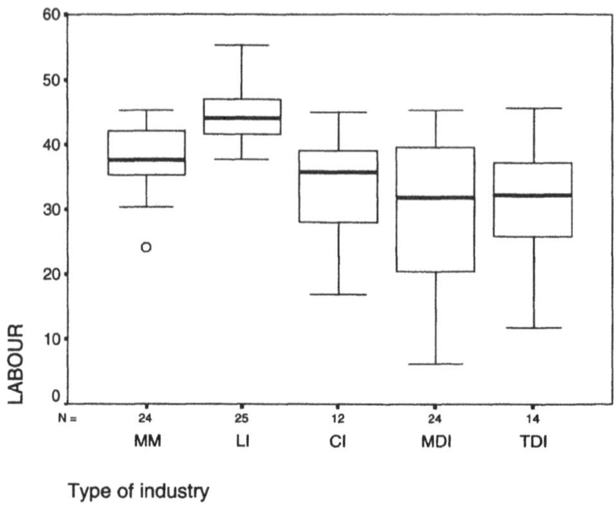

Fig. 2. Boxplot for labour intensity

In contrast to labour intensity, the boxplots for the three other variables tell a different story. As can easily be seen, the isolation of particular *capital* intensive industries within a distinct industry type significantly reduces overall variation. The same applies to expenditures on *advertising* or R&D, where all the industries with the greatest dependence on a particular input have also been classified within their proper category. Additionally, mainstream manufacturing seems to exhibit a kind of minimum threshold for research expenditures, which puts its overall research effort ahead of capital- and labour intensive as well as marketing driven industries.

4 Taxonomy II: labour skills

Analogous to the first taxonomy introduced above, the important aspect of human resources will be captured within a second classification. The exposition will be abbreviated, since the technique is mostly parallel to that used in the preceding section. This time, the taxonomy will be based on occupational data discriminating between two different types and two levels of labour skills. It is assumed that the actual use of certain skills reflects corresponding technological constraints and market opportunities.

The data, which have been published by the OECD (1998), are available at the 2-digit level of ISIC Rev. 2 and distinguish four broad types of occupations, for which shares in total employment can be calculated:

i. *white-collar high-skill* (legislators, senior officials and managers; professionals, technicians and associated professionals);
ii. *white-collar low-skill* (clerks, service workers, shop and sales workers);
iii. *blue-collar high-skill* (skilled agricultural and fishery workers, craft and related trade workers); and finally

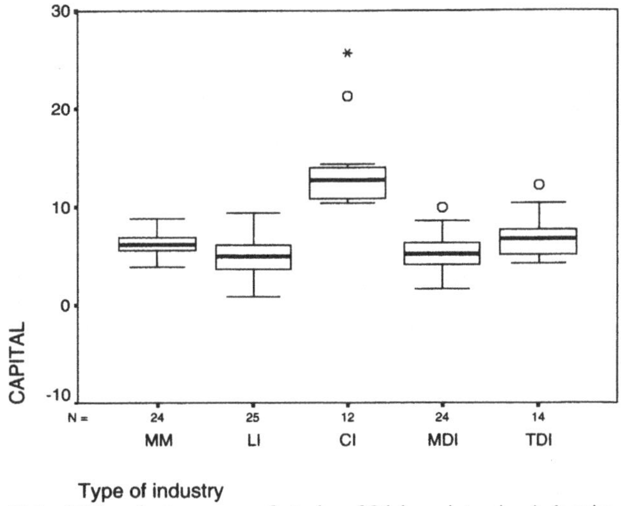

Note: MM mainstream manufacturing; LI labour intensive industries; CI capital intensive industries; MDI marketing driven industries; TDI technology driven industries.

Fig. 3. Boxplot for capital intensity

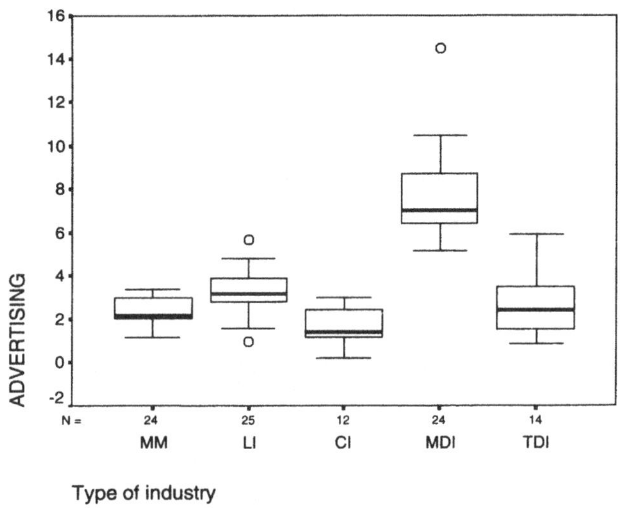

Fig. 4. Boxplot for advertising outlays

iv. *blue-collar low-skill* (plant and machinery operators and assemblers, elementary occupations).

The shares in total employment of blue-collar high-skill and white-collar high-skill were used to maintain a concept of orthonormal space and a linear independence of the respective vectors of the variables. This reflects a conceptualisation of two linearly independent dimensions of blue- versus white-collar occupations, within which the individual scores for each industry illustrate the respective requirements

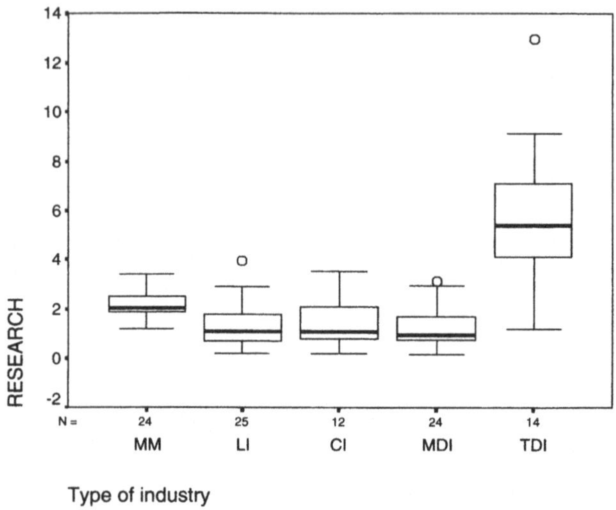

Note: MM mainstream manufacturing; LI labour intensive industries; CI capital intensive industries; MDI marketing driven industries; TDI technology driven industries.

Fig. 5. Boxplot for research expenditures

for skilled labour. Since the number of available observations was much lower than in the previous case, the clustering process was executed in a one-stage algorithm. In order to maintain the linear independence of the basis vectors, only the shares of high-skilled white-collar and high-skilled blue-collar workers were used as discriminatory variables.

The overall clustering algorithm does not work equally as well as in the prior case of typical factor input combinations. This is most probably due to the higher level of aggregation in the available data. As a consequence, the boundaries are more difficult to draw and high-skill industries are mainly defined by outlying cases. Nevertheless, the most pronounced pattern was revealed by the combination of average linkages within groups and the Euclidean distance.

The patterns in Figure 6 primarily reflect the distinction between industries with high shares of blue-collar workers (from the top down to 'rubber & plastic products') and those with a majority of white-collar workers (from 'pharmaceuticals' down to the bottom of the dendrogram).

Finally, four distinct types of industries were constructed, reflecting relative differences in their skill requirements (see Annex). In short, the so-called *high-skill industries* include non-electrical machinery among the more typical blue-collar industries, and pharmaceuticals, computers and office machinery, as well as aircraft, among the more typical white-collar industries. To a certain extent, they are all outlying cases, placed at the outer fringes of their respective clusters. In contrast, the final grouping of particularly *low-skilled* industries is represented by a block of observations, characterised by rather similar occupational structures with particularly low shares of white-collar high-skills and mean to low shares of blue-

```
*** HIERARCHICAL CLUSTER ANALYSIS ***

Rescaled Distance Cluster Combine

C A S E                   0       5       10      15      20      25
Label                     +-------+-------+-------+-------+-------+
Motor vehicles    MSBC    -+
Other transport   MSBC    -+------+
Other manufac.    MSBC    -+      +----+
Machinery         HS      --------+    I
Food, drink       LS      -+--------+  +---+
Textiles          LS      -+        I  I   I
Stone,clay,etc.   LS      ---+      +--+   I
Other metals      LS      ---+-+    I      +-+
Ferrous metals    LS      ---+ +----+      I I
Nonferr. metals   LS      -----+           I +---+
Wood products     MSBC    --------+--------+ I   I
Fabric. Metal     MSBC    --------+          I   +----------------+
Shipbuilding      MSBC    -------------------+   I                I
Rubber, plastic   LS      -----------------------+                I
Pharmaceuticals   HS      ---------------------+----------+       I
Office machin.    HS      ---------------------+          I       I
Petroleum ref.    MSWC    ------+---+                     +-------+
Electronics       MSWC    ------+   +-----------+         I
Basic chemicals   MSWC    ----------+           I         I
Electrical equ.   MSWC    ----+-----+           +---------+
Instruments       MSWC    ----+     +------+    I
Paper, printing   MSWC    ----------+      +----+
Aircraft          HS      -----------------+
```

Note: MSBC medium-skilled blue-collar industries; HS high-skilled industries; LS low-skilled; MSWC medium-skilled white-collar industries.

Fig. 6. Manufacturing sectors grouped according to similarities in labour skills (average linkage within group; squared Euclidean distance)

collar high-skills. All the remaining cases were labelled *medium-skilled*, belonging to the groups of either typical *blue-* or *white-collar* industries.

5 On the complementarity between intangible investments and human resources

In the new taxonomy presented in Section 3, the empirical concept of the firm was based on a broad notion of the production function, focusing on factor inputs to the generation of revenues rather than on pure physical output. This approach brought into perspective expenditures on R&D and advertising as intangible investments, intended to raise the consumer's willingness to pay for the specific output of the firm. In contrast, the second taxonomy was based purely on human resources,

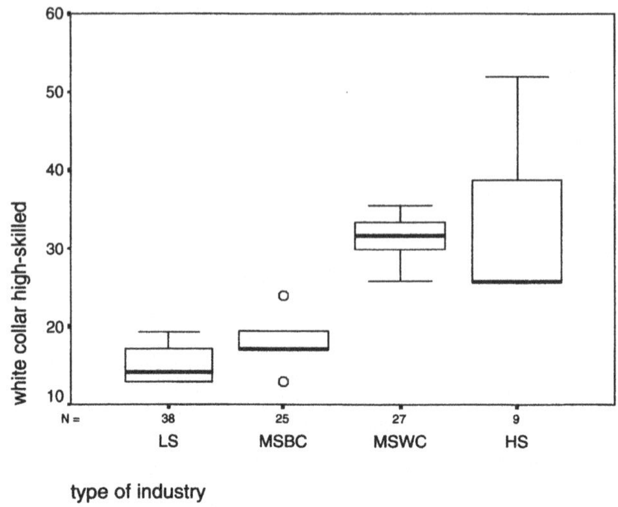

Fig. 7. Shares in total employment: white-collar high-skilled

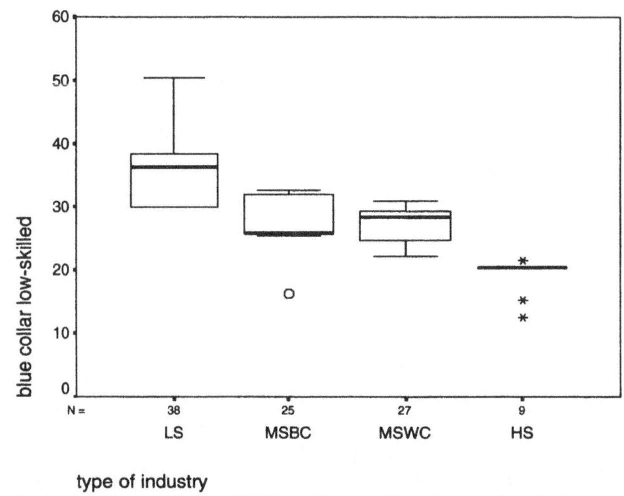

Note: LS low-skilled; MSBC medium-skilled blue-collar industries; MSWC medium-skilled white-collar industries; HS high-skilled industries.

Fig. 8. Shares in total employment: blue-collar low-skilled

focusing on two different kinds and two different levels of labour skills. In the respective statistical cluster analysis, both classifications had to be treated separately, enabling the formulation of a reasonable concept for measuring similarities or differences between industries. Otherwise, a reasonable assumption of orthonormal space would not have been possible, and correlations between the input and skill variables would have introduced major distortions.

The purpose of this section is to investigate whether and to what extent these two different analytic layers and the respective taxonomies interlock. The interpretative framework starts with the assumption that human resources and factor inputs form two different layers in the 'anatomy' of the firm. Any inputs must be drawn as specific services from the pool of resources, which are either externally or internally available to the firm. Thus for each of the four dimensions of factor inputs, it is in principle possible to define a corresponding resource base. This refers to tangible assets such as e.g. the physical stocks of plant and machinery, as well as intangible knowledge-based resources, as, e.g., the basic technological and marketing know-how for launching a new research programme or advertising campaign. A part of these knowledge-based resources might be purchased on external markets – specialised consulting services, for example. Other parts can be generated by the accumulation of past experiences within the firm. Finally, these knowledge-based resources are also dependent on the available qualifications of the people employed. Because of a lack of appropriate data on the other dimensions, it is only this last aspect which can be integrated into our empirical setting. Despite their contributions to the general stock of accumulated knowledge, any investments e.g. in advertising and R&D are exclusively attributed to the analytic layer of actual service inputs. Any positive feedback on the own knowledge base are thought of as indirect effects, stemming from the accumulation of past experiences.

Investigating the relationship between labour skills and typical factor input combinations, no clear causal structure can be specified. As, for example, research activities might be dependent on the availability of skilled labour, the causal link can also be precisely the reverse – in the sense that the demand for skilled labour positively depends on corresponding investments in R&D. Therefore, the true question of interest is, whether the different types and different degrees of human resources, as well as intangible investments, can be characterised by any statistically significant complementary relationships. Thus, we are not concerned with any detailed specification of functional forms and causal links, but concentrate only on the observable patterns of co-movement. This task bears some similarity to recent theoretical (Milgrom–Roberts, 1990, 1995) and empirical (Ichniowski, Shaw and Prennushi, 1997) research on firm organization and the complementarity of different work practices.

Beginning with a simple correlation analysis, the following relationships between the continuous variables on factor inputs and occupations appear to be significant at the 5% level: capital intensity is negatively correlated with blue-collar high-skills, whereas labour intensity is negatively correlated with both types of white-collar occupations and positively correlated with blue-collar low-skills (+0.480). Advertising outlays are positively correlated with white-collar high-skills (+0.310), but negatively so with white-collar high-skills. Among the four input variables, intangible investments in R&D are most strongly related to the dimension of human resources, with significant correlations to all different types of occupations. Strong positive correlations occur with regard to both types of white-collar labour, and are particularly strong for high-skills (+0.678) and somewhat less so for low-skills (+0.324). In contrast, research expenditures are negatively correlated to shares of blue-collar workers in total employment.

A more rigorous testing can be provided by non parametric tests on the significance of observable differences. This corresponds essentially to an F test, in which an estimate of the *between-groups variance* is compared with an estimate of the *within-groups variance* by dividing the former by the latter. If this ratio exceeds a critical value in the F distribution table, the hypothesis that the two groups exhibit the same mean values can be rejected at the given level of significance. The essential reasoning behind the analysis of variance is, that if the groups come from the same population, then the between-groups variance should be similar to the within-groups variance. The higher the F ratio, the more unlikely it is that the differences between the means are merely due to chance.

Because the groupings vary in size and lack homogenous variances (heteroscedasticity), the following four non-parametric tests, which were designed for unrelated samples and non categorical data, were applied. The *Median test* reveals whether the differences in the median of two or more samples are significant. Similarly based on the Chi-square statistics, the *Kruskal–Wallis H* statistic is used to test for the significance of differences across all five industry types, by comparing the number of times a score from one industry grouping is ranked higher than a score from another. The *Mann–Whitney U* test statistic is also based on the mean ranks of different samples, whereas the *Kolmogorov–Smirnov Z* statistic compares the distribution of values in each pair of industry types. While the Kruskal–Wallis H statistic and the Median test reveal whether the typology as such discriminates significantly according to the particular variables, both the Mann–Whitney U test and the Kolmogorov–Smirnov Z statistic are additionally applied in the comparison of individual pairs of industry types.

In short, the non-parametric tests on differences in the employment shares of various types of occupations across industries, classified according to their typical factor input combinations, indicate significant relationships between intangible investments and human resources (Table 3):

- *Technology driven* industries characterised by typically high intangible investments in research and development significantly show the lowest shares in both blue-collar low- and high-skill labour, but the highest shares in white-collar low- and high-skill labour.
- In contrast, for *marketing driven* industries, the shares of blue-collar low-skill labour are significantly lower and those of blue-collar high-skill labour are higher than in most other groupings. The reverse applies with regard to white-collar occupations, where low-skills have high shares and high-skills hold low shares.
- As far as the other groupings are concerned, the most pronounced observation concerns *labour intensive industries*, which typically employ many more blue-collar than white-collar workers of both types. In contrast, mainstream manufacturing and capital intensive industries each reveal a mixed balance.

Finally, the twofold application of multinomial logit regressions is used in a further exploration of the presumed complementary relationship between the two distinct analytic layers. For lack of a more profound specification of the causal model, a simple regression was calculated first, taking taxonomy I as a dependent variable and the respective shares in employment of white-collar low-, white-collar high- and blue-collar high-skilled labour as independent variables. Mainstream manufacturing is used as the comparison group. The underlying hypothesis is that the final classification of a particular industry within either technology driven, marketing driven, capital- or labour-intensive production can be predicted, at least in part, on the basis of its known human resource base. Assuming complementarity between labour skills and intangible inputs to production, the most simple prediction would suggest positive coefficients for both types of high-skilled labour and negative coefficients for both types of low-skilled labour for marketing and technology driven industries, and the reverse for labour intensive industries. No specific hypothesis on different labour requirements relative to the comparison group seems obvious for capital intensive industries.

Since we are already familiar with prior test statistics, we must expect that these simple predictions will not be confirmed in their entirety. Corresponding to our lack of any causal hypothesis, an industry's probability of being particularly *capital intensive* cannot be related to any of the occupational categories. In contrast, e.g. more shares of blue-collar high-skill labour have a significant impact on the probability of being grouped among *labour intensive industries*. Similarly, the only positive and strong effect on the probability of belonging to *marketing driven industries* depends on the share of low-skill white-collar workers, whereas higher shares in white-collar high-skill labour significantly decrease this probability. The regression analysis nevertheless confirms at least one of the prior predictions:

- A larger share of high-skilled white-collar labour has a significant positive impact on the probability of belonging to the *technology driven industries*.

The Likelihood-Ratio-Index (Pseudo R^2) is 0.37 (in multinomial logit regressions, values between 0.2 and 0.4 are generally considered as indicative of a good fit). Out of 99 cases, each offering 5 alternative realisations, the regression model achieves a score of 61 correct predictions (correct scores are those on the vertical axis of Table 5). Hence, the model supports the claim that the two dimensions of intangible investments on the one hand, and human resources on the other, are strongly interlocked.

Since we tend to imagine a kind of interdependent causation and do not think in terms of a unidimensional causal specification, it might be reasonable to test also for the analogous hypothesis that the relative size of factor inputs has a significant impact on the overall level of required skills. Certainly, this does not eliminate the problem posed by the endogeneity of the variables. Nevertheless it does provide an additional test of whether and how intangible investments and human resources interlink.

Compared to the grouping of industries characterised by particularly high shares of low-skilled (mostly blue-collar) workers, the results indicate the positive impact of higher labour intensities on the probability of belonging to the grouping of medium-skilled blue-collar industries. Besides this significant, albeit weak effect, research expenditure is the only variable which exerts a meaningful impact on the demand for particular types of occupations:

Table 3. Non parametric tests for significant differences in labour skills

Shares in total employment	Cases	Mean value	Mean rank	Taxo I	Mann–Whitney U test/ Kolmogorov–Smirnov Z test				
					MM	LI	CI	MDI	TDI
Blue collar/low skill	25	31.39	52	MM	–	–	–	–	*
Median test: ***	25	31.37	59	LI	–	–	–	***	***
Kruskal-Wallis	11	33.42	67	CI	–	–	–	***	**
H test: ***	24	28.93	44	MDI	–	**	***	–	**
	14	24.67	28	TDI	**	***	***	*	–
White collar/low skill	25	12.78	47	MM	–	***	–	***	***
Median test: ***	25	11.16	25	LI	***	–	–	***	***
Kruskal-Wallis	11	13.60	48	CI	–	*	–	–	**
H test: ***	24	15.49	69	MDI	***	***	–	–	–
	14	15.64	70	TDI	***	***	–	–	–
Blue collar/high skill	25	34.49	46	MM	–	**	–	**	***
Median test: ***	25	40.49	69	LI	***	–	***	**	***
Kruskal-Wallis	11	29.79	30	CI	–	***	–	***	**
H test: ***	24	37.18	62	MDI	*	–	***	–	***
	14	24.90	18	TDI	***	***	*	***	–
White collar/high skill	25	22.13	54	MM	–	**	–	**	***
Median test: ***	25	16.69	30	LI	***	–	**	–	***
Kruskal-Wallis	11	22.84	60	CI	–	***	–	**	**
H test: ***	24	18.08	40	MDI	**	–	**	–	***
	14	34.35	88	TDI	***	***	***	***	–

Note: *** significant at the 1% level; ** 5% level; * 10% level.

MM mainstream manufacturing; LI labour intensive; CI capital intensive; MDI marketing driven; TDI technology driven industries.

- Higher intangible investments in R&D have a significant positive effect on the probability of belonging either to the grouping of *high-skilled* industries or *white-collar medium-skilled* industries.

Again, with a value of 0.29, the Likelihood-Ratio-Index is satisfactory. Based on the data for typical factor intensities, the model generates 57 correct predictions from 99 cases (each offering 4 alternative realisations).

In short, this final section proves the existence of strong empirical regularities between the distinct analytic levels of human resources on the one hand, and intangible investments on the other. The interpretative framework follows the resource-based view of the firm in its distinction of general resources available to the firm, and the specific bundles of services, which can be drawn from it. Seen from this perspective, human knowledge-based resources form a separate analytic layer, which is nevertheless interlocked with the observed variables on factor inputs – the actual

Table 4. Do skill requirements explain factor inputs (taxonomy I)?

Multinomial regression						
Occupations	Coef.	Std. Err.	z	P > z	(95% Conf. Interval)	
Labour intensive industries						
WC/HS	−0.082	0.091	−0.902	0.367	−0.259	0.096
WC/LS	−0.087	0.284	−0.306	0.760	−0.643	0.470
BC/HS	0.164	0.071	2.319	0.020	0.025	0.302
Constant	−3.638	4.202	−0.866	0.387	−11.875	4.599
Capital intensive industries						
WC/HS	−0.081	0.089	−0.910	0.363	−0.255	0.093
WC/LS	0.150	0.235	0.637	0.542	−0.311	0.611
BC/HS	−0.064	0.062	−1.033	0.302	−0.185	0.057
Constant	1.049	3.943	0.266	0.790	−6.679	8.777
Marketing driven industries						
WC/HS	−0.296	0.118	−2.503	0.012	−0.528	−0.064
WC/LS	0.970	0.247	3.916	0.000	0.484	1.454
BC/HS	0.114	0.081	1.402	0.161	−0.045	0.272
Constant	−11.443	4.872	−2.349	0.019	−20.992	−0.895
Technology driven industries						
WC/HS	0.365	0.169	2.166	0.030	0.035	0.695
WC/LS	−0.027	0.252	−0.107	0.915	−0.521	0.467
BC/HS	0.059	0.094	0.629	0.530	−0.126	0.441
Constant	−12.391	8.022	−1.545	0.122	−28.114	0.317
Log likelihood = -96.571596				Observations		= 99
				Lrchi2(12)		= 115.61
				Prob>chi2		= 0.0000
				Pseudo R2		= 0.3744

Note: Mainstream manufacturing is the comparison group.

Table 5. Predicting taxonomy I (factor inputs) according to labour skills

Taxonomy I (factor inputs)	MM	LI	CI	MDI	TDI	Total (obs.)
MM	*14*	7	2	1	1	25
LI	6	*18*	0	1	0	25
CI	5	2	*0*	1	3	11
MDI	0	5	0	*17*	2	24
TDI	1	1	0	0	*12*	14
Total (pred.)	26	33	2	20	18	99

Table 6. Do factor inputs explain skill requirements (taxonomy II)?

Multinomial regression						
Factor inputs	Coef.	Std. Err.	z	P > z	(95% Conf. Interval)	
Medium-skilled blue-collar industries						
Capital	-0.102	0.125	-0.813	0.416	-0.348	0.144
Labour	0.135	0.050	2.648	0.007	0.036	0.234
R&D	0.527	0.358	1.474	0.140	-0.174	1.228
Advertising	0.156	0.132	1.181	0.238	-0.103	0.414
Constant	-6.560	2.809	-2.336	0.020	-12.065	-1.056
Medium-skilled white-collar industries						
Capital	0.113	0.085	1.335	0.182	-0.053	0.280
Labour	-0.050	0.038	-1.338	0.181	-0.124	0.023
R&D	1.453	0.369	3.942	0.000	0.731	2.176
Advertising	-0.031	0.143	-0.219	0.826	-0.312	0.250
Constant	-2.145	2.094	-1.024	0.306 -	6.249	1.960
High-skilled industries						
Capital	-0.096	0.191	-0.505	0.614	-0.471	0.278
Labour	0.069	0.077	0.897	0.370	-0.082	0.220
R&D	1.724	0.405	4.257	0.000	0.930	2.517
Advertising	-0.104	0.249	-0.416	0.677	-0.592	0.384
Constant	-6.906	4.135	-1.670	0.095	-15.011	1.199
Log likelihood = -90.3076				Observations	= 98	
				Lrchi2(12)	= 72.36	
				Prob>chi2	= 0.0000	
				Pseudo R2	= 0.2860	

Note: Low skilled industries is the comparison group.

stream of long term investments in such areas as physical capital, advertising or R&D included.

The overall results confirm a particularly strong relationship between intangible investments in R&D and the underlying type of human knowledge-based resources, which emphasise the complementary nature of white-collar high-skill, and to a somewhat lesser degree, white-collar low-skill occupations. Intangible investments in advertising and brand creation cannot be associated with any significant complementary relationship to either of the two types of high-skilled labour. On the contrary, their overall pattern exhibits a strong complementarity with respect to white-collar low-skill occupations.

Table 7. Predicting taxonomy II (labour skills) by means of factor input combinations

Taxonomy II (labour skills)	Low skill	Medium skill/ blue-collar	white-collar	High skill	Total (obs.)
Low skill	*25*	9	3	1	38
Med skill/blue-collar	9	*14*	2	0	25
Med skill/white-collar	4	3	*17*	3	27
High skill	2	3	3	*1*	9
Total (pred.)	40	29	25	5	99

6 Summary and conclusions

This paper is centred around three primary tasks. First of all, the analysis set out to test the hypothesis that the importance of intangible investments and specific skill requirements differs across industries reflecting distinct characteristics of the market process. The results from the statistical cluster analysis convincingly demonstrate that industries do indeed differ in their propensity to undertake intangible investments in advertising or R&D – and that they do so in a remarkably systematic and pronounced way. The same applies to human resources, albeit in a less pronounced and clear-cut pattern. This structural dimension of otherwise firm specific investments and capabilities highlights the existence of substantial differences in the characteristics of the market process, with respect both to the technological opportunity set and the responsiveness of demand vis-à-vis such intangible investments. Although these shape the general constraints on firm behaviour imposed by the selective market environment, the relationship between markets and firms is clearly endogenous, as it is firm specific investments and capabilities which simultaneously generate these characteristics.

Secondly, by means of statistical cluster techniques, two new taxonomies of manufacturing industries were created. The first is based upon distinctions in an industry's typical factor input combinations and comprises data on labour inputs, capital investment, as well as intangible investments in advertising and R&D. The second taxonomy focuses on the human resources dimension and is based upon the average shares of different types of occupations, distinguishing between blue- and white-collar, as well as high- and low-skill labour. Allowing serious problems due to the lack of internationally comparable data to be side-stepped, the taxonomies can be used as discriminatory variables for comparative analysis of sectoral data (Peneder, 2000) as well as, e.g., highlighting the interrelationship between firm behaviour and more general market characteristics (Kaniovski-Peneder, 2000). Promising new research on the impact of industrial structure on aggregate growth performance of OECD countries is currently undertaken.

Thirdly, the analysis set out to test the hypothesis that the employment of high-skilled labour varies complementarily to intangible investments in advertising and R&D. The results of the correlation analysis, non parametric tests on the statistical significance of observable differences, as well as multinomial logit regressions, uniformly stressed the dependence of this relationship on the particular types of intangible investments and occupations. For example, no significant complementary relationship between high-skilled labour and advertising outlays could be found. On the contrary, advertising is revealed as having the most complementary relationship to the employment of white-collar low-skilled labour. But for intangible investments in research and development, a strong and highly significant complementarity to the employment of high skilled white-collar workers does exist, confirming our prior expectations in favour of this specific relationship.

Appendix

The taxonomies of manufacturing industry

			Share in value added/net turnover in%			
NACE	Industry	Skill type	Capital	Labour	R&D	Advertising
Mainstream manufacturing (MM)			**6.28**	**37.83**	**2.17**	**2.35**
1730	Finishing of textiles	LS	6.56	40.70	n.v.	n.v.
1770	Knitted and crocheted articles	LS	6.00	43.49	1.98	2.89
1750	Other textiles	LS	7.30	37.22	1.73	1.14
1760	Knitted and crocheted fabrics	LS	8.64	42.41	1.98	2.92
2120	Articles of paper, paperboard	MSWC	6.69	36.01	3.40	3.01
2430	Paints, coatings	MSWC	3.88	24.25	2.66	2.69
2510	Rubber products	LS	6.81	38.67	2.54	2.03
2520	Plastic products	LS	8.86	37.90	2.01	2.98
2610	Glass and glass products	LS	8.84	35.62	2.55	3.37
2660	Concrete, plaster, cement	LS	5.94	41.45	1.21	2.05
2680	Other mineral products	LS	6.41	30.93	1.89	1.82
2720	Tubes	LS	7.40	41.68	2.04	2.01
2870	Other metal products	MSWC	6.07	43.31	1.44	3.03
2910	Machinery f. mech. power	HS	6.22	39.77	2.30	2.57
2920	Other machinery	HS	5.21	43.60	2.01	1.60
2930	Agricultural machinery	HS	4.02	30.41	3.35	1.12
2950	Special purpose machinery	HS	5.33	45.33	2.49	2.68
2960	Weapons and ammunition	HS	6.14	44.11	1.70	2.08
2970	Domestic appliances n. e. c.	MSWC	5.78	31.46	1.51	3.11
3110	Electric motors, generators	MSWC	5.30	41.06	2.65	1.36
3130	Isolated wire and cable	MSWC	6.62	35.18	2.29	2.11
3140	Accumulators, batteries, etc.	MSWC	6.89	32.24	2.29	2.11
3150	Lighting equipment, lamps	MSWC	4.15	35.39	2.29	2.11
3540	Motorcycles and bicycles	MSWC	5.66	36.22	2.06	2.16
3550	Other transport equipment	MSWC	6.32	37.21	1.82	3.37

Labour intensive industries (LI)			**5.00**	**44.75**	**1.44**	**3.30**
1720	Textile weaving	LS	9.33	45.97	0.69	4.79
1740	Made-up textile articles	LS	4.59	44.02	1.60	3.16
1810	Leather clothes	LS	0.83	43.18	2.70	3.40
1820	Other wearing apparel	LS	2.17	40.81	1.45	3.86
1830	Articles of fur	LS	3.23	37.68	3.96	2.95
2010	Sawmilling, etc.	MSWC	7.02	39.97	0.19	3.57
2020	Panels/boards of wood	MSWC	6.30	39.01	0.69	4.37
2030	Carpentry and joinery	MSWC	4.08	47.04	0.67	3.23
2040	Wooden containers	MSWC	4.91	48.18	1.06	3.57
2050	Products of wood; cork, etc.	MSWC	3.37	40.78	2.70	3.11
2620	Ceramic goods	LS	5.60	41.80	1.04	4.45
2640	Construction materials	LS	7.35	44.02	0.22	2.38
2670	Processing of stone	LS	5.18	46.89	1.10	2.74
2810	Structural metal products	MSWC	3.63	46.73	0.44	1.57
2830	Steam generators	MSWC	4.53	47.23	0.92	0.94
2840	Metal processing	MSWC	6.12	47.07	1.59	1.77
2750	Casting of metals	LS	6.84	50.63	0.78	3.08
2850	Treatment, coating of metals	MSWC	6.00	44.67	2.60	4.62
2940	Machine-tools	HS	4.55	43.38	2.31	3.27
3160	Electrical equipment n. e. c.	MSWC	5.58	41.55	2.92	5.66
3420	Bodies for motor vehicles	MSWC	9.31	52.54	0.70	2.53
3510	Ships and boats	HS	2.90	55.25	0.97	3.11
3520	Railway vehicles	MSWC	4.88	43.74	1.48	3.10
3610	Furniture	MSWC	3.94	45.30	1.32	4.62
3620	Jewellery and related articles	LS	2.72	41.22	1.79	2.77
Capital intensive industries (CI)			**14.01**	**33.43**	**1.46**	**1.64**
1710	Textile fibres	LS	12.36	44.98	1.60	2.98
2110	Pulp & paper	MSWC	21.28	30.43	1.05	1.91
2310	Coke oven products	MSWC	13.74	38.98	1.11	1.38
2320	Refined petroleum prod.	MSWC	25.73	16.85	0.68	1.38
2410	Basic chemicals	MSWC	14.33	21.52	3.55	2.49
2470	Man-made fibres	MSWC	12.94	28.83	3.15	1.14
2630	Ceramic tiles and flags	LS	10.65	38.49	0.22	2.38
2650	Cement, lime and plaster	LS	10.53	27.29	0.54	2.74
2710	Basic iron & steel	LS	13.71	39.01	1.10	1.19
2730	Processing of iron & steel	LS	10.41	36.17	0.88	0.18
2740	Basic non-ferrous metals	LS	11.13	35.31	1.04	0.67
3430	Parts for motor vehicles	MSWC	11.33	43.29	2.62	1.28

Marketing driven industries (MDI)			**5.11**	**30.15**	**1.26**	**7.58**
1510	Meat products	LS	6.36	36.33	0.28	5.86
1520	Fish and fish products	LS	7.13	33.19	1.00	7.23
1530	Fruits and vegetables	LS	6.75	21.91	0.78	7.30
1540	Oils & fats	LS	8.55	18.93	0.15	7.09
1550	Dairy products; ice cream	LS	6.27	24.82	1.67	5.46
1560	Grain mill prod., starches	LS	7.18	14.47	0.94	8.72
1570	Prepared animal feeds	LS	5.09	18.28	0.94	8.72
1580	Other food products	LS	5.29	22.39	0.65	6.93
1590	Beverages	LS	5.88	18.40	0.76	6.47
1600	Tobacco products	LS	1.58	6.33	0.47	7.61
1910	Tanning/dressing of leather	LS	5.16	41.86	0.92	6.62
1920	Luggage, handbags, etc.	LS	2.06	39.49	0.92	6.62
1930	Footwear	LS	2.37	39.53	0.92	6.62
2210	Publishing	MSWC	3.93	31.10	3.16	6.41
2220	Printing	MSWC	5.60	40.59	1.36	6.22
2230	Recorded media	MSWC	9.99	27.83	1.58	6.64
2450	Detergents, clean., perfumes	MSWC	4.61	14.58	2.78	9.45
2820	Tanks, reservoirs, radiators	MSWC	4.14	44.11	0.40	5.15
2860	Cutlery, tools, gen. hardware	MSWC	5.53	45.06	1.88	10.49
3350	Watches and clocks	MSWC	3.03	37.70	0.99	9.33
3630	Musical instruments	LS	2.36	45.25	0.87	7.33
3640	Sports goods	LS	4.20	31.89	1.70	5.73
3650	Games and toys	LS	4.96	31.72	2.95	14.48
3660	Miscellaneous manufacturing	LS	4.54	37.90	2.13	9.39
Technology driven industries (TDI)			**6.91**	**31.21**	**5.85**	**2.64**
2420	Agro-chemical products	MSWC	7.63	11.87	1.21	2.73
2440	Pharmaceuticals	HS	7.19	16.35	12.97	5.93
2460	Other chemical products	MSWC	7.71	24.01	3.41	2.98
3000	Office machinery, computers	HS	7.07	31.63	6.91	1.49
3120	Electricity distribution, etc.	MSWC	4.91	37.25	4.63	1.68
3210	Electronic components	MSWC	12.16	33.30	7.12	2.20
3220	Telecoms equipment	MSWC	5.64	33.93	9.15	1.52
3230	Audiovisual apparatus	MSWC	10.42	30.88	5.54	3.48
3310	Medical equipment	MSWC	5.58	32.73	7.15	1.41
3320	Precision instruments	MSWC	4.23	43.82	5.30	2.61
3330	Process control equipment	MSWC	4.95	43.19	4.02	0.83
3340	Optical instruments	MSWC	6.35	26.69	6.09	4.27
3410	Motor vehicles	MSWC	7.86	25.78	4.31	2.03
3530	Aircraft and spacecraft	HS	5.06	45.56	4.14	3.74

Note: LS low-skill; MSBC medium-skilled blue collar; MSWC medium-skilled white collar; HS high skilled industries.

References

Achilladelis B, Schwarzkopf A, Cines M (1987) A study of innovation in the pesticide industry: analysis of the innovation record of an industrial sector. Research Policy 16: 175–212

Andersen HB (1997) Technological change and the evolution of corporate innovation. Doctoral thesis, University of Reading, UK

Arrow KJ (1962) The economic implications of learning by doing. Review of Economic Studies 29: 155–173

Aghion P, Howitt P (1998) Endogenous growth theory. MIT Press, Cambridge, MA

Caves RE (1996) Multinational enterprise and economic analysis, 2nd edn. Cambridge University Press, Cambridge, UK

Davies S, Lyons B (1996) (eds), Industrial organization in the European Union. Structure, strategy, and the competitive mechanism. Clarendon Press, Oxford

Davies S, Rondi L, Sembenelli A (1998) S.E.M. and the structure of EU manufacturing, 1987–1993. CERIS Working Paper 5, Torino

Dorfman R, Steiner PO (1954) Optimal advertising and optimal quality. American Economic Review 44: 826–836

Dosi G, Teece DJ, Chytry J (eds) (1998) Technology, organization, and competitiveness. Perspectives on industrial and corporate change. Oxford University Press, Oxford

Dunning JH (1994) Multinational enterprises and the global economy. Addison-Wesley, Wokingham

Grossman GM, Helpman E (1991) Innovation and growth in the global economy. MIT Press, Cambridge, MA

Hatzichronoglou T (1997) Revision of the high-technology sector and product classification. STI Working Paper 2, OECD, Paris

Ichniowski C, Shaw K, Prennushi G (1997) The effects of human resource management practices on productivity: a study on steel finishing lines. American Economic Review 87 (3): 291–313

Kaniovski S, Peneder M (2000) On the structural dimension of competitive strategy. Vienna, WIFO working papers No. 145

Kirzner IM (1997) Entrepreneurial discovery and the competitive market process: an Austrian approach. Journal of Economic Literature 35: 60–85

Krugman P (1979) Increasing returns, monopolistic competition, and international trade. Journal of International Economics 9: 469–479

Legler H (1982) Zur Positionierung der Bundesrepublik Deutschland im internationalen Wettbewerb. Forschungsberichte des Niedersächsischen Instituts für Wirtschaftsforschung, Bd. 3

Metcalfe JS (1998) Evolutionary economics and creative destruction. The Graz Schumpeter lectures. Routledge, London

Metcalfe JS (1995) The economic foundations of technology policy: equilibrium and evolutionary perspectives. In: Stoneman P (ed) Handbook of the economics of innovation and technological change, pp 409–512. Blackwell, Oxford

Metcalfe JS (1994) Competition, Fisher's principle and increasing returns in the selection process. Journal of Evolutionary Economics 4: 327–346

Milgrom P, Roberts J (1995) The economics of modern manufacturing: reply. American Economic Review 85 (4): 997–999

Milgrom P, Roberts J (1990) The economics of modern manufacturing: technology, strategy, and organization. American Economic Review 80 (3): 511–528

OECD (1998) OECD data on skills: employment by industry and occupation. STI Working Paper 1998/4, OECD, Paris

Peneder M (2001) Entrepreneurial competition and industrial location. Edward Elgar, Cheltenham

Peneder M (2000) Intangible assets and the competitiveness of European industries. In: Buiges P, Jacquemin A, Marchipont F (eds) (2000) Competitiveness and the value of intangible assets. Edward Elgar, Cheltenham

Peneder M (1995) Cluster techniques as a method to analyse industrial competitiveness. International Advances in Economic Research 1 (3): 295–303

Penrose E (1959) The theory of the growth of the firm, 3rd edn. Oxford University Press, Oxford

Schulmeister S (1990) Das technologische Profil des österreichischen Außenhandels. WIFO Monatsberichte 63 (12): 663–675

Sutton J (1998) Technology and market structure. MIT Press, Cambridge, MA

Sutton J (1991) Sunk costs and market structure. Price competition, advertising, and the evolution of concentration. MIT Press, Cambridge, MA

Teece D, Pisano G (1998) In: Dosi G et al. (eds) The dynamic capabilities of firms: an introduction, pp 1–14. Oxford University Press, Oxford

The dynamics of vertically-related industries. Innovation, entry and concentration*

Andrea Bonaccorsi and Paola Giuri

Sant'Anna School of Advanced Studies, via Carducci, 40, I-56127 Pisa, Italy
(e-mail: {bonaccorsi; giuri}@sssup.it)

Abstract. This paper studies the joint dynamics of concentration, number of firms and products in a vertically-related industry, and develops an original approach based on the analysis of the network of vertical relations between individual firms in buyer and supplier industries. The formation and evolution of the network is explained through technology and market factors such as the level of economies of scope, degree of technological co-specialisation between products in upstream and downstream industries, buyers' sourcing strategies and degree of market fragmentation. The paper develops an empirical analysis of the commercial jet and turboprop aircraft and engine industries from 1948 to 1997. An econometric analysis is developed to test some hypotheses on the transmission mechanisms of structural changes in the two pairs of industries.

JEL Classification: L13, L19, L22, L62

Key words: Vertically-related industries – Network – Concentration – Entry – Exit

---

* Reprinted from International Journal of Industrial Organization, Vol. 19, Bonaccorsi A., Giuri P., The long-term evolution of vertically-related industries, 1053–1083, © Copyright (2001), with permission from Elsevier Science. The authors wish to thank Elsevier Science who has kindly given permission for publishing an extended version of the article as a chapter of this book.
We thank Paul Geroski and anonymous referees for useful comments on a previous draft of the paper. We also thank John Sutton, Jean Luc Gaffard, Nick von Tunzelmann and participants to seminars and conferences held in London, Nice, Urbino, Manchester and Lausanne. The financial support of the Italian Ministry of Research (MURST 40% - "Industrial dynamics and interfirm relations") is gratefully acknowledged.

1 Introduction

This paper develops the argument that the long term structural evolution of an industry depends, in predictable ways, on the evolution of a vertically-related, downstream industry. More precisely, the evolutionary dynamics of the downstream industry, in terms of the number of firms and products, entry, exit and concentration, is transmitted to the upstream industry via the structure of the network of vertical exchange relations.

The claim that the vertical structure of an industry is a determinant of its concentration has been made repeatedly in the literature. We develop an original approach which is based on a construct, the network of vertical relations between individual firms in buyer and supplier industries. The formation and evolution of the network is explained through technology and market factors such as the level of economies of scope, the degree of technological co-specialisation between products in the upstream and downstream industries, the buyers' sourcing strategies and the degree of market fragmentation.

This paper analyses two pairs of vertically-related industries, the commercial jet aircraft and engine industries, and the commercial turboprop aircraft and engine industries, from the first introduction of the jet (1958) and turboprop (1948) technologies to 1997.

Turboprop and jet engine industries exhibit different structural dynamics with respect to the pattern of entry and exit of firms and products and the level of industry concentration. We study the evolution of the upstream industry by looking at the evolution of the downstream industry, and make the argument that the transmission of effects from one industry to the other is determined by the structure of the network linking the two.

Technological and market factors shape the emergence of different structures of vertical networks in the jet and turboprop industries. We identify two basic configurations, partitioned and hierarchical, and show the way in which different network configurations in the jet and in the turboprop are responsible for sharply different transmission mechanisms of the changes of the downstream to the upstream industry.

An econometric analysis is developed to test some hypotheses on the transmission mechanisms in the two pairs of industries.

2 Background literature

The idea that the evolution of an industry may crucially depend on what happens to a vertically-related, downstream industry has been repeatedly proposed in industrial organisation. The earliest formulation of this idea was proposed by J.K. Galbraith with the notion of countervailing power (Galbraith, 1952). It was subsequently placed by Bain (1968) and Scherer (1980) within the structure-conduct-performance paradigm (see also Scherer and Ross, 1990), and recently reprised by von Ungern-Sternberg (1996) and Dobson and Waterson (1997). The main purpose in revisiting the theory of countervailing power is the analysis of the effects of bilateral oligopoly on the price for the buyer industry and for the final consumer.

Although appealing, the notion is quite reductive: it is argued that concentration in a downstream industry is followed by parallel processes of concentration in the upstream industry. No implications are drawn on the specific mechanisms of transmission; other parameters of industrial dynamics, such as the number of firms and products, entry and exit, are simply not considered. However, an increase in concentration is the outcome of a variety of processes: incumbent suppliers may survive but face a redistribution of their market shares, or may merge, or there may be exit of less efficient suppliers. Furthermore, there is scarce and contradictory empirical support to the notion (Lustgarten, 1975; LaFrance, 1979; Ravescraft, 1983).

In a different context, a large stream of literature in industrial organisation deals with sourcing decisions. The starting point of this research is the analysis of procurement decisions in the public sector, particularly in defence, with the purpose of designing optimal procurement schemes under imperfect information.

Rogerson (1994) summarises specific features of the defence acquisition process, which affect the decision for dual or sole sourcing: competition of suppliers is desired in the design phase but, given small quantities and discontinuous demand, only one source is selected because buying a defence system from a single supplier allows economies of scale in production. Important issues for government are also the choice of the sourcing solution, which gives the right incentives to invest in innovation and in efficient production, and the possibility of monitoring suppliers' innovative efforts under asymmetric information and uncertainty about technology. A number of theoretical studies have tried to understand the real advantages of single or dual sourcing strategies (Demsky et al. 1987; Riordan and Sappington, 1989; Anton and Yao, 1987, 1989, 1992). Lyon (1996) empirically confirmed the hypothesis of negative effects of dual sourcing.

In summary, this literature provides elements for understanding the costs and benefits of isolated procurement decisions. Understanding the effects of a sequence of decisions regarding multiple products in a dynamic context is beyond its scope, and the implications of sourcing strategies for the structural dynamics of supplier and buyer industries are not addressed.

Existing theories of industrial dynamics posit a direct relation between attributes of technological regimes and stylised dynamic properties of the structural evolution of the industry (Nelson and Winter, 1982; Malerba and Orsenigo, 1993, 1996a). Developments in the modelling of industrial dynamics provide a clear analysis of the relationship between technological regimes, evolution of market demand, and structural evolution of the industry (Dosi et al. 1995, 1997; Malerba et al. 1998; Winter et al. 1997, 1999).

Recently, in different industries, patterns of vertical integration and disintegration, and division of labour have been observed as relevant engines of changes of industry structures. In the chemical industry, the emergence of specialised engineering firms represents an example of the economies of specialisation and division of labour at the industry level, which enormously affected the evolution of the industry structure, through the entry of new firms and the intensification of the competition (Arora and Gambardella, 1998). In the computer industry (Bresnahan and Malerba, 1999; Bresnahan and Greenstein, 1999) and in the semiconductor industry (Langlois and Steinmueller, 1999), the emergence of standards and platforms stimulated the entry and growth of specialised suppliers of components. In these industries, the changing patterns of vertical integration/disintegration strongly influenced the dynamics of competition.

One very recent frontier in evolutionary modelling of industrial dynamics is the explicit consideration of the joint dynamics of two vertically-related sectors. Malerba et al. (1998) developed a 'history friendly' model of the dynamics of vertical integration and disintegration occurring in the computer industry as the result of the dynamics of capability and upstream and downstream market structure, the latter being represented by the presence of a dominant leader which can exploit static and dynamic increasing returns from specialisation or vertical integration.

In his recent major work, John Sutton discusses the notion of interdependence of sub-markets (Sutton, 1998). The extent to which industries are concentrated is limited by the degree of interdependence between markets that address specific needs or use particular technologies within the same industry. Sutton predicts that, in industries with high interdependence of sub-markets, the lower bound to concentration is high, while industries with fragmented sub-markets exhibit a variety of outcomes, some of which may be characterised by low concentration.

In sum, although the need of including the vertical structure in the analysis of long term evolution of industries is clearly recognised, much work must still be done.

We develop an original approach based on a construct, the network of vertical relations between individual firms in buyer and supplier industries, which is used to explain the evolution of vertically-related industries. The formation and evolution of the network is explained through various characteristics of vertical relations drawn from contributions in industrial organisation and evolutionary economics.

There are a number of advantages originating from the use of a network approach to study the coevolution of vertically related industries. First, the unit of analysis is the single transaction, but it is not isolated from all other transactions taking place in the industry. As Holmstrom and Roberts (1998) pointed out, "in market networks, interdependencies are more than bilateral, and how one organises one set of transactions depends on how the other transactions are set up". Second, specific factors featuring vertical relations, such as the presence of asset specificity, technological complementarities, frequency of relations, or pattern of sourcing, are carefully reflected in several network measures at the transaction level, at the firm level (buyer and supplier), and at the overall network level. For example, the adoption of single or multiple sourcing strategies is represented by the number of relations for buyers. Diffuse adoption of multiple sourcing leads to a dense network, while single sourcing shapes a less dense network, which is also partitioned if the supplier industry is not monopolistic. Third, the relationship between the dynamics of vertical relations and industrial dynamics can be studied through the analysis of the relationship between variables describing upstream and downstream industries, i.e. level of concentration, dynamics of market shares, entry and exit, and variables describing the network.

This paper analyses a case of vertically-related industries, i.e. a supplier industry that sells only to one downstream industry, which in turn cannot substitute the products with those of competing sectors. Aircraft and engine industries, in both jet and turboprop, are characterised by a stable pattern of vertical separation, insofar as vertical integration of engine production never occurred over the entire history of these industries[1]. There is also no diversification, either in market demand for the upstream industry, or in supply sources for the downstream one. Of course, there is still the possibility that the survival of firms is subsidised by military sales or government interventions. Although this applies to specific circumstances, it is difficult to accept as an explanation of long term evolution[2].

This is an ideal case for discussing the coevolution of two industries, since external influences are ruled out. In some sense, we are dealing with a quasi laboratory experiment. In addition, we are able to push the analysis of vertical structure to the finest level of detail, namely transactions between individual customers and suppliers over the entire history of the industry.

The study is developed as follows. Next section describes three basic industry variables – i.e. number of firms, concentration and introduction of new products – and demonstrates sharp differences in the jet and turboprop vertically-related industries[3]. The structural dynamics of the industry highlights some relevant stylised facts with respect to the pattern of *entry and exit of vertically-related firms and products*. Section 4 examines in depth the *structural dynamics of the networks* connecting vertically related industries. Finally, in Section 5 we develop an econometric model to test the hypothesis that the *relation between upstream and downstream industry variables depends on the structure of the network.*

3 The long term evolution of the turboprop and jet industries

3.1 A brief history

The propeller-piston engine combination was the prominent aircraft propulsion system until the Second World War. Military needs to operate at higher altitude and speed induced the search for alternative forms of propulsion systems, which resulted in the affirmation of two propulsion systems: *turboprop* and *turbojet* (Constant, 1980). Each propulsion system was designed for specific ranges of operating conditions (aircraft speed, altitude, air density and temperature, passenger capacity). Military needs for turboprop were less pressing, because of the higher per-

[1] There are not specific studies addressing vertical integration of the engine industry. Technological bases are entirely different in the engine and aircraft industries, the costs of development of engines and aircraft are huge and excess correlation of risk has been considered a reason for vertical separation (Klein, 1977).

[2] Although economies of scope are clearly relevant in R&D, commercial engines are developed to be integrated into commercial aircraft, with no easy cross-over with military products. Moreover, in the cross country distribution of aircraft-engine relations, we do not observe any specialisation of supply relations by country; that is, aircraft manufacturers did not necessarily buy engines from national suppliers. On the contrary, global sourcing strategies have occurred since the beginning of the industry. The history of the supply relations offered many examples of the interrelations between US and European engine and aircraft companies: Pratt & Whitney supplied Sud Aviation, while Rolls Royce powered some Boeing and Douglas aircraft by the 1960s; Airbus aircraft have been mostly powered by General Electric and Pratt & Whitney engines.

[3] The description of the data is reported in Appendix 1.

formances promised by the turbojet technology, but the turboprop was of more immediate interest to the airlines than the jet (Miller and Sawers, 1968).

The 1950s represented a transition period for the airliners, as air carriers adopted the jet or turboprop engined designs for some operations, but continued to buy piston engines for others (*Flight International*, 1999a). The turbojet engine was more efficient than a piston propeller engine at speeds over about 450 m.p.h. (Miller and Sawers, 1968). At medium speed and altitudes, the turboprop was generally more efficient than the pure turbojet. This period was characterised by high uncertainty about the cost performances of jet and prop technologies and by intense competition between the two systems.

Competition between jet and prop occurred especially in the segment of aircraft with 51–90 seats. Over time, the markets with smaller seat capacity became dominated by the turboprops, while the markets with larger seat capacity (91–120 to more than 400 seats) were dominated by the jet. As shown by Figure 1, the size of the market for the jet engine was substantially larger than the turboprop. The evolution of the market has been characterised by several peaks. The birth of the turboprop industry is characterised by a continuous and rapid growth until the peak in 1959. After the introduction of the jet, which witnessed a tremendous growth until 1968, the demand for turboprop engines sharply decreased and stayed at low levels, with some oscillations, until the 1980s. The deregulation in the airline industry, more than any other factor, created the market for new turboprops, which has stimulated the development of new technologies for turboprop engines and aircraft, the introduction of many new programs and the entry of new actors (Aerospatiale, 1989; Airbus Industrie, 1991; FAA Forecasts and Technology Plans, 1994). As shown in Figure 1, after 1978 the market for turboprop engines was characterised by remarkable rates of growth. The production of engines more than doubled in the 10 years following deregulation.

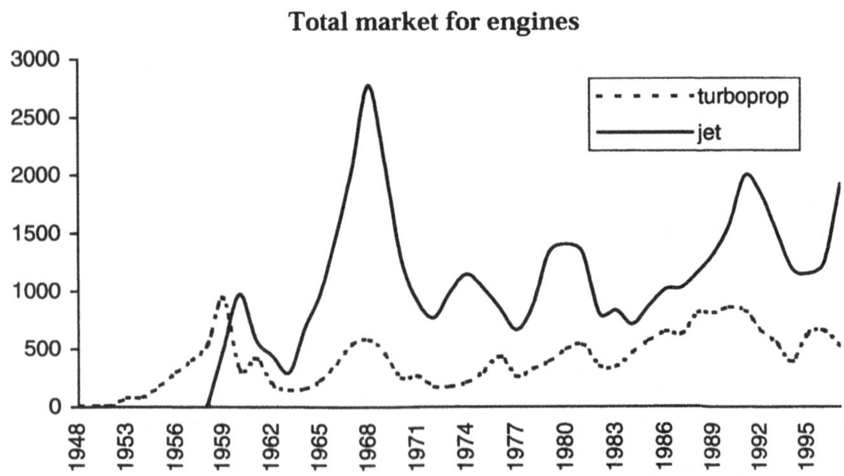

Fig. 1. Evolution of markets for turboprop and jet engines

During the 1990s, the regional aircraft market underwent a new radical change. The market shifted, this time irreversibly, from turboprop to turbofan, and the supply of turboprops faced a dramatically reduced demand (Schaffler, 1991; U.S. International Trade Commission, 1998; *Flight International*, 1997a,b; 1998a,b; 1999b). The process of technological substitution led to a structural decline of the turboprop industry, characterised by a reduction of orders, a process of riconversion to the jet technology undertaken by turboprop aircraft manufacturers, the birth of joint development programs among turboprop aircraft manufacturers (*Flight International*, 1997c; 1998c; 1999c), and the exit of some airframe producers.

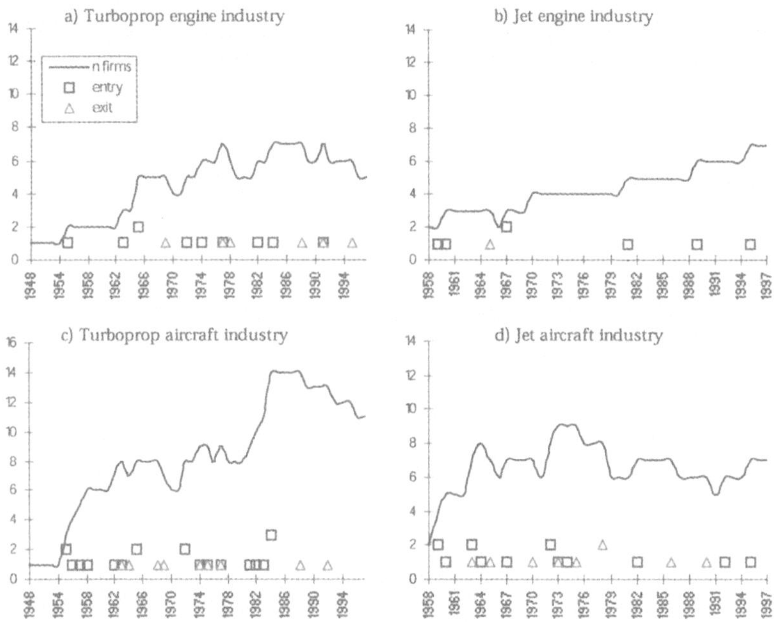

Fig. 2. Number of firms, entry and exit

3.2 Entry, exit and number of firms

The turboprop engine industry is composed of a small number of players, while the number of aircraft manufacturers is higher. Since their birth, the engine industry has counted nine engine manufacturers, while the aircraft industry number is 22. Figures 2a and 2c show the evolution of the number of firms and the process of entry and exit. The peak number of firms competing at the same time is seven in the engine industry and 14 in the aircraft industry; this occurred during the 1980s, when the turboprops registered high rates of development. The two industries did not experience a shakeout during their evolution, although several companies exited the industry in the last decade[4]. The reduction in the number of players can be

[4] It is interesting to note that the patterns observed in the two engine industries are different from the patterns predicted and explained by different streams of research on the dynamics of industry

considered as part of the process of progressive disappearance of the industry, due to technological substitution by the jet.

It is interesting to analyse the relation between the entry and exit of aircraft and engine manufacturers. The history of each entry and exit event highlights some stylised facts.

1) New aircraft manufacturers are either served by existing or new suppliers.

2) Entry of suppliers always occurs to serve new customers. After the first flight of the Vickers Viscount in 1948 and of the Fokker F27 in 1955, powered by the Rolls Royce Dart, in 1957 Allison entered the market to power the Lockheed L-188 Electra. In 1963, the aircraft manufacturer Nord Aviation entered the market, supplied by Turbomeca. In 1965, Pratt & Whitney and General Electric supplied two different aircraft programs introduced by the entrant de Havilland Canada. With the exception of Walter, which entered to substitute Pratt & Whitney to power the Let 41, all the other engine manufacturers entered by supplying new entrants in the aircraft industry: Pilatus and Lycoming in 1973, Casa and Garrett in 1974, Yunshui and Dongan in 1984, IPTN and Allison in 1995.

3) New aircraft programs introduced by existing customers are generally powered by the established engine supplier, unless the new program is introduced in a different segment of the market not served by the existing supplier. Specialisation by market segments and the absence of economies of scale and scope lead to a configuration of the industry in which engine suppliers operate mainly in just one segment of the market. In this industry context, although turboprop aircraft manufacturers operated mainly in single sourcing, the introduction of a program in a new market segment often required the establishment of a supply relation with a second source. An example is provided by Casa, which was supplied by Garrett in the less than 30-seat segments and by General Electric in the segment 31–50 seats.

4) Minor engine manufacturers are induced to leave the industry by the exit of their main customer. This is true for the cases of Allison in 1969, Turbomeca in 1976, Walter in 1995 and Lycoming in 1996. Note that these firms were not owned by aircraft manufacturers, but were independent, privately-held companies. By contrast, Rolls Royce, the leader of the market for many years, progressively lost market shares and left the industry after the exit of its former customers. The last supply relation gained by Rolls Royce was Fairchild in 1968. After the financial crisis of the 1970s, Rolls Royce concentrated its efforts only in the jet market and did not introduce new products in the turboprop industry. This caused its exit from the market in 1988.

The jet engine and aircraft industries are also composed of a small number of players. As with the turboprop, the number of players is higher in the aircraft than in the engine industry. Since their birth, the engine industry has counted seven engine manufacturers, while the aircraft industry has 16 (Figs. 2b and 2d). The peak number of firms is seven in the engine industry and nine in the aircraft industry. The jet engine industry did not experience a shakeout during its evolution, as no firms exited the industry. The aircraft industry experienced a reduction in the number

population (Klepper, 1996, 1997; Hannan and Carroll, 1992; Carroll and Hannan, 2000). Explanations of the diversity have been proposed in Bonaccorsi and Giuri (2000a) for the turboprop and in Bonaccorsi and Giuri (2000b) for the jet.

of firms, which continued in 1998 with the exit of Fokker and the acquisition of McDonnell Douglas by Boeing. However, as evident from Figure 2d, there is no clear evidence of shakeout as new firms entered the small jet segment of the market.

With respect to the stylised facts observed in the turboprop industry, the jet market presents sharp differences.

1) New aircraft manufacturers are always served by existing engine suppliers. The only exception is the entry of Embraer supplied by the new entrant Allison in the segment of the small regional jets.

2) Engine manufacturers do not enter to supply new, but existing aircraft manufacturers, except for the case of Embraer and Allison mentioned above. Entry of suppliers is not fuelled by the entry of new customers or the launch of programs, but by the increasing adoption of multiple sourcing at the aircraft manufacturer and at the program level. Airbus, Boeing and McDonnell Douglas operated with three, and in some periods with four or five, engine manufacturers. Starting from the B747, other programs such as the A330, the B767 and the B777 have been powered by engines from three different engine manufacturers. The affirmation of multiple sourcing strategies implies that the launch of a new program is an opportunity for more than one engine manufacturer, which compete to gain the launch order or a large share of total orders. However, a number of minor aircraft manufacturers operated in single sourcing, but they did not create opportunities for entry, as their aircraft were powered by engines of existing suppliers, in most cases by Rolls Royce engines.

3) There is no established pattern for the choice of the engine supplying the launch of a program by an existing aircraft manufacturers. Obviously, companies operating in single sourcing powered new programs with their established suppliers, although most of them introduced just one program during their life. Companies operating in multiple sourcing used in some cases already involved suppliers, in other cases suppliers that, although active in the market, did not supply them before.

4) As no exit of engine manufacturers occurs during all the history, there is no relation with the exit of aircraft companies.

3.3 Industrial concentration

In the turboprop market, the level of industrial concentration analysed is clearly higher in the supplier than in the buyer industry (Figs. 3a and 3c).

In the engine industry, the CR2 index is very high throughout the period. It is at its maximum value during the first years of the industry life, as there are only two firms. It starts to decrease for the entry of new firms, but at least the 70% of the market is still dominated by two companies. The second half of the time series shows again an increasing trend. The same pattern is displayed by the Herfindahl index. The dynamics of market shares underlying the evolution of the level of concentration reveals that the market is always dominated by a strong leader. The former leader was Rolls Royce, which maintained its position until the end of the 1960s. As mentioned above, the leader progressively decreased its efforts in the turboprop industry and lost market share with the exit of its customers. The leadership position was gained by Pratt & Whitney, which entered in 1965, through

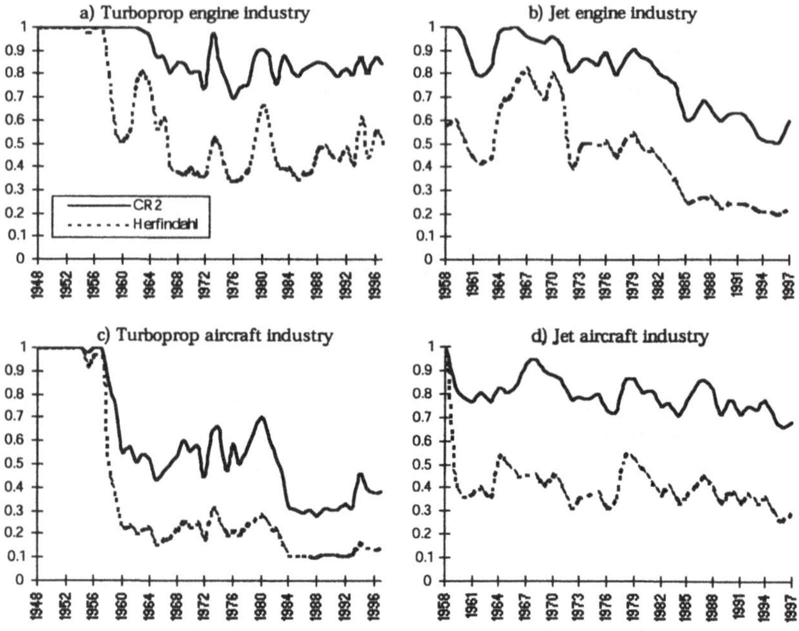

Fig. 3. Level of industrial concentration

the acquisition of several customers in the smaller seat-segment of the market; they witnessed a substantial growth during the 1960s and the 1970s and ended up representing a large share of the total market. However, other companies in the market secured important shares in different market segments.

On the buyer side, the level of industrial concentration was remarkably lower. The turboprop aircraft industry is characterised by a low and decreasing level of concentration, and by the absence of a dominant leader in the total market and in each segment. Until the 1970s a few large companies owned large shares of the markets, and the two largest, de Havilland Canada and Embraer, dominated on average 50% of the market. The level of concentration rapidly decreased after the entry of new competitors, which eroded the positions of the incumbents in all market segments. The industry then became highly fragmented, and was characterised by the complete lack of dominant players.

The dynamics of concentration in the jet engine and aircraft industry is different from that of the turboprop. The jet engine industry shows a remarkable process of decreasing concentration, while the aircraft industry is characterised by a quite high and unstable level of concentration (Figs. 3b and 3d).

In the first decade of the jet engine industry, two firms created a quasi stable duopolistic market structure, in which the first mover (Pratt & Whitney) had a pronounced leadership over the second player (Rolls Royce). The level of concentration slightly decreased due to the subsequent entry of two large competitors (General Electric and Snecma), which rapidly gained significant market shares. During the 1980s, industry concentration decreased very rapidly, and the leader was replaced by the entrant of the second phase, General Electric. The final configuration of the industry was oligopolistic, characterised by strong instability of market shares and turbulence in the players' positions. The entry of minor companies, which acquired small shares of the market, contributed very slightly to the reduction of concentration. By contrast, intense competition among the four larger companies for the same customers and products led to mobility of their market shares.

The jet aircraft industry has been characterised by the persistence of a strong leader, Boeing, for all the industry life. The CR2 index indicates that two companies dominated at least 70% of the market. The oscillation of the two indexes until the end of the 1970s is explained by the entry of new firms, which tried to compete with Boeing and the follower, McDonnell Douglas. The growth of the new entrant, Airbus, which eroded the leader's market share, the exit of Lockheed and other minor companies, and the emergence of the market for small jets, initially dominated by newcomers, explained the slow decline of the concentration. In 1998 the industry concentrated again, as Boeing acquired competitor McDonnell Douglas and Fokker exited the industry.

Comparing the engine and aircraft industries, we may observe that, although the processes of concentration are different in the four industries, the dynamics of the two pairs of vertically-related industries are similar over time. In the turboprop industry, they show simultaneous increasing and decreasing patterns, while in the jet industry they show a decreasing pattern, characterised by different competitive dynamics among major players.

3.4 Introduction of products

Figure 4 displays the evolution of the number of products and the introduction of new products in the four industries analysed[5].

The turboprop aircraft and engine industries shows a strong similarity in the patterns (Figs. 4a and 4c). The number of products increases during the 1950s, shows a weak reduction due to the exit of the first models introduced and tends to stabilise until the 1980s. Then the number of products rapidly increases, following the growth of the market fuelled by the deregulation in the United States and in Europe.

The patterns in the jet engine and aircraft industries are clearly different. In the jet engine market, the number of products grows almost monotonically, except a reduction during the 1990s, which reflects the substitution of many models of the second generation, introduced during the 1970s (Fig. 4b). The number of aircraft increases at very high rates during the 1960s, then decreases and tends to stabilise during the 1980s and the 1990s (Fig. 4d). The sharp growth and exit of products

[5] Our data on engine and aircraft products make a distinction between the program and the different versions of the same program. For example, the B777 is a program, while the B777-100A, B777-100B, B777-200 are versions. A program is based on a core technology platform, while several versions are developed on the same platform by modifying design parameters. In the jet engine technology, a program is launched around a new turbine, while variations in the propulsion system may lead to several versions. The graphs show the number of versions, because programs are used for many applications through the development of different versions. The version is therefore the more fine grained innovative output of firms.

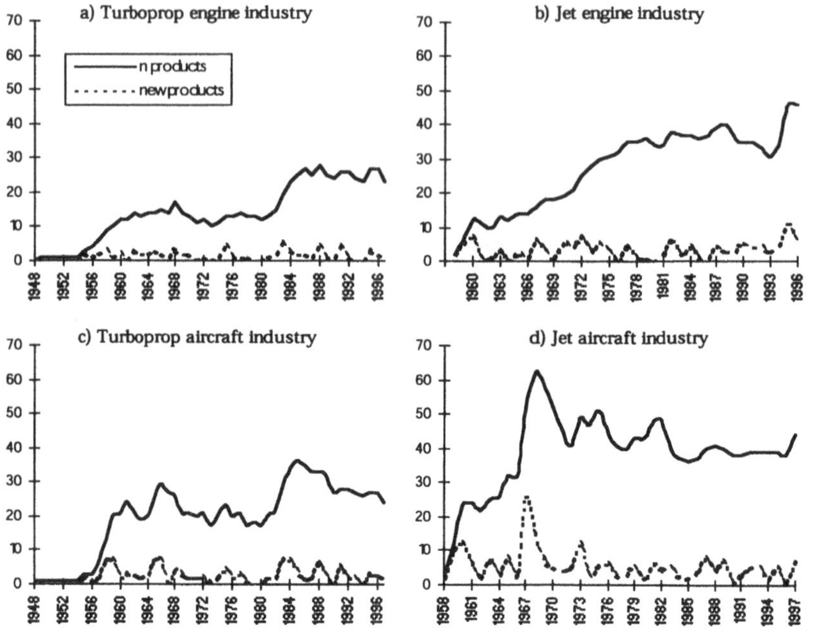

Fig. 4. Total number of products and introduction of new products

during the 1960s is explained by the introduction of a high number of different versions of the first programs introduced, the B707, DC-8 and DC-9, rapidly replaced by improved versions and new aircraft programs.

The analysis of the patterns of introduction of products in the two pairs of industries is explained by the following observations.

In the turboprop engine industry, more than 55% of the engine versions are introduced to power a single aircraft program version. About 40% of the engine versions power a few, in most cases two or three, versions of the same aircraft program. Only five engines power more than one aircraft program of different aircraft manufacturers in the same segment of the market. In the turboprop aircraft industry most of the aircraft, 109 out of 129, are powered by only one engine version, while a small number of aircraft integrate two versions of the same engine program.

In the jet engine industry we may observe three differentiated patterns.

A number of engines power only one aircraft version. This is the case of engines powering aircraft for specific markets (the Rolls Royce M45 which powered the VFW614 in the regional market, the Olympus 593 which powered the Concorde) and the case of the *last generation* engines, which are developed in many versions to power the same aircraft model in order to satisfy the need of aircraft manufacturers to meet the specific requirements of airlines. Examples are a number of versions of the Trent, the PW 4000 series, the IA V2500 series. Different versions of these engines are integrated on the same aircraft version in order to obtain specific operating conditions.

Other engines are integrated in different models and versions of the same program.

Finally, several engines of the *first and second generation* are integrated in many versions and programs of different aircraft manufacturers. To name some examples, in the 1960s the Pratt & Whitney JT3D-3B was integrated into 38 versions of the B707, B720 and DC-8; the JT8D-7 in 18 versions of the Caravelle, B727, B737 and DC-9; while in the 1980s and the 1990s the General Electric CF6-80C2 powered 22 versions of the A300, A310, B747, B767 and MD-11.

An inverse pattern may be noticed in the aircraft industry. Many different versions of the *first generation* of aircraft (B707, B720, DC-8, DC-9 and Caravelle) were powered by only one engine version. The same pattern was found for the regional aircraft market (Fokker, BAe 146, Canadair Regional Jet).

Many aircraft of the *second generation* (B727, B737, B747, MD-80) integrated a number of versions of the same engine program, while the aircraft of the *third generation* (Airbus A300-600, A319, A320, A330, B757, B767, B777) were introduced in multiple sourcing. Each aircraft integrates different versions of programs of different aircraft manufacturers. As an example, each version of the B777 integrated several versions of General Electric, Pratt & Whitney and Rolls Royce engines.

In summary, the pattern observed in the turboprop industry suggests the existence of close technological complementarity between engines and aircraft, which induces the introduction of new engine versions in correspondence with the launch of new aircraft. In the jet industry, aircraft and engines of different generations are designed to operate in many seat-range conditions. Technological factors underlying the differences between jet and turboprop have been discussed by Bonaccorsi and Giuri (2000a, 2001), which provide evidence of the lack of economies of scale and scope in design and production activities in the turboprop industry and of their existence in the jet industry.

4 The structure and dynamics of the network

4.1 Networks

In order to explain the difference in the way the structural dynamics of two pairs of industries are related, we examine the properties of the network of vertical exchanges that connects them.

Network analysis has been applied in many fields of the social sciences, including economics, sociology, and organisation, for analysing different structures of interactions among agents (individuals, firms, groups of actors, technical artefacts). In the analysis of industries, network concepts and techniques are increasingly used in the field of inter-firm agreements (joint ventures, licensing, technological alliances, consortia and the like) (Powell, 1996; Orsenigo et al. 1998, 2000). We apply network analysis to the study of vertical relations between buyers and suppliers.

The unit of analysis is the transaction of engines occurring between an engine and an aircraft manufacturer each year. Relations are characterised by high frequency and stability. The yearly average number of transactions of each relation is 85 in the jet and 37 in the turboprop, while the minimum and the maximum are, respectively, 11 and 436 in the jet and 3.33 and 132 in the turboprop. The range of variation depends on the market shares of the companies and on the size of the market segment. Relations are also very stable over time. In fact, interruptions of relations are very rare, with the exception of cases of the exit of actors, while new relations tend to be added and to persist over time. The total number of relations that compose a network depends on the entry and exit of actors, and on the creation and interruption of relations. The stability of the relations implies that the structural configuration of the network is persistent over time, and it is not subject to short run changes. The way relations are distributed among actors is not random, but depends on structural factors of the industries under study.

The network may assume different topologies, which can be represented by different network measures. For the purposes of this study, we analyse three structural properties of the network: the *relational intensity*, the *distribution* of the relations across actors and the *separability* of the network.

For the analysis of vertically-related industries, we study bipartite graphs, in which edges connect vertices from different sets of actors (buyers and suppliers), and there are no ties within each set (Borgatti and Everett, 1997; Asratian et al. 1999). The edges in the network are determined by the order of an engine placed by an aircraft company to an aero-engine manufacturer at a given date. The structure of the relations is represented for each year by a biadjacency matrix, the cells of which represent the binary variable "a relation exists / does not exist".

We selected the following network measures to study the structural properties of the network: density, k-core, cut-points and bi-components.

The *density* is essentially a count of the number of edges actually present in a graph, divided by the maximum possible number of edges in a graph of the same size. Density is a synthetic measure of network structure that provides information about the group *relational intensity* and the cohesion of a graph, but does not include information about the variability among actor degrees.

We also calculate measures at the sub-graph level, to analyse the *distribution of the relations* across actors. In particular, we study the number and size of k-cores, which give a clear representation of the presence of cohesive sub-groups. A k-core is a connected maximal induced sub-graph which has minimum degree greater than or equal to k (Wasserman and Faust, 1995). The degree of an actor is the number of edges incident with that vertex. Each member of a k-core is related to at least k other actors on the other set. We calculate the number of k-cores in the jet and turboprop networks for every possible value of k and for each year of their life. For $k = 1$, the number of cores indicates the degree of *partition* of a network. The higher the number, the higher the degree of partition of the network, as the network can be separated in n sub-graphs without deleting any vertex. The presence of cores with degree greater than or equal to k denotes cohesiveness of a graph, that is, the distribution of relations across actors is not dispersed but is concentrated in sub-groups of intensely connected actors. In particular, a network characterised by a restricted core and a periphery of disconnected actors can be defined as *hierarchical*.

The *separability* of the network is finally analysed through the number of bi-components and cut-points of the graphs. A *bi-component* (or block) of a graph is a maximally non-separable sub-graph; it requires deletion of at least two vertices to disconnect it. A *cut-point* is a node that is connected to more than one bi-components, therefore its deletion disconnects the graph.

The number of bi-components and cut-points provides an indication of the connectivity of a graph. The higher its number, the higher the probability that a graph will be disconnected for the deletion of a vertex, that is, for the interruption of a relation or the exit of a player. The size of the bi-components, similarly to the analysis of k-cores, gives an indication of the size of highly connected sub-groups. Precisely, bi-components with more than two actors correspond to k-cores with k greater than 1[6].

4.2 Networks in turboprop and jet industries

Figures 5 and 6 exhibit pictures of the network structure in the turboprop and jet markets at four points in time, while Table 1 summarises the network measures at the group and sub-group level for each year of the period. The measures computed are number of actors (*actors*), level of relational density of the network (*density*), number of k-cores with $k = 1$ (*1-core*), number of k-cores with $k = 2$ (*2-core*), number of vertices in the core with $k = 2$ (*2-core size*), number of bi-components (*bi-comp*), number of actors presents in more than one bi-component (*cutpoints*).

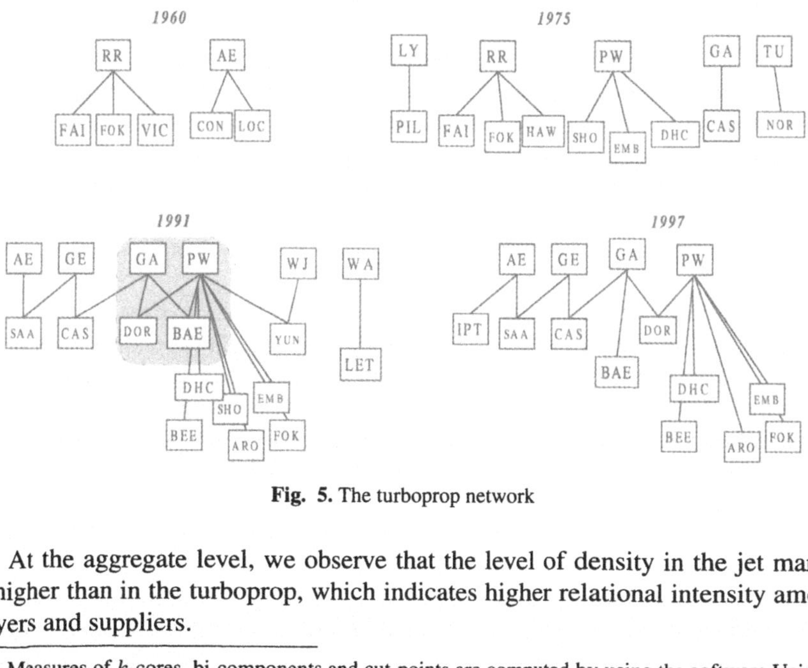

Fig. 5. The turboprop network

At the aggregate level, we observe that the level of density in the jet market is higher than in the turboprop, which indicates higher relational intensity among buyers and suppliers.

[6] Measures of k-cores, bi-components and cut-points are computed by using the software Ucinet 5 (Borgatti et al. 1999).

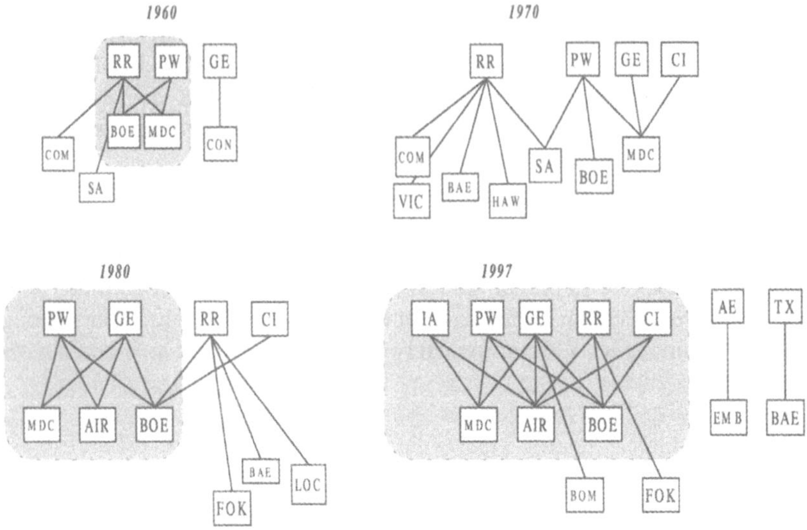

Fig. 6. The jet network

The analysis at the sub-graph level provides details on the distribution of relations across sub-groups of actors. In particular, it gives evidence of the degree of partition and of the formation of a hierarchical structure of the network.

The degree of partition of the network in the jet market is lower than in the turboprop, as the number of k-cores with degree equal to one is smaller. In the turboprop network, the number of 1-cores increases over time and ranges between three and six for a large portion of the time, although it tends to decrease over the last years of the industry evolution. In 1975 there were five partitions while in 1997 there is only one partition, as there are not completely separated sub-graphs. In the jet, the network is composed of only one or two sub-graphs over its life, except in the last three years, in which there are three 1-cores. The larger partition is composed of 10 actors in the market for large commercial aircraft. The other two partitions are two pairs of vertically-related firms in the market for small regional jets.

The number of 2-cores denotes the degree of hierarchisation of the network. The network of the jet assumes a *hierarchical* configuration as it is possible to identify a cohesive core in which the actors have degree greater than or equal to two. The core emerged during the first stage of the industry life and was initially composed of four actors (shaded area in Fig. 6, 1960). The entry of new actors at the end of the 1960s and at the beginning of the 1970s destabilised the network and the core. The intensification of the relational activities of entrants and incumbents led again to the emergence of a core, which expanded during the industry evolution. In fact, as is evident from the table, the number of actors grew from four in 1975 to eight in 1997, therefore a larger part of actors entered the core. The core was composed of the incumbent engine suppliers and of major aircraft manufacturers operating in multiple sourcing. On the other hand, a periphery was also created in the network, which was composed of actors with degree equal to one, that is, aircraft manufac-

Table 1. Density, *k*-cores and bi-components in turboprop and jet networks

	TURBOPROP							JET						
	actors	*density*	*1-core*	*2-core*	*size 2-core*	*bi-comp*	*cut-points*	*actors*	*density*	*1-core*	*2-core*	*size 2-core*	*bi-comp*	*cut-points*
1953	2	1,00	1	-	-	1	0							
1954	2	1,00	1	-	-	1	0							
1955	5	0,50	2	-	-	3	1							
1956	6	1,00	1	-	-	2	1							
1957	7	0,50	2	-	-	3	1							
1958	8	0,50	2	-	-	5	1	4	0,5	2	-	-	2	0
1959	8	0,50	2	-	-	5	1	6	0,5	2	-	-	4	2
1960	8	0,50	2	-	-	5	2	8	0,47	2	1	4	4	1
1961	8	0,50	2	-	-	6	2	8	0,47	2	1	4	4	1
1962	9	0,58	2	-	-	7	2	8	0,53	2	1	5	3	1
1963	11	0,33	3	-	-	8	2	10	0,43	2	1	4	6	1
1964	10	0,38	2	-	-	8	3	11	0,42	2	1	4	7	2
1965	13	0,25	3	-	-	10	4	10	0,43	2	1	4	6	2
1966	13	0,31	2	-	-	10	5	8	0,67	1	1	4	5	2
1967	13	0,29	2	-	-	10	5	10	0,48	1	1	4	7	3
1968	13	0,28	2	-	-	11	4	10	0,43	1	1	4	6	3
1969	12	0,26	3	-	-	9	2	10	0,43	1	-	-	9	4
1970	10	0,30	3	-	-	6	2	11	0,36	1	-	-	10	4
1971	10	0,29	3	-	-	7	3	10	0,38	2	-	-	8	3
1972	13	0,23	4	-	-	9	2	12	0,36	2	-	-	10	4
1973	13	0,25	4	-	-	7	2	13	0,38	2	-	-	9	4
1974	15	0,25	4	-	-	8	2	13	0,38	2	-	-	9	2
1975	15	0,20	5	-	-	9	2	13	0,38	2	1	4	9	2
1976	14	0,21	4	-	-	10	4	12	0,42	2	1	4	7	3
1977	16	0,16	6	-	-	11	3	12	0,46	1	1	6	7	3
1978	14	0,23	4	-	-	8	3	12	0,43	1	1	4	9	2
1979	13	0,23	4	-	-	8	3	10	0,51	1	1	4	6	3
1980	13	0,29	3	-	-	7	3	10	0,52	1	1	5	7	2
1981	14	0,28	3	-	-	9	4	11	0,44	1	1	5	6	3
1982	16	0,24	3	-	-	12	5	12	0,41	1	1	5	8	3
1983	17	0,22	4	-	-	11	4	12	0,35	2	1	4	8	3
1984	21	0,16	5	-	-	12	5	12	0,44	1	1	5	7	3
1985	21	0,21	3	-	-	12	7	12	0,36	2	1	5	8	3
1986	21	0,22	2	-	-	12	7	12	0,35	2	1	4	8	3
1987	21	0,27	1	-	-	12	7	11	0,44	2	1	6	5	2
1988	21	0,22	2	-	-	12	7	11	0,49	2	1	6	4	2
1989	19	0,21	3	-	-	12	6	12	0,49	2	1	6	5	3
1990	19	0,21	3	-	-	12	6	12	0,44	2	1	6	6	3
1991	20	0,24	2	1	4	12	6	11	0,49	2	1	6	5	3
1992	19	0,22	3	1	4	12	5	12	0,44	2	1	6	6	4
1993	18	0,34	1	1	4	12	5	12	0,44	2	1	6	6	4
1994	18	0,34	1	1	4	12	5	12	0,44	2	1	6	6	4
1995	18	0,27	2	1	4	12	6	14	0,39	3	1	8	5	2
1996	16	0,25	2	-	-	12	7	14	0,39	3	1	8	5	2
1997	16	0,33	1	-	-	12	7	14	0,39	3	1	8	5	2

turers in single sourcing and engine suppliers in the regional market. The network was also partitioned in the last three years, as there were three subgroups, two of them in the regional market (Allison-Embraer and Textron-British Aerospace), and the other containing the 2-core.

The network in the turboprop presents a very different structural dynamics. In fact, except for a five-year period, there are no highly connected sub-groups of actors, as indicated by the absence of 2-cores in almost all the industry life. Only from 1991 to 1995 a 2-core was formed by four actors, but it was transitory (see Fig. 5, 1991). This indicates that relations among actors are exclusive and sparse, and are not concentrated around a group of actors.

Further evidence on the degree of connectivity and separability of the networks is given by the number of bi-components and cut-points. In the jet market, these two variables increase until the beginning of the 1970s, when the network is destabilised by the entry of new actors. Subsequently, the indicators decline until 1997, suggesting higher connectivity of the network. In the turboprop, both the number of bi-components and cut-points grow over time, denoting lower connectivity and higher separability of the graph. The difference of the values in the jet and turboprop markets is not very large, especially if compared with the number of actors, which is larger in the latter. This reflects the presence of turboprop engine suppliers, namely Rolls Royce and Pratt & Whitney, which are present in a high number of bi-components. Precisely, during the 1970s the number of cut-points is very low, compared to the number of bi-components, because many aircraft manufacturers in single sourcing are supplied by only two engine companies. During the 1980s and 1990s, although the number of bi-components in which Pratt & Whitney is active is still high, the number of cut-points grows, denoting increasing separability of the network through other nodes.

Combining these results, we observe a *partitioned structure* of the turboprop network and a *hierarchical structure* of the jet network.

4.3 Causes of differing network structures

Where do different structural configurations of the network come from? What caused these different network structures to emerge?

We believe that differences in the network structures in the two pairs of industries depends on the following structural factors: the degree of economies of scope, the sourcing strategy adopted by aircraft manufacturers, and the level of market segmentation. We summarise here the differences between jet and prop with respect to these structural factors, which have been studied in related papers (Bonaccorsi and Giuri, 2000a, 2001).

Economies of scope

We calculated some statistics on the number of versions per product to measure the degree of robustness of basic designs (programs) as an indication of the presence of *economies of scale and scope* in the design and manufacturing activities of aircraft and engine companies. Robustness occurs in terms of adaptability of the basic design to different customers and to different market segments. It allows some degree of economies of scale (high commonality of parts) and scope (high product variety) on the R&D and production side, and offers the possibility of enhanced learning from user experience. In fact users, working with similar platforms, can be more apt to ask for specific modifications of design. This type of design applies not only to the sequential introduction of models, but also to different more or less contemporaneous versions (Rothwell and Gardiner, 1989, 1990).

We found that the average and the maximum number of versions per program in the jet market is higher than in the prop, both in aircraft and engine. In addition, the number of programs in turboprop is higher than in turbojet, although the size of the market is smaller. This suggests that the turboprop market is more fragmented in different programs and that the degree of economies of scale and scope is much lower. Similar results have been observed by calculating indicators on the number of *applications* (aircraft programs and versions) per engine programs.

How did economies of scope influence the network structure?

In the presence of economies of scope, engine programs are potentially applicable to different aircraft programs of different manufacturers. This allows engine companies to relate to many buyers, and potentially to all of them. By contrast, in the absence of economies of scope, turboprop companies may relate to a reduced number of aircraft companies. The only exception is the case of companies producing many different programs, but this would be an extremely expensive strategy which is not plausible in industries characterised by huge costs of development and large break even point.

Buyers' sourcing strategies

By single sourcing we mean that aircraft manufacturers select just one engine supplier for each new programme, possibly by asking the same manufacturer to develop several versions of the basic engine for the corresponding versions of the aircraft programme.

In the turboprop market, the demand for engines takes place within an almost generalised single-sourcing strategy. Out of 22 aircraft manufacturers, 10 operate using multiple sourcing and 12 with single sourcing. Companies operating with dual sourcing still equip the aircraft program through single sourcing.

In the jet market, the pattern of sourcing of aircraft manufacturers changed considerably over the industry evolution. While at the birth of the industry, aircraft makers tended to operate through single sourcing, or in some cases dual sourcing for specific aircraft models, the solution of the major technical problems of the first stage of industry evolution, while reducing the degree of uncertainty, allowed designing aircraft to integrate different engine configurations. Innovation at the design level permitted the increasing interchangeability of engines on newer model aircraft (Bluestone et al. 1981). Airframes started to be built to accept any one of several turbine configurations offered. This trend led to the shift from single to dual and multiple sourcing strategies of large aircraft manufacturers. Multiple sourcing policies began to be adopted at the aircraft program level, and in some cases at the version level.

The differences in the jet and prop economies of scope and sourcing strategies suggest that in the turboprop market there is a technological co-specialisation of engine and aircraft, while in the jet aircraft market product architectures are open and characterised by a standardised interface for engines. Jet engines are optimised for an envelope of operational conditions, while turboprop engines are designed for a narrow range of operational parameters.

With respect to the effects on the network structure, in the turboprop market the diffusion of single sourcing at the firm, at the segment and at the program level induces a partitioned network. In the jet market the shift from single and dual toward multiple sourcing strategies of major aircraft manufacturers led to the formation of connected network structures.

Market segmentation

The structure of demand for engines is represented by the aircraft manufacturers. In the turboprop market, customers operate mainly in just one segment of the market. Out of 22 manufacturers over the life of the industry, 12 were active in one segment, nine in two segments, and just one in three segments. No one covered four segments, including the rapidly disappearing segment of 91–120 seats dominated by the jet technology. If we look at the final structure of the industry, out of 11 still active in 1997, again the largest part (72%) still operate in one segment. Those that operated in more than one segment developed their models at different dates, with the exception of the joint venture Aerospatiale-Alenia, which developed the ATR 42 in 1984 and the ATR 72 in 1988, following the same commonality strategy that is typical of the jet industry.

This is in sharp contrast with the pattern found in the jet aircraft industry (Sutton, 1998; Bonaccorsi, 1996), which is one of declining number of large suppliers operating simultaneously in all segments of the market. Since large jet aircraft manufacturers operate across all segments and plan their engine acquisition strategies in an integrated manner, they favour complete range suppliers. As a result of this enormous pressure, all large engine suppliers operate in all market segments. The only jet engine suppliers that were able to survive in one or a few segments are those that supplied smaller jet engine manufacturers (e.g. British Aerospace or Embraer), which do not target the large transport aircraft market. On the supply side, the presence of economies of scale and scope, reflected in the presence of robust designs of engines which power many aircraft configurations, supports a configuration of interdependence of sub-markets (see Sutton, 1998).

The same pressure is not found in the turboprop industry. Faced with customers who demand engines for just one aircraft size at a time, with no significant interdependence among segments, turboprop engine suppliers could survive with a limited range of products. The low level of economies of scope supports this configuration.

How does this affect the network?

Fragmented markets limit the scope for higher relational intensity across actors. Suppliers (or customers) operating in one or two segments can in practice relate to a reduced number of customers (suppliers), although the potential number of customers is higher. By contrast, in industries characterised by interdependence of sub-markets, each actor can in principle relate to actors on the other side in any segment.

5 Observing the transmission mechanism

The comparison between the two pairs of industries shows an intriguing difference. In the turboprop industry, the dynamics of the upstream industry closely parallel the dynamics of the downstream one. The turboprop aircraft industry exhibits a remarkable growth in the number of firms, a rapidly declining and then stable level of concentration, and an increasing number of products, with some oscillations. Almost the same patterns of change apply to the related engine industry.

This is not true for the jet. In the aircraft industry the number of firms is quite stable, concentration is slightly decreasing and the number of products has a peak and then declines. By contrast, a different pattern is found in the engine industry: the number of firms is always growing, concentration is sharply decreasing, and the number of products has an increasing trend.

We develop the hypothesis that the relation between the structural dynamics in the upstream and downstream industry is regulated by the structure of the network of vertical relations, that is, changes in the downstream industry are transmitted to the upstream industry with an intensity which depends on the structure of the network. In other words, our interpretation of the network as a transmission mechanism of changes from a downstream to an upstream industry is that a hierarchical network *filters* the transmission of effects, while a partitioned network transmits *directly* and entirely the effects. In a hierarchical network, any change on the buyer side (entry of firms, introduction of new products, change in market shares) affects more than one supplier. By contrast, in a partitioned network a change on the buyer side is very likely to affect only one supplier.

To test these hypotheses, we build three regression models in which the dependent variables (*Eng_Var*) are, in turn, concentration, number of firms and number of new products of the downstream industry, while the independent variables (*Air_Var*) are the respective indicators in the downstream industry (see Appendix 2 for the list of variables). The three regressions are carried out for the turboprop and the jet markets.

In order to test the significance of a network variable, in a modified specification we add density as an independent variable in the three regressions. We chose the more synthetic network measure, since we are interested in the explanation of the long run dynamics of the upstream industry.

The ADF test on the variables under study revealed that they are all non stationary and first order integrated, as they become stationary after one difference. We use an unrestricted Error Correction Model to test the short and long run relations between the variables. The specification of the general model is the following:

$$\Delta Eng\_Var_t = \alpha + \beta_0 \Delta Air\_Var_t + \pi_1 Eng\_Var_{t-1} + \pi_2 Air\_Var_{t-1} + \varepsilon_t \quad (1)$$

This equation is a reparametrisation of an ARDL(1) process, where π_1 and π_2 are the long run parameters, and π_2/π_1 represents the speed of adjustment towards the long run solution of the model. The test for the long run solution is based on these hypotheses:

$$H_0 : \quad \pi_1 = 0$$
$$H_1 : \quad \pi_1 = \pi_2 = 0$$

Asymptotic critical values for the t and F distributions are tabulated by Pesaran et al. (2000)[7]. In Appendix 3 we report the t and F critical values for I(1) variables in the case of unrestricted intercept and no trend in the equation.

[7] Although it is more general, we do not use the Johansen method of cointegration because it requires a large number of observations. However, when the number of the variables is no greater than 2, and

The second specification of the model also tests the significance of density in the explanation of the dynamics of the upstream industry.

$$\begin{aligned} \Delta Eng\_Var_t &= \alpha + \beta_0 \Delta Air\_Var_t + \beta_1 \Delta DENS_t \\ &+ \pi_1 Eng\_Var_{t-1} + \pi_2 Air\_Var_{t-1} + \pi_3 DENS_{t-1} + \varepsilon_t \end{aligned} \tag{2}$$

These specifications of the models are built on the idea that causality goes from the downstream to the upstream industry. We believe that this hypothesis is realistic in these industries. In fact, while the aircraft industry represents the whole market for the engine industry, influencing directly its dynamics, a reverse causality direction can be assumed to exist in the case of radical innovation in components or monopolistic structure of the supplier industry. That is, while one can find interesting examples of innovation in components which determined important changes in the downstream industry (ex. computer and microprocessor, see Malerba et al. 1999; Bresnahan et al. 1999), we believe that this is not the case for the jet and turboprop industries. In fact, in these industries, after the transition to the gas turbine technology, which lead to the disappearance of the commercial piston engine and aircraft industries, no other radical innovation occurred which strongly affected the dynamics of the buyer industry. However, we carry out Granger causality tests to check the statistical significance of the hypothesised direction of causality.

Tables 2–4 show the results of the regressions for the three variables in the turboprop industry.

From the two regressions in Table 2, it is clear that there is a positive and significant short run impact of the downstream concentration on the upstream concentration. The F-tests of the restrictions on the coefficients π_1 and π_2 are larger than the critical values in both cases, suggesting cointegration between the three variables. However, the coefficient π_2 for the downstream concentration is significant only in the first regression, while in the second regression density explains upstream concentration in the long run.

Results of the regressions in Table 3 confirm the existence of a short run relation between downstream and upstream number of firms. The variables are not cointegrated and, in the long run, density seems to affect negatively the number of firms. The higher the density, the lower are in fact the opportunities available for supplying incumbent firms, already attached to other suppliers. In addition, in a partitioned network, density acts as a barrier to entry at a lower level compared to a hierarchical network.

Results in Table 4 reveals positive and significant short run and long run relations between the number of new products in the turboprop aircraft and engine industry. The F-statistics are higher than the critical values in both regressions, but in the second the value is smaller as the coefficient for density is not significant.

We carried out a test of Granger causality in the first regression specification for a statistical verification of our hypothesis on the direction of causality. In all

the independent variable is weakly exogenous, results of the unrestricted ECM are considered robust. In the case of more than two variables, the application of the Johansen method to our data produced ambiguous results, probably due to the small number of observations. Therefore, we prefer to use the unrestricted ECM model to study the effects of downstream concentration and density on the upstream concentration, although we cannot check for the number of cointegrating relations.

Table 2. Relation between upstream and downstream concentration – *Turboprop*

Dependent variable: ΔPE_HERF_t	Regression 1		Regression 2	
	Coefficient	t-Statistic	Coefficient	t-Statistic
C	0.111	2.445 **	0.099	2.246 **
ΔPA_HERF$_t$	0.599	3.459 ***	0.400	2.172 **
PE_HERF$_{t-1}$	−0.317	−2.742 ***	−0.457	−3.653 ***
PA_HERF$_{t-1}$	0.213	2.304 **	0.039	0.339
ΔP_DENSITY$_t$	–	–	0.203	1.729*
P_DENSITY$_{t-1}$	–	–	0.397	2.432 **
Adjusted R-squared	0.32		0.38	
F-statistic	7.76 ***		6.30 ***	
F-statistic H_0 : $\pi_1 = \pi_2 = (\pi_3) = 0$	9.31		4.70	
serial correlation LM test (F-stat)	0.03		0.21	
n	44		44	

$^{*}p < 0.10$, $^{**}p < 0.05$, $^{***}p < 0.01$.

Table 3. Relation between upstream and downstream number of firms – *Turboprop*

Dependent variable: ΔPE_NFIRM_t	Regression 1		Regression 2	
	Coefficient	t-Statistic	Coefficient	t-Statistic
C	0.052	0.228	1.369	2.320 **
ΔPA_NFIRM$_t$	0.468	5.248 ***	0.429	4.841 ***
PE_NFIRM$_{t-1}$	−0.128	−1.600	−0.264	−2.761 ***
PA_NFIRM$_{t-1}$	0.062	1.401	0.054	1.291
ΔP_DENSITY$_t$	–	–	−1.298	−2.010*
P_DENSITY$_{t-1}$	–	–	−1.798	−2.434 **
Adjusted R-squared	0.43		0.55	
F-statistic	11.79 ***		9.14 ***	
F-statistic H_0: $\pi_1 = \pi_2 = (\pi_3)$=0	0.33		2.86	
serial correlation LM test (F-stat)	0.03		0.04	
N	44		44	

$^{*}p < 0.10$, $^{**}p < 0.05$, $^{***}p < 0.01$.

three cases, it is the downstream variable which drives the upstream, and not the reverse.

In the jet market, results are quite different. The Chow test for the structural stability of the relation between downstream and upstream concentration revealed

Table 4. Relation between upstream and downstream number of new products – *Turboprop*

Dependent variable: ΔPE_NPRO_t	Regression 1		Regression 2	
	Coefficient	t-Statistic	Coefficient	t-Statistic
C	0.763	2.957 * **	1.014	2.474 * *
ΔPA_NPRO$_t$	0.461	6.682 * **	0.449	6.408 * **
PE_NPRO$_{t-1}$	−1.145	−7.268 * **	−1.157	−7.242 * **
PA_NPRO$_{t-1}$	0.404	3.609 * **	0.398	3.536 * **
ΔP_DENSITY$_t$	–	–	−1.440	−1.289
P_DENSITY$_{t-1}$	–	–	−0.683	−0.865
Adjusted R-squared	0.73		0.73	
F-statistic	39.34 * **		2.38 * **	
F-statistic H$_0$:				
$\pi_1 = \pi_2$=(π_3)=0	10.53		19.11	
serial correlation				
LM test (F-stat)	0.47		0.48	
n	44		44	

$^{*}p < 0.10$, $^{**}p < 0.05$, $^{***}p < 0.01$.

Table 5. Relation between upstream and downstream concentration – *Jet*

Dependent variable: ΔJE_HERF_t	Regression 1		Regression 2	
	Coefficient	t-Statistic	Coefficient	t-Statistic
C	−0.173	−2.353 * *	−0.324	−3.297 * **
C2	0.055	0.508	0.087	0.822
ΔJA_HERF$_t$	1.270	5.872 * **	1.322	6.340 * **
ΔJA_HERFT2$_t$	−0.551	−2.220 * *	−0.628	−2.608 * *
JE_HERF$_{t-1}$	−0.534	−4.808 * **	−0.545	−4.293 * **
JA_HERF$_{t-1}$	1.298	4.930 * **	1.290	5.011 * **
JA_HERFT2$_{t-1}$	−0.569	−2.040 * *	−0.659	−2.427 * *
ΔJ_DENSITY$_t$	–	–	1.015	1.015
J_DENSITY$_{t-1}$	–	–	0.366	2.197 * *
Adjusted R-squared	0.53		0.57	
F-statistic	8.11 * **		7.26 * **	
F-statistic H$_0$:				
$\pi_1 = \pi_2$=(π_3)=0	9.43		8.65	
serial correlation				
LM test (F-stat)	1.01		0.42	
n	39		39	

$^{*}p < 0.10$, $^{**}p < 0.05$, $^{***}p < 0.01$.

a structural break in 1978[8]. We carried out the above regressions by including a

[8] The test for the structural stability was carried out in all regressions, but only for the concentration in the jet industry it was significant.

dummy variable for the intercept and a dummy variable for the coefficient of the downstream concentration (Table 5). Results are very interesting, as they show the existence of a short and a long run relation between the two variables, but the sign of the relation is different in the two periods. Before 1978, when the density of the network is smaller, and the core of the network is not stable, the sign of the relation is positive. In the second period, the hierarchical network filters the transmissions of the changes of the downstream to the upstream industry. In this case, the coefficient for the downstream concentration is negative. In fact, larger market shares of the downstream leaders are distributed among suppliers in the core almost equally, leading to a reduction of the upstream concentration. The coefficient for density is significant only in the long run, and this reflects the hierarchisation of the network which is not captured by an aggregate indicator such as density. The tests for Granger causality indicate that both directions are significant, suggesting also influence of the upstream on the downstream industry.

The regressions for number of firms and number of new products show that there is no relation between upstream and downstream variables (Table 6 and 7). Density seems to affect negatively the number of firms in the short run, suggesting again the role of density as a barrier to entry. In Table 7, the F-values are largely higher than the critical value, but this reflects the higher significance of the lagged dependent variable.

Table 6. Relation between upstream and downstream number of firms – *Jet*

Dependent variable: ΔJE_NFIRM_t	Regression 1		Regression 2	
	Coefficient	t-Statistic	Coefficient	t-Statistic
C	0.441	1.005	0.701	0.696
ΔJANFIRM$_t$	0.081	0.870	−0.125	−1.317
JENFIRM$_{t-1}$	−0.020	−0.389	−0.027	−0.599
JANFIRM$_{t-1}$	−0.036	−0.626	−0.064	−1.060
ΔJ_DENSITY$_t$	–	–	−4.230	−3.485 * **
J_DENSITY$_{t-1}$	–	–	−0.077	−0.055
Adjusted R-squared	0.00		0.33	
F-statistic	0.98		4.80 * **	
F-statistic H$_0$: $\pi_1 = \pi_2 = (\pi_3) = 0$	0.74		0.73	
serial correlation LM test (F-stat)	1.98		3.05*	
N	39		39	

*$p < 0.10$, **$p < 0.05$, ***$p < 0.01$.

Summarising, we observe positive relations between each couple of variables in the turboprop market, while in the jet market the relation is significant only for the concentration index, with different sign of the coefficients in the two periods,

Table 7. Relation between upstream and downstream number of new products – *Jet*

Dependent variable: ΔJE_NPRO_t	Regression 1		Regression 2	
	Coefficient	t-Statistic	Coefficient	t-Statistic
C	2.407	2.853 * **	4.841	1.330
ΔJA_NPRO$_t$	0.137	1.635	0.111	1.118
JE_NPRO$_{t-1}$	−0.749	−4.582 * **	−0.812	−4.576 * **
JA_NPRO$_{t-1}$	0.025	0.241	0.019	0.164
ΔJ_DENSITY$_t$	–	–	−8.522	−1.219
J_DENSITY$_{t-1}$	–	–	−5.043	−0.625
Adjusted R-squared	0.41		0.40	
F-statistic	9.92 * **		6.16 * **	
F-statistic H$_0$: $\pi_1 = \pi_2$=(π_3)=0	28.97		7.38	
serial correlation LM test (F-stat)	0.46		0.37	
N	39		39	

$^{*}p < 0.10$, $^{**}p < 0.05$, $^{***}p < 0.01$.

and it is not significant for the number of firms and of new products. How can this difference be explained?

As shown in Section 3, the network linking the two industries assumes a sharply different configuration in the two cases. We argue that networks matter, and precisely that the partitioned network directly transmits the effects to the upstream industry, while the hierarchical network filters the effects. How does the network act as a transmission mechanism?

With respect to the *number of firms*, we observed in the turboprop industry that the entry of suppliers always follows the entry of an aircraft manufacturer. In a context characterised by stability of relations, high costs of switching suppliers, and preference towards single sourcing, the opportunities for entry come from the entry of a new unattached buyer and much less frequently from the introduction of a new program. Specialisation by market segments and the absence of economies of scale and scope lead to a configuration of the turboprop industry in which engine suppliers operate mainly in just one segment of the market. In this industry context, although turboprop aircraft manufacturers operated mainly in single sourcing, the introduction of a program in a new market segment often require the establishment of a supply relation with a second source. An example is provided by Casa, which was supplied by Garrett in the less than 30-seat segments and by General Electric in the segment 31–50 seats.

In addition, in such a fragmented market, the exit of a customer means, in the short run, the disappearance of the market for the supplier, which is therefore forced to exit the industry. In fact, the interruption of relations due to the exit of aircraft manufacturers represents a loss of market share that cannot be compensated by the acquisition of supply relations with existing customers, already attached to competing suppliers in single sourcing. In partitioned networks characterised by the presence of quasi exclusive relations, the higher the density, the lower the probability that a new firm can join the network. The network becomes in fact saturated at a lower level of density.

However, in the jet industry, the entry of jet aircraft manufacturers may produce opportunities for the entry of engine suppliers only indirectly through the growth of the market. Entry of suppliers is not fuelled by the entry of new customers or the launch of programs, but by the increasing adoption of multiple sourcing at the aircraft manufacturer and at the program level. Airbus, Boeing and McDonnell Douglas operated with three, and in some periods with four or five, engine manufacturers. Starting from the B747, other programs such as the A330, the B767 and the B777 have been powered by engines of three different engine manufacturers. The affirmation of multiple sourcing strategies implies that the launch of a new program is an opportunity for more than one engine manufacturer, which compete to gain the launch order or a large share of total orders. However, a number of minor aircraft manufacturers operate in single sourcing, but they have not created opportunities for entry, as their aircraft have been powered by engines of existing suppliers, in most cases by Rolls Royce engines. The coexistence of multiple sourcing within the core and single sourcing at the periphery contributed to the emergence of a cohesive and hierarchical structure of the network. The openness of the network structure, that is the reachability of central nodes of the network, reduces the barriers to entry of new suppliers, which may enter for supplying existing and already attached buyers.

With respect to the level of *concentration*, in the turboprop industry, although the levels of concentration upstream and downstream are very different, their dynamics are highly correlated in the short run. Again the partitioned structure of the network, mirroring an extremely fragmented market structure in which the customer may represent the market of only one supplier, transmits directly the effects of changes in market shares and, therefore, in the level of downstream concentration to the upstream industry. In other words, the growth or the decline of the market share of a customer, or the exit or entry of a customer, cause a growing or declining concentration, respectively. In a fragmented structure, this change impacts on the market share of only one supplier, and induces a change in the level of upstream concentration of the same direction and intensity.

The results of the regressions on total *number of products, and number of new products* in the aircraft and engine industries confirm even more sharply the previous results. In the jet industry, there is no relation between number of new products upstream and downstream, while in the turboprop the two variables are cointegrated. In the prop, the technological co-specialisation of engine and aircraft, which is also reflected in the high number of bi-components and of k-cores with minimum degree equal to 1, explains the entry of isolated couples of vertically-related firms and products. This means that new engines are realised for specific new aircraft and are very rarely used for other applications. On the other side, a new aircraft is designed to integrate a specific engine. Therefore, the dynamics of the number of products is highly related.

The same pattern of cospecialisation is not observed in the jet market.

6 Conclusions and further research

The main message of the paper is that networks matter in the explanation of the evolution of industries. Depending on technology and market factors, networks of vertical relations assume a variety of structural configurations and change over time. Once formed, networks evolve themselves and act as constraints to the evolution of industries, transmitting effects from related industries according to their configuration.

We have shown that specific characteristics of the industries, which refer to economies of scope, technological cospecialisation, buyers' sourcing strategies and market segmentation, are reflected in *partitioned* network structures in the turboprop industry and in *hierarchical* network structures in the jet.

The econometric analysis demonstrated that the structural dynamics of downstream and upstream industries, measured in terms of number of firms, industrial concentration and the introduction of new products, are positively related in the turboprop markets, while in the jet market there are not significant relations, except for concentration in the first period, when the core is not stabilised and the relation is positive, while in the second period the relation is negative. We propose that partitioned network structures *transmit directly* the changes of downstream to upstream industries, while hierarchical networks *filter* the effects.

The analysis developed in this paper may be applied in different industrial contexts. Further research will aim to study the relations between airline companies and aircraft manufacturers. Two important factors may have affected the sourcing of airlines: the role of commonality across different engines and aircraft, which allows important cost savings for airlines, and the trend toward outsourcing of maintenance activities, which reduces the cost savings of having a single supplier. The analysis of the vertical relations between engine, aircraft and airline industries will offer different cases characterised by different structures of upstream and downstream industries. While in this paper we analysed *small-number buyer and supplier industries*, the analysis of the airline industry will remove this restriction on the downstream side, as the industry is composed of hundreds of companies and is characterised by very turbulent dynamics, also fuelled by the deregulation process that has occurred in the United States and in Europe. Again, the comparison between markets for jet and turboprop aircraft will allow further specifications of market structures and of demand regimes, as the development of air carriers in the markets for turboprop and jet followed differentiated dynamics.

A further development of this work will have as its object of analysis the dynamics of the network of vertical relations between avionics and aircraft manufacturers. In that case we will extend the application of the theory to a *supplier industry* characterised by the presence of a *large number of firms*. The avionics industry is also composed of a number of market segments characterised by different technologies which witnessed strong changes in the last few decades.

The analysis of the network in different industrial contexts will provide cases which will enrich and enlarge the general applicability of the proposed approach.

Appendix 1. Data

Empirical analysis is carried out in the turboprop and jet aircraft-engine industries from their birth to 1997, by using a *proprietary database* built upon several sources of data.

Specifically, we use the *Atlas Aviation* and *Jane's All the World Aircraft* databases, IATA publications, technical press and literature on the history and technological development of the aviation industry.[9] The *Atlas Aviation Database* contains all the transactions occurring from 1948 to 1997 between aircraft manufacturers and airline companies (orders) in the market for large commercial aircraft. The data distinguish the engine technology adopted, jet and turboprop, and for each transaction it is possible to identify the engine model integrated into the aircraft ordered. The jet industry includes all turbojet and turbofan engines, from the first Pratt & Whitney JT3 introduced in 1958. The turboprop includes all turbine propeller engines from the Rolls Royce Conway in the Vickers Viscount in 1948.

The database provides data on more than 85,000 transactions, carried out by 5,900 operators, 27 aircraft companies and 11 engine manufacturers, and involving 102 aircraft models (more than 450 versions) and 260 engine types. For each transaction, the database provide three monthly dates: contract, first flight (also indicated as production date), and delivery. We use the first flight as unit of analysis as it is subject to less fluctuation. To reduce discontinuity in the data, monthly dates are transformed into annual dates. Data on three aircraft programs not included in Atlas have been added by using Aerospatiale (1990) data on orders and deliveries.

Transactions also include second-hand transfers between operators. As we are interested in the relations between engine and aircraft manufacturers, we consider only the first introduction of the product and do not consider each subsequent transaction occurring between airline companies. The final number of transactions used in the analysis is 27,000.

We integrated the *Atlas* database with data on the number of engines powering each aircraft, by using other sources: *Jane's All the World Aircraft* publications and the technical press (in particular, *Flight International* and *Aviation Week and Space Technology*). Data on seat capacity of aircraft, and information about segmentation by seat are provided by company reports (in particular Boeing, Airbus, Aerospatiale).

Russian aircraft and engine transactions are excluded from this analysis, because of some incompleteness and uncertainty about data in the version of the database used for this research. This is not a problem with respect to the objectives of this thesis, since historically Russian engines have been exclusively integrated into aeroplanes produced in Russia, and thus the relational dynamics in the engine industry of the rest of the world are not influenced very much.

Entry is defined as the first date an engine manufacturer supplies an engine to an aircraft manufacturer (indicated by the date of production). A firm experiences

[9] Among others, Miller and Sawers, 1968; Phillips, 1971; Klein, 1977; Constant, 1980; Bluestone et al. 1981; Bright, 1981; Mowery and Rosenberg, 1982, 1989; Hayward, 1986, 1994; Vincenti, 1990; World Aerospace Technology, 1993; Norris and Wagner, 1997; Sutton, 1998; U.S. International Trade Commission, 1998.

exit when it does not supply engines for at least 5 consecutive years. Entry and exit of companies are analysed simply by counting the number of companies in the engine industry and their life cycle. The calculation of rates of entry and exit is not significant given the small number of players.

Data on which concentration measures are computed are based on total sales of commercial aircraft manufacturers over the entire period of observation, expressed in physical quantities (orders). To take into consideration sales of aero-engine firms, aircraft orders are multiplied by the number of engines installed in the model, as described in the technical literature. No consideration is given to the spare units sold in the maintenance and repair market. Market shares are therefore defined in terms of quantities rather than turnover, since there is no such detailed information available at the level of individual aircraft and engine programs.

Appendix 2. List of Variables

Eng_Var	Dependent variable (engine industry)
Air_Var	Independent variable (aircraft industry)
PE_HERF	Herfindahl index – turboprop engine industry
PE_NFIRM	Number of firms – turboprop engine industry
PE_NPRO	Number of new products – turboprop engine industry
PA_HERF	Herfindahl index – turboprop aircraft industry
PA_NFIRM	Number of firms – turboprop aircraft industry
PA_NPRO	Number of new products – turboprop aircraft industry
P_DENSITY	Density of the turboprop network
JE_HERF	Herfindahl index – jet engine industry
JE_NFIRM	Number of firms– jet engine industry
JE_NPRO	Number of new products– jet engine industry
JA_HERF	Herfindahl index – jet aircraft industry
JA_HERFT2	Dummy for Herfindahl index (1978–1997) – jet aircraft industry
JA_NFIRM	Number of firms – jet aircraft industry
JA_NPRO	Number of new products – jet aircraft industry
J_DENSITY	Density of the jet network
C2	Dummy for constant (1978–1997)

Appendix 3. Critical values for F and t

	F		t	
	$k = 1$ *	$k = 2$	$k = 1$	$k = 2$
p¡0.10	4.78	−4.14	2.91	−3.21
p¡0.05	5.73	−4.85	3.22	−3.53
p¡0.01	6.68	−6.36	3.82	−4.10

Sources: Pesaran et al. (2000)
* k is the number of independent variables in the regressions. $k = 1$ corresponds to the first regression specification, while $k = 2$ to the second specification.

References

Aerospatiale (1989) Etude de Marchè. Transport Regional. Planning Strategique, Suresnes

Airbus Industrie (1991) Market perspective for civil aircraft

Anton JJ, Yao DA (1987) Second sourcing and the experience curve: price competition in defence procurement. Rand Journal of Economics 18, 57–76

Anton JJ, Yao DA (1989) Split awards, procurement, and innovation. Rand Journal of Economics 20, 538–552

Anton JJ, Yao DA (1992) Coordination in split award auctions. Quarterly Journal of Economics 107, 681–707

Arora A, Gambardella A (1998) Evolution of Industry Structure in the Chemical Industry. In: Arora A, Landau R, Rosenberg N. (Eds.) Chemicals and Long-Term Economic Growth: Insights form the Chemical Industry. Wiley, New York

Asratian AS, Denley TMJ, Haggkvist R (1998) Bipartite Graphs and their Applications. Cambridge University Press, Cambridge, MA

Bain JS (1968) Industrial Organisation. Wiley, New York

Bluestone B, Jordan P, Sullivan M (1981) Aircraft Industry Dynamics. An Analysis of Competition, Capital and Labor. Auburn House Publishing Company, Boston

Bonaccorsi A (1996) Cambiamento Tecnologico e Competizione nell'Industria Aeronautica Civile. Integrazione delle Conoscenze e Incertezza. Guerini e Associati, Milano

Bonaccorsi A, Giuri P (2001) Increasing returns and network structure in the evolutionary dynamics of industries. In: Saviotti PP (ed.) Applied Evolutionary Economics: New Empirical Methods and Simulation Techniques, Edward Elgar, (forthcoming)

Bonaccorsi A, Giuri P (2000a) Non shakeout patterns of industry evolution. The case of the turboprop engine industry. Research Policy 29, 847–870

Bonaccorsi A, Giuri P, (2000b) Industry life cycle and the dynamics of the industry network. Working Paper (2000)/04, Sant'Anna School of Advanced Studies, Pisa

Borgatti SP, Everett MG (1997) Network Analysis of 2-Mode Data Social Networks 19, 243–269

Borgatti SP, Everett MG, Freeman LC (1999) Ucinet 5 for Windows: Software for Social Network Analysis, Natick, Analytic Technologies

Bresnahan TF, Greenstein S (1999) Technological competition and the structure of the computer industry. Journal of Industrial Economics 37, 1–40

Bresnahan TF, Malerba F (1999) The Computer Industry. In: Mowery D, Nelson RR (eds.) The sources of industrial leadership. Cambridge University Press, Cambridge, MA

Bright, CD (1978) The Jet Makers. Regence Press of Kansas, Lawrence

Carroll GR, Hannan MT (2000) The Demography of Corporations and Industries. Princeton University Press, Princeton

Constant EW (1980) The origin of the turbojet revolution. John Hopkins University Press, Baltimore and London

Demski JM, Sappington DEM, Spiller P (1987) Managing supplier switching. Rand Journal of Economics 18, 77–97
Dobson PW, Waterson M (1997) Countervailing power and consumer prices. Economic Journal 107, 418–430
Dosi G, Malerba F, Marsili O, Orsenigo L (1997) Industrial structures and dynamics: evidence, interpretations and puzzles Industrial and Corporate Change 6, 3–24
Dosi G, Marsili O, Orsenigo L, Salvatore R (1995) Learning, market selection and the evolution of industrial structures. Small Business Economics 7, 411–436
FAA Forecasts and Technology Plans (1994) Avionics Communication Inc. Leesburg
Farrell J, Monroe HL, Saloner G (1998) The Vertical Organisation of Industry: Systems Competition versus Component Competition. Journal of Economics and Management Strategy, 7, 143–182
Flight International (1997a) Regional power struggle, 11–17 June, 133–134
Flight International (1997b) Regional aircraft risks, 12–18 November, 36
Flight International (1997c) AIR and Embraer start talks on joint 70-seat regional jet, 25 June–1 July, 8
Flight International (1998a) Jet age dawns for 328, 11–17 February, 36–37
Flight International (1998b) Regional jam, 23–29 September, 28–29
Flight International (1998c) ATR holds 728JET partnership discussion, 9–15 December, 4
Flight International (1999a) A celebration of flight (1908)–(1998), 11–21
Flight International (1999b) Jet there soon, 10–16 February, 30
Flight International (1999c) ATR team sets tight deadline for regional jet project talks, 13–19 January, 7
Galbraith JK (1952) American Capitalism. The Concept of Countervailing Power. Basil Blackwell, Oxford
Hannan MT, Carroll GR (1992) Dynamics of Organisational Populations. Oxford University Press, Oxford
Hart O, Tirole J (1990) Vertical Integration and Market Foreclosure, Brookings Paper on Economic Activity: Microeconomics. Special Issue, 205–286
Hayward K (1986) International Collaboration in Civil Aerospace. Frances Pinter Publishers, London
Hayward K (1994) The World Aerospace Industry. Collaboration and Competition. Duckworth & RUSI, London
Holmstrom B, Roberts J (1998) The Boundaries of the Firm Revisited. Journal of Economic Perspectives 12, 73–94
IATA (1956)–(1994) World Air Transport Statistics. Montreal, Canada
Jane's All the World's Aircraft (1940)–(1998) Jane's Information Group, Sentinel House
Klein B, Crawford RG, Alchian AA (1978) Vertical Integration, Appropriable Rents, and the Competitive Contracting Process. Journal of Economic Behaviour and Organization, 21, 297–326
Klein BH (1977) Dynamic Economics. Harvard University Press, Cambridge
Klemperer P (1987) The Competitiveness of Markets with Switching Costs. Rand Journal of Economics, 18, 138–150
Klepper S (1996) Entry, exit and growth over the product life cycle. American Economic Review 86, 562–583
Klepper S (1997) Industry life cycles. Industrial and Corporate Change 6, 145–181
LaFrance VA (1979) The impact of buyer concentration – An extension. Review of Economics and Statistics 61, 475–476
Langlois RN, Steinmueller WE (1999) The Evolution of Competitive Advantage in the Worldwide Semiconductor Industry (1947)–(1996) In: Mowery D, Nelson RR (eds.) The sources of industrial leadership. Cambridge University Press
Lustgarten SH (1975) The impact of buyer concentration in manufacturing industries. Review of Economics and Statistics 57, 125–132
Lyon TP (1996) Competition and Learning Curves in Defence Procurement: An empirical analysis. Paper prepared for the Logistic Management Institute, McLean
Malerba F, Nelson R, Orsenigo L, Winter S (1998) Vertical integration and specialisation in the evolution of the computer industry: towards a history friendly model. Paper presented at the 7th Conference of the International Joseph Schumpeter Society, Vienna, June, 13–16
Malerba F, Orsenigo L (1993) Technological regimes and firm behaviour. Industrial and Corporate Change 2, 45–71

Malerba F, Orsenigo L (1996) The dynamics and evolution of industries. Industrial and Corporate Change 5, 51–87
Miller R, Sawers D (1968) The Technical Development of Modern Aviation. Routledge & Kegan Paul, London
Monteverde K, Teece DJ (1982) Supplier Switching Costs and Vertical Integration in the Automobile Industry. Bell Journal of Economics, 13, 206–213
Mowery DC, Rosenberg N (1982) The Commercial Aircraft Industry. In: Nelson R (ed.) Government and Technological Progress. Pergamon Press, New York
Mowery DC, Rosenberg N (1989) Technology and the Pursuit of Economic Growth. Cambridge University Press, Cambridge
Nelson R, Winter S (1982) An Evolutionary Theory of Economic Change. The Belknap Press of Harvard University Press, Cambridge
Norris G, Wagner M (1997) Giant Jetliners. Motorbooks International Publishers and Wholesalers, Osceola
Orsenigo L, Pammolli F, Riccaboni M (2000) Technological change and network dynamics. The case of the bio-pharmaceutical industry. Research Policy, 30, 485–508
Orsenigo L, Pammolli F, Riccaboni M, Bonaccorsi A, Turchetti G (1998) The dynamics of knowledge and the evolution of an industry network. Journal of Management and Governance 1, 147–175
Pesaran MH, Shin Y, Smith RJ (2000) Bounds testing approaches to the analysis of level relationships, Mimeo
Phillips A (1971) Technology and Market Structure. A study of the Aircraft Industry. Heath Lexington Books, Lexington, Massachusetts
Powell WW, Koput KW, Smith-Doerr L (1996) Collaboration and the locus of Innovation: Networks of learning in biotechnology. Administrative Science Quarterly 41, 116–145
Ravescraft DJ (1983) Structure-profit relationships at the line of business and industry level. Review of Economics and Statistics 65, 22–31
Riordan MH (1998) Anticompetitive Vertical Integration by a Dominant Firm. American Economic Review, 88, 1232–1248
Riordan MH, Sappington EM (1989) Second sourcing. Rand Journal of Economics 20, 1, 41–58
Rogerson WP (1994) Economic Incentives and the Defence Procurement Process. Journal of Economic Perspectives 8, 65–90
Rothwell R, Gardiner P (1989) The strategic management of re-innovation. R&D Management, 147–160
Rothwell R, Gardiner P (1990) Robustness and Product Design Families. In Oakley M (ed.) Design Management, Blackwell Reference, Oxford
Scherer FM (1980) Industrial market structure and economic performance. Rand McNally, Chicago
Scherer FM, Ross D (1990) Industrial Market Structure and Economic Performance. Houghton Mifflin Company, Boston
Sutton J (1998) Technology and Industry Structure, MIT Press, Cambridge, MA
US International Trade Commission (1998) The Changing Structure of the Global Large Civil Aircraft Industry and Market: Implications for the Competitiveness of the US Industry. Washington DC
Vincenti WG (1990) What Engineers Know and How They Know It. Analytical Studies from Aeronautical History. The John Hopkins University Press, London
von Ungern-Sternberg T (1996) Countervailing power revisited. International Journal of Industrial Organisation 14, 507–520
Wasserman S, Faust K (1995) Social Network Analysis: Methods and Applications. Cambridge University Press, Cambridge, UK
Williamson OE (1985) The Economic Institutions of Capitalism. The Free Press, New York
Williamson OE (1999) Strategy Research: Governance and Competence Perspectives, Mimeo
Winter SG, Kaniovski YM, Dosi G (1997) A baseline model of industry evolution. IIASA Interim Report, IR-97-013, March
Winter SG, Kaniovski YM, Dosi G (2000) Modeling industrial dynamics with innovative entrants. Structural Change and Economic Dynamics 11, 255–293
World Aerospace Technology (1993) A personal view of five decades of jet engines, 55–58

The role of innovation and quality change in Japanese economic growth

Derek Bosworth, Silvia Massini, and Masako Nakayama

Manchester School of Management, UMIST, PO Box 88, Manchester M60 1QD, UK
(e-mail: {derek.bosworth,silvia.massini}@umist.ac.uk)

Abstract. This paper explores the use of time series data to isolate quality change in the Japanese economy using a hedonic procedure. We argue that the traditional approach to hedonic estimation based upon panel data sets of different brands in a given product area is extremely resource intensive and, thus, unlikely to be adopted by official statistical bodies outside of key areas, such as computers. This paper adopts a "top-down" approach to see whether more traditional measures of technical change, such as patents, can be used to separate pure inflation from quality change. If this is possible, it offers a much simpler route to estimate the role of quality change in economic growth and performance. In practice, we extend the analysis not only to include patents, but other forms of intellectual property that might reflect technology and attribute changes, such as designs, utility models and trademarks. We begin by taking a longer-term historical perspective, exploring the development of indigenous inventive capacity in Japan during the early years when R&D data are not available. It is possible to show that the rise in utility models pre-dates the main growth in patenting activity, suggesting the development in more low-level indigenous creative work prior to higher level inventive activity. The principal aim of this paper, however, is to demonstrate that it is possible to develop robust models to explain changes in the producer price index in Japan, which can then be used to re-examine Japanese growth performance over the period from about 1960. If the official Japanese statistical body has fully accounted for quality change in the price indices (i.e. produced fully quality-constant price deflators), then the official estimates of growth will be correct. However, we provide strong evidence that this is not the case. Changes in quality, proxied by the IP variables, are important determinants of prices in Japan over the period 1960 to 1995 as a whole. Indeed, we provide evidence that the true rate of growth of the Japanese economy, taking into account the rate of quality change, is significantly higher than that suggested in official statistics.

Key words: Hedonic regressions – Economic growth – Japan

JEL Classification: O47, O30, N15

Correspondence to: D. Bosworth

1 Introduction

This paper focuses on the role played by innovation and quality change as drivers of Japanese growth. Even using official data, the Japanese economic growth record stands out compared with other industrialised countries over the period studied in this paper. However, we will attempt to demonstrate that the official statistics not only conceal an even more dramatic growth, but also the key roles played by innovation and quality change. In order to do this, we develop a time series hedonic procedure that draws upon both standard and novel measures of innovative activity. Our interest in this work stems from an earlier project, which not only indicated the deficiencies of standard methods of accounting for quality change in official statistics, but also the problems of operationalising hedonic procedures as the principal alternative approach (Bosworth et al., 1993).

In essence, the issue concerns the isolation of quality constant price deflators (i.e. "pure price" deflators). In particular, if quality is constant, then real output growth is simply nominal output growth minus the rate of change in prices. If quality is changing over time, however, real output growth has both quality and volume change components. Derivation of this real growth should be calculated as the rate of change in nominal output minus the purely inflationary component of price change. This requires the construction of a pure inflation index, which is obtained by purging the price index of any effects arising from increases (or decreases) in product quality. If the price index is not quality adjusted, the growth in real output reflects only the changes in volume and underestimates the true growth whenever quality is rising.

The USA has led the way in the exploration and use of hedonic procedures for obtaining estimates of quality constant price indexes, as well as the implications for economic growth and productivity measurement. Gordon (1992), for example, argues that,

> "...it is likely that measurement issues bias downwards the growth rate of manufacturing output in every country, due to inadequate adjustments of price indexes for quality change, but more outside the US due to the absence of a computer price deflator."

Gordon has estimated that the bias for consumer durables may be around 1.5% per annum and, for producer durables, about 3% per annum. This suggests that measured increases in real output will be underestimated (as are real inputs, although the latter is more downward biased than the former). The results of work by Lichtenberg and Griliches (1989) suggests that, over a period of five years in the 1970s, the failure of official statistics to quality adjust the US *PPI*, led to an underestimate of productivity growth of about one-third of its total.

Nevertheless, the standard hedonic procedure is extremely resource intensive, both in the sense of the need to collect and collate technical characteristics/attribute data for each product area, and in the need to periodically re-estimate the empirical specification. Then there is the issue of aggregation from the bottom up, in order to obtain economy-wide estimates. In other areas of work involving innovation, the literature has suggested the use of proxy measures such as R&D expenditure, patent counts, etc. (Schmookler, 1966). Thus, we present an alternative hedonic route, using time series data first suggested by Bosworth (1976). In the present work, however, we further extend this idea by not only utilising traditional intellectual property (IP) measures, such as patents, but other measures thought to reflect product innovation, such as trademarks, designs and utility models (Bosworth et al., 2000). Assuming that we can demonstrate to the readers' satisfaction that this approach works, then, given the relative ease of collecting these types of intellectual property statistics, it opens up the possibility of the much wider application of hedonic techniques.

Section 2 explores the underlying issues in more depth, providing some evidence from the hedonics literature. Section 3 continues with a longer term, historical examination of the Japanese experience in terms of the intellectual property measures that we use as proxies for various dimensions of innovation. Section 4 develops the hedonic specification and reports the results of controlling for quality in the main price deflator, as well as the results of estimating a knowledge production function for the Japanese economy. Section 5 then discusses the implications for growth in the Japanese economy. Finally, Section 6 provides the main conclusions of the present paper.

2 Role of quality

2.1 Underlying problem

First we turn to the reason for needing quality constant prices. This lies in the most appropriate measure of output and, thereby, economic growth. The basic idea is that pure volume measures of output or growth (i.e. number of cars, tons of steel, etc.), give a highly imperfect impression of economic performance when quality is changing significantly. Thus, the quality element needs to be "forced" into the output or growth measure. In principle, this is achieved by deflating nominal output by a quality constant price index, $p(\overline{Q})$. Take the construction of real output, RY, from nominal output, NY,

$$RY = p(Q)Y = \frac{NY}{p(\overline{Q})} \tag{1}$$

where $p(Q)$is an index of quality based upon the price that consumers are willing to pay for that level of quality per unit of the volume of output, Y. If, on the other hand, quality change is not accounted for in the form of a quality-constant price index, then, the output corresponds to a pure volume measure,

$$Y = \frac{NY}{p(\overline{Q})p(Q)} \tag{2}$$

The implication is that, if price deflators are not appropriately quality adjusted, then the resulting growth estimates which rely on them will be biased, and the degree of bias is the rate of unaccounted quality change, $p(\overset{o}{Q})$.

2.2 Matched products versus hedonic approaches

Where quality change appears to be important, most official statistical bodies adopt a "matched products" rather than a hedonic approach. The matched products method attempts to construct a representative basket of brands for a given product. The price of each element of the basket is then tracked and an overall price index is formed as the weighted sum of prices, where the weights reflect the quantities of each of the brands sold. To maintain quality constant, as one brand disappears, an attempt is made to replace it with another of equal quality. The problems that arise when this approach is operationalised are immediately apparent, for example, new brands may appear on the market with quite distinct attributes and technical characteristics. In addition, by definition, brands are to some degree distinct (i.e. have different attributes and characteristics), and it is impossible to replace a disappearing brand by an exact equivalent. Most worrying is that this approach is less dependable the more rapid is the rate of technological and product change.

The hedonic approach has been carried out on a wide variety of products (cars, housing, etc.)[1]. With the exception of computers in the USA, however, it does not appear to have filtered into official use. The basic idea as operationalised to date can be illustrated using a price-attribute relationship,

$$p_{i,t} = p(Q)_{i,t}p(\overline{Q})_{i,t} = f(T(1)_{i,t}, T(2)_{i,t}, ..., T(m)_{i,t}, D_t) \quad (3)$$

where the price and quality variables have already been defined, the subscripts i and t denote the ith brand and the tth time period, T(j) is the jth technical characteristic, and D is a time dummy. D is the mechanism by which pure-inflation effects are picked up for the product area in question. In principle, the coefficients on each of the attributes are the shadow prices of that characteristic (i.e. what customers are willing to pay at the margin for a small change in that characteristic). The strength of the approach lies in the ability to account for new characteristics, which makes it potentially very valuable during periods of more rapid and radical technical change. On the other hand, the approach does have a number of conceptual and practical weaknesses. One weakness is that the shadow prices only reflect this marginal valuation under quasi-competitive conditions of monopolistic competition with a wide range of competing brands (Rosen, 1974). In addition, there are practical issues of the effects of multicollinearity between the T(j), which affects the derivation of meaningful shadow prices, α_j (though this does not affect the construction of a composite quality index, $p(Q) = \alpha_1 T(1) + \alpha_1 T(2) + \ldots + \alpha_1 T(m)$). More problematic is the omission of relevant characteristics. In addition, it can be seen that the data demands are enormous when this is operationalised for a large number of brands at the product level.

The matched products approach appears at its strongest in situations where technical change is slowest, and weakest when technical change is rapid, which is probably exactly the time when quality change is most important. Our experience based on the UK matching of computers is that it is extremely difficult to undertake such a matching process, and this is presumably the reason why the Bureau

[1] The original study was of asparagus sold on the Boston vegetable market (Waugh, 1928). For a review, see Berndt (1991).

of Economic Analysis in the USA has switched to a hedonic approach in the case of computers (Bosworth et al., 1993). However, the practical difficulties of operationalising the hedonic procedures in a bottom-up fashion across all product groups explains their very limited adoption by official statistical bodies.

2.3 Evidence of the extent of underestimation of real output

We now turn to the evidence available as to whether the likely bias caused by the use of matched products rather than hedonic procedures is of sufficient magnitude to worry about. Berndt (1991, p. 126) compares the results of the "matched models" and the hedonic regression approaches for processors, disk drives and computers as a whole. The results indicate just how misleading the matched models might be. In the case of processors, shown in Figure 1, the matched products quality constant price index exhibits an average annual growth rate of −8.5% compared with the corresponding hedonic measure of −19.2%. The results for disk drives, shown in Figure 2, indicate an average annual rate of growth of -6.9% and -16.8% for the two measures respectively. This implies that the original official Bureau of Economic Analysis matched products measure underestimated the real rate of growth of processor output by around two thirds each year. Of course, computers are a special case,

> "...if the automobile and airplane businesses had developed like the computer business, a Rolls Royce would cost $2.75 and run for 3 million miles on one gallon of gas. And a Boeing 767 would just cost $500 and circle the globe in 30 minutes on five gallons of gas." (Tom Forrester, quoted in Berndt, 1991, p. 102).

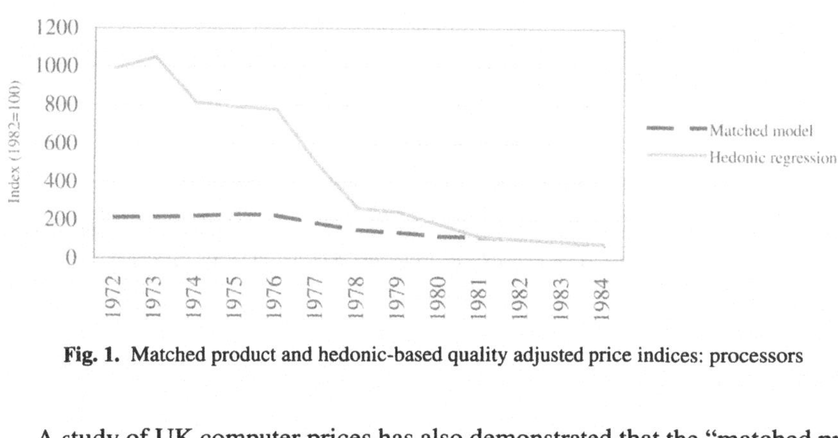

Fig. 1. Matched product and hedonic-based quality adjusted price indices: processors

A study of UK computer prices has also demonstrated that the "matched products" approach fails to capture the true quality change at all well (Bosworth et al., 1993). On a quality adjusted basis, using hedonic techniques, the price of computers in the UK fell by about 70% between December 1986 and May 1992, while the official series showed a fall of only 25% over the same period. This suggests that

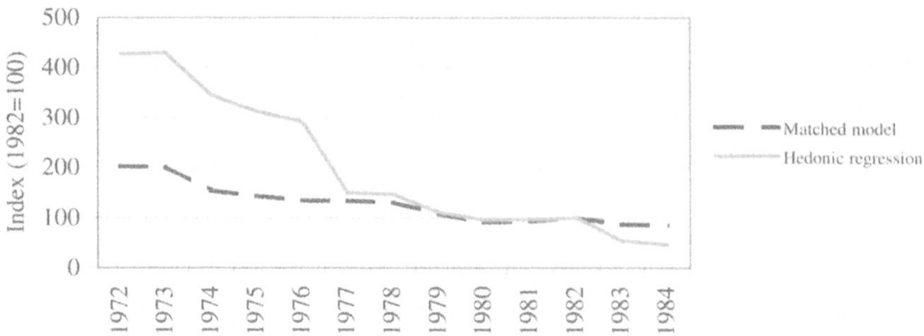

Fig. 2. Mathced product and hedonic-based quality adjusted price indices: disk drives

the real growth in computers was underestimated by about 66%, similar to the US study reported above.

While computers are a special case (Berndt, 1991 and Denison, 1989), the differences in quality change and the extent to which they are accounted for in the statistics are potentially very important for the measurement of productivity change and growth. Gordon (1992) reviews a large part of the appropriate literature. He concludes that the traditional problem of substitution bias in the use of fixed weight price indexes is of minor importance compared to the bias introduced by the introduction of new products, changes in the quality of existing products and outlet substitution bias. Gordon estimates that the quality bias for consumer durables in the US may be around 1.5% per annum and, for producer durables, about 3% per annum. This implies that a standard price index approach will yield measures of both real output and capital inputs that are too low, but the latter will be more inaccurate than the former. Lichtenberg and Griliches (1989) estimate that the US *PPI* adjusts for only about two thirds of actual quality change, resulting in an underestimate of total factor productivity growth of 34 per cent over the 1972-77 period. Later work, on data for 1977-82, also suggests the two-thirds estimate is about correct, and that there is considerable under recording of productivity growth.

Table 1. Rates of price change in machine tools (per cent per annum)

Machine type	*Annual average rate of price change*	*Quality constant rate of price change*	*Proportion accounted for by quality change (%)*
Boring and drilling	11.4	7.5	33.8
Abrading	15.1	8.4	44.4
Planing, shaping and slotting	10.9	7.6	29.7
Milling	6.5	1.6	75.7
Screwing and threading	8.3	2.5	69.7
Turning	5.5	3.3	39.3
Punching and shearing	19.2	6.3	67.4
All machine tools	10.5	6.9	34.4

Source: Bosworth (1976)

The vast majority of the work on hedonic functions uses some form of brand-level panel data set, containing detailed attribute or characteristics data. Bosworth (1976), however, argues that this is not always possible because of the non-availability of brand data, the lack of information about attributes and the resource intense nature of hedonic estimation. This earlier article proposed the use of time series data, incorporating patent data to reflect the extent of changes in technical characteristics. This was applied at a very detailed level to explain changes in machine tool prices, by type of machine. Intuitively it seemed that the average rates of change of machine tool prices over the post-War period up to the mid-1970s could not have been solely the product of inflation, but were also a reflection of the major shifts in machine tool technology (i.e. the introduction of CNC machines). Given the absence of any systematic data on technical characteristics, these were proxied by changes in weight (reflecting size and complexity) and patent grants by type of machine. Table 1 sets out the main results, suggesting that, overall, just over one-third of price change could be attributed to changes in quality, rather than simply inflation. It is noticeable, however, that some of the changes are closer to the differences between the matched products and hedonic estimates for computers, reported above.

2.4 Further problems

Insofar as statistical offices are able to *partially* hold quality constant in the deflators, some of the quality change is forced into the output measure. This is not an issue if all we want is some estimate of real output growth, as long as we can add the estimated rate of quality growth not accounted for (i.e. still in the official deflator), $p(\overset{o}{Q})$ from above, to the rate of growth of the official estimates of real output. However, it is an issue if we want to say something about the relative roles of quality *versus* volume in economic growth, then ideally we also need to decompose the official real output growth estimates into their quality and price components.

The problem is that the official statistical body might be partially successful in holding quality constant. If the overall level of quality is $p(Q)$, then $p(Q)_1$ is taken to represent the part accounted for and $p(Q)_2$ represents the unaccounted part. By implication, the official estimates of real output can be written as,

$$p(Q)_1 Y = \frac{NY}{p(Q)_2 p(\overline{Q})} \tag{4}$$

The official deflator, $p = p(Q)_2 p(\overline{Q})$ now forces a part of the quality change into the real output measure, while a further part is treated as if it is pure inflation. If we wish to fully isolate the "quality" from volume effects, equation (4) suggests that we also need to decompose the $p(Q)_1 Y$ measure into its two component parts and reconstitute $p(Q) = p(Q)_1 p(Q)_2$. Interestingly, while the matching procedure has controlled for $p(Q)_1/p(Q)_1 p(Q)_2$ of the quality change, it does not provide explicit estimates of $p(Q)_1$. One clue to the way forward is in the production function work where official measures of output are regressed not only on physical capital and labour, but on past research and development (R&D), where the role played by R&D is in explaining the quality of output and not its volume (Mairesse and Sassenou, 1991; Mairesse and Mohnen, 1995). This is essentially the approach we adopt below, although again using time series data rather than panel data sets.

3 Japanese innovation and growth: a historical perspective

3.1 Evidence from IP statistics

In this section we explore the historical IP data in an attempt to see whether it throws any light on the evolution of indigenous innovative activity in Japan. We focus on four categories of IP: patents, designs, trademarks and utility models. Each of these reflects a somewhat different aspect of the innovation process. Patents represent more fundamental inventions that have an industrial application and are, generally, the outcome of formal research and development. Utility models are often called "petty patents" and, as this suggests, relate to more minor modifications and adjustments that would not constitute an invention. Designs relate to industrial designs, which reflect the configuration of a product. Trademark activity is taken to reflect the extent of product modification and new product launch. We begin by presenting the total IP activity data (i.e. resident and foreign combined), because the resident data is incomplete for part of our historical period.

3.2 Total IP activity in Japan

Figure 3 shows the long-term trends in IP activity in Japan from 1885 up to 1982. It is clear that there is a long-term upward trend in all areas of activity which dates back to the 19th Century, punctuated by interruptions caused by the two World Wars.

Figure 4 shows the proportions of patents, utility models and designs (note that the three sum to 100 per cent – to make the diagram clear, we have omitted trademarks). It can be seen that utility models and, to a lesser extent, designs increased their share of the total up to the beginning of the Second World War. Not too much should be made of the War period itself, as the volume of activity was very low with respect to all areas of IP. Equally, the post-War period is one when patents increased their share of the total, primarily at the expense of utility models.

One line of investigation is to examine the ratio of resident (indigenous) to total IP activity.[2] This is based on the hypothesis that, given the speed of Japanese development, this should be reflected in a rise in the ratio of resident to non-resident activity. There are two problems with this, first that the data are incomplete and second that the Japanese economy was largely closed to the outside world for many years. Figure 5 shows the highly incomplete WIPO (World Interloctual Property Organisation) data for the period from 1923. Figure 6 demonstrates the equivalent

[2] Some care needs to be taken in interpreting such information. In particular, the levels of the ratios will depend on the width of patent scope (Japanese patents are generally believed to be narrower in scope than, say UK and especially US patents). However, the trends may be less affected by such factors.

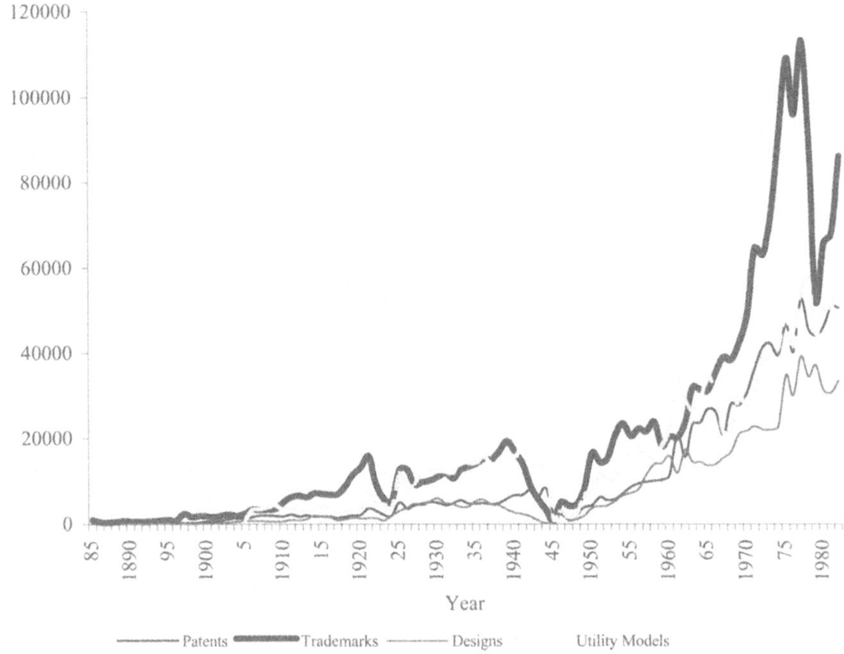

Fig. 3. Japanese IP data, historical series

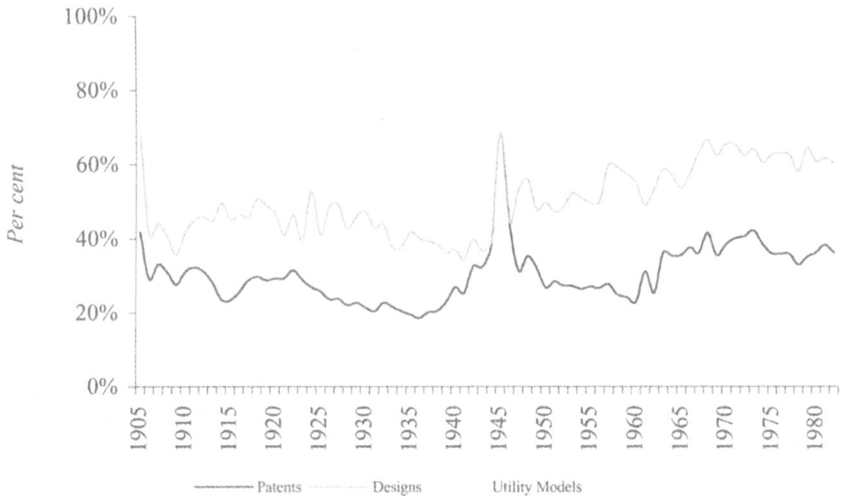

Fig. 4. Japanese IP data, share of total

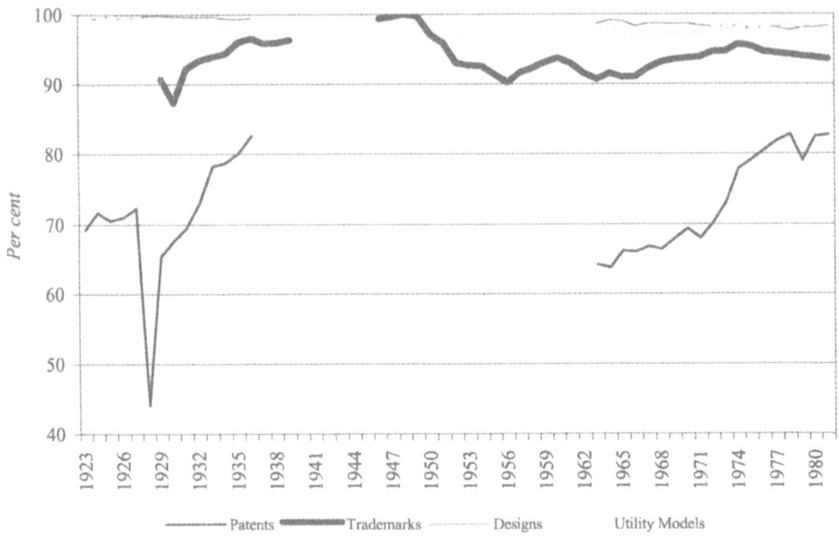

Fig. 5. Proportion of resident to total IP activity

series from the *Japanese Yearbook*, which gives two early post-War observations and then a continuous annual series from 1970.

Figure 5 shows the overall growth in the proportion of resident to total patents in the inter-War period and, again, reflects the overall growth in patenting activity in Japan at that time. However, the ratio starts at a very much lower level in the WIPO data in 1963. While domestic activity increased from 3994 in 1936 to 14937 in 1963 (an almost four fold increase), the corresponding increase for foreign activity was from 842 to 8366 (an almost ten fold increase). Foreign patenting activity in the inter-War period was very low, indeed, having risen from a very small base in the early to late 1920s, it declined back almost to its 1923 level by 1936. This presumably reflects the effects of the recession in the world as a whole. In Japan, however, while the resident patenting activity grew strongly in the 1920s, it did not decline in the 1930s, although it levelled off.

Figure 6, taken from the *Japanese Yearbook*, throws a little more light on the subject, as it provides evidence for 1960 and 1965, as well as continuing the series on to more recent years. The first feature is that there was a decline in the ratio of resident to total patenting activity from 1960 (when it was about 70 per cent – roughly equal to its 1923 value) and 1965 (when it reached a low of 62 per cent), before growing almost monotonically to just under 90 per cent in 1994. This persistent growth in the ratio of domestic to total activity in patents seems to be unique amongst the major industrial countries. Based upon this evidence, it appears that there was a surge of foreign patenting activity during the post-War reconstruction period, which was eventually overtaken by the growth in indigenous inventive activity in Japan.

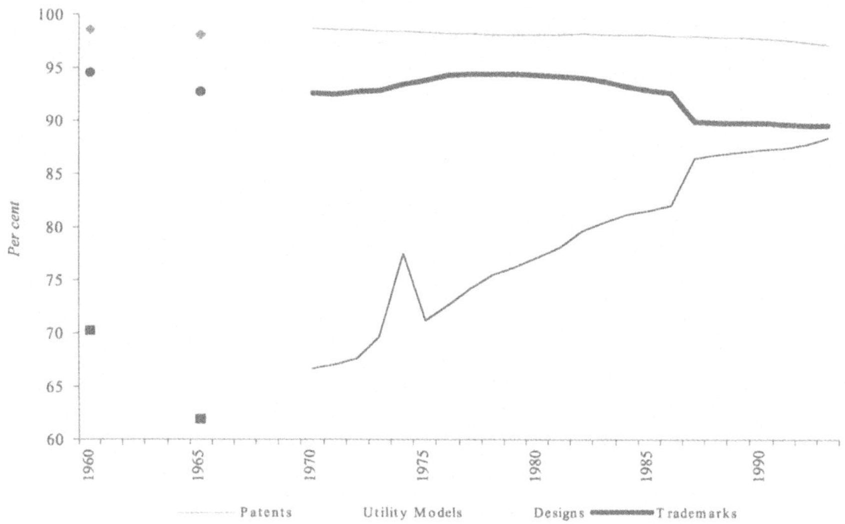

Fig. 6. Japanese IP, domestic origin

3.3 Resident Japanese IP activity

The lack of information about the activities of residents *versus* foreigners in Japan during the early post-War period is particularly unfortunate. Figures 7 and 8, constructed from WIPO, *Historical Statistics*, provide greater insights about the relative importance of patents, designs and trademarks and utility models over the period from 1923 to 1982. Each line shows a ratio of the particular type of IP in question to that of patents, and each has been smoothed in order to highlight the longer term trends. The data we now examine are for Japanese residents only, in order to try and see the changing indigenous efforts. While no data are published around the time of the Second World War, the contrast between the pre- and post-War periods throws some light on the nature of the early industrial development of Japan. Utility models (petty patents), associated with more minor improvements, have always been important in Japan, but they exhibit a major surge in the inter-War period, rising from a ratio of 0.5 *vis a vis* domestic patents to a high of 1.8 in 1930, and then levelling off. Similarly, design activity showed an upward trend relative to patents throughout the 1920s and 1930s, rising from a low of 2.5 in the early 1920s to a high of 4.1 towards the mid-1930s. After an initial fairly significant fall in the trademark to patent ratio, the ratio then remained roughly constant. The trends in the patent and design series show the growing relative importance of minor invention and design activity during the inter-War period.

Figure 8 shows that these upward trends in utility models to patents were eventually reversed in the post-War period. While the gaps in the data make hard and fast conclusions difficult, Figure 7 suggests that the peak of design activity to patents appears to have been around 1930, while that of utility models was around 1934. The ratio of utility models to patents, having risen from 2.5 to 4.1 over the interwar period, fell from 4.1 in 1934 to 1.9 in 1965, and continued to fall to a level of 1.2 by 1982. A broadly similar picture can be seen in the case of design activity, which,

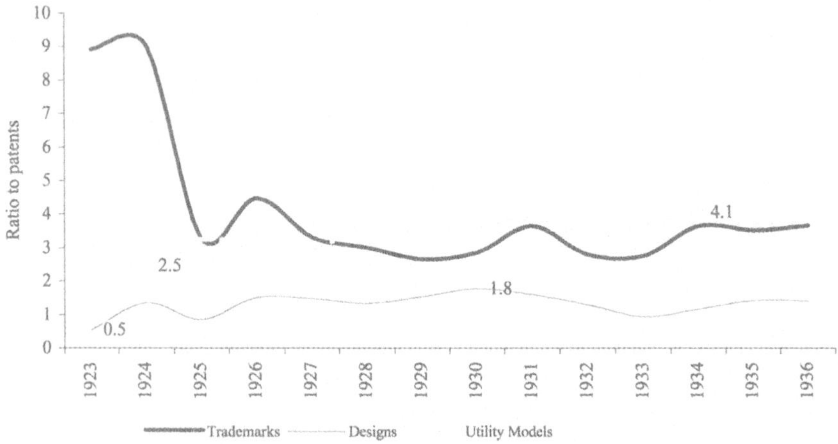

Fig. 7. Japanese IP trends pre-war resident grants and registrations

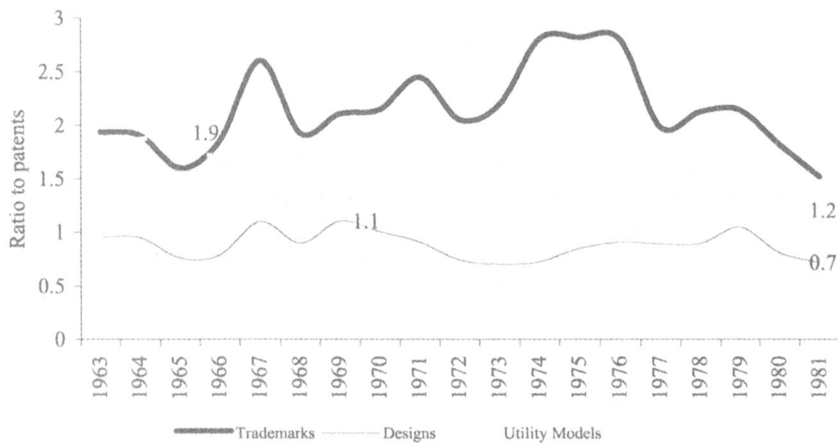

Fig. 8. Japanese IP trends post-war resident grants and registrations

having risen from a ratio of 0.5 to 1.8, had fallen to 1.1 by the late 1960s/early 1970s, and fell to 0.7 by the end of the sample period (Fig. 8). The long term trends suggest that the Japanese had laid the foundations for minor modifications in the early inter-War period, while indigenous inventive activity became increasingly important from about the mid-1930s. Again, it is difficult to pin down the exact "turning point" given the gap in the data.

4 Empirical specification and results

4.1 Approach to modeling quality change

Following the theoretical discussion of the location of the accounted and unaccounted quality change, our empirical specifications are as follows. The pure hedonic regression explains the producer price index, p,

$$p_t = f(X_t, P_{t-i}, U_{t-i}, D_{t-i}, T_{t-i}) \tag{5}$$

where X_t denotes the inflationary component of price change, and the variables P_t, U_t, D_t and T_t represent the product quality or attributes, proxied by patents, utility models, designs and trademarks respectively. The IP variables are appropriately lagged, i periods (where i may differ across the IP variables), to show the effect of technical change on the price deflator. The knowledge production function, which is a quasi-hedonic regression equation, explains the official index of real output (value added, Y),

$$Y_t = f(K_t, E_t, P_{t-i}, U_{t-i}, D_{t-i}, T_{t-i}) \tag{6}$$

where, in addition to above, K_t denotes the tangible capital stock and E_t is the number of employees. As noted earlier, $Y_t = p(Q)_{1t}V_t$ has both a quality and volume component, of which, V_t is explained by changes in tangible inputs and $p(Q)_{1t}$ of is determined by the proxies for technology and attributes (i.e., P, U, D and T). Given the large number of explanatory variables, in order to maximise degrees of freedom, we first estimate a standard production function,

$$Y_t = f(K_t, E_t) \tag{7}$$

and take the residual as a measure of the unaccounted price change,

$$p(Q)_{1t} = Y_t / \hat{Y}_t = f(P_{t-i}, U_{t-i}, D_{t-i}, T_{t-i}) \tag{8}$$

where $\hat{Y}$ denotes the predicted value of Y. We return to the precise functional forms used in the estimation below.

4.2 Variables used in the estimation

The variables, which are set out in Table 2, are taken from the *Japanese Yearbook of Statistics*. The dependent variable, Y_t, is measured as the value added of incorporated businesses, while K_t and E_t represent the tangible capital stock and total employees (including directors) in incorporated businesses. All variables in nominal value are deflated by official deflators. The variable p_t is the producer price index (*PPI*), while the attempt to capture purely inflationary pressures, X_t, was represented by the fuel price index (*FPI*) and capacity utilisation (*KU*), although we did investigate other proxies for this influence. Of these two measures, the *FPI* was always preferred on statistical grounds, largely, we believe, because Japan's economy is highly dependent on foreign sources of fuel, the price of which is largely exogenously determined.

In Figure 9 we show the profile of the Producer Price Index (*PPI*), as well as the Fuel Price Index (*FPI*) and Capacity Utilisation Index, used to reflect the inflationary component of the price index. The profiles of *PPI* and *FPI* are quite similar. The *FPI* shows the well-known peaks due to the two oil shocks in the early and late 1970s, which are reflected to a smaller extent in the *PPI* profile. Related to the oil shocks, we note that the Capacity Utilisation index follows an opposite trend compared to the *PPI* and *FPI* series, at least until the late 1980s, when the Japanese economy was booming. In the 1990s the three series show a similar pattern, this time because of the economic problems that hit the East Asian region and, after a short recovery, the financial crisis in the late 1990s.

Table 2. Descriptive statistics for variables used in the estimation, 1961–1997

Variables	Definitions	Means	Standard Deviations
PPI	Producer price index	81.54	21.09
FPI	Fuel price index	110.00	53.19
PA	Patent applications	216.27	116.47
UA	Utility model applications	134.46	58.79
DA	Design applications	46.10	8.92
TA	Trademark applications	139.87	52.62
PR	Patent grants	54.56	38.02
UR	Utility model registrations	44.03	14.48
DR	Design registrations	28.23	8.79
TR	Trademark registrations	91.55	50.75

It should be noted that the means and standard deviations vary with the sample period.

The IP variables, P_t, U_t, D_t and T_t are taken from the *Science and Technology* section of the same source. There are two choices here, between: applications and grants/registrations; domestic, foreign and total IP. *A priori*, we would have preferred grants and registrations to applications, as those that do not enter into force can be argued not to represent an improvement on the existing state of the art. However, applications appear to work much better in a number of our regressions for reasons that we outline below. In some of our specifications, it appears *a priori* more sensible to have a measure of the rate of change of the IP stock. This is calculated using, for example, patents in force, as a measure of the patent stock, PS. The patent stock is obviously comprised of patent grants and, thus, the rate of change of the stock, $\overset{o}{PS}$, would most naturally be calculated as, PR_t/PS_t (where *PR* denotes patent registrations). There are *a priori* grounds in preferring total IP to either domestic or foreign. We will demonstrate later that the total measure out-performed the alternatives.

In Figures 10 and 11 we show the patterns of the IP variables, patents, utility models and designs, respectively as applications and registrations, in relation to the price indexes, *PPI* and *FPI*. The patterns of applications appear to be smoother than the equivalent registrations. Among the IP variables, utility models and designs follow a pattern more similar to the price index series, showing an upward trend until the early 1980s, and then declining. An exception is the pattern of patents applications, which shows an upward trend over the whole period, including the 1980s and 1990s. The dip in patent applications in the early-mid 1990s reflects the economic recession mentioned above.

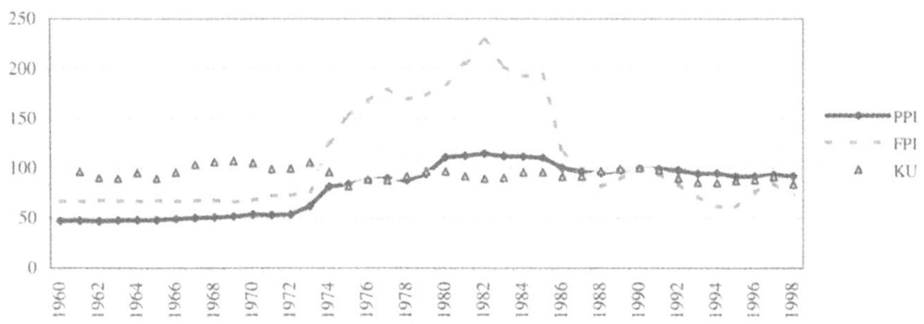

Fig. 9. Producer price, fuel price and capital utilisation

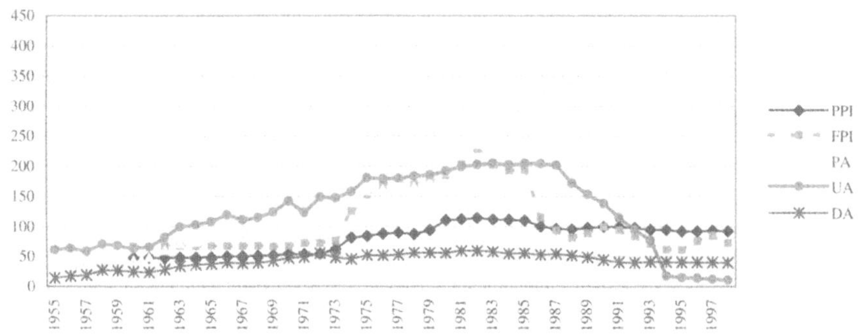

Fig. 10. Price indexes and IP variables – applications

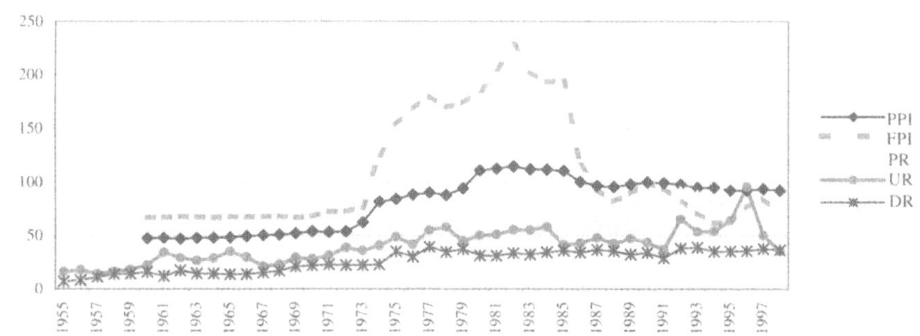

Fig. 11. Price indexes and IP variables – registrations

With regard to the patents registrations we can identify two main "jumps": the first between 1991 and 1992 and the second between 1994 and 1996. The first break in the trend is due to the introduction of electronic applications by the Japanese Patent Office. The JPO expected to see a big improvement in the patenting process, but the transition to the new system proved to be more difficult than they had expected. However, by 1992 the learning process using the electronic applications was completed and the number of patent and utility model registrations increased. The second break appears to be linked to the introduction of the *Patents and Utility Models Law, 1993* (Granstrand, 1999). In addition, the faster procedures for utility models were introduced in 1994 in response to a general shortening of the product life cycle. Utility models could be registered soon after the examination of the formal applications, without a proper examination of the utility model. Since a patent can be only be registered 18 months after the application, the effects of changes in the laws and procedures are likely to be seen in 1995–1996 – the years showing dramatic increases in patent grants and utility model registrations.

We were able to use both applications and registrations (grants) for our IP variables. Thus, while registrations are generally preferred to applications in technological change studies (i.e. because they have been screened by the IP examiners, with only those with sufficient novelty eligible for IP protection), the changes in registration behaviour induced by the new laws and administrative procedures suggest that the reverse might be true in this instance. In practice, we are able to show that both applications and registrations give very similar results if the sample period is restricted to end in 1991. However, if more recent years are included, applications continue to work in the regressions but registrations or grants do not.

Before presenting the econometric results, we show the correlations between the IP variables in Tables 3 and 4. The corresponding correlations between applications and registrations for the same IP variable are as follows: patents: 0.714; utility models: −0.109; designs: 0.581; and trademarks: 0.662. Note that applications and

Table 3. Partial correlations, application variables

	PPI	FPI	PA	UA	DA	TA
PPI	1.000					
FPI	0.682	1.000				
PA	0.779	0.099	1.000			
UA	0.365	0.728	−0.121	1.000		
DA	0.633	0.768	0.233	0.796	1.000	
TMA	0.584	0.100	0.727	0.078	0.390	1.000

Table 4. Partial correlations, registration variables

	PPI	FPI	PR	UR	DR	TR
PPI	1.000					
FPI	0.682	1.000				
PR	0.458	−0.049	1.000			
UR	0.686	0.338	0.832	1.000		
DR	0.879	0.488	0.599	0.745	1.000	
TR	0.663	0.105	0.835	0.715	0.802	1.000

registrations appear in separate regressions and, hence, strong correlations between these do not give rise to any adverse effects. However, some of the correlations reported in Tables 3 and 4 are quite high, especially between trademarks and patents, although there are other examples. These are likely to introduce multicollinearity problems in the estimation of our models and it is likely that including highly correlated variables in the same model will not provide robust coefficient estimates. However, in terms of the separation of price and quality effects, the key correlation in our specifications concerns the fuel price and the IP variables. Here we generally see much lower correlations, especially in the registration data, but also in the case of patent and trademark applications. These correlations are further weakened when we move to the current *FPI* and lagged IP, as used in the models below.

4.3 Empirical results

4.3.1 Order of integration. The Dickey-Fuller unit root tests show all our variables in the dataset are integrated of order 1, I(1), except the ratio of designs to capacity utilisation which is I(0). From a visual analysis of our series we expected them to be I(1), hence we undertook the test on the series differenced twice as suggested by Dickey and Pantula (1987). In general, when using non-stationary series, it is possible to falsely conclude that a causal long-run relationship exists between the two series. However in our study, given the different nature of the processes measured by the variables, we expect to find long run relationships (i.e. in levels) between the price deflator and the lagged technological change variables, rather than the sort of short run relationships that are analysed by differentiating the series. This is also reflected in the nature of the time series data we use. As we expect the effects of technological change to occur over a longer time span, we utilize annual data for prices and IP, rather than monthly or quarterly series. The official annual price data are an average of the monthly series for that year, which avoids the sometimes highly erratic changes over shorter time periods.

4.3.2 Producer price index regressions. The first stage is to attempt to separate the effects of pure price and quality change from the producer price index. Based upon a number of exploratory estimations we found that *PPI* and *FPI* cointegrate, that is their movements over time are well represented as short term changes around a long term relationship, with the inclusion of some impulse dummies corresponding to the oil shocks. Column (1) in Table 5 provides an example based on an error correction model. In practice, it is a fairly common feature of all of the results that one or more dummy variables relating to the oil shocks are needed in the specifications of the quality adjusted price index models in order to ensure the diagnostics are acceptable.

Given this finding, therefore, we started from a very simple model exploring the effects on *PPI* of lagged patent grants and the current *FPI*. As we noted above, our *a priori* expectation was that patent grants ought to provide a better measure of technical change than applications. Hence, we began with the grant and registration data. Although our feasible dataset for this model is 1961–1998, we only obtain acceptable results for the shorter period of time 1961–1991, after which the patent registration series shows a pattern that diverges from the *PPI*. Column (2) provides the results in levels for the period 1961–1991, when patent registrations are included. It can be seen that, in addition to *FPI*, the patent registration variable lagged one period is also highly significant. The R^2 is high, as might be expected in a "levels" regression, and the F statistic is significant at the 1% level. The diagnostics suggest that there are no problems of heteroskedasticity or normality.

As we have already noted, the significance of the patent variable falls away if the data period is extended to 1998, for the reasons outlined above. Hence, we also explored the role of patent applications. The results in column (3) suggest that patent applications work just as well as patent grants over the period 1961-91, and column (4) demonstrates that the results are sustained when the sample period is extended to 1998. The results now suggest a somewhat longer lag on the patent variable, as might be expected given the delay between application and registration. In addition, the model includes an impulse dummy variable for 1980, following the oil crisis of 1979. Some slight residual autocorrelation emerges but not to a worrying degree (see both *DW* in the indecisive interval and the AR F test). Correcting for residual autocorrelation using autoregressive least squares, however, provides similar coefficients, as shown in column (5). The lagged residual is significant at around the 10% level, but the similarity of the coefficients suggests autocorrelation was not a major problem in the previous regression.

There are some issues to take into account in models that introduce additional IP variables. First, the short sample is not long enough to build models with a high number of explanatory variables. Second, the correlation between our IP variables may cause some problems in terms of obtaining robust coefficient estimates. We will demonstrate below, for example, that patents and trademarks do broadly the same job in these regressions, and we found it difficult to obtain meaningful coefficient estimates on these variables when they were both included in the model. Further evidence of this problem can be found in the fact that the best two models include either patents applications lagged two periods and designs applications lagged three periods, or patents applications lagged two periods and utility model applications lagged one period. These results are shown in columns (6) and (8) respectively. The coefficients on *FPI* (at around 0.25-0.26) and lagged patent applications (at around 0.14–0.15) are largely unchanged between the two alternative models. The first model, which includes designs, shows evidence of slight autocorrelation, which can be corrected by using autoregressive least squares, as shown in column (7). While the lagged residual in the ALS version is significant at the 1% level, there is little change in the parameter estimates, although the coefficient on lagged designs rises slightly. The impulse dummy relating to 1980 continues to be significant.

We continued to check whether patent registrations out-performed applications in any of these models. However, the registration version of the model only had good diagnostics over the shorter period, 1961-1991, and consistently poor diagnostics when extended to the full data period. Examples of the disruptive effects of the legal and administrative changes are illustrated in columns (9) and (10). Neither the introduction of a dummy variable nor a trend variable for the post-1991 period corrects the problem with the registration model, giving unsatisfactory diagnostics

Table 5. Hedonic regression results, explanation of PPI

	(1)	(2)	(3)	(4)	(5)	(6)	(7)	(8)	(9)	(10)	(11)
DV	DPPI	PPI	PPI	PPI	PPI	PPI	PPI	PPI	PPI	PPI	PPI
Sample	1961–98	1961–91	1961-90	1961–98	1961–98	1961–98	1961–98	1961–98	1961–91	1961–98	1961–98
Constant	0.0187	10.469	19.366	19.533	21.083	14.9620	13.785	17.712	9.583	11.648	10.717
	(0.098)	(2.785)	(12.076)	(12.184)	(7.200)	(5.867)	(3.425)	(10.768)	(3.898)	(3.322)	(3.860)
DFPI	0.1216										
	(11.320)										
FPI		0.2236	0.2797	0.2912	0.2813	0.2619	0.2489	0.2662	0.1845	0.1702	0.1929
		(8.139)	(22.437)	(26.848)	(15.658)	(15.680)	(10.821)	(19.020)	(9.313)	(7.044)	(9.917)
PR–1		1.0415									
		(10.343)									
PR–2									0.5611	0.0513	
									(4.943)	(1.262)	
PA–2			0.1602	0.1484	0.1458	0.1394	0.1358	0.1502			
			(21.449)	(31.297)	(17.397)	(23.009)	(15.495)	(33.789)			
DA–3						0.2146	0.2861				
						(2.230)	(2.133)				
DR–3									1.0368	1.8328	1.6167
								(4.881)	(10.654)	(9.982)	
UA–1								0.0318			
								(2.574)			

Note: t values are given in parentheses.

Table 5 (continued)

	(1)	(2)	(3)	(4)	(5)	(6)	(7)	(8)	(9)	(10)	(11)
DV	DPPI	PPI	PPI	PPI	PPI	PPI	PPI	PPI	PPI	PPI	PPI
Sample	1961–98	1961–91	1961-90	1961–98	1961–98	1961–98	1961–98	1961–98	1961–91	1961–98	1961–98
TR–1											0.0818
											(3.099)
ECM-1	−0.0201										
	(−1.913)										
ID73	7.6612										
	(6.742)										
ID74	13.253										
	(10.627)										
ID78											−21.1900
											(−3.941)
ID79	5.7110										
	(5.067)										
ID80	15.0980		13.1900	12.839	10.087	12.947	10.273	13.122			
	(13.352)		(3.729)	(3.630)	(3.495)	(3.868)	(3.711)	(4.003)			
$\hat{U}-1$					0.4655		0.4280				
					(2.693)		(2.610)				

Table 5 (continued)

	(1)	(2)	(3)	(4)	(5)	(6)	(7)	(8)	(9)	(10)	(11)
DV	DPPI	PPI	PPI	PPI	PPI	PPI	PPI	PPI	PPI	PPI	PPI
Sample	1961–98	1961–91	1961-90	1961–98	1961–98	1961–98	1961–98	1961–98	1961–91	1961–98	1961–98
R^2	0.9592	0.9318	0.9846	0.9814		0.9839		0.9850	0.9714	0.9290	0.9576
F	121.54	191.3	554.45	599.53		503.43		525.67	305.52	148.23	186.35
DW	2.20	1.41	1.54	1.26		1.31		1.34	1.36	1.36	1.38
RSS	37.81	1343.85	296.74	389.51		338.50		324.40	563.91	1491.14	890.06
AR 1-2	0.925	0.927	1.560	3.0471		2.391		3.143	1.589	2.045	1.808
ARCH 1	1.445	0.516	0.149	1.491	3.847	0.004	4.066	0.151	0.693	0.173	0.264
Normality	0.084	0.428	2.118	0.339	3.338	1.233	0.732	3.25	4.094	3.791	2.222
Xi^2	0.709	0.561	0.984	1.263	1.375	0.603	1.014	1.106	2.433	0.651	3.414*
XiXj	0.616	0.576	0.723	1.336	1.206	0.483	1.609	0.925	1.493	2.142	2.365*
RESET	0.688	0.263	0.079	0.004		1.326		0.476	3.300	7.629**	3.791

and an adverse affect on the patent registration coefficient. In general we conclude that applications tend to better represent the impact of technological change on price indexes over the longer period.[3]

Finally, we discuss the introduction of trademarks into the model. As noted above, trademarks tend to do broadly the same job as patents, and it is not possible to obtain a satisfactory specification when both variables are included together. Interestingly, trademarks were not subject to legal and administrative changes in the early 1990s, and it is possible to find a specification for the entire sample period through to 1998, in which trademark and design registrations (appropriately lagged) are both significant and the diagnostics are acceptable. This suggests that, if this exercise were undertaken at a sectoral level, trademarks might well play a useful role for those areas of activity where there is little or no patenting activity.[4]

4.3.3 Knowledge production function estimates. The interpretation of the total factor productivity based estimates are more difficult for a number of reasons: (i) if we find no relationship between *TFP* and intellectual property, this may reflect the fact that the Japanese statistical authorities have failed to account for quality change at all, or it may be a failure on our part to specify the relationship properly; (ii) if we find a significant relationship between *TFP* and IP, this might be a consequence of the degree to which quality has been included in the real output measure, or it might reflect the role of IP in driving non-quality related output per unit of input. In addition, at this stage, we cannot account for changes in the quality of inputs in driving real value added. Nevertheless, it is interesting to see if our initial estimates suggest that this might be a useful line of investigation. We will demonstrate below that this at least gives some upper-bound estimates of the overall contribution of quality change in Japanese growth.

The first step in this process is to obtain estimates of the rate of change in *TFP*. To do this, we have regressed real value added on the real tangible capital and total employment. The function is assumed to be Cobb-Douglas, as the log-linear functional form is easily interpreted in terms of *TFP*. Column (1) of Table 6 clearly shows that a simple linear relationship of this type has a number of problems. The relationship exhibits strong autocorrelation and heteroskedasticity. The resulting coefficient estimates have little meaning and reflect the bias caused by the associated statistical problems. The use of ALS, shown in column (2), improves the estimates significantly, although there is a problem with normality. The latter is corrected by the inclusion of a dummy variable, as shown in column (3), suggesting some shift in the relationship around 1975 (i.e. shortly after the first oil crisis). These estimates now have the typical characteristics of Cobb-Douglas functions, with a capital share

[3] We undertook a variety of other experiments with regard to the functional form, such as estimating the models with the variables in logs. The results continued to be largely consistent where just the two price indexes were in logs, but when the IP variables were in logarithmic form the models were not satisfactory. We also found similar results as in the models in levels using patent applications and registrations: patent registrations only appear to provide good models in the shorter time period. Our preferred models are those reported in the main text.

[4] Indeed, when using either Japanese or Foreign IP series, rather than the total of both, we only found meaningful results when using trademarks and designs.

of 0.44 and labour share of 0.67, suggesting increasing returns to scale, with a scale factor of 1.11. It is possible to duplicate this result directly using the rate of change version of the equation, shown in column (4). The statistical problems with this relationship are corrected by a dummy variable in 1975, as shown in column (5). Note, however, that this estimate exhibits returns to scale of almost 1.5, which appears intuitively implausible. We believe that this may be a consequence of not adjusting for the changing quality of inputs, as suggested by the growth accounting literature.

Table 6. Production function estimates, 1961–1997

	(1)	(2)	(3)	(4)	(5)
Sample	1961–97	1961–97	1961–97	1961–97	1961–97
DV	LRVA	LRVA	LRVA	$\overset{\circ}{RVA}$	$\overset{\circ}{RVA}$
Constant	−25.9860	−5.3323	−7.7075	0.0198	−0.0138
	(−4.158)	(−1.158)	(−1.817)	(1.094)	(−0.958)
LK	0.2687	0.3109	0.4351		
	(2.049)	(2.092)	(3.804)		
LE	1.8045	0.6095	0.6714		
	(0.2918)	(2.309)	(2.743)		
$\overset{\circ}{K}$				0.4506	0.9007
				(2.749)	(6.285)
$\overset{\circ}{E}$				0.6463	0.5748
				(2.303)	(2.818)
ID75			−0.1082		−0.2933
			(−3.018)		(−5.624)
$\hat{U}-1$		0.9218	0.8922		
		(29.06)	(17.975)		
R^2	0.9839			0.3530	0.6697
F	1070.4			9.277	22.301
DW	0.358			1.34	1.58
RSS	0.489			0.116	0.059
AR 1–2	28.627**			2.338	1.344
ARCH 1	20.47**	1.746	0.470	0.175	0.466
Normality	4.637	7.121*	2.387	8.543*	4.225
Xi^2	2.170	1.064	0.830	15.447**	0.355
XiXj	2.638*	0.838	0.907	12.31**	0.304
RESET	56.38**			6.581*	0.003

Note: t values given in parenthesis.

The results shown in column 3 of Table 6 were used to construct an index of *TFP*, while column (5) from that table was used to estimate annual rates of change in *TFP*. The former was regressed on the IP variables in levels (see column (1) in Table 7), and the latter was regressed on rates of change in IP (see column (2) of Table 7, i.e. $\overset{\circ}{TFP}$), measured as new IP divided by the stock of IP in force. The

Table 7. Role of IP in explaining TFP

	(1)	(2)
Sample	1961–1997	1961–1997
DV	$\mathring{TFP}$	$\mathring{TFP}$
Constant	−7.9497	−0.0645
	(−41.571)	
DA–1	0.0050	
	(1.966)	
PA–3	0.0007	
	(1.608)	
$\mathring{PA}$ -1		0.0032
		(2.605)
ID75	−0.0874	−0.2875
	(−2.528)	(−7.436)
$\hat{U}-1$	0.8622	
	(16.985)	
R^2		0.6549
F		32.262
DW		1.85
RSS		0.049
AR 1–2		0.055
ARCH 1	0.035	1.231
Normality	2.458	3.991
Xi^2	1.467	0.188
XiXj	1.325	0.188
RESET		3.540

Note: t values given in parenthesis.

results shown in Table 7 indicate that the role of IP does not emerge with the same clarity in the case of *TFP*, as in the more direct hedonic regressions outlined above. However, there is some evidence that patenting activity has a positive impact on *TFP*. In column (2) of Table 7, for example, a 1% rise in the patent stock, lagged 1 period, produces a 0.003% increase in the rate of *TFP* growth.

5 Implications for Japanese growth

We now trace the implications of the results for real Japanese growth. In the case of the hedonic results, we chose to work with equation (8) from Table 4, which not only has typical parameter estimates, but also satisfactory diagnostics. These results suggest that a 1% increase in the fuel price index causes a 0.33% rise in the *PPI*, while a 1% increase in patent activity produces a 0.37% increase in prices

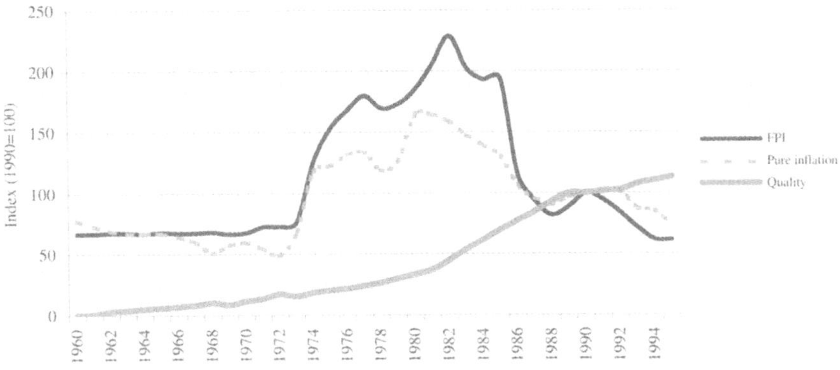

Fig. 12. Pure inflation and quality change in the Japanese economy

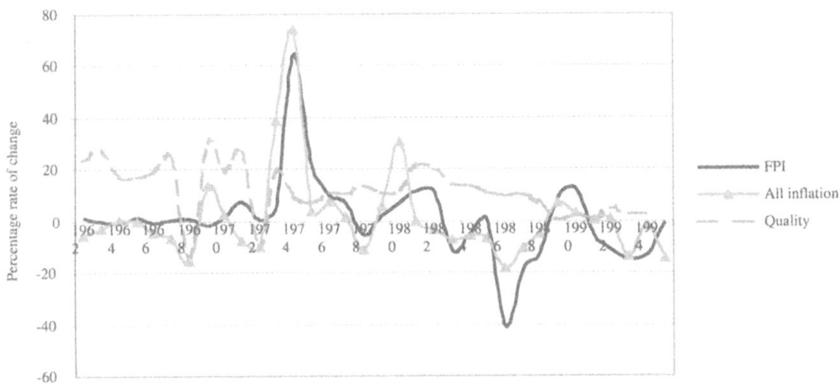

Fig. 13. Rates of change in inflation and quality

and a 1% rise in design activity causes a 0.16% rise in the *PPI*. Figure 11 shows the overall pattern of the fuel price index, *FPI*, which demonstrates the effects of the two fuel price jumps in the early and late 1970s. The quality index is the weighted sum of the patent and design variables, where the weights are the shadow prices from the hedonic regression. This index appears to reflect the strength of the economy, for example, flatter in the 1970s, rising significantly in the 1980s, before leveling off in the late 1980s-early 1990s, and then some recovery at the very end of the period. The pure inflation index, shown in Figure 12 is constructed as the difference between the *PPI* and the quality index.

The changes over time are better expressed as rates of change of the three series, as shown in Figure 13. The *FPI* and all inflation indices are highly cyclical, although, on balance, showing negative rates of price change for substantial periods (the main exceptions being the two oil crises). The rate of change in the quality index, on the other hand, is generally positive and much more stable, but, if anything appears to show a long term decrease over time. The rate of improvement falls from over 20% per annum at the start of the period to close to zero by the end.

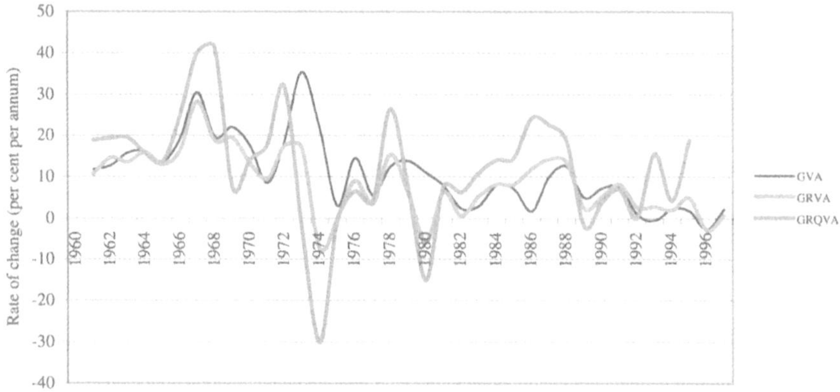

Fig. 14. Growth in nominal and quality unadjusted and adjusted real output

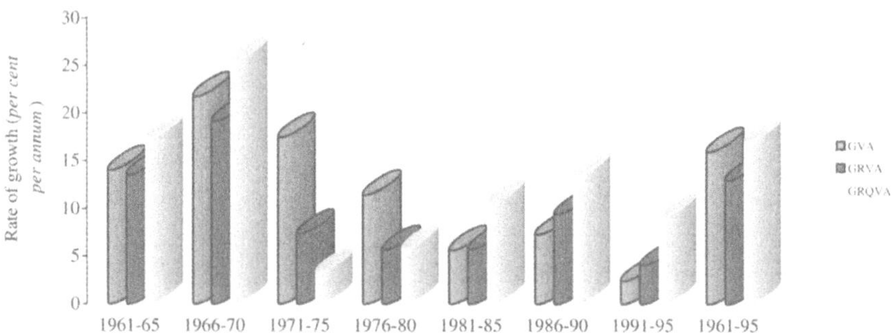

Fig. 15. Comparisons of nominal, quality unadjusted and quality adjusted growth rates

It can be seen from Figure 14 that, at various times, real value added rises more rapidly than nominal value added (compare *GRVA* and *GVA* respectively). This is due to the effects of a falling *PPI* during these periods. It can also be seen that the real value added constructed using the pure-inflation index grows at least as quickly and often more quickly than the official measure, with the exception of the 1970s (compare *GRQVA* and *GRVA* respectively). Figure 14 therefore reports on the rates of change in these indexes. While on a number of occasions the real value added constructed with the pure inflation index falls below one or both of the other series, on balance it remains above them, consistent with the underestimation of growth in the official statistics.

Finally, Figure 15 provides a summary of the three measures of Japanese growth. It can be seen that the value added deflated by the pure inflation index (*GRQVA*) indicates more rapid Japanese growth in every five year period than the official real index (*GRVA*), with the exception of 1971-75. Indeed, in all but two of the five-year periods, the *GRQVA* series grows more quickly than the growth in nominal value added. The overall position is shown by the final trio of bars, which report the overall annual average rates of change for the period 1961–1995. The average annual rates of growth are 16% (*GVA*), 13% (*GRVA*) and 17% (*GRQVA*) respectively. This result occurs because the rate of increase in quality is not only higher than the rate of growth in the official estimates of real value added, but also higher than the rate of growth in nominal value added.

6 Conclusions

In this paper we have investigated whether official statistics underestimate the true rate of growth of the Japanese economy. In doing so, we have explored the possible use of more aggregate time series data for this exercise, rather than the detailed, brand and product level information that is often used in hedonic studies. Rather than attribute information, this study has used a number of IP variables to represent technological and attribute change. In particular, we have explored the use of patents, designs, utility models and trademarks. While, *a priori*, it was anticipated that the IP registration data would form a better proxy in this exercise, in practice, changes in the law and in the administration of the patent system in particular produce a break in the relationship after 1991 when using the registration information. This discontinuity is not present when application data are used, and the resulting relationships appear stable, significant and economically plausible.

It seemed important to extend the IP data beyond patents, particularly when investigating price changes at an aggregate level. While patents are extremely important in the Japanese economy, they nevertheless still remain most relevant to the manufacturing sector and to larger firms. Other IP, such as trademarks (which have been argued to be linked to new product launch), have a much wider sectoral significance. In addition, utility models may be relatively more important for small than for large firms. In practice, while we indicated a preference for a hedonic specification incorporating patents and designs, based on the statistical tests of the empirical results, specifications including trademark activity in place of patents also performed well. Again, this suggests it may be possible to use trademark activity where patent data are either unavailable or not relevant to a particular sector.

This paper provides evidence that the official Japanese *PPI* does not appear to be a quality-constant index. In particular, we have shown that the price index is influenced not only by external and exogenously given factors, such as fuel prices, but also technological and attribute changes in Japanese products. There is a robust and significant statistical relationship between the *PPI*, the current fuel price index, patent applications lagged two periods and design applications lagged three periods. Using the results to construct a quality constant deflator, it is possible to show that official estimates of Japanese growth based upon nominal value added divided by the published *PPI* appear to underestimate the true rate of growth by about 4 percentage points *per annum* over the period 1961-1995. In other words, Japanese growth is underestimated in the official statistics by about one-third, a finding that appears broadly in line with a number of studies that have looked at this problem in the context of the USA.

Finally, we draw a number of conclusions about the longer term changes in inventive activity and quality change in Japan. We began by exploring the development of indigenous inventive capacity in Japan during the early years when R&D data were not collected. We demonstrated that the rise in utility models pre-dates the main growth in patenting activity, suggesting the development in more low-level indigenous creative work prior to higher level inventive activity. Our best estimate based upon the relative importance of patents is that the rise in high-level indigenous creative activity appears to begin in the mid-1930s. Then, using the results of the regression analysis based on data for the post-War period, we have demonstrated that the rate of quality improvement has not only changed year on year, but appears to have fallen from a high of over 20 per cent improvement per annum around 1960 to marginally above zero by the mid-1990s.

References

Berndt, ER (1991) The practice of econometrics: classical and contemporary. Addison Wesley, Reading, MA

Bosworth, DL (1976) Price and quality changes in metal working machine tools. Applied Economics 8: 283–288

Bosworth, D, Stoneman, P, McCausland, D (1993) Quality adjusting the PPI for computers. A Report to the CSO, Warwick University, Coventry

Denison, EF (1989) Estimates of productivity change by industry: an evaluation and an alternative. Brookings Institute, Washington, DC

Dickey, DA, Pantula, SG (1987) Determining the order of differencing in autoregressive processes. Journal of Business and Economic Statistics 15: 455–461

Gordon, RJ (1992) Discussion. Economic Policy October: 414–421

Granstrand, O (1999) The economics and management of intellectual property. Edward Elgar, Cheltenham

Lichtenberg, F, Griliches, Z (1989) Errors of measurement in output deflators. Journal of Business and Economic Statistics 17: 1–9

Mairesse, J, Mohnen, P (1995) Research & development and productivity: a survey of the literature. INSEE, Paris

Mairesse, J, Sassenou, M (1991) R&D and productivity: a survey of econometric studies at the firm level. Science, Technology, Industry Review 8: 9–43. OECD, Paris

Rosen, S M (1974) Hedonic prices and implicit markets: product differentiation in pure competition. Journal of Political Economy 82 (1): 34–55

Schmookler, J (1966) Invention and economic growth. Harvard U.P., Cambridge, MA

Waugh, F V (1928) Quality factors influencing vegetable prices. Journal of Farm Economics 10 (2): 185–196

Entrepreneurs, innovations and market processes in the evolution of the Swedish mobile telecommunications industry

Staffan Hultén[1] **and Bengt Mölleryd**[2]

[1] Stockholm School of Economics and Ecole Centrale Paris, France
[2] Stockholm School of Economics and Evli Securities, Sweden

Abstract. This paper analyses the entrepreneurial actions that were critically important for the development of mobile telecommunications in Sweden. Three types of theoretically determined entrepreneurial actions were identified and analysed: a) innovative in the Schumpeterian sense that create disequilibrium, b) the resolution of structural tensions or reverse salients, and c) the restoration of equilibrium by capturing possibilities already available. Our analysis shows that the Schumpeterian innovations were most important in giving Sweden the position as a first-mover in the development of the mobile telephone industry. But, on some occasions the other types of entrepreneurship gave momentous force to the expansion of the mobile telecommunication business.

JEL Classification: L96, M13, O31, O33

Key words: Entrepreneurs – Innovations – Commitments – Mobile Telecommunications

1 Introduction

When mobile telephony appeared few persons could envisage the benefits of the new technology. Mobile telephone usage had to be invented in the same way as mobile telephone technology. A substantial part of the entrepreneurial activities were oriented towards the interpretation of the technology. Eventually the mobile telephone of the 1990s emerged – a device for co-ordination and easy talk.

In a few years time the third generation (3G) of mobile telephone systems will be launched in Europe. The total investments will amount to hundreds of billion USD.

The licenses will in some countries cost more than the accumulated investments in the second generation of mobile telephone systems. In Germany the auction of 3G licenses brought in 98.9 billion DEM, in the United Kingdom the figure was £ 23 billion.[1] The capitalistic system seems to be willing to commit enormous resources to this future merger of internet and mobile telecommunications. What consumers will gain from this is today of course nearly unknown and the mobile telecommunication operators have not established any business models – how to charge the customers and for what. Ultimately, the established usage of mobile telephones will be replaced or complemented by data transfers over mobile telecommunication networks. This is a classical case of ex ante Schumpeterian creative destruction: an old technology will be prematurely replaced (from a technical point of view) by a new technology.

We suspect that two "truths" in modern economics underpin this situation. The first one is the historically successful track record of firms committing themselves fully to communication technologies with substantial network externalities. It seems as if it is always advantageous to be a first-mover. Everyone involved in the betting on the third generation mobile telecommunications knows the success stories of Bill Gates and Microsoft, Nokia, Cisco, Ericsson and other information technology firms. [2]

The second truth is less obvious. When we teach strategy to future top managers we have been struck by the fact that nearly every student opts for a tough strategy in a strategic game based on the Prisoner's Dilemma, despite the fact that a mutual soft strategy gives the highest total payoffs.[3] However, the way the game is constructed, if one actor is tough and the other is soft, toughness pays off. As a consequence, MBA students in general play tough and get low total payoffs in the hypothetical game. This fact puzzled us until a student enlightened us with a simple piece of information. The reason students tend to choose toughness is that they are taught that different humans are predestined to select tough or soft. Some humans are by nature soft. So, if in one game the actors cannot collude there is a high probability that a tough actor meets a soft competitor. The students learn that in the real world they are certain to meet soft competitors.

The combination of these two truths gives the simple lesson that a reputation of toughness and willingness to commit will make you a winner. What is missing in this simplified reality is of course that success in the business world takes something more than toughness and commitment. Otherwise brutes and criminals would run business firms.

A Schumpeterian assumption would be that the missing explanatory factor in planned creative destruction is entrepreneurship. Of course it is conceivable that entrepreneurship without toughness and commitment is not enough to succeed in building system technologies.

---

[1] The UMTS Forum web site: www.umts-forum.org August 5 2001.

[2] Codified in popular business books like Information Rules by C. Shapiro and H.R. Varian (1999).

[3] This is of course the basic problem of the Prisoner's Dilemma (R. Axelrod, 1984). But in chapter 7 in Axelrod's book toughness can be replaced by cooperation in a repeated game through four different mechanisms: a. enlarge the shadow of the future, b. change the payoffs, c. altruism, and d. teach reciprocity.

To explore the hypothesis that entrepreneurship is a decisive factor in system building, we will play the tape again of the technical and commercial break-through of mobile telecommunications in Sweden. We analyse how different entrepreneurs through their activities developed the mobile telecommunication business in Sweden. In particular we study two aspects of the entrepreneurial activities - how they were organised and their effects on the development of the technology and market expansion. The goal of this exercise is to find out what kind of entrepreneurship was crucial along the mobile telecommunication trajectory. The reason we chose Sweden is that Sweden and Swedish firms consistently were first-movers in mobile telecommunications from 1950–1995.

2 Theoretical framework

In a paper by Volberda and Cheah (1993) it is suggested that different types of entrepreneurs play different roles in an entrepreneurial process. Volberda and Cheah (1993) put forward the view that the entrepreneurial process consisted of a dynamic alternation between the Schumpeterian and the Austrian entrepreneur. Austrian entrepreneurs promote equilibrium, which results in change within an existing situation as entrepreneurs strive to discover gaps, increase the knowledge of the situation and reduce the general level of uncertainty. The short-run processes, which Kirzner perceives as being comprised of arbitrage and speculative activities, are based on the fact that, at a given point in time, a market economy is not fully co-ordinated. According to this view, entrepreneurship could be regarded as a co-ordinating mechanism.

In contrast, the Schumpeterian entrepreneur disturbs an equilibrium by introducing an innovation into the circular flow of the economy. After that an innovation has created a new market we can envisage the entry of many different Schumpeterian entrepreneurs conducting different entrepreneurial activities – creation of new products, new production processes, new distribution channels, etc.

If we expand this classification we can add a third type of entrepreneurship: The Hughesian entrepreneurs that are system builders, characterised by their ability to solve reverse salients, socio-technological bottlenecks emerging as a system technology advances.

2.1 Schumpeterian entrepreneurship

The interaction between entrepreneurial activities and technology, according to Schumpeter (1947), is a vehicle for economic development. Schumpeter portrays economic development as a perennial gale of restructuring and expansion resulting from innovative re-combinations of resources. Witt (1995) points out that the evolutionary element in Schumpeter's approach is that economic change is produced endogenously in the economy. The entrepreneurs have the function of carrying out new combinations based on existing knowledge, artefacts, services etc. The entrepreneur thereby reforms or revolutionises the pattern of production and distribution by exploiting an invention. Only the most gifted entrepreneurs are assumed to be the pioneers capable of overcoming the hurdles facing an entirely new venture. The motives for entrepreneurial conduct, according to Schumpeter, are will to conquer, proving oneself superior to others, the joy of creating, otherwise simply exercising one's energy and ingenuity.

For Schumpeter (1949) entrepreneurship consists in "doing things that are not generally done in the ordinary course of business routine, it is essentially a phenomenon that comes under a wider aspect of leadership". Schumpeter (1947, p. 150) distinguishes between adaptive and creative responses. By adaptive response Schumpeter refers to adjustment and adaptation in the form of more people, bigger quantities, and expansion through existing practices. The creative response occurs whenever the economy or an industry or some firms in an industry do something outside of the range of existing practices leading to a new social and economic situation.

Hérbert and Link (1982) argue that the Schumpeterian entrepreneur is a construct, like Weber's charismatic leader, which is introduced to disrupt the self-perpetuating equilibrium. This is in accordance with Schumpeter (1949) who underscores that the entrepreneur should not be interpreted as being the equivalent of a single physical person so much as a function, and that every social environment has its own ways of filling the entrepreneurial function. It could for example be fulfilled through co-operation between actors (Schumpeter 1949).

2.2 Austrian entrepreneurship

Mises postulates that knowledge is never complete or perfect – which explains why markets constantly are in a state of disequilibrium. Human action, according to Mises (1963), is founded on subjective knowledge of the environment and human action is one of the forces that creates change in the economic system. Therefore, knowledge about people, local conditions and particular circumstances are equally important for economic success as scientific facts. Entrepreneurs, according to Mises, are the first to understand that there is a discrepancy between what currently is done and what ought to be done.

The basic idea in Kirzner's (1973, p. 17) notion of entrepreneurship is that the market process essentially is entrepreneurial. For Kirzner (1973, p. 81) entrepreneurship is about to "perceive new opportunities which others have not yet noticed" and an ability to see where new products have become unexpectedly valuable to consumers. Moreover, Kirzner (1979, p. 115) regards entrepreneurship as the grasping of opportunities that somehow have escaped notice, and that the entrepreneurs "fulfill the potential for economic development that a society already possesses". Kirzner criticises neoclassical theory as it does not state how equilibrium is accomplished from an initial state of disequilibrium, and that it leaves no room for purposeful human action. In Kirzner's theory market participants learn about what other market participants are likely to do and capture possibilities that already are available, and it is through this discovery process that entrepreneurs move economic markets in the direction of equilibrium. The market is in Kirzner's (1973, p. 9) view made up of interacting decisions of consumers, entrepreneur-producers, and resource owners. The market process is set in motion by the results of the initial market-ignorance of the actors. Taken over time the "... systematic changes in the interconnected network of market decisions constitutes the market process" (Kirzner 1973, p. 10).

2.3 Development blocks, structural tensions and reverse salients

A particular kind of entrepreneurial problem is the structural tensions or bottlenecks that develop after the initial innovation. [4] The internal combustion car evolved in such a pattern. As the technology advanced new entrepreneurial problems emerged, stretching from technical matters such as the invention of starting-lighting-ignition, to economic problems such as mass production and institutional issues such as who should pay for road construction. These problems represented structural tensions that were opportunities for entrepreneurs.

Dahmén (1950) developed the notion of the development block, that is constituted by "... a sequence of complementarities which by way of a series of structural tensions, i.e., disequilibria, may result in a balanced situation" (Dahmén 1989, p. 111). The unbalanced situation could be reflected in price and cost signals in markets, which are noted by firms and may give rise to new techniques and new products. Incomplete development blocks generate both difficulties and opportunities for firms and entrepreneurs (Dahmén 1989, p. 109).

But, in certain cases reverse salients may develop as the technological system expands.[5] Reverse salients are components in the system that have fallen behind or are out of phase with the others (Hughes 1987). Hughes underscores that when a reverse salient cannot be corrected within an existing system, the problem becomes a radical one, the solution to which may bring a new and competing system. The origin of the reverse salient is often the accumulated actions of the decision makers in the technological system. This is due to the fact that management in a technological system chooses technical components or technological advances that support the structure, or organisational form of management.[6] The components of technological systems are consequently socially constructed artefacts invented and developed by system builders and their associates (Hughes, 1987).

2.4 Entrepreneurial actions along the technological trajectory

We can now construct a simple taxonomy of entrepreneurial actions that develop along a technological trajectory. These actions are of three classes: a) innovative in the Schumpeterian sense that create disequilibrium, b) the resolution of structural tensions or reverse salients, and c) the restoration of equilibrium by capturing possibilities that already available.

[4] "Structural tensions" is a central concept in Dahmén's approach, by which he refers to the fact that depressive pressure is predominant as long as complementary factors are missing. (Dahmén 1989, p. 111).

[5] A salient is a protrusion in a geometric figure.

[6] On this point see also C. M. Christensen (1997) Chapter 2.

We can envisage that the different types of entrepreneurs dominate sequential phases along the evolution of a technological trajectory. The Schumpeterian entrepreneur disturbs an equilibrium by introducing innovations creating "structural tensions" that sometimes result in reverse salients attracting the interest of Hughesian entrepreneurs. As the technology develops and the industry matures Kirznerian entrepreneurs handle the imbalances created by the Schumpeterian and Hughesian entrepreneurs. Eventually an equilibrium is reached which in its turn will be overthrown by a new innovation.

3 Entrepreneurship along the mobile telecommunication trajectory[7]

This case records the pre-mass market phase of mobile telecommunications. This phase roughly started in 1967 when Swedish Telecom started to investigate the building of a national mobile telephone network and ended when handportable mobile telephones were introduced in 1987. The case will focus on the activities of the operators and to a lesser extent on the activities of telecommunication suppliers, standardisation organisations and politicians.

3.1 The Land Mobile Radio Survey 1967

In August 1967, Carl-Gösta Åsdal, chief engineer at Swedish Telecom Radio, submitted a report regarding the future of mobile telecommunications in Sweden.[8] The report proposed that Swedish Telecom should supply a nation-wide mobile telephone network and paging networks. Although the report was optimistic concerning the future prospects for mobile telephony, the service was thought to be limited in terms of size and lacking in profitability.

The investigation suggested that Swedish Telecom should develop a fully automated national mobile telephone system, and that MTB, the second Swedish automatic mobile system, should be expanded on a regional level, pending the advent of a national system. Moreover, Swedish Telecom should initiate and conduct the development and the design of the system, since the manufacturing of mobile telephone switches and other products was not expected to reach such volumes that the industry would be willing to invest sufficient capital.

The conclusive argument behind Åsdal's proposition to establish a national system was the positive market reactions from subscribers in trade and industry. The report regarded it as indisputable that the mobile telephone service should be provided by Swedish Telecom, since the service would be integrated with the public telephone network. The report further considered Swedish Telecom to be best suited to own the mobile telephones, which could then be leased out to subscribers. The mobile telephones were expected to be of standard types, obtainable from any telephone supplier after modifications.

Despite the fact that most of the mobile telephone systems in the world at that time were manual, Åsdal forecasted that the future was headed towards fully

[7] The case study draws on B. Mölleryd (1997) and B. Mölleryd (1999).

[8] Swedish Telecom, Landmobil radiokommunikation (1967).

automated systems. But none of the existing, fully automated mobile telephone systems in the world, such as the Improved Mobile Telephone System (IMTS) in the US, were suitable for expansion nationally. Neither was it regarded as realistic to prepare expansion of a manual system.

One result of Åsdal's investigation was that Swedish Telecom's radio laboratory began development work, under the supervision of Ragnar Berglund and Östen Mäkitalo. They conducted a feasibility study in 1967, which showed that a substantial amount of technological progress was needed in order to develop a system according to Åsdal's principles. Mäkitalo suggested it was preferable to await the technological development of computing capacity in electronics since computers were needed to operate a sophisticated mobile telephone system with advanced features such as roaming and handover. This position was endorsed by Åsdal, who actively supported the development work at the radio laboratory. [9] Partly based on these insights Åsdal reached the conclusion that it would be better for Swedish Telecom to work with the other Nordic telecommunication administrations in order to develop a common standard. Åsdal saw the advantages of a Scandinavian system, a market with 23 million inhabitants was big enough for the industry to consider it profitable to develop systems and mobile telephones.

3.2 Development of a Nordic system

The other Nordic administrations were also active in the field of mobile telephony and discussed the possible introduction of fully automated mobile systems. It was therefore not difficult for Åsdal to launch and get support for the idea of a joint Nordic mobile telephone system at the Nordic telecommunications conference in 1969.

The Nordic administrations decided at the 1969 conference to develop a fully automated common Nordic mobile telephone system, and a working group was set up and named the Nordic Mobile Telephone Group, the NMT-Group. [10] The Group would meet up to seven times a year. Meetings could last up to five days and more than 100 such meetings were held. The development began at the beginning of 1970, and the group was working on principles for signalling between mobile telephone switches, radio base stations and mobile telephones. However, the NMT-Group's first assignment was to develop a manual system, ready to be used immediately, and to work out a frequency plan.[11]

3.3 Development and opening of a manual system

The reason the NMT-Group had to work on a manual system was that it was anticipated that it would take quite a long time to develop an automated system,

[9] Interview Carl-Gösta Åsdal, former manager Swedish Telecom Radio 16 April 1991, and interview Östen Mäkitalo Telia Research, 24 February, 1993.

[10] Memorandum Swedish Telecom, Introduction of a national automatic mobile telephone network MTC, July 1975.

[11] In the frequency band 453–455 MHz and 463–465 MHz.

while it was considered important to offer a national mobile telephone service immediately.

In 1971, the Nordic telecommunication conference approved the plans of a manual system, and decided on new rules which allowed the cross-border use of mobile telephones in Scandinavia. Manual Mobile Telephone System D (MTD) networks were established in Sweden, Denmark and Norway.

The Swedish MTD network was introduced in December 1971. Subscribers were assisted by operators from cord operated switchboards at six service centres. Each operator filled in a form regarding the subscriber's number and length of the call.[12] The deployment of MTD began in the Mälardalen region close to Stockholm and expanded gradually throughout the country. The system's radio parts were interconnected with the public telephone network at the service centres. The system had 80 channels and when fully extended 110 radio base stations. The system lay in the 460 MHz band. Aerials to radio base stations were located on TV and radio masts, which gave an effective range. MTD did not provide roaming or handover. No particular mobile telephone switches were required since it was a manual system.

To place a call to a MTD telephone, the operator had to know roughly where the subscriber was located in order to direct the call over the nearest radio base station. It was an open system at first. The subscribers were called by their telephone numbers, consequently everyone had to listen to the calling channel. This meant that other subscribers could listen to calls in progress. When selective calls were introduced in 1974, no one had to wait for the calling channel; instead an optical signal turned on a lamp at the receiver.[13] As far as calls from the mobile telephone were concerned, the operator responded to tone signalling by activating the calling channel.

The MTD network created opportunities for growth. Swedish Telecom's automatic systems had together less than 1000 subscribers in the late 1960s. The influx of subscribers to the MTD network was around 2300 annually; 200 000 calls were exchanged every month and the operating revenue was about SKr 20 million.[14] Profitability was satisfactory when the number of subscribers reached 10,000 but the costs increased immensely when the number came up to around 20,000 since at that level more than 400 telephone operators were needed, accounting for about 60–70 per cent of the network's total cost.

As mentioned earlier, the Land Mobile Radio Survey Commission of 1967 proposed that Swedish Telecom should own the mobile telephones and lease them to the subscribers. This was the established model to organise telecommunication services, and was also used for the first two mobile telephone systems, the telephones being considered an integrated part of the system. The NMT-Group suggested this model for the new manual system in January 1970. However, it required a considerable investment on the part of Swedish Telecom over and above the SKr 20 million per year that was to be invested in network expansion. The necessary investment for the procurement of mobile telephones was estimated at SKr 40 million. Swedish Telecom considered it unfeasible to obtain sufficient capital to purchase

[12] Håkan Bokstam, "Televerket landsomfattande mobiltelefonsystem", *Tele* 1/1972.

[13] Håkan Bokstam, "Televerket landsomfattande mobiltelefonsystem", Tele 1/ 1972.

[14] Carl-Gösta Åsdal, "Televerket radioverksamhet, Landmobil radio", *Tele* 2-3 1977.

mobile telephones through the state budget and time was also limited. Therefore Swedish Telecom began discussing the possibility of breaking with the established convention - that telecommunication operators should control all parts of the telephone system - and liberalising the market. Suppliers could then market the phones directly to end-users, thereby also promoting the mobile telephone service. This move was also inspired by experiences from Denmark and Norway, where the mobile telephone national operators had opened the market for mobile telephones.[15]

In 1971, Swedish Telecom liberalised the mobile telephone market and allowed mobile telephone suppliers to market their products directly to end-users. Subscribers could purchase or lease the telephones from a retailer or distributor selling mobile telephones.[16] This step changed the definition of the mobile telephone system as it created a separate market for mobile telephones, and resulted in the emergence of independent distributors. However, the mobile telephones had to be approved by Swedish Telecom.

The liberalisation of the mobile telephone market meant that Swedish Telecom could not assign Teli, Swedish Telecom's mechanical shop, the development and manufacture of proprietary mobile telephones. Swedish Telecom considered that Teli lacked the appropriate competence and preferred to have external suppliers. The operator thought it sufficient with type approvals to have control over which mobile telephones were used.[17] The range of products increased, and among the suppliers of mobile telephones to MTD were: AP, Handic, Mitsubishi (Gadelius), Salora, Storno, Sonab and Svenska Radioaktiebolaget (SRA). Competition between suppliers intensified and the marketing of the mobile telephone service was stimulated, since mobile telephone suppliers - through sales – participated in the promotion of the mobile telephone service.[18]

3.4 NMT development from 1972

From mid 1972, working groups were established to address different aspects of the development of the Nordic system. In 1972, the NMT-Group commissioned the Danish company Storno (Motorola acquired Storno in 1985), the dominating radio company in Scandinavia at that time, to perform a signalling study and to carry out research on three different signalling methods. The study formed the basis for decisions and resulted in the group deciding on binary signalling.

Åke Lundqvist, SRA, in 1971 had expressed to Östen Mäkitalo that it was necessary to select tone signalling (computer or digital signalling) instead of the five tone signalling, according to a CCIR standard, proposed by Storno. The principal reason for Lundqvist taking that position was that five-tone signalling was primarily aimed at mobile radio, and it restricted the number of subscribers to 100,000.[19]

A group of 10-15 people at the Swedish Telecom radio laboratory in Stockholm as well as two or three persons in Norway were in charge of the NMT project. The

[15] M. Karlsson (1998), p. 223.

[16] Håkan Bokstam, "Televerket landsomfattande mobiltelefonsystem", *Tele* 1/1972.

[17] Interview Carl-Gösta Åsdal, former Swedish Telecom Radio, 16 April 1991.

[18] Carl-Gösta Åsdal, "Televerket radioverksamhet, Landmobil radio", *Tele* 2-3 1977.

[19] Interview Åke Lundqvist, Ericsson, 22 February 1994.

NMT-Group wanted to find a cost effective and flexible system, which was not too demanding of computer power. The mobile telephones required program memory of only a couple of kilobytes.[20] A necessary requirement was that the system should be able to handle 180 channels. Many of the suppliers were reluctant and doubtful whether this was possible.[21]

The NMT-Group introduced a proposition, in time for the Nordic telecommunications conference in 1975, for a fully automated mobile telephone system.[22] The system was so designed that subscribers would be treated in a similar way as in the public telephone network. The system's requirements were:

1) automatic switching and charging – to and from the mobile telephone,
2) an ability to call any permanent telephone subscriber or other mobile telephones,
3) calls should work at home radio base stations as well as at other radio base stations,
4) the subscriber capacity should be adequate to handle future growth,
5) the system should automatically give access to roaming and automatic switch between base stations (handover).

Östen Mäkitalo was involved in the NMT-Group from 1975. He had worked on the Mobile Telephone System C until the mid 1970s which, in practice, was the Swedish contribution to the Nordic standard. Mäkitalo applied his own version of Moore's Law and predicted that the performance increases of computers and microprocessors should continue and that by the end of 1970's the necessary computing power would be available at a low enough cost. He often presented his prediction in the form of a curve – called "Östen's curve"[23] Despite the bearish forecasts of subscriber growth Mäkitalo wanted to enlarge the capacity and improve the frequency efficiency by using a small cell technique. Mäkitalo had discovered that the subscribers only moved a couple of kilometres at normal speed during a mobile call lasting approximately two minutes. It was not necessary for a mobile system cell to have a range of some ten or twenty kilometres in the areas of big cities - a few kilometres were enough. By having tightly packed small radio base stations, a frequency could be repeated more often which was a considerable improvement in economising on frequencies. Still, it was necessary to have access to processing capacity to look after how the frequencies were used. True, small cell techniques did bring the need for additional radio base stations, but the higher capacity would compensate the extra investment.[24]

It was necessary to conduct a pilot test, since the NMT-Group counted on the need for a sophisticated technique to handle the interface between the mobile telephone and the radio base station. The developers considered that one of the administrations should handle the test system, in preference to the industry, so as to utilise the knowledge that the NMT-Group had built-up and so as to maintain control

[20] Interview Östen Mäkitalo, Telia Research, 23 February 1993.

[21] Interview Åke Lundqvist, Ericsson, 22 February 1994.

[22] The system's detailed specification was nearly completed in 1975, and revised 1977–1978. NMT-Group, Nordic Mobile Telephone System Description, NMT Doc. 1 1977, Revised February 1978.

[23] Interview Östen Mäkitalo, Telia Research, 23 February 1993.

[24] Östen Mäkitalo, *Frekvensekonomi i mobilradiosystem*, Tele 4/1975.

over the standard.[25] The NMT-Group considered the Swedish administration, with Swedish Telecom's radio laboratory, best suited for the test. The pilot system, with comprehensive tests of all switches, radio base stations and mobile telephones ran for two and a half years and was completed in early 1978. Swedish Telecom used ten converted MTD- telephones with software developed by Swedish Telecom. The cost was SKr 1.2 million, which was shared between the administrations according the number of fixed telephones in the respective countries.[26] The cost was split in the following way: Sweden 52 per cent, Denmark 20 per cent, Norway 14 per cent and Finland 14 per cent.[27]

The NMT-Group suggested that the subscribers should purchase or lease the mobile telephones from radio suppliers, after the telephones had been accepted according to type. This proposition was supported by the Nordic Telecommunications conference.[28] No administration was then forced to invest in either mobile telephones, distribution or service networks. The end-users could freely choose equipment and would be able to receive service from private firms when travelling to another Nordic country.[29]

3.5 Launch of Nordic Mobile Telephone - NMT

In October 1981, NMT 450 was inaugurated in Sweden. Sales were fairly modest at first; the range of mobile telephones was limited since the type approval was delayed, and few manufacturers had the capacity to deliver. But a year after the start, the number of subscribers had increased to more than 35,000 in Scandinavia, and traffic growth exceeded projections. As from October 1982, roaming began to work between Denmark, Norway and Sweden. According to the original plans, NMT 450 would cover the need until a European system was introduced. However, capacity problems soon emerged in the network. In 1984 it was difficult at peak hours to get through on the network in Stockholm, so in order to increase the capacity, the network was modified into a partial small cell system in 1985, with a large number of radio base stations with a short range. This resulted in a capacity ceiling of 250,000 subscribers for NMT 450. However, the NMT-Group did not believe that the 180 radio channels in NMT could handle the growing traffic in larger cities, despite the small cell technique, which was why a decision was made to extend NMT to the 900 MHz band.[30]

The specifications for the new standard, NMT 900, were completed in 1985. They drew on NMT 450, but the system was in a higher frequency band and had more channels, and featured a few new components such as noise limiter and compander. The system was a small cell system, which gave higher capacity, suitable

[25] NMT-Group, memorandum to the Nordic Telecommunication conference 1975.

[26] According to the budget 1800 working days SKr 400-500, and purchase of material SKr 310,000.

[27] NMT-Group, memorandum to the Nordic Telecommunication conference 1975.

[28] Ibid.

[29] Swedish Telecom, Memorandum July 1975, Introduction of a national automatic mobile telephony, MTC.

[30] NMT-Group, memorandum 12 January 1983.

for handportable or pocket telephones, which had not been permitted in the NMT 450 system, due to their low transmitter output power. At first, the plan was to expand NMT 900 only in the urban regions and on connecting European highways.

During 1984—1985 it was discussed whether it would be possible for Swedish Telecom to purchase a fully developed system, such as the American AMPS, or the British TACS, which would give end-users access to a considerably larger mobile telephone market. Advocates of this principle could be found within industry, but Swedish Telecom decided to concentrate on NMT. Åke Lundqvist at Ericsson Radio tried, without success, to convince Swedish Telecom to select a standard that was already developed. But Carl-Gösta Åsdal, responsible at Swedish Telecom Radio, responded that in such a case the operator could purchase equipment from an American company.[31] Lars Ramqvist, Ericsson Radio's managing director, also tried to convince the Director General at Swedish Telecom, Tony Hagström, to select an AMPS system, but Swedish Telecom rejected that proposition.

In August 1986 the NMT 900 was opened for traffic in Sweden. System growth was sluggish at first due to the fact that the new system did not offer subscribers anything extra in comparison with NMT 450. Not until the launch of handportable telephones, and the expansion of the network outside urban regions, did the market grow. The increasing number of subscribers also motivated an expansion of the network in the whole of Sweden.

3.6 Co-operation between the industry and the NMT-Group

During the development of NMT, the industry was continuously sharing the specifications and proposing changes. This gave the NMT-Group an opportunity to find technically and economically realistic solutions while the suppliers could effect improvements to radio base stations, switches and mobile telephones. When the NMT-Group held its first information meeting at the end of 1971, some 40 companies expressed their interest in developing equipment to the Nordic standard. In this period, the NMT-Group met representatives from a large number of Swedish as well as international companies, such as Tekade, ITT, Martin Marietta, Motorola, AP Radiotelefon, Sonab, SRA, Storno, and Ericsson. Several Japanese companies, such as Mitsubishi and NEC, also showed interest, and were prepared to participate in the new technical development.[32]

In 1977, the NMT-Group invited tenders from a number of companies. In competition with firms, like Fujitsu, Hitachi, Motorola and NEC, Ericsson obtained in September 1978 the order for Denmark, Norway and Sweden with its AXE switches.

At first, Ericsson's intention was to offer the AKE-13 exchange, which had been developed during the 1960s, and which had a computer controlled cross bar switch system. But Swedish Telecom did not consider the system to be sophisticated enough which was why it prescribed Ericsson to adapt the AXE switch to mobile telecommunications. The digital AXE switch was developed at Ellemtel

---

[31] Interview Åke Lundqvist, Ericsson, 22 February 1994.

[32] Interview Östen Mäkitalo, Telia Research, 23 February 1993.

Utvecklings AB, which was Ericsson's and Swedish Telecom's joint development unit.[33]

3.7 Expansion of NMT internationally

It was considered important from the beginning that NMT had to expand beyond the Nordic markets, so that industrial enterprises would become interested in further development of mobile systems and telephones. Swedish Telecom therefore presented the advantages of NMT in various contexts, to convince foreign operators to invest in NMT.[34]

NMT 450 was one of the alternatives considered when the Department of Trade and Industry and the two network operators in the United Kingdom were going through the process of deciding which available standard to chose. Other alternatives were a Japanese standard from Nippon Telephone & Telegraph (NTT), the German system C450, a system developed by Alcatel and Philips called MATS-E and the US standard AMPS.

The Japanese system was considered to be technically acceptable but was only supplied by one company (NTT) and was therefore not an alternative. The same was the case for the German C450 system, which was considered as elegant but very expensive and only available from Siemens. The MATS-E system developed by Alcatel and Philips was technically attractive but unproven. NMT 450 was not selected because of insufficient capacity for the centre of London and a relatively slow signalling speed. The AMPS standard was tested and met the general requirements. It was available from several suppliers and operated at a frequency band only 70 MHz below the 900 MHz band, which was why it was considered to be the best alternative to be used in the UK. The two appointed operators and the Department of Trade and Industry in 1983 decided to modify the American standard Advanced Mobile Phone System (AMPS) and name it Total Access Communication System (TACS).[35]

A difference between NMT and TACS is that NMT has an open interface between base stations and mobile telephone switches which enables supplier independence. Although not chosen by the UK authorities, the NMT standard succeeded in getting established in a number of other countries. As NMT was not patented it was open to any supplier interested in building systems or in mobile telephones. This contributed to the continuous growth in the installed base of subscribers, pressing down prices on system and telephones.[36]

3.8 Competitors in the network operators' market

A number of companies operated mobile telephone networks in Sweden until 1981, when Swedish Telecom got its first major competitor in Comvik. The majority of

[33] See J. Meurling and R. Jeans (1985) and B-A Vedin (1992).

[34] Interview Bo Magnusson, Swedish Telecom International/Telia, 20 September, 1991.

[35] G. A. Garrard (1998), p 98.

[36] Interview Jan Sverup, Ericsson Radio, 7 May 1991.

these companies were local or regional operators, but some covered relatively large parts of the country. According to Swedish Telecom, the common problem with the private operators was their financial weakness, which Swedish Telecom had to solve by keeping the companies alive, in order to avoid the subscribers from being affected.[37]

In 1971, a total of 13 operators offered mobile services and together they had 45 private base stations. The three largest operators were Telelarm AB, AB Svenska Sambandscentralen, and Nordiska Radiocentralen. The mobile terminals were leased or privately owned and each customer needed a permit from Swedish Telecom to operate a radio transmitter.[38] In 1970, Telelarm received permission to operate in the 400 MHz band, at which time it had 151 subscribers, a number which had grown two years later to 800.[39]

Swedish Telecom presented a new and restrictive policy in November 1979 with the aim of: a) protecting the public network from interference, b) limiting the number of private networks in order to provide a rational solution for less profitable areas, c) maintaining frequency economy and d) creating a more pleasant environment by limiting the number of antennas and radio installations. Swedish Telecom's standpoint was that mobile telephone networks with manual connection to the public network could be established in areas where Swedish Telecom's network was yet not extended. Automatic mobile telephone traffic was not permitted according to the directive.[40]

Svensk Kommunikationskonsult AB, the general agent for Salora mobile telephones, acquired Telelarm in 1979, and changed the company's name to Företagstelefon AB. This firm purchased the only remaining private competitor Nordiska Biltelefonväxeln AB in 1980 giving it access to an additional number of frequencies and bringing its installed base to 1,900 subscribers.

In October 1980 Företagstelefon applied for a licence to operate a fully automated mobile telephone system with mobile telephone switches supplied by Rydax Inc and with interconnection to the public network. [41] Swedish Telecom rejected the application. Företagstelefon's managing director Bo Hammarstedt appealed the decision to the Director General of Swedish Telecom. A series of exchanges took place between the two parties. Swedish Telecom consistently refused to give permission for an automatic exchange. The private company emphasised the necessity of taking advantage of more advanced technology to improve the efficiency of the mobile telecommunication operation. Swedish Telecom argued that mobile telephone equipment fell under the regulation of voice communication over the public network, which according to the telecommunication policy set by the Swedish Parliament and Swedish Telecom's Directive should be included in the monopoly area. However, the Director General of Swedish Telecom announced that he was willing

[37] Swedish Telecom, Comment General Director, 14 October 1981.

[38] M. Karlsson (1998) p. 228.

[39] M. Karlsson (1998) p. 228.

[40] M. Karlsson (1998) p. 229.

[41] Communication from Företagstelefon to Swedish Telecom, 13 October, 1980.

to allow Företagstelefon to connect its system to the public telephone network, provided that the mobile network was operated manually.[42]

In March 1981, Företagstelefon applied for a type approval to operate their radio switches manually, which was approved by Swedish Telecom at the end of May. The two parties decided to co-operate and Swedish Telecom would assist Företagstelefon to improve its manual system. The switches would be modified and approved for manual connection, frequencies in the 450 MHz band would be allocated, a method for transferring certain customer categories from the MTD system would be discussed and Företagstelefon would be able to establish an integrated secretary service within the NMT system.[43]

3.9 The growth of Kinnevik and Comvik in mobile telephony

In September 1981, Industriförvaltnings AB Kinnevik acquired Företagstelefon and reorganised it into Comvik. The company then had a network with frequencies spread over different frequency bands and had the ambition to consolidate a network in the 460 MHz band. [44] Soon afterwards the company introduced its mobile telephone system, which consisted of six Rydax mobile telephone switches from E.F. Johnson in the US.

The main owner of Kinnevik, Jan Stenbeck, was engaged in mobile telephony projects in the US through his American company Millicom Inc., which he founded in March 1979 together with Shelby Bryan. They had a vision that mobile communication would become a major market. Millicom's business concept was to take advantage of the deregulation within telecommunications by applying for licences, as well as operating mobile telephone networks together with local partners and investors internationally. [45] But the allocation of operators' licences in the US was turning out to be a lengthy process, which made Stenbeck move his attention to Swedish mobile telephony.[46]

In the late 1970s, mobile telephony was tested at three locations in the US. Millicom succeeded to get one of the three development authorisations to establish a test system in Raleigh-Durham in North Carolina. Millicom's objective was to evaluate the market for new handportable telephones, which was then a radical innovation since all phones at the that time were quite large and could only be mounted in vehicles.[47] The plan was that Millicom together with E.F Johnson and Racal and other partners would develop a handportable mobile telephone in the US.

[42] Communication from Swedish Telecom to Företagstelefon, 21 November, 1980, Directive 5 September, 1980, Communication from Företagstelefon to Swedish Telecom, 11 November, 1980, 12 December, 1980.

[43] Communication from Företagstelefon to Swedish Telecom, 24 March, 1981, Communication from Swedish Telecom to Företagstelefon, 26 May, 1981, Swedish Telecom report, 3 June, 1981, M. Karlsson (1998) p. 232.

[44] Industriförvaltnings AB Kinnevik Annual Report 1989.

[45] Millicom Annual Report 1985.

[46] J. Meurling and R. Jeans (1994). M. von Platen (1993).

[47] G. A. Garrard (1998).

The project did not result in any new products and was dismantled after a year.[48] Nevertheless, an outcome of the project was that Millicom succeeded in concluding a deal with Racal, helping them to submit a successful bid in December 1982 for one of the two UK mobile telephone network licences and establish the network operator Racal-Vodafone.[49]

In 1986, Racal bought out Millicom and Hambros from Racal-Millicom Ltd in a deal that valued the company at $ 80 million. It gave Millicom shares in Racal Telecom and US$ 30 million in cash as a transfer of its 10 per cent pre-tax profit royalty for profits for the subsequent 15 years.[50] Millicom gradually sold its holdings to finance investments in mobile telephone networks in developing countries.[51] This meant that Kinnevik, which partly owned Millicom, had an extensive international mobile network operation as well as a mobile telephone business in Sweden.

Comvik pursued the plans Företagstelefon had outlined for the modernisation of its network. In September 1981, Swedish Telecom discovered that Comvik violated the permit by using an automatic switch. According to Swedish Telecom, there was an apparent risk of serious interference to the public telephone network, since it was unclear how the signalling was worked out.[52] Swedish Telecom threatened to disconnect Comvik's system from the public telephone network.[53]

Comvik appealed the disconnection to the Director General of Swedish Telecom, and claimed that the company's 15-year-old mobile telephone business was threatened.[54] But the Director General found no reasons to alter Swedish Telecom's decision, since mobile telephony was protected by monopoly, which concerned "equipment for duplex voice communication over the public telephone net".[55]

Swedish Telecom argued that an exemption from the monopoly would set a precedent, resulting in more companies wishing to operate private mobile networks. Thereby there was a clear risk of not being able to expand the new NMT system to remote regions of the country. Swedish Telecom claimed that Comvik's aim was primarily to cover areas with a potential for high volume traffic, leaving less profitable areas to the government owned operator.[56] At this time, Comvik attracted about 30 per cent of the new mobile telephone subscribers.

The significance of this case and the reluctance to open the market to competition is underscored by a communication from LM Ericsson to the Government. Björn Svedberg, Chief Executive Officer, LM Ericsson supported Swedish Telecom's restrictive policy and argued that Swedish Telecom should be able to establish a national network without competition from a private network operator. It could

---

[48] Millicom Annual Report 1985 and 1986.

[49] G. A. Garrard (1998) p. 32.

[50] Millicom Annual Report 1985 and 1986.

[51] G. A. Garrard (1998). Interview Håkan Ledin, Millicom International, 12 March, 1991.

[52] Interview Carl-Gösta Åsdal, Swedish Telecom Radio, 16 April 1991.

[53] Communication from Swedish Telecom's Radio Control station to Swedish Telecom's Radio Division 25 September 1981.

[54] Communication from Comvik to Swedish Telecom 30 September 1981.

[55] Communication from Swedish Telecom to Comvik 21 November 1981.

[56] Communication from Swedish Telecom to Comvik 10 October 1981, Communication from Swedish Telecom to the Government 14 October 1981.

challenge NMT's expansion since it was anticipated that the private operator primarily would expand in profitable urban areas. Svedberg emphasised that a rapid deployment of NMT throughout Sweden was a prerequisite for NMT's as well as Ericsson's success on international mobile telephone markets, thereby securing employment in Sweden.[57]

Comvik appealed to the Swedish Government and asked for a licence to connect its automatic system to the public telephone network, arguing that it was specialised in customer related services. Comvik also claimed that a rejection of their request would make customers suffer economically, left holding worthless mobile telephones that had cost almost SKr 10,000 each. The company would have to terminate its business and employees would lose their jobs.[58] Comvik emphasised that it was not going to change its business since the company and its predecessors had been operating a licence for 15 years. So, according to Comvik, the case could not be considered as setting a precedent. Besides, Swedish Telecom would have no problem in competing with Comvik, since the NMT system was technically considerably more sophisticated.[59]

In December 1981, the Government decided to grant Comvik the licence. However, it was referred to as an exemption. The Government did not question whether mobile telephony was protected by Swedish Telecom's monopoly, since it had to do with voice communication over the public network, but argued that specific circumstances applied and that it was one way to increase the competition in the market. The decision was not about allocating frequencies, which meant that Comvik had to settle for the 26 frequencies it already had at its disposal.

After the first expansion of the network for SKr 24 million, Comvik's network covered the midth and south of Sweden, but the network was not cellular from the start. Comvik leased and sold telephones under their own logotype. E.F. Johnson supplied the first one and from 1983 Comvik also offered a model from Nils Mårtensson's newly founded company Technophone.

During 1982, the company invested SKr 53 million in the network. In early 1983, the network was almost nation-wide with 140 radio base stations, rented telephone lines, six automatic switches and six staffed centrals. In 1984, when Comvik asked for another twelve frequencies, the Director General of Swedish Telecom denied this, and Comvik appealed to the Government. Swedish Telecom responded to the appeal in September 1984. [60] In April 1985 Industriförvaltnings AB Kinnevik sent an official letter to the Government promising to create 50 to 100 new openings at their plant in Fagersta if they were allocated another 12 frequencies. [61] In June 1985, the Government allocated another eight frequencies to Comvik.[62]

---

[57] M. Karlsson (1998) p. 235.

[58] Communication from Comvik to the Government, 6 October 1981.

[59] Communication from Comvik to the Government, 6 October, 1981, 26 October 1981.

[60] Communication from Comvik to Swedish Telecom, 22 March, 1984, Communication from Swedish Telecom to Comvik, 22 May, 1984, Communication from Swedish Telecom to the Government 25 September, 1984.

[61] Communication from Industriförvaltnings AB Kinnevik to the Government, 16 April, 1985.

[62] Government Decision, 27 June, 1985, II 1153/84.

Comvik in 1985 dropped the plans to construct a complete new system. According to the management, the company would have built a TACS system if it had been allocated sufficient frequencies.[63] In April 1986, Comvik requested the Government to annul the existing limit regarding frequencies, as well as to clarify that there were no main obstacles in allocating Comvik 120 frequencies in the 900 MHz band.[64] The original system was improved, and during 1986, roaming was introduced between switches. In Stockholm the system was reconstructed into a small cell system, which tripled call capacity. The coverage was improved, particularly in northern Sweden. In June 1987, the Government decided that the exemption regulation was to continue, but that Swedish Telecom should allocate another 16 frequencies to Comvik, giving the firm 50 frequencies at its disposal.[65]

3.10 Ericsson adopts mobile telephony

Ericsson in the 1970s had a rather guarded approach towards the concept of a public mobile telephone system, as such, since the company's aims were more in line with mobile radio systems used by emergency services and in the transport sector.[67] But orders from the Nordic telecommunication administrations, as well as from Saudi Arabia, demonstrated mobile telephony's market potential. Accordingly, Ericsson secured a place in the domestic market and obtained access to a mobile telephone network to exhibit to potential customers.

Ericsson was soon the dominating enterprise in mobile telephony, a position it succeeded in retaining even when the world market expanded. Ericsson was aware from the outset that the AXE switches could more than cope with the modest subscriber growth projected. When the influx of subscribers grew considerably, the switches could easily handle the growth and generated a positive revenue stream for the mobile telephone network operators.

Prior to 1982, LM Ericsson and its subsidiary SRA tendered for different contracts which concerned mobile telephone systems: LM Ericsson offered mobile telephone switches while SRA offered radio base stations, i.e. system integration was the buyers' function. But from 1982 onwards, the firm's objective was to sell integrated systems. A contract awarded by the telecommunication administration in the Netherlands for the expansion of an NMT network there triggered this change in policy. LM Ericsson, as usual, offered the switches and SRA the radio base stations but the Dutch telecommunication administration was only interested in buying switches from Ericsson, and intended to buy radio base stations from Motorola. Ericsson's reaction was positive initially, but Åke Lundqvist, Managing Director of SRA, objected. He managed to stop the deal and to force Ericsson

[63] Interview Thomas Julin, Comvik GSM AB, 11 April 1991.

[64] Communication from Comvik to the Government, 18 April 1986.

[65] Government Decision, 12 June, 1987, II 678/86.

[66] In addition to Ericsson a large number of Swedish and Nordic firms benefited from the early standardisation of NMT and the creation of a Nordic manual mobile telephone network in the 1970s. Nokia is of course the outstanding example. For more details on the Swedish firms see Mölleryd (1999).

[67] Interview Bengt Dahlman, Ericsson (Magnetic) 3 September 1991.

not to supply the switches unless the Ericsson Group supplied the radio base stations as well. Lundqvist was convinced that if Motorola got access to the AXE technology, Ericsson's position would weaken considerably. In the final event, the Dutch telecommunication administration decided to purchase the equipment from Ericsson/SRA, but stipulated that the network should consist of the small cell technique. Ericsson's experts advised it would take two to three years to develop such a technique. Lundqvist, through an American friend, turned to Chandos Rypinski, an American expert in cell structure, who not only played an important role in the expansion of the Dutch system but also in Ericsson getting established in the US.[68]

3.11 GSM development

The market for mobile telephony in Europe was fragmented and underdeveloped in the 1980s, with numbers of incompatible analogue mobile telephone standards in operation, with no particular market leader. The infrastructure equipment and the mobile telephones themselves were quite expensive and the products limited in variety in the majority of the European countries, apart from the United Kingdom and the Nordic countries.

As early as in 1970, the NMT-Group discussed a future European mobile telephone system, although the group anticipated difficulties in reaching an agreement regarding the standard. The NMT-Group realised that at that time it would have been too time-consuming to try to convince Europe that a mutual standard was advantageous. On the other hand, the group saw the possibility of achieving a limited level of compatibility between different European standards.[69]

In the 1980s, the Commission of the European Communities actively promoted the development of a pan-European mobile telephone standard. It was anticipated that it would contribute to a positive economic development in Europe, primarily in two respects. Firstly, inter-country and inter-personal communication would improve, generating positive effects on business life. It would be possible, for instance, to use mobile telephones when travelling throughout Europe. Secondly, by creating a single market for mobile telephone systems as well as mobile telephones it would strengthen the European telecommunication industry, something that was considered to be essential. The Commission also considered it as essential to introduce competition into the sector and separate the regulatory duties from the operational activities in the telecommunications administrations.

A first step towards a mutual European system was taken in 1982, when the Conference on European Posts and Telecommunications (CEPT), consisting of national telecommunication administrations from 26 member states, decided to assemble a group called Groupe Spéciale Mobile (GSM) - in the early 1990s the name was changed to Global System for Mobile Telecommunications - which was commissioned to develop a mobile telephone standard. The Nordic countries were instrumental in promoting this initiative.

---

[68] Ibid.

[69] NMT-Group, minutes meeting number 5, 20–22 January 1971.

Frequencies between 862 MHz and 960 MHz had been reserved at the World Administrative Radio Conference of the International Telecommunications Union (ITU) in 1978.[70] Subsequently the conference of European Posts and Telecommunications Administrations decided to allocate this frequency band to mobile telephony.

The inaugural meeting of GSM took place in Stockholm in December 1982, where representatives from eleven countries met under the chairmanship of Thomas Haug from Swedish Telecom. Thomas Haug had also chaired the NMT-Group from 1976. The GSM-Group was to design a number of interfaces in the mobile telephone system, to facilitate communication between switches and radio base stations, human beings and machines.[71] The GSM-Group met regularly, and the standardisation work expanded, involving more and more people. In 1985, the detailed specifications were approved and communicated throughout the industry.

It was already assumed from outset that the GSM system would be based on digital transmission, even though it was not officially decided until 1987. According to Mäkitalo, a digital GSM had several advantages such as:

1) improved speech quality,
2) improved combined services,
3) higher capacity, and
4) extended security through encryption.[72]

During 1985–1986, the GSM-Group explored different alternatives for handling the radio transmission. A wideband solution was first discussed, but Swedish Telecom decided at an early stage to concentrate on narrowband Time Division Multiple Access (TDMA), which divides the frequency spectrum into a number of time slots. At first, Ericsson tried the Frequency Division Multiple Access (FDMA) technique, which divides the frequency spectrum into a number of frequencies, before the company decided on TDMA.

France and West Germany were in favour of wideband TDMA, and their respective national operators had invested about $50 million in development work and, together with a handful of firms, developed a prototype for wideband transmission. Moreover, Italy and the United Kingdom joined France and Germany in this venture.[73]

In 1986, a decisive test for the selection of radio transmission technique was performed in Paris by Center National d'Etudes Télécommunications (CNET), supervised by the GSM-Group. Altogether, eight prototypes were tested: four resulting from the Franco-German alliance and four originating from the Nordic region: one from Swedish Telecom, one from Ericsson, one from Nokia, and one from Trondhiems Technical University in Norway. The results of the test were presented at a plenary session held by CEPT in Madeira in February 1987 that decided that the system should be based on TDMA. However, it was more intricate to determine whether GSM should use wide- or narrow band TDMA. Finally, it

---

[70] G. A. Garrard (1998) p. 63

[71] Interview Thomas Haug, Swedish Telecom, 11 September 1990.

[72] Ibid.

[73] G. A. Garrard (1998) p. 129.

was decided that GSM should be based on narrow band TDMA - which the Nordic systems were based on - as this enabled faster hand-offs, smaller cells, down to 100 meter radius, and was compatible with existing spectrum planning. The main argument against the wide band solution was that it demanded considerably larger investments in densely populated areas.[74] In 1987, the Commission of the European Communities made several decisions, in the form of directives, related to the pan-European standard. The plan for the introduction and establishment of a digital cellular system throughout Europe was taken in recommendation 87/371/EEC. Directive 87/372/EEC required national frequency regulators to co-ordinate the allocation of 2 x 9 MHz of spectrum in the frequency band reserved for mobile telephony. The directive also stated that the allocated frequency band should be made available for GSM in accordance with demand, implying that analogue systems, such as NMT 900, should be dismantled from the year 2000.[75]

As the projections for the future growth of mobile telephony in the latter part of the 1980s were modest and analogue networks were concurrently expanded throughout Europe, it was considered necessary by the Commission that the European network operators make a commitment to implement GSM-networks. This would create a sufficient market to convince the industry to make vast investments in research and development for the pan-European GSM standard. In May 1987, ministers from France, Italy, the United Kingdom and West Germany called for an agreement between network operators in Europe to be formalised in a Memorandum of Understanding (MoU). The MoU stated that the signatories were committed to introduce GSM networks by January 1, 1991, later put back to July 1, 1991. The MoU was signed in Copenhagen on September 7, 1987 by operators and regulators from thirteen countries: Belgium, Denmark, Finland, France, Ireland, Italy, the Netherlands, Norway, Portugal, Spain, Sweden, the United Kingdom, and West Germany. In line with European Commission's ambition to liberalise the telecommunications market, a number of directives were initiated under article 90 of the Treaty of Rome. The first of these, issued on May 16, 1988, concerned competition in markets for telecommunications equipment and ensured liberalisation of the equipment market including mobile telephones.[76] The second, usually referred to as the service directive, was issued in 1990 and ensured the separation of the telecommunications operation and regulation. These initiatives played an important role in shaping the development of mobile telephony.

With the ambition of liberalising the mobile telephony market, it was not consistent to develop GSM within the CEPT organisation, as it was only opened for national telecommunication administrations. Therefore, the responsibility for the development of the GSM standard was transferred to the newly founded European Telecommunications Standards Institute (ETSI) in 1989. ETSI is open to any organisation based in Europe involved in the telecommunications industry, enabling suppliers and other industry participants to take an active part in the standardisation process.

---

[74] Ibid. Interview Åke Lundqvist, Ericsson, 22 February 1994.

[75] G. A. Garrard (1998).

[76] 88/301/EEC.

The specifications for the GSM standard, comprising some 5,000 pages, were completed in 1989. After 1989, further development of standardisation was initiated, including supplementary services and speech codes. By the end of 1990 it was estimated that ten manufacturers had invested 5000 man-years of effort in the development of GSM at a total cost of $350 million.[77]

4 Conclusion

Sweden has consistently been a first mover in mobile telecommunications. The position as a first mover has been underpinned by entrepreneurial actions of different types. In the case study we have identified three different organisations of the entrepreneurial activities: by individuals, in firms, and in networks. In some cases individual entrepreneurs acted also as members of an organisation and in other cases organisations acted both alone and in co-operation with other organisations.

In the late 1960's and early 1970's a number of important entrepreneurial actions took place.

1. The decision by Åsdal, acting for Swedish Telecom, to start a co-operation between Sweden and the other Nordic countries.
2. The decision by the Nordic telecommunications operators to continue to plan a cellular network with roaming and handover so it could be launched when computing capacity became available at a reasonable price.
3. The decisions by the Nordic telecommunications operators to satisfy the growing demand for mobile telephony with a manual system before an automatic system could be built.
4. The setting free of the market for mobile telephones by Swedish Telecom, which resulted in more investments in mobile telephones and the launch of different telephone designs. This strengthened the growth of the market.
   In the early 1980's two more important entrepreneurial actions occurred.
5. Swedish Telecom took a relatively soft approach to Comvik's attempts to grow. The Swedish government supported Comvik by giving the firm more frequencies. These actions were based on a flexible interpretation of the regulatory framework.
6. When the market for mobile telephone systems increased, the individual entrepreneur (Lundqvist, working at Ericsson) forced the firm to adopt a systems approach to mobile telecommunication technology.
   Later in the 1980's two entrepreneurial actions were important for the growth of mobile telephony.
7. The continuing pressure of Comvik for new frequencies put pressure on Swedish Telecom to more aggressively expand the sales of mobile telephones.
8. When work started on a European mobile telecommunications system, the Nordic model with committee work was adopted by Europe.

[77] G. A. Garrard (1998) p. 129.

How can we classify these entrepreneurial actions in terms of innovations that: a) create disequilibrium, b) resolve structural tensions or reverse salients, or c) restore equilibrium by capturing possibilities that already are available?

a) The committee work on a Nordic and a European level aimed at innovating the mobile telephone industry and in creating new business opportunities. Hence, this is an example of Schumpeterian entrepreneurship. Two other examples of this type of entrepreneurship are Ericsson's decision to sell systems instead of system components and Comvik's attacks on the regulatory system. Ericsson in selling mobile telecommunications systems changed the way mobile telephony was marketed to operators. This Schumpeterian innovation created huge problems for Ericsson's competitors and the firm supplied more than 50 % of all mobile telephone systems at the end of the 1980s. Comvik's successful attacks on the legal structure of the Swedish telecommunication industry helped the firm to expand its network and resulted also in that Sweden before any other country issued three GSM licenses. The politicians believed so much in competition in 1990 that they decided that the competitive advantages of three GSM operators out-weighted the lower frequency efficiencies.
b) The strategic decision to build a manual network while an improved automatic cellular network was developed, and the strategic decision to postpone the launch of the new automatic network until computing capacity was affordable are two examples of ex ante avoidance of reverse salients. The entrepreneurs saw the dangers of continuing on the historical technological path and made decisions about the future based on simple models – Moore's Law and demand expanding in accordance with a product life cycle. The decisions gave the wanted results despite the fact that the demand forecasts underestimated the subscriber growth.
c) The decision to let market forces decide on prices and the design of mobile telephones was something that facilitated the system to reach equilibrium. The network builders could focus on network expansion while consumers and mobile telephone producers could develop the mobile telephones. Truly an example of capturing possibilities that already were available. A similar case could be made about the leniency towards the small operators and Comvik. The firms were already active in the market and were actually contributing towards a better functioning of the market process. For example by contesting the monopoly and by providing differentiated products geographically and feature wise.

A striking fact about the totality of these entrepreneurial actions is that they follow the logic of co-operation rather than toughness and stubborn commitment. We can note that many of the individual entrepreneurial activities aimed at co-operation, either inside an organisation or in a network of organisations.

When the architect of Swedish mobile telephony, Åsdal, identified obstacles, he invited the other Nordic operators to participate in the development. When the builders of the manual MTD network became aware of the huge costs of supplying mobile telephones, they gave away that market to private firms. When Mäkitalo found out that a model similar to Moore's Law was applicable and that a small cell structure was advantageous he informed his colleagues and wrote articles. When Lundqvist noted that it was better to sell a mobile telephone system rather than components he made Ericsson change strategy. Few of the entrepreneurs in this case study got rich, with the exception of the owner of Comvik. At the end of the day they became heroes of Swedish technology and two of them were awarded the Swedish Royal Institute of Technology's Great Prize in 1994.

References

Axelrod R (1984) The evolution of cooperation. Basic Books

Bokstam, H (1972) Televerkets landsomfattande mobiltelefonsystem. Tele 1

Christensen CM (1997) The innovator's dilemma. Boston Mass: Harvard Business School Press

Dahmén, Erik (1950) Svensk Industriell företagarverksamhet. Stockholm: Industrins Utredningsinstitut.

Dahmén, Erik (1989) Development blocks in industrial economics. In: Industrial Dynamics. edited by B. Carlsson, Norwell, MA: Kluwer Academic Publishers.

Garrard Garry A (1998) Cellular communications: Worldwide market development. Norwood, MA: Artech House.

Hérbert RF, Link An (1982) The Entrepreneur – mainstream views and radical critiques. New York: Praeger Publishers.

Hughes PT (1987) The evolution of large technical systems. In: Social Construction of Technological Systems. Bijker W, Hughes T, Pinch T (eds.) Cambridge, Ma: MIT Press

Hughes PT (1994) Technological momentum. In: Does Technology Drive history? (ed) Smith MR, Marx L Cambridge, MA: The MIT Press

Industriförvaltnings AB Kinnevik, Annual Report 1989

Karlsson Magnus (1998) The liberalisation of telecommunications in Sweden. Linköping: Linköpings University

Kirzner IM (1973) Competition and entrepreneurship. Chicago: University of Chicago Press

Kirzner IM (1979) Perception, Opportunity and Profit. Chicago: University of Chicago Press

Landmobil Radiokommunikation (1967) betänkande angivet av arbetsgruppen för mobiltelefonsystem, Televerket

Mäkitalo Ö (1975) Frekvensekonomi i mobilradiosystem. Tele nr 4

Meurling J, Jeans R (1985) A switch in time. Chicago: Telephony

Meurling J, Jeans R (1994) The mobile phone book. London: CommunicationsWeek International

Millicom Annual Report 1985–1986

Mises von L (1963) Human action. New Haven: Yale University Press

Mölleryd BG (1997) The building of a world industry – the impact of entrepreneurship on swedish mobile telephony. via Teldok 28 E

Mölleryd BG (1999) Entrepreneurship in technological systems – the development of mobile telephony in Sweden, Dissertation, Stockholm: The Economic Research Institute

Office of Technology Assessment (1995) Mobility and the implications of wireless technologies.

PA Computers and telecommunications (1984) Swedish telecom mobile communications market study phase 2. Final Report, March

Platen v M (1993) Boken om Stenbeck. Stockholm: Dagens Industris Förlag

Schumpeter A J (1947) The creative response in Economic history. Journal of Economic History 149–159

Schumpeter AJ (1949) Economic theory and entrepreneurial history. In: Change and the entrepreneur: postulates and patterns for entrepreneurial history. Cambridge: Harvard University Press

Shapiro C, Varian HR (1999) Information rules - A strategic guide to the network economy. Boston Mass: Harvard Business School Press

UMTS Forum web site www.umts-forum.org

U.S. Congress, office of technology assessment (1995) Wireless technologies and the national information infrastructure. (Mobility and the Implications of Wireless Technologies) OTA-ITC-622, Washington, DC: U.S. Government Printing Office, July

Vedin B-A (1992) Teknisk revolt. Stockholm: Atlantis

Volberda WH, Cheah H-B (1993) Entrepreneurship and business development. In: Entrepreneurship and Business Development. Klandt H (ed.) Aldershot: Avebury.Ashgate Publishing

Witt U (1995) Introduction. In: Evolutionary Economics. Witt U (ed.) Aldershot: Elgar

Åsdal K-G (1977) Swedish Telecoms radioverksamhet. Landmobil radio. Tele 2–3

The new geography of corporate research in Information and Communications Technology (ICT)

John Cantwell[1] **and Grazia D. Santangelo**[2]

[1] Department of Economics, Whiteknights, PO Box 218, Reading RG6 6AA, UK (e-mail: J.A.Cantwell@reading.ac.uk)
[2] Facoltà di Giurisprudenza, Università degli Studi di Catania, Via Gallo, 24, 95124 Catania, Italy (e-mail: gsantangelo@lex.unict.it)

Abstract. In the new ICT-based paradigm MNCs have increasingly locationally dispersed competence-creating activities. Using patent data granted in the US to the largest European-owned electronic corporations and all the largest companies in other industries for their ICT research in the European regions, this paper investigates the regional dispersion of such research. We find that co-specialised electronic companies do not tend to develop related R&D in the same regional location, but non-electronic firms undertake related ICT development in a common centre of excellence. Thus, intra-industry competition encourages the geographical separation of co-specialised research, while inter-industry cooperation entails the co-location of related research.

Key words: Co-specialisation – Co-location – Innovation – International dispersion

JEL classification: O30, L10, R10

1 Introduction

In the ICT-based paradigm, the role of innovation in promoting change, development and transformation of the underlying socio-economic capitalist system has been clearly enhanced. Historically, the impact of innovation on the socio-economic system has taken place through a continuous and incremental adaptation

Correspondence to: J. Cantwell

of economic and social organisation. This can be observed by comparing the old paradigm, based on energy and oil-related technologies, and the new paradigm, based on ICT and related technologies. The former was grounded on mass production with its economies of scale and specialised corporate research and development (R&D), while the latter is characterised by economies of scope, derived from the interaction between flexible but linked production facilities and a great diversity of search in R&D. A distinctive feature of the current modes and forms through which innovation generates major systemic techno-socio-economic changes can be identified in the increasingly complex character of technology, the consequent rise in technological interrelatedness, and the great technological combinations made feasible by information and communications technologies (ICT). If these specific features have emphasised the role of the firm as the main actor in the development of new knowledge, in the most recent literature they have also led to the re-discovery of the significance of the local dimension in the creation of new knowledge. As highlighted by Nooteboom (1999), new distant-shrinking technologies are unlikely to undermine the value of proximity because the diffusion of codified knowledge amplifies rather than devalues the significance of local tacit knowledge.

Due to the increasingly complex nature of technology, the boundaries of the firm cannot fully encompass any longer the entirety of new knowledge generation required by an innovative company. This has promoted the adoption of corporate sectoral and spatial strategies aimed at outsourcing new knowledge creation from complementary technological sources and absorbing them into the firm's technological portfolio through inter-firm relationships. With the ultimate aim of enhancing competencies for the sake of corporate competitiveness, firms are extremely sensitive in seeking out foreign centres of excellence for the creation of knowledge outside the primary fields of their own industry, while tending to avoid locating technology creation in their primary fields in the home centres of their major international competitors (Cantwell and Santangelo, 1999, 2000; Cantwell and Kosmopoulou, 2002). On the grounds of this strategic distinction, the relationships between firms and territorial units follow different patterns. The selection of locations by multinational corporations (MNCs) for the purpose of siting their corporate research and development (R&D) laboratories is highly influenced by the complementary capabilities of indigenous institutions and technological traditions, especially when local firms are not major competitors in the same final product markets. MNCs seem to be attracted most by the overall economic structure of regional systems and by the consequent opportunities for successful corporate performance. Thus, they may be targeted locations by MNCs not just for the purpose of knowledge exchange, but also because they have historically hosted established financial centres, corporate headquarters and therefore provide opportunities for corporate control. In any case, local economic growth and external corporate spatial attraction are interdependent and mutually reinforcing phenomena as their interplay reproduces geographical hierarchies and corporate competitive performances over time through *vicious* and *virtuous* cycles.

Similarly, due to the complex character of technology and the increasing innovation potential from the fusion of formerly unrelated types of technologies through ICT, the link between the local and corporate dimension has gained a strategic importance in the generation of technologies that are core to the current paradigm (notably ICT). The need to develop these technologies is shared by the firms of all industries, and the knowledge spillovers between MNCs and local firms in this case may be inter-industry in character. Thus, ICT development in centres of excellence is not the prerequisite of firms of the ICT industries themselves, for which such activities are primary and generally concentrated at home, but instead involves the efforts of the MNCs of other industries in these common locations (Cantwell, 1999). However, firms of the ICT industries may distinguish the varying degree of centrality to themselves of different technologies from within the whole basket of ICT fields when targeting centres of excellence for the purpose of siting research activity. ICT firms may locate abroad (in other centres for the industry) the development of some ICT fields which lay outside their own primary interests among the ICT sectors. In the learning process, the attraction to locationally separate and differentiated sites for the creation of different ICT fields is due to the complexity of combinations required within cutting-edge technology, the consequent uncertainty and risks linked to high R&D costs, and the need to cooperate between firms within a region while increasing knowledge flows within MNCs across regions or national boundaries in order to design the multi-technology systems which are brought together in part through ICT.

In this context, the aim of this paper is to analyse the spatial dispersion or concentration of ICT corporate innovative activity within the European ICT industry[1] at a sub-national (i.e. regional) level. Using patent data granted in the US to the largest European-owned electronic corporations for their research located in the most developed European regions, the present study is the latest development of a sequence of related studies building upon one another. The initial starting-point of the research were the results obtained in a previous analysis (Santangelo, 1998), from which clusters of European-owned electronic companies specialising in equivalent ICT technologies were identified over the period 1969–1995. On these grounds, a further investigation tested the hypothesis re-stated below in the context of three national groups of regions (Santangelo, 2000). Beginning from these results, the present study explores further the following hypothesis by taking into account a wider geographical span of European regions than in the earlier work just referred to.

Hypothesis: in each of the three sub-periods under analysis (1969–1977, 1978–1986 and 1987–1995) the technological clusters of electronic companies identified as being grouped together in terms of common profiles of specialisation in ICT technology (co-specialisation)[2] do not locate their related R&D in some common regions (co-location)[3] because of mutual deterrence due to region-specific capa-

[1] The reasons for focusing on the European ICT industry are twofold: first the strategic significance of this industry in the current techno-socio-economic paradigm, and second the specific importance of the European ICT industry for historical and political evolution as well as its market structure (e.g. the conditions under which the industry has developed and operated and its oligopolistic structure).

[2] As can be inferred from the text, the term co-specialisation refers to the co-presence of technological expertise of two (or more) firms in the same technological field/s.

[3] The term co-location refers to the co-presence of R&D activity of two (or more) firms in the same territorial space.

bilities. In other words, we hypothesise that regions provide a source of expertise for corporate technological development as they are co-identified with companies which focus regional technological specialisation in some fields rather than others. Therefore, we hold that close co-specialisation within any industry leads to inter-regional interchange as well as inter-firm competitive relationships. However, in contrast, when looking at the location of ICT technological development by the firms of other industries (not electronics) we hypothesise that at least in the most important regional centres there will be a pattern of intra-regional inter-industry cooperation in the absence of direct competition for sector-specific resources or (of course) in downstream output markets. Thus, within the electronics industry the development of a particular type of technology is locationally separated in accordance with ownership (each leading firm has its own favoured home location), while for the firms of other industries we expect geographical agglomeration in each of the relevant foreign centres (the leading firms of other industries may each develop ICT technologies in one of the major foreign centres of excellence for the chosen field).

The paper is organised in 7 main sections. The following section elaborates upon the framework for analysing the implications of sectoral and spatial corporate strategies for competence accumulation. Section 3 describes the data used. Section 4 discusses the geographical evolution of ICT corporate research activity in the European regions. The empirical evidence on the hypothesis of co-location and co-specialisation is presented and discussed in the Section 5. Section 6 reviews the analytical framework in the light of the findings, and investigates whether the co-location of technological activity is likelier when firms do not compete in output markets. Some conclusions are drawn in Section 7.

2 Strategic implications of sectoral and geographical aspects of corporate R&D: a preliminary theoretical framework

The construction of a complete economic theory of technological change grounded on the evolutionary tradition calls for a theory of the firm able to fully understand and explain the production, development and re-generation of corporate technological capabilities over time. This need arises from the evolutionary concept of technology, characterised by both a public (or codified) and a private (or tacit) element (Nelson, 1992). The former includes free and widely available knowledge (e.g. blueprints, books, etc.), while the latter attains to knowledge embedded in corporate *routines* upon which the firm's specific capabilities are built and reproduced over time. Therefore, firms play a crucial role in the process of knowledge creation, characterised by the path-dependency of firms' specific technological trajectories (Cantwell and Fai, 1999). Nonetheless, it should be emphasised that tacit knowledge is specific to organisations as well as geographic locations, this increasing its internal circulation but impeding its external accessibility (Amin and Wilkinson, 1999).

If the specificity of corporate capabilities in generating innovation defines the firm's own trajectory of technological development, some general trends concerning corporate technological diversification and internationalisation have been identified in the literature (Cantwell and Piscitello, 2000). As far as diversification is concerned, the complex character of technology, increasingly emphasised by interrelatedness and fusion between formerly separated technologies (Kodama, 1992), has promoted the rise of the model of the "multi-technology" firm (Granstrand and Oskarsson, 1994; Granstrand, 1996; Granstrand et al., 1997; Patel and Pavitt, 1997). This corporate "model" allows the firm to adapt in the new techno-socio-economic conditions, in which major new opportunities for innovation based upon corporate technological diversification have arisen. The technological base of the company needs to be wider than (and is now less well related to) its product base, as the development of corporate capabilities in a wider spectrum of technologies has become an essential condition for exploiting potential technological combinations made feasible by the current paradigm, and thereby to facilitate successful output performance. The reason lies in the growing number of technologies entering into the production of a single product. Nonetheless, if corporate multi-technological portfolios are a necessary requirement to perform successfully, the firm finds it difficult to develop competencies across an increasingly wide range of technologies. In this sense, technological complexity, interrelatedness and fusion affects inter-firm relationships which create opportunities for knowledge outsourcing and technology-based alliances.

The corporate outsourcing of new knowledge has also a spatial dimension, since the firm may enhance its capabilities by increasingly dispersing geographically its intra-firm networks across a range of locations. This internationalisation strategy is substantially different from the international strategy adopted in the early post-war period, in which the primary aim was the conquest of new markets through the adaptation of products to local consumer preferences. Conversely, the closer international corporate integration that has occurred in the leading MNCs since the 1960s, aims to establish geographically dispersed networks for the purpose of the transfer of technology, skills and assets across national borders between the parent company and its affiliates (Cantwell, 1995). The sustainable competitive advantage built on this transfer lies in the two-way interaction between parent and subsidiaries. The transfer goes from the parent to subsidiaries, but also from the subsidiaries to the parent company. Local laboratories played a new role within the whole corporate structure by sourcing new knowledge from the local environment rather than carrying out merely demand-oriented activities (Papanastassiou and Pearce, 1994; Barlett and Ghoshal, 1995; Zanfei, 2000). Moving from the idea that increasing returns are essentially a regional and local phenomenon arising from regional economic agglomeration and specialisation (Krugman, 1991a,b), different approaches[4] emphasising the role of local *spatial areas* for the purpose of global competitiveness, have flourished in recent economic theory. In analysing MNCs' internationalisation strategy, it emerges clearly that multinationals target local *spatial areas* where they can enjoy externalities and knowledge spillovers as well as corporate control and a dynamic environment. This promotes, in turn, a ranking of the attractiveness of territorial units, which consequently engage in competition to attract foreign direct investment (FDI) as well as to encourage a concentration of

[4] For a critical overview see Boschma and Lambooy (1999) and Martin (1999).

corporate activity in "higher" order territorial units. In this context, the distinctive technological specialisation of regional systems of innovations (RIS) plays a great role, as corporate activity is more and more sensitive to the conditions of localised cumulative processes. Due to the multinationals' aim to outsource knowledge by tapping into local expertise, the historical heritage of RIS (understood in terms of technological capabilities) becomes a main factor driving MNCs' location strategy across national borders. Following Lawson's (1999) competence perspective on the regional systems, like firms regions can be regarded as ensembles of competencies that emerge from, but are not reducible to, social interaction.

Therefore, the sectoral and spatial composition of corporate R&D impacts on the firm's strategic goal of competence accumulation. As the issue of the extent of co-location and co-specialisation with other firms attains to inter-company relationships, significant implications arise for inter-firm strategic dynamics which affect the pattern and degree of competence accumulation over time. The distinction between strong and loose co-specialisation between two (or more) companies helps to drive their spatial interaction in terms of either co-location or non-co-location, as well as to shape their corporate strategies with respect to one another in terms of either competition or cooperation respectively. Corporate competitiveness, understood as the outcome of a process of competence accumulation, can be derived from either competitive or cooperative linkages, the balance between which is dictated by the degree of overlap in firms' specialisation profiles and reflected in their spatial strategies. In this sense, the relationship between co-location and co-specialisation and its implications for corporate strategy can be summarised in Figure 1, as a framework for analysis. Figure 1 plots the "degree of co-location"[5] on the y-axis and the "balance between inter-firm competition and cooperation" on the x-axis. The former variable runs from no co-location to co-location, while the latter variable runs from high corporate competition to high corporate cooperation. Therefore, the closer to the origin, the greater the extent of no co-location (and the intensity of competition) between companies, while the further away from the origin the greater is the degree of co-location (and the intensity of inter-firm cooperation). The hypothesised relationship between the two variables is shown by the downward sloping trend line in Figure 1.

This implies that if firms do not co-locate their research in a given field/s, the reason may be found in the strong co-specialisation of their technological profiles in the field/s in question which renders them principally as competitors to one another. The strong overlapping of the capabilities of two (or more) firms in a particular field/s acts as a deterrent to the common development of their related R&D in the same location. In this sense, few complementarities can be enjoyed by firms as they are competing around an equivalent spectrum of activity. Conversely, a pattern of co-location of R&D activity in a particular field/s is likely to be associated with a loose co-specialisation between two (or more) firms in the field/s in question. The lack of a complete technological overlap (and rather the presence of some areas of expertise that distinguish one firm from another) creates the conditions

[5] The variable refers to the co-location of R&D activity carried out in fields of (either close or loose) corporate co-specialisation.

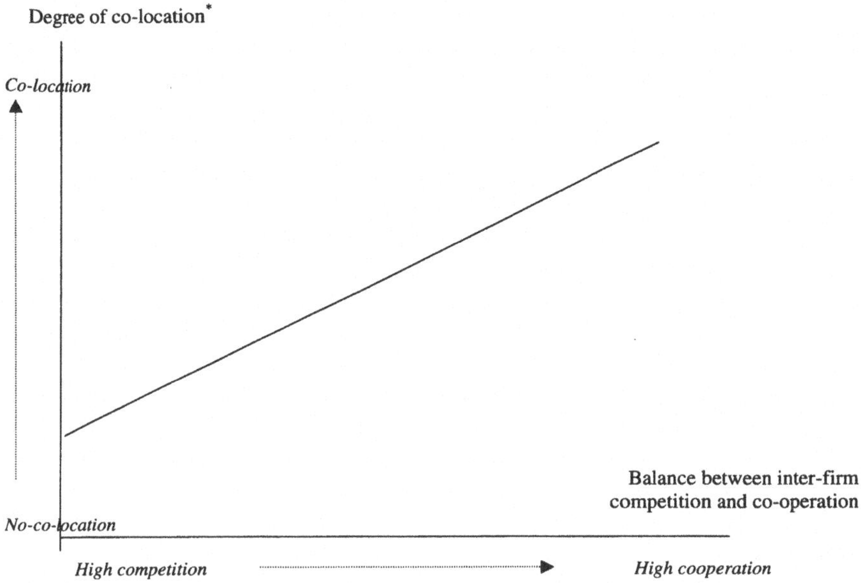

* The variable refers of the co-location of R&D activity carried out in fields of (either close or loose) corporate co-specialisation

Fig. 1. The theoretical framework

for inter-firm cooperation, through which complementarities can be enjoyed and competencies related to the firm's own technological profile can be absorbed from the external environment into its corporate technological path.

3 Some notes on the data

The data, upon which the empirical evidence provided in this paper is based, concern US patents granted to the world's largest firms, drawn from a database held at the University of Reading. For the purpose of the present paper, it is worth noting that the patent document records (among other useful information) the address of both the inventor and owner (assignee) of the invention. In other words, the location in which the R&D activity was carried out is provided by the patent records, and can be separated from the location of the headquarters of the parent company to which the patent is ultimately assigned, which can be inferred through the corporate consolidation of patents.[6]

At least two major advantages are associated with the use of these data. First, the data refer to patenting in a common third country – i.e. the US – which allows

[6] Mergers and acquisitions are largely recognised in the data through the practice in most groups of centralising the patent application procedure in the parent company. In other important cases affecting the ultimate ownership of significant numbers of patents, the change in ownership structure has been incorporated into the organisation of the data, which involves in some cases the creation of a new corporate group and, in others, the expanded consolidation of groups with newly acquired subsidiaries.

a more reliable international comparison on a commonly imposed standard and under a common legal framework. Second, foreign-owned patents (in this instance European-owned patents due to European-located research facilities) are expected to be of a higher average quality than domestic patents (i.e. US-owned from US-located facilities) as it is reasonable to assume that only patents that have survived preliminary testing in the home or host country will be extended abroad. The wider advantages and disadvantages of the statistical use of patent data have been covered extensively in the literature (see Pavitt, 1985; Griliches, 1990), and will not be discussed further here. However, in an analysis of ICT a drawback in the use of patent data lies in the fact that software innovations have started to be patented in a major way only since the mid-1990s. Nonetheless, for the purpose of the present study, the significance of this omission from the range of ICT fields included is somewhat limited by the weak European performance in software technology by comparison with their US competitors (Malerba et al., 1997).

Going further into the detail of the structure of the Reading database, each patent is classified by the type of technological activity with which it is primarily associated. The 399 original classes identified in this way by the US Patent and Trademark Office can be collected together into 56 technological sectors, of which 6 sectors comprise the main fields of ICT.[7] The data used refer to all 23 European-owned firms classified in the broad electrical corporate industrial group among the world's 784 largest firms. For the purpose of this paper, the empirical analysis was carried out at the level of the original 399 patent classes by selecting those 30 classes which make up the 6 ICT technological sectors.[8] The broad electrical[9] corporate industrial group includes the electrical equipment (communications) and the office equipment (computing) industries.[10] In this context, the corporate geographical distribution of ICT technological development is investigated across Europe over the period 1969–1995.

The spatial analysis of large corporate research activity in the European ICT industry is carried out at a sub-national level. The geographical distribution of European electronic corporate patenting activity in the ICT technological sectors is investigated across Belgium, Germany, France, Italy, the Netherlands, Sweden, Switzerland and the UK, for which regionalised patent data are available in the Reading database. For each of these countries, the sub-national entities identified correspond to territorial units as classified by the European Nomenclature of Territorial Units of Statistics (NUTS). In order to ensure as much comparability as possible, the NUTS 1 level is used to identify Belgian, Dutch, German and UK re-

[7] A list of the 6 ICT technological sectors is provided in the appendix (Table A1).

[8] A list of the ICT original technological patent classes is provided in the appendix (Tables A1a and A1b).

[9] Hereafter, this corporate industrial group will be named as "electronic" in order to take into account the technological historical development. The definition of "electrical" attains to a historical classification of the data.

[10] Some of the 23 electronic corporations were dropped from the analysis on the grounds of the relatively small number of patents in the technological patent classes considered, the outcome being that the firms in the sample vary from one sub-period to another. As listed in Table A2, 20 firms were examined in the sub-period 1969–1977, whilst 21 and 19 were taken into account in the two later sub-periods respectively.

gions, while as far as French, Italian and Swedish regions are concerned, the NUTS 2 level is adopted. In the case of Switzerland, no NUTS subdivision is available for the Swiss territory as it is a non-European Union (EU) member. Therefore, in the Reading database, Switzerland is geographically subdivided in 12 regions according to proximity to big cities. As pointed out by Eurostat (1995) and Dunford (1996), despite the aim of ensuring that comparable regions appear at the same NUTS level, the same level of disaggregation in various countries still implies considerable differences between regions in terms of area, population, economic weight or administrative power, and so it is necessary to choose the most appropriate NUTS level in each case to reduce the effect of inter-country differences in the classification scheme. Thus, the 3 Belgian *régions*, the 4 Dutch *landsdelen*, the 22 French *régions*, the 16 German *länder*, the 20 Italian *regioni*, the 8 Swedish *riksområden* and the 11 UK *standard regions* seem to allow some comparability as far as innovative activity is concerned (see Table A3).[11]

In addition, the sub-national units identified host almost completely the entire bulk of research activity conducted within the European electronic industry in Europe as shown by Table 1. In Table 1, the European regions are arranged by national groups and ranked according to the share of research activity carried out by European electronic companies relative to Europe as a whole (as measured by the corporate patents sourced from such research facilities).

4 The regional hierarchy in the corporate development of ICT research activity

In order to provide some general background on the regional geography of corporate development of ICT research activity within the European electronic industry, we begin by drawing a picture of the spatial dynamics concerning the distribution of R&D laboratories across European regions. The aim is to gather some of the major elements which can help us to analyse whether technologically co-specialised groups of firms are co-located in the regions considered.

As shown in Table 1, the regional areas under analysis account for almost 98% of the entire ICT research activity carried out in Europe by the largest European-owned electronic companies.

Similarly, the figures in the Table provide us with some preliminary understanding of the geographical hierarchy in terms of the location of the corporate development of ICT research activity as shown by the top-ranked position of German, French, Dutch and UK regions. In order to have a more detailed overview of this trend, Table 2 ranks the regions considered by the percentage of European ICT technological activity located in each of them relative to Europe as a whole over the period 1969–1995.

The figures reveal a pattern of strong concentration in a few major centres of ICT corporate technological development in the industry in question. Almost

[11] These territorial levels of analysis have also been adopted by Cantwell and Iammarino (1998, 2000) in the case of Italy and the UK respectively, by Cantwell and Noonan (2002) in the case of Germany and by Cantwell and Santangelo (1999) in the case of Belgium, the Netherlands, Sweden and Switzerland.

Table 1. Distribution of US patents attributed in European electronic companies in all ICT technological sectors relative to Europe (%), by nationality of the region, 1969–1995

German regions	31.72%
French regions	21.44%
Dutch regions	19.78%
UK regions	14.47%
Swedish regions	4.45%
Italian regions	2.77%
Swiss regions	1.96%
Belgian regions	1.14%
Total European regions	**97.73%**

70% of the research activity European-owned electronic companies conducted in Europe is due to facilities in six regions: Bayern, Ile de France, Zuid-Nederland, the South-East (UK), Baden Württemberg and Stockholm. Therefore, the pattern of a strong concentration of R&D in Europe, suggested by Caniël's findings (2000) is also confirmed in the context of the European electronic industry. It is interesting to mention that the German, Dutch, Swedish and UK regions identified maintain their top position in the geographical hierarchy by comparison with the results obtained in previous empirical studies on this industry, focusing on an either small or different span of European regions (Cantwell and Santangelo, 2002; Santangelo, 2000). This suggests that the inclusion of new regional data now available does not undermine the strength of these top-ranked regional units overall.

The six locations top-ranked in Table 2 hold also a leading position as major *spatial areas* for ICT knowledge creation across the ICT technological sectors considered individually as illustrated in Table 3, in which all the regional locations considered are ranked by the size of the European-owned research in each of the ICT technological sectors relative to Europe as a whole.

From a glance at the figures, it emerges clearly that the six regional centres are at the top of the hierarchy in each of the technological fields analysed, allowing for some variance in their ranking across sectors. Nonetheless, a few exceptions should be mentioned as illustrated by the case of Zuid-Nederland in "special radio systems", and Stockholm in "image and sound equipment" and "semiconductors". The two regions seem to lose position in these sectors in the overall ranking, although they do not slip down much. It should also be mentioned that the six regional locations identified can be labelled "higher" order regions as well in terms of production. Figure 2 illustrates their per capita GDP for 1994 by comparing them with the "Europe 15" total.

On average this cross-sectoral intra-industry analysis seems to provide further empirical support for the idea of "regional clubs" within Europe (Verspagen,

Table 2. Distribution of US patents attributed to European electronic companies in all ICT technological sectors relative to Europe (%), by region, 1969–1995

Region		Region	
Bayern	20.49%	Luzern	0.13%
Ile de France	17.02%	Pays de la Loire	0.13%
Zuid-Nederland	16.53%	Yorkshire and Humberside	0.12%
South-East (UK)	9.90%	Bremen	0.11%
Baden-Württemberg	3.22%	Bourgogne	0.11%
Stockholm	2.52%	Wales	0.10%
Niedersachsen	1.78%	Franche-Comté	0.10%
Oost-Nederland	1.76%	Småland med öarna	0.10%
Schleswig-Holstein	1.59%	Sicily	0.09%
West-Nedeland	1.43%	Alsace	0.09%
Hamburg	1.36%	Emilia Romagna	0.09%
Lombardy	1.21%	Aquitaine	0.08%
Nordrhein-Westfalen	1.21%	Poitou-Charentes	0.07%
Piedmont	1.20%	Nord-Pas-de-Calais	0.06%
Rhône-Alpes	1.10%	Noord-Nederland	0.06%
South West (UK)	1.03%	Thüringen	0.06%
Hessen	1.00%	Brandenburg	0.05%
North West (UK)	0.96%	North (UK)	0.05%
Basel	0.94%	Fruli-Venezia Giulia	0.05%
Vlaams Gewest	0.85%	Lazio	0.05%
East Midlands	0.80%	Région Wallonne	0.05%
Bretagne	0.74%	Mecklenburg-Vorpommern	0.04%
Östra Mellansverige	0.74%	Midi-Pyénées	0.04%
Basse-Normandie	0.71%	Champagne-Ardenne	0.04%
West Midlands	0.68%	Sachsen-Anhalt	0.03%
East Anglia	0.67%	Toscana	0.03%
Provence-Alpes-Côte d'Azur	0.64%	Picardie	0.03%
Zürich	0.60%	Övre Norrland	0.03%
Berlin	0.45%	Sachsen	0.02%
Västsverige	0.39%	Lorraine	0.02%
Sydesveige	0.38%	Bern	0.02%
Rheinland-Pfalz	0.31%	Fribourg	0.02%
Norra Mellnsverige	0.30%	Calabria	0.02%
Centre, Bassin Parisien	0.29%	Limousin	0.02%
Bruxelles	0.24%	Liguria	0.01%
St. Gallen	0.24%	Campania	0.01%
Scotland	0.15%	Languedoc-Roussillon	0.01%
Haute-Normandie	0.15%	**Total 97,73%**	

Table 3. Distribution of US patents attributed to European electronic companies in each of the ICT technological sectors relative to Europe (%), by region and ICT sector, 1969–1995

Telecommunications		Other electrical communications systems		Special radio systems	
Bayern	27.16%	Bayern	22.13%	Ile de France	40.76%
Ile de France	16.60%	Ile de France	14.92%	South-East (UK)	23.85%
Zuid-Nederland	9.03%	South-East (UK)	14.75%	Bayern	9.07%
South-East (UK)	8.07%	Zuid-Nederland	12.49%	Oost-Nederland	7.95%
Stockholm	5.66%	Baden-Württemberg	2.85%	Stockholm	1.90%
West Nederland	3.22%	Niedersachsen	2.60%	Baden-Württemberg	1.79%
Baden-Württemberg	3.09%	North West (UK)	2.10%	Västsverige	1.79%
Niedersachsen	2.09%	Piedmont	1.93%	East Midlands	1.34%
South West (UK)	1.88%	West Nederland	1.76%	South West (UK)	1.34%
Vlaams Gewest	1.66%	Stockholm	1.76%	Zürich	1.01%
Bretagne	1.59%	Östra Mellansverige	1.59%	Basel	1.01%
Östra Mellansverige	1.31%	Hamburg	1.42%	Schleswig-Holstein	0.56%
Nordrhein-Westfalen	1.22%	Schleswig-Holstein	1.26%	East Anglia	0.56%
East Anglia	1.06%	West Midlands	1.01%	West Nederland	0.56%
West Midlands	0.97%	Bretagne	1.01%	Sydsverige	0.56%
Sydsverige	0.94%	Basel	1.01%	Nordrhein-Westfalen	0.45%
East Midlands	0.88%	Nordrhein-Westfalen	0.92%	Haute-Normandie	0.45%
Zürich	0.81%	Rhône-Alpes	0.92%	Niedersachsen	0.34%
Hamburg	0.75%	East Midlands	0.84%	Zuid-Nederland	0.34%
Piedmont	0.75%	South West (UK)	0.84%	Östra Mellansverige	0.34%
Basel	0.75%	Hessen	0.67%	West Midlands	0.22%
Hessen	0.72%	Oost-Nederland	0.67%	North West (UK)	0.22%
North West (UK)	0.72%	Zürich	0.67%	Wales	0.22%
Västsverige	0.59%	Västsverige	0.59%	Scotland	0.22%
Norra Mellansverige	0.59%	Berlin	0.50%	Picardie	0.22%
Lombardy	0.56%	East Anglia	0.50%	Nord-Pas-de-Calais	0.22%
Provence-Alpes-Côte d'Azur	0.56%	Scotland	0.50%	Bretagne	0.22%
Rheinland-Pfalz	0.53%	St. Gallen	0.50%	Poitou-Charentes	0.22%
Bruxelles	0.50%	Bremen	0.42%	Aquitaine	0.22%
Schleswig-Holstein	0.47%	Norra Mellansverige	0.42%	Brandenburg	0.11%
Rhône-Alpes	0.47%	Yorkshire and Humberside	0.34%	Bremen	0.11%
Oost-Nederland	0.38%	Lombardy	0.34%	Hessen	0.11%
Centre, Bassin Parisien	0.25%	Aquitaine	0.34%	Thüringen	0.11%
Wales	0.22%	Rheinland-Pfalz	0.25%	Yorkshire and Humberside	0.11%
Småland med öarna	0.22%	Thüringen	0.25%	Liguria	0.11%
Yorkshire and Humberside	0.16%	Centre, Bassin Parisien	0.25%	Lombardy	0.11%
Haute-Normandie	0.16%	Basse-Normandie	0.25%	Centre, Bassin Parisien	0.11%
Basse-Normandie	0.16%	Franche-Comté	0.25%	Bourgogne	0.11%
Alsace	0.16%	Noord-Nederland	0.25%	Alsace	0.11%
St. Gallen	0.16%	Vlaams Gewest	0.25%	Pays de la Loire	0.11%

Table 3 (continued)

Telecommunications		Other electrical communications systems		Special radio systems	
Bremen	0.13%	Sydsverige	0.25%	Midi-Pyrénées	0.11%
North (UK)	0.13%	Haute-Normandie	0.17%	Bern	0.11%
Scotland	0.13%	Pays de la Loire	0.17%	Småland med öarna	0.11%
Pays de la Loire	0.13%	Provence-Alpes-Côte d'Azur	0.17%		
Region Wallonne	0.13%	Brandenburg	0.08%		
Champagne -Ardenne	0.13%	Wales	0.08%		
Lazio	0.09%	Friuli-Venezia Giulia	0.08%		
Sachsen-Anhalt	0.06%	Emilia-Romagna	0.08%		
Emilia-Romagna	0.06%	Calabria	0.08%		
Bourgogne	0.06%	Picardie	0.08%		
Nord-Pas-de-Calais	0.06%	Nord-Pas-de-Calais	0.08%		
Lorraine	0.06%	Lorraine	0.08%		
Poitou-Charentes	0.06%	Bruxelles	0.08%		
Limousin	0.06%				
Övre Norrland	0.06%				
Brandenburg	0.03%				
Mecklenburg-Vorpommern	0.03%				
Franche-Comté	0.03%				
Midi-Pyrénées	0.03%				
Noord-Nederland	0.03%				
Bern	0.03%				
Luzern	0.03%				
Total	**98.62%**		**97.82%**		**99.55%**
Zuid-Nederland	30.28%	Bayern	27.11%	Zuid-Nederland	22.05%
Ile de France	14.58%	Zuid-Nederland	15.33%	Bayern	17.56%
Bayern	12.35%	Ile de France	11.26%	Ile de France	16.48%
South-East (UK)	9.72%	South-East (UK)	8.72%	South-East (UK)	7.30%
Schleswig-Holstein	6.15%	Baden-Württemberg	3.65%	Baden-Württemberg	3.67%
Hamburg	3.24%	Basse-Normandie	3.65%	Piedmont	2.35%
Baden-Württemberg	3.02%	Oost-Nederland	3.07%	Lombardy	1.87%
Niedersachsen	2.96%	Lombardy	2.96%	Stockholm	1.82%
Rhône-Alpes	2.12%	Basel	2.75%	Niedersachsen	1.71%
Oost-Nederland	1.34%	Rhône-Alpes	1.69%	Hessen	1.64%
Nordrhein-Westfalen	1.28%	Hamburg	1.27%	Nordrhein-Westfalen	1.45%
East Anglia	1.23%	Nordrhein-Westfalen	1.16%	Oost-Nederland	1.37%
Bretagne	1.17%	East Midlands	1.06%	Hamburg	1.32%
Hessen	1.01%	Provence-Alpes-Côte d'Azur	1.00%	North West (UK)	1.32%
Stockholm	0.89%	North West (UK)	0.95%	Rhône-Alpes	1.16%
Piedmont	0.84%	Centre, Bassin Parisien	0.95%	Berlin	1.13%
West Nederland	0.61%	Schleswig-Holstein	0.90%	Schleswig-Holstein	1.08%
Vlaams Gewest	0.50%	Stockholm	0.90%	Vlaams Gewest	1.00%
Provence-Alpes-Côte d'Azur	0.39%	Hessen	0.79%	South West (UK)	0.95%
Rheinland-Pfalz	0.34%	West Midlands	0.74%	Provence-Alpes-Côte d'Azur	0.95%
Zürich	0.34%	St. Gallen	0.74%	West Nederland	0.95%

Table 3 (continued)

Image and sound equipment		Semiconductors		Office equipment and data processing systems	
East Midlands	0.28%	Luzern	0.74%	East Midlands	0.71%
South West (UK)	0.28%	Sicily	0.63%	West Midlands	0,66%
Lombardy	0.28%	Rheinland-Pfalz	0.53%	Östra Mellansverige	0.50%
Alsace	0.28%	Zürich	0.53%	Basel	0.47%
Basel	0.28%	Östra Mellansverige	0.53%	Zürich	0.45%
Mecklenburg-Vorpommern	0.22%	East Anglia	0.48%	Basse-Normandie	0.34%
North West (UK)	0.22%	South West (UK)	0.48%	East Anglia	0.26%
Haute-Normandie	0.22%	Norra Mellansverige	0.48%	Bruxelles	0.24%
Berlin	0.17%	Niedersachsen	0.42%	Franche-Comté	0.21%
Brandenburg	0.17%	West Nederland	0.32%	Bretagne	0,21%
Bremen	0.17%	Vlaams Gewest	0.32%	Scotland	0,16%
West Midlands	0.17%	Berlin	0.26%	Sydsverige	0.16%
Wales	0.17%	Bourgogne	0.21%	Emilia-Romagna	0.13%
Friuli-Venezia Giulia	0.17%	Pays de la Loire	0.21%	Centre, Bassin Parisien	0.13%
Bruxelles	0.17%	Västsverige	0.21%	Bourgogne	0.13%
St. Gallen	0.17%	Emilia-Romagna	0.16%	Pays de la Loire	0.13%
Yorkshire and Humberside	0.11%	Sydsverige	0.16%	Norra Mellansverige	0.13%
Lazio	0.11%	Piedmont	0.11%	Rheinland-Pfalz	0.11%
Centre, Bassin Parisien	0.11%	Toscana	0.11%	Småland med öarna	0.11%
Bourgogne	0.11%	Poitou-Charentes	0.11%	Thüringen	0.08%
Västsverige	0.11%	Noord-Nederland	0.11%	Yorkshire and Humberside	0.08%
Sachsen	0.06%	Bruxelles	0.11%	Friuli-Venezia Giulia	0.08%
Thüringen	0.06%	Brandenburg	0.05%	Haute-Normandie	0.08%
Scotland	0.06%	Sachsen	0.05%	St. Gallen	0.08%
Toscana	0.06%	North (UK)	0.05%	Sachsen-Anhalt	0.05%
Poitou-Charentes	0.06%	Campania	0.05%	North (UK)	0.05%
Aquitaine	0.06%	Calabria	0.05%	Lazio	0.05%
Midi-Pyrénées	0.06%	Haute-Normandie	0.05%	Nord-Pas-de-Calais	0.05%
Region Wallonne	0.06%	Nord-Pas-de-Calais	0.05%	Poitou-Charentes	0.05%
Luzern	0.06%	Franche-Comté	0.05%	Aquitaine	0.05%
Östra Mellansverige	0.06%	Aquitaine	0.05%	Noord-Nederland	0.05%
Småland med öarna	0.06%	Midi-Pyrénées	0.05%	Fribourg	0.05%
Sydsverige	0.06%	Languedoc-Roussillon	0.05%	Västsverige	0.05%
Övre Norrland	0.06%	Fribourg	0.05%	Bremen	0.03%
		Sachsen	0.03%		
		Toscana	0.03%		
		Champagne-Ardenne	0.03%		
		Picardie	0.03%		
		Alsace	0.03%		
		Midi-Pyrénées	0.03%		
		Region Wallonne	0.03%		
		Bern	0.03%		
		Luzern	0.03%		
		Övre Norrland	0.03%		
Total	**99.05%**		**98.47%**		**95.54%**

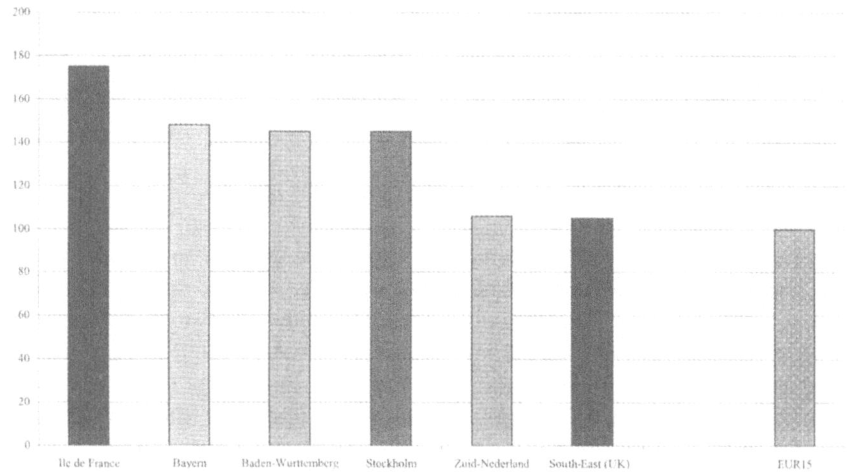

Fig. 2. Per capita GDP (ECU) – 1994, by "higher" order region

1997). If the geographical ranking across sectors is rather stable at the top of the hierarchy, stability also characterises to some extent lower ranked regions across ICT sectors. This stability over time may be interpreted as a result of cumulative causation mechanisms promoting *vicious* and *virtuous* circles and reinforcing geographical inequalities. A strong path-dependent character is found in the local ability to develop an entrepreneurial environment through expertise accumulation and, consequently, to attract the quality investments of MNCs (Metcalfe, 1996). In this sense, the global-local nexus shapes both the geographical distribution of corporate activity and the hierarchies of local areas. Thus, local (regional) systems of innovation do not undermine the globalisation process in terms of the production and diffusion of technology, but rather they reinforce it (Howells, 1999). Nonetheless, despite the stability of "higher" and "lower" order regions, there is still room for policy in infrastructure and education in order to facilitate the process of local growth through FDI strategies (Cantwell and Iammarino, 2000). On the host region perspective, Vence-Deza (1996) places great emphasis on the importance of diversity and complementarity in boosting local expertise in lagging regions.

5 Co-specialisation and co-location

By considering a span of regions wider than in the earlier work mentioned above, in order to analyse whether in each of the three sub-periods – 1969–1977, 1978–1986 and 1987–1995 – the technological clusters of co-specialised firms identified in the study referred to above (Santangelo, 1998) are co-located, for each firm a cross-regional RTA index was calculated at the level of the ICT patent classes[12] as a proxy for the geographical division of labour within the firm. For each European electronic firm (i), the index is defined as the share of its US patents granted in a

[12] A selection of ICT patent classes is considered in this paper. The analysis focuses on the patent classes labelling the technological clusters in Figure 3 in each of the sub-periods (1969–1977, 1978–1986 and 1987–1995) rather than all ICT patent classes recorded in the Reading database.

patent class (c) in all European regions which are attributable to research in region (r), relative to its share of patents granted in all ICT patent classes over all European regions which is due to its activity in the same region (r). Thus, the index can be formalised as:

$$RTA_{irc} = \left(P_{irc} / \sum_{r} P_{irc} \right) / \left(\sum_{c} P_{irc} / \sum_{r} \sum_{c} P_{irc} \right) \tag{1}$$

where P_{irc} is the total number of patents granted to firm (i) in region (r) in a patent class (c). It is worth emphasising that while we confine our attention here to just some regions, the RTA index has been calculated relative to activity in all European regions. As the index is a comparative measure, high (low) values of RTA_{irc} indicate corporate advantage (disadvantage) in locating research activity in a specific patent class in the region in question. Therefore, the index enables one to evaluate for each European electronic firm the significance of the regional location in a patent class in Europe relative to the overall European significance of the same region in all ICT patent classes. As such, it maps the profile of corporate technological specialisation across the ICT fields in each region.

The analysis tests whether co-specialised firms co-locate their research activity in the technological fields of co-specialisation. As in the previous study (Santangelo, 2000), the criterion adopted to identify eventual co-location of corporate co-specialised research in ICT requires that, in each of the three sub-periods, at least 50% of the firms in each technological cluster show the highest RTA_{irc} value in the relevant patent class (around which the cluster is formed) in the same regional location, and that this regional location is common to all other firms in the cluster (which all conduct at least some activity there).

Figure 3 depicts the clusters of ICT European-owned corporate groups which were found to be specialised in the same technological classes (see Santangelo, 1998). In the present study, technological clusters of firms which were found to be also co-located are reported through the use of a grey background. By comparison with the earlier work conducted on a small number of European regions (Santangelo, 2000), the number of co-located groups of co-specialised firms is reduced to cluster B2.2, still hosted in the South-East (UK). Conversely, the co-location of the clusters C, C1 and C2, in the South-East (UK) has not been confirmed by the present analysis.

Therefore, despite the high concentration of European ICT technological development revealed in Tables 2 and 3, the average trend emerging from our earlier study that European electronic firms specialised in the same patent class/es carry out the related R&D in geographically separate regional locations is confirmed and further reinforced by the present analysis, based on a wider geographical span of European regions. Given the stability of the regional hierarchy discussed in the previous section, these results seem to suggest that, although firms do not locate R&D activity in fields of co-specialisation in the same region, on average the production of ICT innovative activity is concentrated in a small number of regional centres. Therefore, if R&D in ICT as a whole is concentrated in a few regions, ICT fields of corporate co-specialisation follow a pattern of geographical dispersion.

As illustrated elsewhere by both of us (Cantwell and Santangelo, 1999, 2000; Cantwell and Kosmopoulou, 2002), MNCs are likely to disperse geographically the

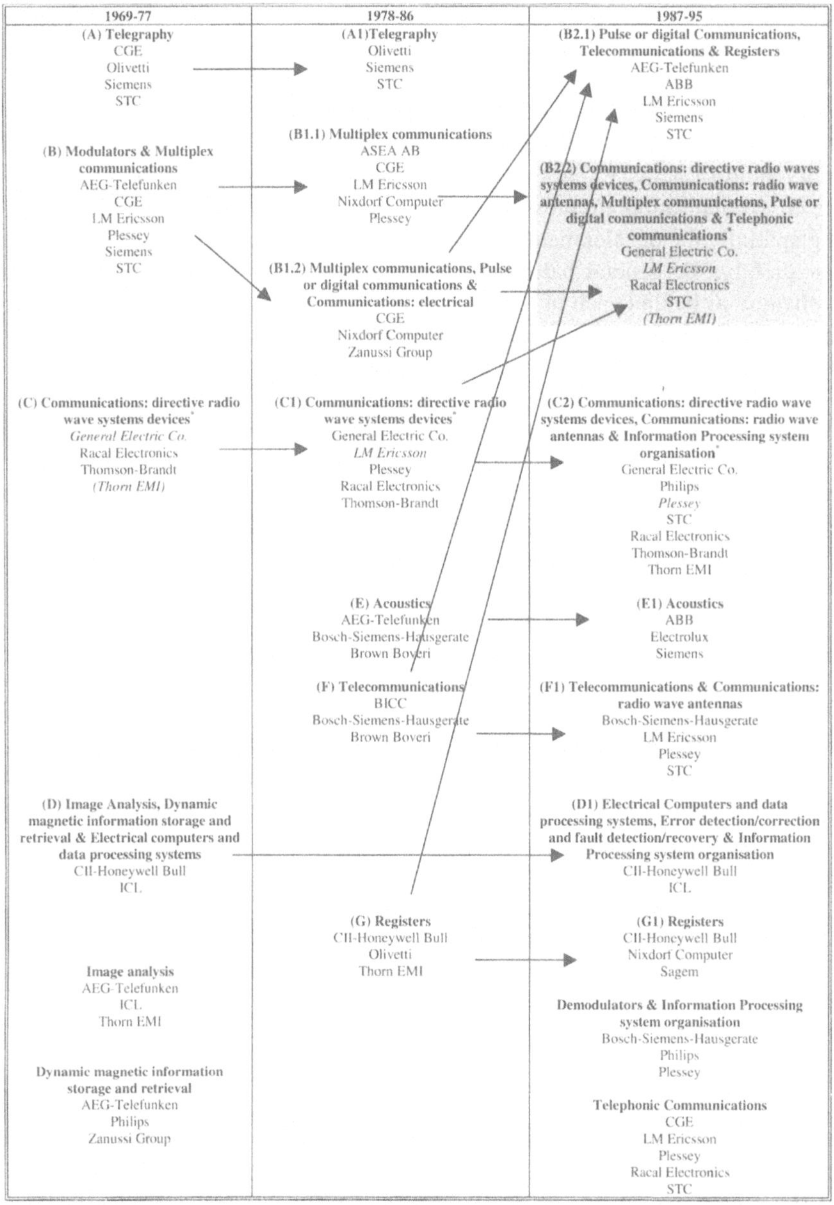

* Italics denote consideration of sector-specific factors, which create correlation problems between each of the companies, whose name is in italic, and all the others and between each others in the case of the 1969–1977 cluster

Fig. 3. European electrical firms specialised in the same patent class/es and located in the same region (reported in grey background), by sub-period, 1969–1995

development of technologies which lie mainly outside their core fields of competence, as the creation of technology in which tacit knowledge carries a greater weight is harder to coordinate across long distances. Therefore, firms tend to concentrate at home industry-specific core technologies in which they are heavily dependent on tacit knowledge, and to disperse the development of technologies outside their primary field of activity. This implies that firms (or groups of firms) that show loose co-specialisation are most likely to cooperate in fringe sectors, while separating spatially the development of core technologies owing to mutual deterrence. Consequently, the general pattern emerging can be defined as an inter-regional interchange, as behind each of these regions there are companies which focus regional technological specialisation in some fields rather than others. In these cases, co-location is deterred by region-specific capabilities.

The relationship between regions follows trajectories of either cooperation or competition. In the latter case, very close co-specialisation generates competition, which in turn leads to an absence of corporate co-location. Companies strong in the same field/s do not co-locate their most related R&D activity because of mutual deterrence. Since in this case indigenous firms and regions are technologically co-identified, a dynamic of inter-regional interchange develops by emphasising the relationship between regional corporate clusters. In the former case, weaker co-specialisation generates cooperation, which may lead to a pattern of spatial co-location. Thus, in this event firms need not be synonymous with regions, and regions can gain an identity which goes beyond that of indigenous firms. This applies to the exception to the general trends found in the empirical analysis and represented by the cluster B2.2 hosted in South-East England (UK). In this case, sectoral and spatial corporate interactions follow a pattern of intra-regional cooperation between co-specialised firms in a common locational pole of attraction. In this special case, the common location – i.e. the South-East (UK) – does gain an identity of its own with respect to localised interaction between indigenous and foreign European-owned firms as a major centre for the development of the technologies of common specialisation. Nonetheless, as far as the South-East (UK) is concerned, it should be also taken into account the presence in the region of a world's major financial centre when assessing corporate location strategy.

6 The theoretical framework revisited

The results of the empirical analysis allow us to elaborate upon the theoretical framework proposed in Section 2 and sketched in Figure 1. Graphically this has been done in Figure 4, in which clusters (illustrating the relationship between the "degree of co-location" and the "balance between inter-firm competition and cooperation") have been mapped by taking into account the patterns of sectoral and spatial corporate interaction corresponding to the two antithetical forms of inter-firm relationship (competition and cooperation).

As discussed above, competitive corporate relationships, dictated by a strong technological co-specialisation, are likely to promote inter-regional competition in equivalent lines of technological development as discussed in the case of the general trend revealed by the empirical analysis. Conversely, cooperative corporate

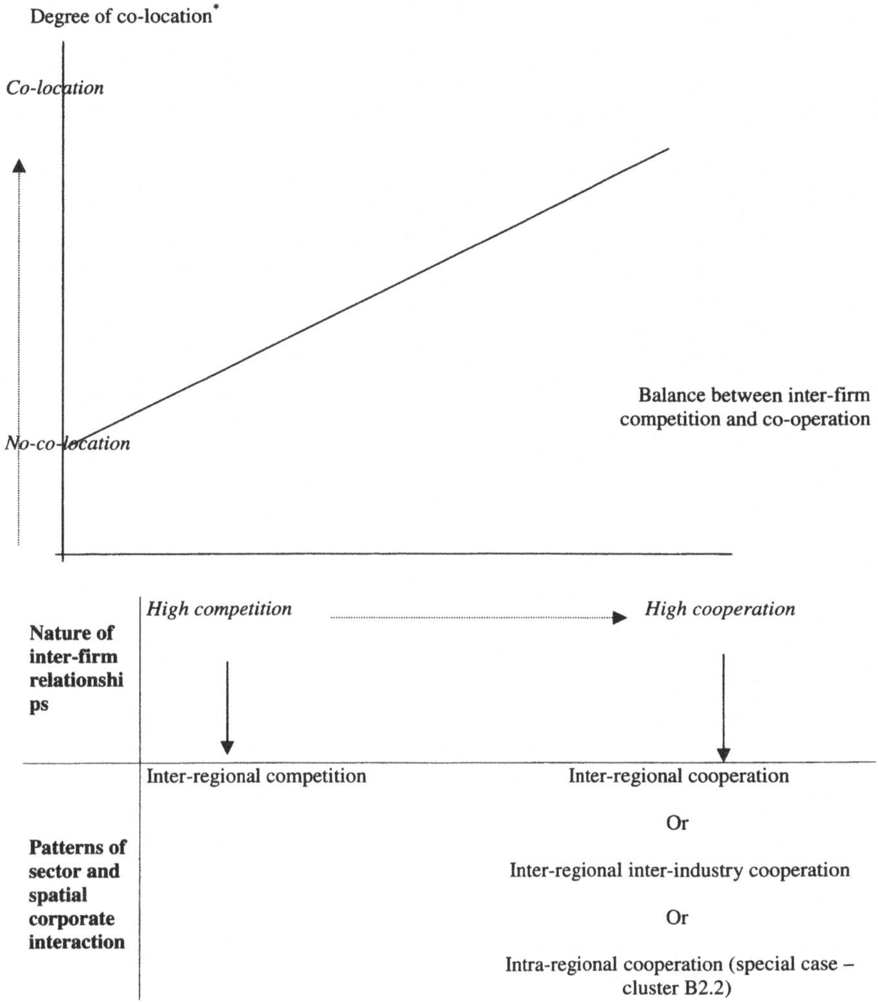

* The variable refers to the co-location of R&D activity carried out in fields of (either close or loose) corporate co-specialisation

Fig. 4. The theoretical framework revisted

relationships, generated by loose technological co-specialisation, may lead conceptually to three alternative potential patterns of sectoral and spatial corporate interactions:

- inter-regional cooperation,
- inter-regional inter-industry cooperation and
- intra-regional cooperation.

An example of the latter has been identified in the cluster B2.2 which is co-located in the South-East (UK), the first can be conceptually understood as the counterpart of the competitive pattern of sectoral and spatial corporate interactions discussed as the general trend captured by the empirical analysis in Section 5. Its distinctive feature lies in the fact that non-co-specialised firms locally cooperate in technical development in fields outside the primary sector of one of the (two or more) cooperating firms. The second alternative conceptually identified concerns sectoral and spatial corporate interactions characterised by a concentration of research in specific regional locations to benefit from untraded externalities, which may well go beyond intra-industry knowledge spillovers as in the Marshall-Arrow-Romer (MAR) model, involving inter-industry and industry-university knowledge spillovers (Feldman, 1993; Audretsch and Feldman, 1996; Audretsch and Stephan, 1996; Anselin et al., 1997; Audretsch, 2000). Therefore, what is proposed is a different concept of agglomeration by comparison with other streams of literature. Unlike Baptista and Swann (1998), who focus on intra-industry agglomeration and knowledge spillovers, we adopt a different level of analysis by looking at inter-industry knowledge flows across European regions.

In order to test whether non-electronic firms locate R&D in relevant "centres of excellence" (i.e. the six selected regions) to enjoy knowledge spillovers from the local environment as well as from indigenous electronic companies, in each of the six selected regions we calculate a RTA index across ICT patent classes. For both European-owned electronic and all non-electronic firms (k=1 or 2), the RTA index is calculated as the share of research carried out in a particular ICT class (c) in all European regions which is attributable to research in region (r), relative to their share of activity in all ICT classes over all European regions which is due to activity in the same region. Thus, for the whole period 1969-95 the index can be formalised as follows:

$$RTA_{krc} = \left(P_{krc} / \sum_{r} P_{krc} \right) / \left(\sum_{c} P_{krc} / \sum_{rc} P_{krc} \right) \qquad (2)$$

where P_{krc} is the total number of patents granted to either European-owned electronic or all non-electronic firms (k) in region (r) in a patent class (c). It is worth emphasising that while our attention is confined here to just some regions, the RTA index has been calculated relative to activity in all European regions. In each of the six selected regions for both European electronic firms and non-electronic firms the figures are reported in Table 4.

At a first glance at the figures, and noting the weak correlation coefficient with the exception of South-East (UK), electronic and non-electronic firms do not tend to co-locate the directly equivalent category of R&D across the whole spectrum of ICT fields, but in each case they tap into just a sub-set of local strengths. Non-electronic companies do not engage in direct competition for highly sector-specific resources with indigenous electronic firms. Rather, inter-industry cooperative relationships develop, amplified by the local strength in the fields of attraction and related areas, and in the dynamic local environment. In this sense, inter-firm interaction follows a regional inter-industry cooperation pattern, characterised by the absence of direct competition for sector-specific resources or (of course) in downstream output markets. In this context, the fact that the only correlation was found in the South-East

(UK) provides further evidence of the broader intra-regional cooperation pattern in that region, which stands out as a more distinctive all-round 'centre of excellence'.

As illustrated in Table 4, for each distribution we also compute the standard deviation and the coefficient of variation (which measure the degree of concentration of technological specialisation across fields). In all six regions, the degree of concentration is greater for non-ICT firms than for European ICT firms. This reflects the fact that, as expected, European ICT firms have on average a much wider breath of specialisation than do non-ICT companies across the ICT fields in each of the selected regions. In each of the six selected regions, non-electronic companies have a much higher degree of focus in what they do within the ICT fields (which for them lie outside their fields of primary activity) than do European-owned electronic firms. However, the fact that despite this the non-electronic companies still conduct quite a wide dispersion of ICT research in these centres supports the argument that research activity in firm- or industry-non-core fields (which ICT are for the non-electronic companies) is geographically dispersed and often localised in "centres of excellence" (such as the six regions identified in Tables 2 and 3). Non-electronic companies need to develop some expertise in ICT (although it lies outside their core technologies) as this is the pervasive technology of the new paradigm. Therefore, they locate a significant part of their R&D effort in the relevant "higher" order regions for these activities in order to enjoy knowledge spillovers from the local environment including the activities of indigenous electronic companies. As Feldman and Audretsch (1999) argue, overlapping diversity between complementary industries showing a common base is a source of greater innovation potential. Following Camagni (1988), this seems to be all the more true in the current techno-socio-economic paradigm in which the creation of spatial synergies is amplified by the new complexity of technological combinations adopted. Innovation potential is enhanced by spillovers to and from firms that have their principal fields of specialisation in other technologies and which are not deterred so much by the presence of strong electronic indigenous companies.

The analysis goes beyond the calculation of simple correlation coefficients between the RTA distributions in each region of European electronic firms and non-electronic firms, by assessing the co-presence or absence of a matching degree of specialisation between the two groups of firms in each patent class. Drawing on the measure of intra-industry trade across sectors used in international trade literature (Grubel and Lloyd, 1975), in each of the six selected regions for both European electronic firms and non-electronic firms we compute the following measure:

$$C_{rc} = 2min(\mathrm{RTA}_{1rc}, \mathrm{RTA}_{2rc})/(\mathrm{RTA}_{1rc} + \mathrm{RTA}_{2rc}) \qquad (3)$$

where C stands for co-presence of a matching degree of specialisation between the two groups of firms (k) (where k = 1 or 2) in a particular ICT patent class (c) in each region (r). This measures varies between 0 and 1, rising with the extent of co-presence.

In each region, we distinguish between ICT classes showing a co-presence of a matching degree of specialisation and ICT classes showing an absence of such. The distinction is made by dividing each regional distribution into values greater than 0.5 and those less than 0.5 (which is taken as a critical value): values above this critical value identified classes showing co-presence, while values below identified

Table 4. RTA values of European-owned electronic and non-electronic companies, by patent class and selected region, 1969–1995

ICT patent classes	Selected regions											
	Baden-Württemberg		Bayern		South-East (UK)		Ile de France		Zuid-Nederland		Stockholm	
	EE	N-E	EE	N-E	EE	N-E	EE	N-E	EE	N-E	EE	N-E
Music	0.000	0.000	0.718	1.370	0.000	0.000	0.287	0.000	2.365	0.000	0.000	5.987
Telegraphy	0.399	0.649	***2.770***	***2.436***	0.585	2.488	0.642	1.895	0.156	0.000	0.198	0.000
Acoustics	0.778	0.314	1.104	0.884	0.759	0.421	0.147	2.430	0.758	3.347	***2.311***	***1.287***
Registers	0.855	0.834	0.741	1.566	1.391	0.320	2.102	0.731	0.167	0.000	***1.270***	***4.105***
Active solid state devices (before 1990)	0.333	3.982	1.192	0.996	0.615	0.509	0.358	0.000	0.975	0.000	0.440	0.000
Electrical transmission or interconnection systems	0.829	1.928	1.331	0.517	0.905	1.902	0.803	0.563	0.989	0.000	0.483	1.355
Electrical digital logic circuits	0.562	2.336	1.949	0.000	0.000	0.000	0.319	3.412	0.547	0.000	0.000	0.000
Demodulators	0.601	0.000	0.711	0.000	0.684	3.359	1.762	0.000	0.820	0.000	0.000	0.000
Modulators	2.723	0.000	0.552	0.000	***2.658***	***7.465***	***1.251***	***2.843***	0.531	0.000	1.541	0.000
Communications: electrical	0.427	1.090	***1.355***	***1.022***	1.190	0.607	0.916	0.870	0.509	0.000	***1.411***	***1.039***
Code data generation or conversion	0.783	0.174	0.942	1.145	***1.370***	***3.677***	0.870	0.764	1.068	0.000	0.582	0.715
Communications: directive radio wave systems devices	0.294	0.470	0.541	1.260	***2.393***	***1.931***	***2.207***	***1.176***	0.019	0.000	0.583	4.404
Communications: radio wave antennas	1.124	0.094	0.195	1.945	2.304	0.722	2.679	0.688	0.022	0.000	0.890	0.000
Selective visual display systems	0.963	1.752	0.152	4.385	4.231	0.000	0.182	1.706	1.689	0.000	0.477	0.000
Television	2.457	0.701	0.349	1.754	0.879	2.687	0.837	1.023	1.652	0.000	0.000	0.000
Active solid state devices (after 1990)	***1.593***	***1.790***	***1.309***	***1.061***	0.983	1.264	0.666	0.138	0.880	0.000	0.427	0.772
Pictorial communication: television	0.709	0.416	0.499	1.525	***1.154***	***1.160***	0.997	0.166	1.767	0.000	0.125	2.015

Table 4 (continued)

ICT patent classes	Selected regions											
	Baden-Württemberg		Bayern		South-East (UK)		Ile de France		Zuid-Nederland		Stockholm	
Dynamic magnetic information storage or retrieval	1.350	0.812	0.615	1.641	0.488	0.060	0.946	0.547	2.192	0.000	0.050	0.000
Electrical computers and data processing systems	0.719	1.762	1.000	0.911	***1.036***	***1.066***	1.137	0.800	0.640	2.504	1.373	0.963
Static information storage and retrieval	0.342	0.136	***1.526***	***3.569***	0.648	2.083	0.765	0.000	0.998	0.000	0.169	0.000
Communications, electrical:												
acoustic wave systems and devices	0.201	0.024	1.394	0.045	0.262	0.592	***1.711***	***3.398***	0.235	0.000	0.597	0.779
Dynamic information storage and retrieval	3.116	0.417	0.069	0.391	0.285	0.400	0.786	0.000	3.143	0.000	0.054	0.000
Multiplex communications	1.107	0.449	1.581	0.422	0.731	1.723	0.825	1.968	0.368	0.000	2.289	0.000
Error detection/correction												
and fault detection/recovery	***1.194***	***2.305***	1.177	0.577	1.004	0.884	0.941	0.449	0.756	16.384	1.675	0.000
Pulse or digital communications	0.602	0.220	1.001	0.620	0.859	1.268	***1.245***	***1.448***	0.903	0.000	1.973	0.000
Electrical pulse counters, pulse dividers or												
shift register circuits and Systems	0.372	1.864	1.204	0.700	0.666	1.668	0.528	0.544	1.415	0.000	0.184	0.000
Telephonic communications	0.811	0.701	1.493	0.000	0.581	0.000	0.686	1.706	0.401	0.000	3.482	0.000
Electrical audio signal processing system and devices	0.564	0.531	1.174	1.993	0.330	1.018	0.320	0.000	2.242	0.000	1.005	0.000
Image analysis	2.936	0.876	0.617	0.000	***1.847***	***1.120***	1.000	1.279	0.763	0.000	0.582	0.000
Information processing system organisation	0.733	2.655	0.832	0.332	0.954	1.018	1.470	0.517	0.514	0.000	1.161	0.000
Telecommunications	1.445	0.347	0.775	0.434	1.091	0.998	0.853	0.169	0.698	0.000	2.060	0.474
Total mean	***0.997***	**0.956**	**0.996**	**1.081**	**1.061**	**1.368**	**0.975**	**1.007**	**0.974**	**0.717**	**0.884**	**0.771**
Standard deviation	**0.801**	**0.966**	**0.561**	**1.027**	**0.867**	**1.484**	**0.605**	**0.993**	**0.755**	**3.000**	**0.872**	**1.474**
Coefficient of variation*	**0.803**	**1.011**	**0.564**	**0.950**	**0.818**	**1.085**	**0.620**	**0.986**	**0.775**	**4.182**	**0.986**	**1.912**
Correlation coeffcient	**−0.130**		**−0.084**		**0.258**		**−0.011**		**−0.075**		**−0.116**	

EE: European-owned firms; N-E: non-electronic firms.

Bold Italics denotes the matching cases of high specialisation between European-owned electronic and all non-electronic firms.

* The coefficient of variation (CV) is equal to σ_{RTA}/μ_{RTA}.

classes showing a lack of co-presence. For each region, Tables 5 and 6 illustrate the two cases respectively.

The picture emerging from Tables 5 and 6 seems to suggest that, although the overlapping fields of specialisation between European electronic firms and non-electronic firms do not drive on average the correlation between the two distributions in the six regions (as shown by the correlation coefficients in Table 4), there is still a proximity between their ICT specialisation profiles in the four largest regions of the six selected at a more detailed level of analysis, that is when we look at each patent class individually.

A majority of the ICT technological fields demonstrate a local co-presence of European electronic and non-electronic firms in Bayern, Baden-Württemberg, South-East (UK) and Ile de France, but this is not the case in Stockholm or Zuid Nederland. Thus, co-presence is greater in the largest regions but tends to be lacking in all others as shown by the C_{rc} regional means reported in Tables 5 and 6, which are greater than 0.5 for Bayern, Baden-Württemberg and South-East (UK) (again), and almost equal to 0.5 for Ile de France. This seems to suggest that in the "centres of excellence" identified non-electronic firms tend to replicate in most but not all cases the ICT specialisation of European-owned electronic firms. The reason might be found in the fact that the two corporate groups compete in different downstream output markets and, therefore, there is no scope for mutual deterrence in the creation of new knowledge in the fields under analysis. These fields are primary fields of activity for electronic companies, whilst they lie outside the core competencies of non-electronic firms, which, however, need to develop some ICT expertise. Non-electronic firms strategically pursue this goal by locating ICT research in centres of excellence where they can enjoy localised knowledge spillovers generated by electronic corporate specialists.

However, it is interesting to note that there seems to be some classes in which the co-presence of a matching degree of specialisation between European electronic firms and non-electronic firms is more consistently observed across regions. This is the case of 'acoustics', 'registers', 'electrical transmission or interconnection systems', 'communications electrical', 'electrical computers and data processing systems' and 'error detection correction and fault detection/recovery', in which both groups tend towards a common pattern of representation in terms of specialisation across regions. Therefore, these fields may be regarded as cases in which inter-firm co-location is not deterred by direct competition. Conversely, it is possible to identify classes showing an absence of a matching degree of specialisation (i.e. 'electrical digital logic circuits', 'demodulators', 'communications radio waves antennas' and 'selective visual display systems'). Thus, these are fields in which some degree of competition between the two groups for resources appears to exist, deterring the co-location of R&D by the two groups of firms.

7 Conclusions

The attempt to adopt a broader approach to the study of innovation is currently on the agenda due to a wider awareness of the advances that have recently been made from the perspective of evolutionary economics in our understanding of this essential feature of capitalism. The great emphasis placed on the firm as major actor in knowledge creation as well as on the local dimension – in which spatially defined networks and infrastructure generate localised knowledge spillovers – can be defined as a distinctive feature of the current techno-socio-economic conditions. This calls for a broader research approach to corporate spatial organisation of innovation in order to provide a more coherent interdisciplinary social science, geographical or historical analysis of innovation as the engine of the change, development and transformation of capitalism. In this context, the empirical analysis carried out in this paper as well as our analytical framework, suggested by the literature and clarified by our findings, attempts to make a contribution to the construction of a more comprehensive theory of technological change, where interactive learning and geographical proximity/distance play a major role. In this spirit, this paper proposes a theoretical structure in which to situate the implications of the co-location and co-specialisation of innovation for inter-firm relationships that follow either mainly competitive or cooperative forms.

The results are consistent with previous empirical investigations. European-owned electronic firms do not tend to locate in the same region R&D activity in field/s in which they are co-specialised. Nonetheless, what is proposed here is a different concept of agglomeration by comparison with other streams of literature as discussed above. The average tendency revealed by the empirical analysis suggests there must be instead an inter-regional interchange of technological development, which defines the competitive pattern of sectoral and spatial corporate interaction as firms become co-identified with regions in terms of their respective profiles of technological specialisation. Therefore, the complete overlap of corporate technological portfolios acts as a deterrent to concentrate geographically research activity in the fields of common specialisation. Conversely, the special case represented in our analysis by the only cluster co-specialised and co-located reveals an intra-regional cooperation pattern of sectoral and spatial corporate interactions as in this case, the region of common location – i.e. the South-East (UK) – gains its own identity in the creation of new knowledge in the technologies of common corporate co-specialisation. As argued by Maskell and Malmberg (1999), this confirms the significance of the institutional embodiment of relevant tacit knowledge in corporate learning. Conceptually, cooperative inter-firm relationships can be also shaped in terms of inter-regional, and inter-regional and inter-industry patterns. The former is characterised by the cooperative co-location of non-co-specialised firms for the purpose of outsourcing new complementary knowledge. The latter refers to regional concentration of loosely co-specialised firms in order to enjoy *untraded* externalities. In both cases, the development of the internal MNC network relies on the grounds of external, locally embedded networks, which implies an increasing degree of autonomy for the subsidiaries (Zanfei, 2000).

Therefore, the argument of agglomeration and knowledge spillovers as major factors of corporate attraction is strong in the case of intra-industry co-location when considering all firms as in the analysis of Baptista and Swann (1998). Conversely, when looking at the activity of large firms in centres of excellence, as we have done in this paper, a greater degree of location separation emerges within industries, but

Table 5. ICT classes showing co-presence of a matching degree of specialisation between European-owned electronic firms and non-electronic firms, by selected regions

ICT patent classes	Selected regions					
	Baden-Württemberg	Bayern	South-East (UK)	Ile de France	Zuid-Nederland	Stockholm
Music		0.688				
Telegraphy	0.762	0.936		0.506		
Acoustics	0.575	0.889	0.714			0.715
Registers	0.988	0.643		0.516		
Active solid state devices (before 1990)		0.911	0.906			
Electrical transmission or interconnection systems	0.601	0.560	0.645	0.824		0.526
Electrical digital logic circuits						
Demodulators						
Modulators			0.525	0.611		
Communications: electrical	0.563	0.860	0.676	0.974		0.848
Code data generation or conversion		0.902	0.543	0.935		0.897
Communications: directive radio wave systems devices	0.770	0.601	0.893	0.695		
Communications: radio wave antennas						
Selective visual display systems	0.709					
Television				0.900		
Active solid state devices (after 1990)	0.942	0.895	0.875			0.712
Pictorial communication: television	0.739		0.997			

Table 5 (continued)

ICT patent classes	Selected regions					
	Baden-Württemberg	Bayern	South-East (UK)	Ile de France	Zuid-Nederland	Stockholm
Dynamic magnetic information storage or retrieval	0.751	0.545		0.733		
Electrical computers and data processing systems	0.580	0.953	0.986	0.826		0.825
Static information storage and retrieval	0.569	0.599				
Communications, electrical: acoustic wave systems and devices			0.613	0.670		0.868
Dynamic information storage and retrieval			0.832			
Multiplex communications	0.577		0.596	0.591		
Error detection/correction and fault detection/recovery	0.682	0.658	0.937	0.646		
Pulse or digital communications	0.536	0.765	0.808	0.924		
Electrical pulse counters, pulse dividers or shift register circuits and systems		0.735	0.571	0.985		
Telephonic communications	0.927			0.573		
Electrical audio signal processing system and devices	0.970	0.741				
Image analysis			0.755	0.878		
Information processing system organisation		0.571	0.968	0.520		
Telecommunications		0.718	0.956			0.374
C_{rc} mean	**0.527**	**0.517**	**0.622**	**0.489**	**0.028**	**0.235**

Table 6. ICT classes showing a lack of a matching degree of specialisation between European-owned electronic firms and non-electronic firms, by selected regions

ICT patent classes	Selected regions					
	Baden-Württemberg	Bayern	South-East (UK)	Ile de France	Zuid-Nederland	Stockholm
Music				0.000	0.000	0.000
Telegraphy			0.380		0.000	0.000
Acoustics				0.114	0.369	
Registers			0.374		0.000	0.472
Active solid state devices (before 1990)	0.155			0.000	0.000	0.000
Electrical transmission or interconnection systems					0.000	
Electrical digital logic circuits	0.388	0.000		0.171	0.000	
Demodulators	0.000	0.000	0.338	0.000	0.000	
Modulators	0.000	0.000			0.000	0.000
Communications: electrical					0.000	
Code data generation or conversion	0.364				0.000	
Communications: directive radio wave systems devices					0.000	0.234
Communications: radio wave antennas	0.155	0.182	0.477	0.409	0.000	0.000
Selective visual display systems		0.067	0.000	0.193	0.000	0.000
Television	0.444	0.332	0.493		0.000	
Active solid state devices (after 1990)				0.342	0.000	
Pictorial communication: television		0.493		0.285	0.000	0.117
Dynamic magnetic information storage or retrieval			0.219		0.000	0.000

Table 6 (continued)

ICT patent classes	Selected regions					
	Baden-Württemberg	Bayern	South-East (UK)	Ile de France	Zuid-Nederland	Stockholm
Electrical computers and data processing systems					0.407	
Static information storage and retrieval			0.475	0.000	0.000	0.000
Communications, electrical: acoustic wave systems and devices	0.211	0.062			0.000	
Dynamic information storage and retrieval	0.236	0.300		0.000	0.000	0.000
Multiplex communications		0.421			0.000	0.000
Error detection/correction and fault detection/recovery					0.088	0.000
Pulse or digital communications					0.000	0.000
Electrical pulse counters, pulse dividers or shift register circuits and systems	0.333				0.000	0.000
Telephonic communications		0.000	0.000		0.000	0.000
Electrical audio signal processing system and devices			0.490	0.000	0.000	0.000
Image analysis	0.460	0.000			0.000	0.000
Information processing system organisation	0.433				0.000	0.000
Telecommunications	0.387			0.331	0.000	
C_{rc} mean	**0.527**	**0.517**	**0.622**	**0.489**	**0.028**	**0.235**

when industries overlap in their technological profiles, geographically localised knowledge flows emerge within relevant centres in areas of common specialisation between several industries.

Appendix

Table A1. List of the ICT technological sectors, 1969–1995

Telecommunications
Other electrical communications systems
Special radio systems
Image and sound equipment
Semiconductors
Office equipment and data processing systems

Table A1a. List of the technological patent classes in the sub-periods 1969–1977 and 1978–1986

Telegraphy	1
Demodulators	2
Modulators	3
Communications, electrical: acoustic wave systems and devices	4
Multiplex communications	5
Pulse or digital communications	6
Telephonic communications	7
Telecommunications	8
Communications: electrical	9
Code data generation or conversion	10
Image analysis	11
Communications: directive radio wave systems devices	12
Communications: radio wave antennas	13
Music	14
Acoustics	15
Pictorial communication: television	16
Electrical audio signal processing system and devices	17
Electrical transmission or interconnection systems	18
Active solid state devices	19
Register	20
Dynamic magnetic information storage or retrieval	21
Electrical computers and data processing systems	22
Static information storage and retrieval	23
Dynamic information storage and retrieval	24
Error detection/correction and fault detection/recovery	25
Electrical pulse counters, pulse dividers or shift register circuits and systems	26

Table A1b. List of the technological patent classes in the sub-period 1987–1995*

Telegraphy	1
Demodulators	2
Modulators	3
Communications, electrical: acoustic wave systems and devices	4
Multiplex communications	5
Pulse or digital communications	6
Telephonic communications	7
Telecommunications	8
Communications: electrical	9
Code data generation or conversion	10
Image analysis	11
Selective visual display systems	12
Communications: directive radio wave systems devices	13
Communications: radio wave antennas	14
Music	15
Acoustics	16
Pictorial communication: television	17
Electrical audio signal processing system and devices	18
Television	19
Electrical transmission or interconnection systems	20
Active solid state devices	21
Electrical digital logic circuit	22
Register	23
Dynamic magnetic information storage or retrieval	24
Electrical computers and data processing systems	25
Static information storage and retrieval	26
Dynamic information storage and retrieval	27
Error detection/correction and fault detection/recovery	28
Electrical pulse counters, pulse dividers or shift register circuits and systems	29
Information processing system organisation	30

* The difference in the number of technological patent classes between the first two sub-periods and the last is due to a post-1990 re-classification.

Table A2. List of European electronic firms in the sample in each of the three sub-periods

1969–1977	1978–1986	1987–1995
AEG-Telefunken	AEG-Telefunken	AEG-Telefunken
BICC	ASEA AB	ABB ASEA Brown Boveri
Bosch-Siemens Hausgeräte	BICC	Bosch-Siemens Hausgeräte
Brown Boveri	Bosch-Siemens-Hausgeräte	CII-Honeywell Bull
CII-Honeywell Bull	Brown Boveri	Electrolux
Electrolux	CII-Honeywell Bull	Compagnie General d'Electricité (CGE)
LM Ericsson	Electrolux	General Electric Co.
Compagnie General d'Electricité (CGE)	LM Ericsson	ICL
General Electric Co.	Compagnie General d'Electricité (CGE)	LM Ericsson
ICL	General Electric Co.	Nixdorf Computer
Nixdorf Computer	ICL	Olivetti
Olivetti	Nixdorf Computer	Plessey
Philips	Olivetti	Philips
Plessey	Philips	Racal Electronics
Racal Electronics	Plessey	Sagem
Siemens	Racal Electronics	Siemens
Standard Telephones and Cables (STC)	Siemens	Standard Telephones and Cables (STC)
Thomson-Brandt	Standard Telephones and Cables (STC)	Thomson-Brandt
Thorn EMI	Thomson-Brandt	Thorn EMI
Zanussi Group	Thorn EMI	
	Zanussi Group	

Table A3. Regional locations where European-owned research activity in ICT technological sectors is carried out, 1969–1995

Belgian regions (NUTS 1)	*Dutch landsdelen* (NUTS 1)	*French régions*	*German Länder* (NUTS 1)
		Ile de France	
Bruxelles/Brussel	Noord-Nederland	Champagne-Ardenne	Baden-Württemberg
Vlaams Gewest	Oost-Nederland	Picardie	Bayern
Region Wallon	West-Nederland	Houte-Normandie	Berlin
	Zuid-Nederland	Centre	Brandenburg
		Basse-Normandie	Bremen
		Bourgogne	Hamburg
		Nord-Pas-de-Calais	Hessen
		Lorraine	Mecklenburg-Vorpommern
		Alsace	Niedersachsen
		Franche Comté	Nordrhein-Westfalen
		Pays de la Loire	Rheinland-Pfalz
		Bretagne	Sachsen
		Poitou-Charantes	Sachsen-Anhalt
		Aquitaine	Schelswig-Holstein
		Midi-Pyrénées	Thüringen
		Limousin	
		Rhône-Alpes	
		Auvergne	
		Languedoc-Roussilon	
		Provence-Alpes-Côte d'Azur	

Table A3 (continued)

Italian regioni (NUTS 2)	*Swedish riksområden* (NUTS 2)	*Swiss regions*
Calabria	Stockholm	Zürich
Campania	Östra Mellansverige	St. Gallen
Emilia Romagna	Småland med öarna	Basel
Friuli-Venezia Giulia	Sydsverige	Bern
Lazio	Västsverige	Thun
Liguria	Norra Mellansverige	Luzern
Lombardy	Mellersta Norrland	Davos
Piedmont	Övre Norrland	Lugano
Tuscany		Sion
Sicily		Lausane
		Fribourg
		Geneve

UK standard regions (NUTS 1)
East Anglia
East Midlands
North
North West
Scotland
South-East
South West
Wales
West Midlands
Yorkshire & Humberside

References

Amin A, Wilkinson F (1999) Learning, proximity and industrial performance: an introduction. Cambridge Journal of Economics 23 (2): 121–126

Anselin L, Varga A, Acs Z (1997) Entrepreneurship, geographic spillovers and university research: a spatial econometric analysis. ESRC Centre for Business Research, University of Cambridge Working Papers No. 59

Audretsch DB (2000) Knowledge, globalisation and regions. In: Dunning JH (ed) Regions, globalization and the knowledge based economy. Oxford University Press, Oxford

Audretsch D, Feldman MP (1996) R&D spillovers and the geography of innovation and production. American Economic Review 86 (3): 630–640

Audretsch DB, Stephan PE (1996) Company scientist locational links: the case of biotechnology. American Economic Review 86 (3): 641–652

Baptista R, Swann GMP (1998) Do firms in clusters innovate more?. Research Policy 27 (5): 527–542

Bartlett CA, Ghoshal S (1995) Transnational management – text, cases and reading in cross-border management. IRWIN, Chicago

Boschma RA, Lambooy JG (1999) Evolutionary economics and economic geography. Journal of Evolutionary Economics 9 (4): 411–430

Camagni R (1988) Functional integration and locational shifts in new technology industry. In: Aydalot P, Keeble D (eds) High technology industry and innovative environments: the European experience. Routledge, London

Caniëls M (2000) Knowledge spillovers and economic growth: regional growth differentials across Europe. Edward Elgar, Cheltenham

Cantwell JA (1995) The globalisation of technology: what remains of the product cycle model? Cambridge Journal of Economics 19: 155–174

Cantwell JA (1999) Introduction. In: Cantwell JA (ed) Foreign direct investment and technological change. Edward Elgar, Cheltenham

Cantwell JA, Fai FM (1999) Firms as the source of innovation and growth: the evolution of technological competence. Journal of Evolutionary Economics 9 (3): 331–366

Cantwell JA, Iammarino S (1998) MNCs, technological innovation and regional systems in the EU; some evidence in the Italian case. International Journal of the Economics of Business 5 (3): 383–407

Cantwell JA, Iammarino S (2000) Multinational corporations and the location of technological innovation in the UK regions. Regional Science 34 (3): 317–332

Cantwell JA, Kosmopoulou E (2002) What determines the internationalisation of corporate technology? In: Forsgren M, Håkanson H, Havila V (eds) Critical perspectives on internationalisation. Pergamon, Oxford (forthcoming)

Cantwell JA, Noonan CA (2002) The regional distribution of technological development: evidence from foreign-owned firms in Germany. In: Feldman MP, Massard N (eds) Institutions and systems in the geography of innovation. Kluwer, Dordrecht

Cantwell JA, Piscitello L (2000) Accumulating technological competence - its changing impact on corporate diversification and internationalisation. Industrial and Corporate Change 9 (1): 21–51

Cantwell JA, Santangelo GD (1999) The frontiers of international technology networks: sourcing abroad the most highly tacit capabilities. Information Economics and Policy 11: 101–123

Cantwell JA, Santangelo GD (2000) Capitalism, profits and innovation in the new techno-economic paradigm. Journal of Evolutionary Economics 10 (1–2): 131–157

Cantwell JA, Santangelo GD (2002) The significance of European small country regions in the geographical division of labour of European information and communications technology (ICT) corporations. International Journal of Technology Management 25 (forthcoming)

Dunford M (1996) Regional disparities in the European Community: evidence from the REGIO databank. Regional Studies 30 (1): 31–40

Eurostat (1997) Regioni – annuario statistico, Luxembourg

Eurostat (1995) Nomenclature of Territorial Units Statistics, Luxembourg

Feldman MP (1993) An examination of the geography of innovation. Industrial and Corporate Change 2: 451–470

Feldman MP, Audretsch DB (1999) Innovation in cities: science-based diversity, spacialization and localized competition. European Economic Review 43: 409–429

Granstrand O (1996) International diversification and multitechnology corporations. Paper presented at the EIBA Annual Conference, Stockholm

Granstrand O, Oskarsson C (1994) Technological diversification in 'multi-tech' corporations. IEEE Transactions on Engineering Management 41 (4): 355–364

Granstrand O, Patel P, Pavitt KLR (1997) Multi-technology corporation: why they have 'distributed' rather than 'distinctive core' competencies. California Management Review 39: 8–25

Griliches Z (1990) Patent statistics as economic indicators. Journal of Economic Literature XXVIII: 1661–1707

Howells J (1999) Regional systems of innovation. In: Archibugi D, Howells J, Michie J (eds) Innovation policy in a global economy. Cambridge University Press, Cambridge, MA

Kodama F (1992) Technology fusion and the new R&D. Harvard Business Review 70 (4): 70–78

Krugman P (1991a) Geography and trade. The MIT Press, Cambridge, MA

Krugman P (1991b) Increasing returns and economic geography. Journal of Political Economy 99 (31): 483–499

Lawson C (1999) Towards a competence theory of the region. Cambridge Journal of Economics 23 (2): 151–166

Malerba F, Lissani F, Torrisi S (1997) Computer and office machinery – firms external growth & technological diversification. EIMS Publications, European Commission

Martin R (1999) Critical survey: the new 'geographical turn' in economics – some critical reflections. Cambridge Journal of Economics 23: 65–91

Maskell P, Malmberg A (1999) localised learning and industrial competitiveness. Cambridge Journal of Economics 23 (2): 167–186

Metcalfe JS (1996) Economic dynamics and regional diversity – some evolutionary ideas. In: Vence-Deza X, Metcalfe JS (eds) Wealth from diversity. Kluwer, Dordrecht

Nelson RR (1992) What is 'commercial' and what is 'public' about technology, and what should be? In: Rosenberg N, Landau R, Mowery DC (eds) Technology and the wealth of nations. Stanford University Press, Stanford

Nooteboom B (1999) Innovation, learning and industrial organisation. Cambridge Journal of Economics 23 (2): 127–150

Papanastassiou M, Pearce RD (1994) Host-country determinants of the market strategies of US companies' overseas subsidiaries. Journal of the Economics of Business 1 (2): 199–217

Patel P, Pavitt KLR (1997) The technological competencies of the world's largest firms: complex and path-dependent but not much variety. Research Policy 26: 141–156

Pavitt KLR (1985) Patent statistics as indicators of innovative activity: possibilities and problems. Scientometrics 7 (1–2): 77–99

Santangelo GD (1998) Corporate technological specialisation in the European information and communications technology industry. International Journal of Innovation Management 2 (3): 339–366

Santangelo GD (2000) Inter-European regional dispersion of corporate ICT research activity: the case of German, Italian and UK regions. International Journal of Economics of Business 7 (3): 275–295

Vence-Deza X (1996) Innovation, regional development and technology policy: new spatial trends in industrialization and the emergence of regionalization of technology policy. In: Vence-Deza X, Metcalfe JS (eds) Wealth from diversity. Kluwer, Dordrecht

Verspagen B (1997) European 'regional clubs': do they exist, and where are they heading? On economic and technological differences between European region. MERIT Working Papers No. 2/97–010

Zanfei A (2000) Transnational firms and the changing organisation of innovative activities. Cambridge Journal of Economics 24: 515–542

Technology transfer in United States universities

A survey and statistical analysis*

Bo Carlsson[1] **and Ann-Charlotte Fridh**[2]

[1] Department of Economics, Weatherhead School of Management, Case Western Reserve University, Cleveland, OH 44106-7206, USA (e-mail: bxc4@po.cwru.edu)
[2] Department of Industrial Economics, Royal Institute of Technology, 10044 Stockholm, Sweden (e-mail: annfr@lector.kth.se)

Abstract. This paper examines the role of offices of technology transfer (OTT) in 12 U.S. universities in 1998 in commercializing research results in the form of patents, licenses, and start-ups of new companies. We study the organization and place of OTTs within the university structure, the process of technology transfer, and the staffing and funding of the office. Data were collected through a mail questionnaire followed up through telephone interviews. We also conducted a statistical analysis of data for 170 U.S. universities, hospitals, and research institutes for the period 1991–1996.

Our findings suggest that technology transfer from universities to the commercial sector needs to be understood in its broader context. The primary purpose of a technology transfer program is for the university to assist its researchers in disseminating research results for the public good. Success in this endeavor is only partially reflected in income generated for the university or the number of business start-ups. The degree of success depends not only on the nature of the interface between the university and the business community but also on the receptivity in the surrounding community as well as the culture, organization, and incentives within the universities themselves.

Key words: Technology transfer – Commercialization – Licensing – Patenting – Start-ups

JEL Classification: L3, O3

* This paper is the result of a project initiated by the Edison BioTechnology Center (EBTC). Financial support from the EBTC and Case Western Reserve University is gratefully acknowledged.

Correspondence to: B. Carlsson

1 Introduction

1.1 Technology transfer

Technology transfer may be defined generally as "the transfer of the results of research from universities to the commercial sector" (Bremer, 1999, p. 2). It may also be more narrowly defined as "the process whereby inventions or intellectual property from academic research is licensed or conveyed through use rights to industry" (AUTM, 1998, p. 3).

Technology dissemination or transfer can occur in many different forms. The publication of research results in scientific journals and books is the most common form of dissemination. In some cases the transfer may occur only if the intellectual property is protected and then commercialized. The issue dealt with in this paper is the narrower one of transfer from universities to industry of intellectual property rights in the form of patents or licenses and via start-ups of new companies.

Technology transfer involves at least two parties. As Lawrence Dubois of DARPA puts it, "technology transfer is a contact sport!" The quality and quantity of interaction are determined not just by the interface (the rules of the game, as it were) between the two parties but also by what each of the players brings to the game. The knowledge, preparedness, organization, culture, and attitudes of both sides are important for successful interaction. The motivations of the two sides are often quite different. The main objective of basic research is almost never inventions. "If inventions do flow from that research activity, it is largely a fortuitous happening that takes place because the researcher, or perhaps, an associate, has the ability to see some special relationship between his scholarly work product and the public need" (Bremer, 1999, p. 4). On the commercial side, the main objective, of course, is profitable exploitation of an innovation or an idea.

1.2 The Bayh-Dole act

The recent increase in university patenting and licensing activity is, at least in part, a consequence of the Bayh-Dole Act, which was enacted by the U.S. Congress in 1980 and became effective on July 1, 1981 (see The Council on Government Relations, 1993). The Act transferred the rights to intellectual property generated under federal grants from the funding agencies to the universities, thus providing the latter opportunities to exploit research results commercially. One of the major arguments for the Act was that a stronger protection of publicly funded research would lead to a faster and stronger technology transfer and hence benefit the taxpayers. The reason behind this argument was that companies need intellectual property rights to pick up, develop, and commercialize the results of university research.

Before the establishment of the Bayh-Dole Act, not many universities found it worthwhile to get into the patenting business since this was connected with a high fixed cost. The Act opened up the possibilities for universities to explore their technology transfer to a larger extent.

There has indeed been an increase in patenting and licensing activity on the part of U.S. universities after the establishment of the Bayh-Dole Act. From 1979 to 1984 the number of patents issued annually doubled (from 177 to 408) and between 1984 and 1989 it doubled again (to 1,008) (Mowery et al., 1999). However, some universities such as Stanford and the University of California were active in technology transfer well before the passage of the Bayh-Dole Act. This means that technology transfer from the universities to industry can not be explained solely by the Bayh-Dole Act. Several other policy decisions were made during this period. For example, in 1980 Diamond v. Chakrabarty upheld a broad patent in the new biotechnology industry which opened the door for many patents in this area. Also, the Court of Appeals for Federal Circuit (CAFC) was established in 1982. The CAFC emerged as a strong protector of the rights of patent holders such as universities.

Following the Bayh-Dole Act, the number of technology transfer offices at U.S. universities increased dramatically. There were 25 such offices in 1980 (before the new legislation), but by 1990 the number had increased to 200. A recent study by Mowery et al. (1998) compares Stanford, the University of California, and Columbia University. The study finds that even without the Bayh-Dole Act, both Stanford and the University of California would have expanded their patenting and licensing activities. Columbia University also made some steps in this direction prior to the Bayh-Dole Act. The study also finds that there has been a change in the attitude and policies regarding the value of research and the potential revenue and profit that it can bring the university. The expanded licensing activities have led to both enthusiasm and resentment over the effect that it might have on the culture and norms of academic research. There is concern that there will be a change in the character of university research towards applied and away from basic research. Yet another study by Mowery et al. (1999) shows the difficulty of managing a technology transfer office with the view of maximizing income. The goals need to be broader than that and be integrated with the entire mission of the university.

1.3 Benefits of technology transfer to academic institutions

Technology transfer programs are important to the academic institutions' mission of education, research, and public service in that they provide:

- A mechanism for important research results to be transferred to the public;
- Service to faculty and inventors in dealing with industry arrangements and technology transfer issues;
- A method to facilitate and encourage additional industrial research support;
- A source of unrestricted funds for additional research;
- A source of expertise in licensing and industrial contract negotiations;
- A method by which the institution can comply with the requirements of laws such as the Bayh-Dole Act (AUTM Licensing Survey FY 1991 - FY 1995, p. 6). Also, it may be used as
- A marketing tool to attract students, faculty, and external research funding.

Thus, the primary purpose of a technology transfer program is to assist the institution, on behalf of its faculty and inventors, in the dissemination of research results for the public good. The income generated through this mechanism is important but is only a part of the total benefit to the institution. According to the 1996 AUTM survey, the gross license income received by the reporting U.S. universities, hospitals, and research institutions amounted to $500 million. While this is a large amount, it represented only about 2.3 percent of the total sponsored research expenditures in the same institutions. However, among all the reporting institutions the percentage ranges from zero to 11 percent.

It is clear, therefore, that while license income is neither a likely major source of research funding nor the sole (or even the most important) benefit to the institution, some universities are much more successful than others in generating such income. This is one of the reasons for this study: to examine the technology transfer function at U.S. universities with respect to organization, risk management, funding, staffing, and professional competence in order to better understand what successful practices and strategies are.

1.4 Benefits to the community

It may well be argued that technology transfer is even more important to the surrounding communities than to the universities because of the benefits it creates to the rest of the society. In an economic impact model developed by the Association of University Technology Managers (AUTM), it has been shown that nearly $25 billion of the economic activity in U.S. can be attributed to the results of academic licensing, supporting 212,500 jobs in fiscal year 1996. For FY 1995, the comparable figures were $21 billion and 180,000 jobs (AUTM, 1998).

In an article investigating the technology transfer function at Stanford University, Fisher (1998) discusses the many benefits to society of a well-functioning transfer of technologies from the research laboratory to the commercial sector. The income that the technology transfer generates can offset the shrinkage of federal funding in comparison with other sources which has been observed in recent years. The AUTM data show that the share of federal funding in the reporting institutions was reduced from 72 to 66 percent between 1992 and 1996.

The creation of wealth, new jobs and new solutions to problems in the society is another benefit. Fisher describes Stanford's success as a leader in technology transfer and the effect that it has had on northern California's Silicon Valley and the biotechnology industry and its role in providing a model for many other universities across the country as well as internationally. "Stanford continues to show the way, providing creative solutions to new challenges as the need for university research becomes even more urgent... And for the great research universities, income from patent licenses can offset the shrinkage of federal funding" (Fisher, 1998, p.76).

It should be noted, however, that Stanford's success in generating new businesses has occurred in spite of a university policy not to specifically promote business start-ups. Thus, success depends not only on university policies and strategies but also on the institutions, entrepreneurial climate, and fertility of the economic soil (including access to venture capital) in the recipient community.

"The real measure of technology transfer is not, of course, the number of patents which the university sector holds, but the amount of technology, represented in and by those patents which has been transferred to the private sector for further development into products and processes useful to mankind" (Bremer, 1999, p. 4). As the Bayh-Dole Act puts it, "the mission of university technology transfer offices is to transfer research results to commercial application for public use and benefit... The major effort of the office is to find companies which have the capability, interest and resources to develop embryonic technologies into useful products" (The Council on Governmental Relations, 1993, p. 2).

Thus, a full evaluation of the output of the technology transfer process is a complex matter. The income from licenses and the number of start-ups are at best only partial measures. Not much is known about this, although a recent study finds that license income increases more than proportionally with R&D expenditures (Siegel, Waldman, and Link, 1999, p. 20).

1.5 Focus and organization of the paper

The focus in this report is on various activities and parameters associated with the technology transfer process on the university side. In the AUTM report several different parameters are presented that can be used to measure technology transfer. The general stimulus for the basic process is research expenditure. Most universities use this variable on the input side to evaluate the technology transfer activity. The most common output variable is number of licenses. It is logical to expect that the larger are the resources spent on research and development, the more licenses and options will be generated. Indeed, recent studies (Adams and Griliches, 1996; Siegel, Waldman and Link, 1999) have shown constant returns to scale in this respect, i.e., that the number of licensing agreements increases proportionally with R&D expenditures. The size, distribution and significance of the parameters describing the technology transfer function are also of interest and will be explored further in our statistical analysis.

The fact that the history, organization, and performance of technology transfer vary from one institution to the next provides a major motivation for this study. Are there certain practices that are more successful than others? How do offices of technology transfer (OTTs) fit within the organization (and overall mission) of the university? What differences can we observe in levels and sources of funding, staffing, and professional expertise? What role do differences in attitudes and procedures with respect to risk management play? How widely do attitudes and procedures with respect to licensing and start-ups differ? Is there a "model" that can be emulated? These are the general questions which prompted the Edison BioTechnology Center (EBTC) to initiate and fund this study.[1] They are also the main questions addressed in the in-depth survey we conducted with twelve U.S. universities, reported in the next section.

In addition to answering these questions concerning the technology transfer organization and process with the help of the questionnaire survey, we have used the insights gained from this analysis in carrying out a statistical analysis of the annual

[1] EBTC is a part of the Ohio Department of Development's Thomas Edison Program which constitutes the backbone of Ohio's technology policy.

survey data collected by the Association of University Technology Managers, Inc. (AUTM). Our aim is to highlight the technology transfer process and some of the most common performance measures that the universities themselves use to evaluate their activity. We have also examined the correlations among the variables and how this can help us understand the technology transfer function. In addition, we have built a simple model, based on our own survey results, which can be used for multivariate econometric analysis. We then used this model in a regression analysis.

It is important to keep in mind that this report focuses primarily on the practices in universities with respect to technology transfer. Thus, it covers only one side of the equation. A more comprehensive report would also cover the absorptive capacity of the actors on the commercial side, as well as the environment supporting the transfer activity. We have not found any such study in our literature review. Such a report would also have to deal with the further complexities arising from the fact that the interaction is different in different technology areas and in different environments. Many studies show that the benefits of technology transfer (often referred to in the economic literature as technological spillovers) are largely local (Bania, Eberts and Fogarty, 1993; Jaffe, Trajtenberg, and Henderson, 1993; Zucker, Darby, and Brewer, 1998; Zucker and Darby, 1996; Audretsch and Stephan, 1996).

The paper is organized as follows. The next section analyzes the results of our survey of the technology transfer function in twelve United States universities. Using the insights gained from this survey we then proceed to a statistical analysis of AUTM data. The paper concludes with a discussion of the results and their implications.

2 Questionnaire survey of technology transfer in U.S. universities

2.1 Information on the survey data collected

We chose 12 universities ranging from top research universities to some regional universities. The universities are The California Institute of Technology, Carnegie-Mellon University, Case Western Reserve University, Emory University, The Ohio State University, Stanford University, The University of Cincinnati, The University of Michigan, The University of Pennsylvania, The University of Texas Southwestern Medical Center, Vanderbilt University, and Washington University.

The data collection was done through a mail questionnaire followed up through telephone interviews. Nine universities responded to the questionnaire; thus, the response rate was as high as 75 percent. The questionnaire included 31 questions divided into three parts.[2]

The data reported here include both the survey data and the background data for each university obtained through the AUTM reports. Thus, some data are available also for the universities which did not respond to our survey, as well as an additional university in Ohio not included in the survey.[3]

[2] The questionnaire is available from the authors upon request.

[3] We will refer to this data set as the subset n, and we will refer to the data set collected by the AUTM as the total sample, N.

2.2 Organization, staffing, and funding

Table 1 provides overview data on the organization, staffing, and funding of technology transfer in the universities covered by the survey. In the table, the universities are ranked in descending order of their annual research budget. Two of the universities had set up their office of technology transfer (OTT) prior to the Bayh-Dole Act. In the other cases, the OTT was established somewhere between 1982 and 1990 with 1986 as the median for the sample as a whole.

In all of the universities surveyed, the OTT is set up as a unit within the university, not as a corporation or other entity separate from the university. In all but two cases the director of the OTT reports to the Vice President for Research (or equivalent position). In the two exceptional cases, the director reports to the Provost and in one case also to the Chancellor for Health Affairs. Thus, there is no doubt that technology transfer is regarded as a matter of strategic and policy concern at the highest level within the universities. How high a priority is placed on these activities varies from one university to another, however.

There is only one technology transfer office at each university, except that in two cases there is also a branch office in the medical school. In one of the universities the OTT also has responsibility for handling industrial grants and in another case works closely with an office handling such grants (as distinct from federal funding).

2.3 Staffing and expertise

The size of the OTT in terms of number of staff members ranges from 2 fulltime equivalents (FTEs) in the smallest institution to18 in the largest, 4.5 being the median number. In most cases, the number of professionals is significantly larger than the number of support staff. In the schools with a small OTT, basic science is the dominant type of expertise. As one would expect, the larger the OTT, the broader is the range of in-house expertise. In three cases there is legal expertise within the OTT itself; in one of these cases that is the only professional expertise represented.

2.4 Risk management and legal expertise

The management of the risk exposure associated with technology transfer is a sensitive and serious issue at most universities. A common method is to accept no liability resulting from technology transfer but to require licensees to indemnify the university if liability should occur (or to obtain the necessary indemnification insurance coverage). This is the method used in four of the surveyed universities. However, this may not work in cases when the potential licensee is a small start-up. In such cases, other solutions have to be found (including not issuing a license at all).

For risk management purposes but also for handling of intellectual property issues, as well as other reasons, legal expertise is often required in connection with technology transfer. This expertise may reside in a variety of places. Six of the

Table 1. Technology transfer in U.S. universities – organization, staffing and funding (AUTM-data from FY 1996, (*) indicates data from interview material)

University	Year OTT started	Reports to: (*)	Staffing (FTEs)	Staffing (FTEs) (*)	Expertise (*)	Legal expertise in OTT (*)	Outside legal advice from (*)	Annual budget for licensing & patenting (*)	University annual research budget ($M)
5	1982	VP Research	15.25	7 professionals, 10 support staff	Basic science; legal (1); mgmt (1)	Yes	Univ. Risk Mgr.	Total budget $2M; patenting $500K	441.3
4	1970	VP Research	18	20 professionals, 5 support staff	2 biotech; 2 physical science; 2 both engineering; industrial experience	No	Univ. Risk Mgr; Outside IP firm	Total budget: $2.2M For patenting: $2M. License income $700K	395.5
8	1978	VP Research	4	3,5 professionals, 2,5 support staff	MBA (1); PhD in science (2); legal (1)	Yes	Univ. IP office	Total budget: $4M; for patenting: $750K	218.0
6	1990	VP Research	4	4 professionals, 2 support staff	Basic & pol science (1); genetics(2); Ind. Engin. (1)	No	Univ. Office of Legal Affairs IP: outside firm	Patenting budget: $400–500K License income $200-300K	207.7
3	1992	Provost	5	4 professionals, 2 support staff	Engineering; basic science; MBA	No	Outside firm	N.A.	165.0
9	1985	VP Research	4.5	3 professionals, 2 support staff	Basic science, business	No	Univ. Office of General Councel	Patenting budget: $855K. License income $580K	146.0

Table 1 (continued)

University	Year OTT started	Reports to: (*)	Staffing (FTEs)	Staffing (FTEs) (*)	Expertise (*)	Legal expertise in OTT (*)	Outside legal advice from (*)	Annual budget for licensing & patenting (*)	University annual research budget ($M)
1	1986	VP Research	3.5	2 professionals, 1 support staff	Legal (2)	Yes	Univ. Attorney & Univ. Risk Mgr.	Total budget: $320K; for patenting: $90K	140.6
7	1990	Provost & V. Chanc.-Health Affairs	2.25	2.5 professionals, 4 support staff	Elec. engin.; life sciences; business	No	Univ. Attorney	Total budget $580K; for patenting: $150K	128.5
2	1983	VP Research	3.5	2 professionals, 1 support staff	Basic science (1); legal (1)	No	Univ. Risk Mgr.	Total budget: $389K; for patenting: $154.2K; license income $136.5K)	64.4
10	1986		11						328.0
11	1985		6.3						250.0
13	1990		6.5						144.1
12	1991		2						13.5
Median	1986		4.5						165.0

nine responding universities have no legal expertise at all in the OTT. Of these, four require indemnification by licensees. All of the universities have a university attorney's office and/or risk management office. Four of the universities obtain legal advice from outside firms concerning matters of intellectual property, while one university has its own intellectual property office.

2.5 Annual budget and research funding

The annual budget for licensing and patenting activities within the OTT ranges from $320,000 to over $2 million. A substantial portion of the difference is explained by the size of license income. Apparently, license income is a major source of funding of technology transfer activities in most universities.

The university-wide annual research budget in the reporting institutions ranges from $13.5 million to $441.3 million, with a median of $165 million. This should be compared with a median of $96 million for all the schools included in the AUTM report. The correlation between the annual budget for licensing and patenting on the one hand and the annual research budget on the other is very low.

2.6 The technology transfer process

The steps to transfer or commercialize a technology are basically the same at all universities. Typically, the process starts with the faculty/researcher/inventor submitting an invention disclosure form (a standard form specified by the university) to the OTT. After reviewing the disclosure, investigating the potential market, and estimating whether or not the expected return warrants the cost of seeking intellectual property protection (patent, copyright, trademark, or other form of protection), the OTT initiates the requisite application. All the universities surveyed claim to actively facilitate patenting, but always within the constraints set by the budget (the patenting cost typically being somewhere in the $15,000 – $20,000 range per application). In all but one of the universities, patent applications are handled by the OTT; in the remaining case an outside firm is used.

Once intellectual property rights have been obtained, technology licenses are typically developed in several stages:

a) **Confidentiality or Non-Disclosure Agreement (NDA)** – If confidential matters need to be disclosed by either party to the other in order to permit substantive discussions, a Non-Disclosure Agreement between the potential licensee and the university will be developed. In some categories of cases a similar arrangement is obtained through a **Material Transfer Agreement (MTA)**.

The typical process for a nondisclosure agreement (NDA) or confidentiality agreement is to start with a standard agreement form. If the standard form is acceptable, the process is usually completed within a day or two. If a non-standard agreement must be negotiated, the process varies from case to case as to its duration and the personnel involved. In most universities, the OTT officers or director have sign-off authority, but in one case the provost has to sign and in another legal counsel has to be sought.

The process for material transfer agreements (MTAs) is similar to that for NDAs. In standard cases, the process can usually be completed within a day or two and is handled entirely within the OTT. But in some universities, other offices process MTAs. In one case, the provost's signature is required (see Table 2).

b) **Business Plan** – In order to have a substantive discussion of a potential license, the university will also need to understand the nature of the Licensee's current business, his/her future business plans, and the specific plans for utilizing, developing and commercializing the technology expected to be licensed ("Licensed Technology").

In the case of well established, robust companies with demonstrated record of revenues, profits, technologies and products, such a business plan will concentrate on the specific development of the licensed technology, the resources to be used, expected development milestones, and the economic results expected, to demonstrate that the licensee intends to develop that technology to the fullest and has the know-how and the resources to do so. In the case of start-ups or newly established firms, the business plan may involve the development of the whole company and its business strategy.

c) **License Term Sheet** – Based on the above, the next and typically the most important step will be the discussions of the key economic terms of the proposed license, resulting in a tentative agreement on terms as defined in a "License Term Sheet".

d) **License Agreement** – Once a basic agreement on economic terms has been reached, the next step is to draft the actual license agreement, i.e., the legal document which incorporates both the economic and other terms of the terms sheet as well as the university's general licensing terms and conditions (commonly also called "legal boiler plate").

In contrast to NDAs and MTAs, there is no standard procedure for license agreements. The process usually starts with a standard form, but modifications and therefore negotiations are always necessary. The time required to reach an agreement varies from a few weeks to several years.

The university personnel involved in these negotiations are the licensing associates of the OTT, plus the inventor or researcher. In one case the OTT director is also involved, along with the university's legal counsel. In another case, in addition to these, the university risk manager also takes part in the negotiation. The authority to sign off on licensing agreements resides in the OTT in all but two cases, the remaining two requiring the signature of the V.P for research or even the provost.

How active a role the university plays in finding potential licensees depends largely on the resources available, particularly the number and capabilities of the OTT staff. In one case, the OTT explores many avenues in trying to locate licensees, including contacting existing licensees and other corporate contacts, searching corporate technology directories, and conducting online literature searches. However, by far the best source of potential licensees is the inventors themselves, since they are likely to know who in industry is doing work related to their inventions. They are often aware of companies that might successfully commercialize their inventions. The OTT encourages companies to indicate what areas are of interest to them; thus

Table 2. Technology transfer in U.S. universities – licensing procedures (data form interview material)

University	NDA nego-tiated by	Duration of process for standard agreement	NDA sign-off by	MTA sign-off by	Licensing: Personnel involved	Licensing sign-off authority	Role in finding licensees	Number of professional staff	Indemnific required
1	OTT, researcher, Univ. attorney, Univ. risk mgr	1-2 days	Provost	Provost	OTT, researcher, Univ. attorney, Univ. risk mgr	Provost	Modest role (limited by resources)	2	Sought but not always required
2	OTT officers	1 day	OTT officers	OTT officers	OTT officers	OTT officers	?	2	Yes
3	OTT officers	Varies	OTT after consultation with legal counsel	N/A	OTT director, licensing officer, Univ. legal counsel	OTT after consultation with legal counsel?	?	4	Yes
4	OTT officers	1 day	OTT officers	OTT officers	OTT officers	OTT officers	Active search via existing licensees, corporate contacts, directories, on-line searches	20	Yes?
5	OTT officers	1 week	OTT director?	Incoming: Dept. of Research Admin.; outgoing: OTT officers?	OTT officers	OTT director?	Fairly active	7	?
6	OTT officers, with input from legal counsel	A few days	OTT director	OTT director? with input from legal counsel and inventor	OTT officers,	OTT director	?	4	Yes
7	OTT officers?	?	?	Div. of Sponsored Research in Provost's office	Inventor; OTT officers	?	Relies on existing network and researcher's contacts	2.5	?
8	OTT officers	?	OTT officers	OTT officers	OTT officers	OTT officers?	Relies on existing network and researcher's contacts	3.5	?
9	OTT officers?	Varies	OTT officers	OTT officers	OTT officers	VP Research	?	3	?

providing a "wish list" of technologies. Often the assessment of the commercial potential of an invention begins by conducting literature searches and by asking contacts in industry for input.

2.7 Patenting and licensing activities

The patenting and licensing activities in the universities included in our survey are summarized in Table 3. With only two exceptions, medicine is mentioned as the school or field most frequently served by the OTT at universities which have a medical school. Engineering is the second most frequently mentioned field. The number of invention disclosures reported for FY 1996 varies from 22 to 300, with a median of 63. The number of patent applications ranged from 4 to 130, with a median of 34. The median number of U.S. patents issued in FY 1996 was 16, i.e., about half the number of applications. In other words, ignoring the time lags between disclosure and application and between application and issuing of patents, only about half of the invention disclosures resulted in patent applications, and only half of the applications resulted in actual patents. Furthermore, only a fraction of patents yield license income. Of the 1,747 active licenses and options in the surveyed universities in FY 1996, 682 (39 %) yielded income. The average license income to the university was $5.9 million, meaning that the average income per license was $180,000. However, the mean was only $80,000. This means that the distribution of income-yielding licenses is highly skewed; with the exception of only one university, the number of patents generating more than $100,000 per year is six or less.

2.8 Policies and procedures for start-ups

The number of start-ups is another indicator of technology transfer performance. The policies and procedures for start-ups depend in large measure on the university's attitude towards risk and the capabilities in the surrounding business community. As already noted, one university (Stanford) has a policy of not creating or helping to create spin-offs (Roberts and Malone, 1996) – and yet has more start-ups to its credit than any other university (AUTM Survey). Thus, the university's policies are not the only determinant of spin-offs and their degree of success; the history, culture, attitudes, industry affiliation, market orientation, etc., of existing businesses, and the presence or absence of venture capital, as well as the vigour and diversity of supporting organizations and institutions are also important. At other universities included in our survey, the attitude towards start-ups ranges from reluctance to play a role in spin-offs at one extreme, to refraining from initiating spin-offs but helping them once underway, and to active encouragement of and involvement in creating spin-offs (including incubation services, financial support, and information/networking services) at the other extreme.

The survey data on start-ups are summarized in Table 4. The table shows little correlation – or perhaps even a negative one – between the university's attitude toward start-ups and performance as indicated by the number of new companies formed.

Table 3. Technology transfer in U.S. universities – patenting and licensing activities (AUTM-data from FY 1996, (*) indicates data from interview material)

University	Medical School	Schools and departments most frequently served by OTT (*)	Invention disclosures received FY1996 (AUTM)	FY1996 U.S. patent applications (AUTM)	FY 1996 U.S. Patents Issued (AUTM)	FY1996 licenses & options yielding license income	Total active licenses & options 1996	FY1996 Gross License income (thousands)	FY 1996 income per license (thousands)	FY 1998 Patents generating more than $ 100,000 (*)
8	No	Biology, chemistry, chemical engineering, applied engineering, applied physics, computers	300	120	31	35	N.A.	$3,900	$111	3
4	Yes	Engineering: electrical. Medicine: various. Various laboratories.	160	130	56	259	903	$43,752	$169	20
10	Yes		136	54	46	29	94	$782	$27	
5	Yes	Medicine. Engineering.	112	65	23	82	142	$1,074	$13	6
3	No	Computer science, robotics. Engineering: electrical & computer. Arts & sciences: chemistry	83	18	8	13	43	$7,135	$549	4
13	Yes		73	23	16	49	82	$2,940	$60	
6	Yes	Engineering: electrical, mechanical. Arts & sciences: physics, chemistry. Pharmacy. Medicine: HLI, surgery, internal medicine, OB/GYN. Agriculture & veterinary: food animal health, veterinary biosceinces, horticulture & crop science, food science.	63	49	21	25	78	$1,097	$44	3

Table 3 (continued)

University	Medical School	Schools and departments most frequently served by OTT (*)	Invention disclosures received FY1996 (AUTM)	FY1996 U.S. patent applications (AUTM)	FY 1996 U.S. Patents Issued (AUTM)	FY1996 licenses & options yielding license income	Total active licenses & options 1996	FY1996 Gross License income (thousands)	FY 1996 income per license (thousands)	FY 1998 Patents generating more than $ 100,000 (*)
9	Yes	*Medicine*	63	34	11	17	38	$2,580	$152	5
7	Yes	*Medicine* (70–75 %). *Engineering* (10–15 %).	58	10	7	27	67	$640	$24	4
1	Yes	*Medicine:* various. *Engineering:* biomedical, materials science, electrical. *Arts & sciences:* physics, biology, chemistry	43	15	8	17	37	$544	$32	3
2	Yes	*Medicine:* molecular genetics, internal medicine, surgery, many others. *Arts & sciences:* chemistry. *Engineering:* chemical eng., civil & environmental, electrical & computer, mtrls science	38	23	8	11	39	$2,208	$201	1
11	Yes		22	44	20	117	218	$9,413	$80	
12	Yes		22	4	5	1	6	$874	$874	
Median			63	34	16	27	73	$2,208	$80	4

Table 4. Technology transfer in U.S. universities – start-up activities (AUTM-data from FY 1996, (*) indicates data from interview material)

University	Attitude to start-ups (*)	FY 1996 Start-up companies formed	Start-up companies formed 94–96
1	University plays no active role in spin-offs, but affiliated organizations sometimes do. The university prefers licensing to large companies rather than to get involved with start-ups.	1	5
2	N.A.	0	1
3	Actively involved in the start-up process, helping out in the commercialization process via documentation package, Incubation service, financial support (cash $20K and credit $30K) and contacts with several organizations.	2	3
4	OTT is not in the business of creating or helping with spin-off or start-up companies.	14	25
5	Actively involved and encourages start-ups. OTT has one person working full time with this. The state provides information and contacts with venture capitalists but offers no direct financial help.	8	9
6	OTT plays an active role in securing technical and financial support. Spin-offs also have access to a local incubator.	0	0
7	Actively involved with the start-up process. The university has an organization specifically for this. There is also an investment fund of $10K available. Has hired a person with commersialization experience for this matter.	1	1
8	Actively involved in the start-up process, helping out in the commercialization process. About 1/3 of all license agreements involve start-ups.	10	16
9	OTT helps with start-ups but does not initiate them.	0	3
10		4	16
11		1	4
12		0	1
13		0	3

2.9 Industry-sponsored research

As shown in Table 5, industry-sponsored research amounted to $192 million in FY 1996, or 7.3 percent of the total research budget in the surveyed institutions. The share of industry-sponsored research remained constant between FY 1991 and FY 1996. (Meanwhile, as noted earlier, the share of industry-sponsored research increased in the entire population of universities included in the AUTM surveys.)

Table 5. Technology transfer in U.S. universities – industry sponsorship (AUTM-data from FY 1996, (*) Indicates data from interview material)

University	Annual budget for licensing & patenting (*)	Annual research budget 1996 , $M	Industry-sponsored research 1996, $M	Industry-sponsored research in % of total FY96	Annual research budget 1991, $M	Industry-sponsored research 1991, $M	Industry-sponsored research in % of total FY91
5	Total budget $2M; patenting $500K	441.3	35.1	8.0	324.1	23.4	7.2
4	Total budget: $2.2M. For patenting: $2M. License income $700K	395.5	19.6	5.0	280.1	14.1	5.0
10		328.0	20.0	6.1	238.0	4.5	1.9
11		250.0	22.3	8.9	181.0	17.6	9.7
8	Total budget: $4M; for patenting: $750K	218.0	4.0	1.8	115.0	2.7	2.3
6	Patenting budget: $400–500K. License income $200-300K	207.7	17.3	8.3	154.0	10.0	6.5
3	N.A.	165.0	21.5	13.0	N.A.	N.A.	N.A.
9	Patenting budget: $855K. License income $580K	146.0	16.4	11.2	82.3	1.0	1.2
13		144.1	13.2	9.2	87.2	2.8	3.2
1	Total budget: $320K; for patenting: $90K	140.6	5.4	3.9	106.2	16.4	15.4
7	Total budget $580K; for patenting: $150K	128.5	10.6	8.2	109.1	24.5	22.5
2	Total budget: $389K; for patenting: $154.2K; license income $136.5K	64.4	4.1	6.4	66.0	5.1	7.8
12		13.5	2.7	20.3	10.2	1.6	15.3
Total		2642.6	192.3	7.3	1753.2	123.7	7.1

2.10 Policy with respect to exclusive vs. non-exclusive licensing

Six of the universities included in our survey report that they have no policy with respect to exclusive or non-exclusive licensing agreements. One university grants only exclusive licenses, while another one usually offers exclusive licenses. One university offers exclusive license options to industry sponsors.

2.11 Patent ownership and sharing of license income

In the event that the university is not interested in pursuing a patent, the ownership rights are usually reassigned to the inventor. In a few cases the university retains patent ownership but licenses it back to the inventor. License income is shared between the university, the management center of the inventor(s), and the inventor(s). As shown in Table 6, the formulae pertaining to income sharing vary. The inventor's share is typically 30 – 50 percent of the net income (i.e., gross income minus legal fees and other expenses).

2.12 Monitoring of royalty agreements

Several of the surveyed universities have effective procedures in place for monitoring royalty agreements, including computerized methods. However, a few of the respondents indicate dissatisfaction with current procedures, and several universities are taking steps to improve the monitoring. Staffing levels and the relatively low priority placed on this activity seem to be the main constraints.

2.13 Proposed success indicators

When asked what they regard as good indicators of performance or success in technology transfer, technology managers come up with a number of suggestions. These include the measures incorporated in this study, particularly the number of patents and licenses and the amount of royalty income. The number of invention disclosures is also often mentioned. A few of the respondents also mention the broader and harder-to-measure aspects of performance, such as the number of spin-offs, faculty satisfaction, and general indicators of overall transfer activity.

We can now summarize the technology transfer process given by the results of the study as shown in Figure 1.

2.14 Statistical analysis of the questionnaire survey data

The technology transfer process can be described in terms of an input-output model in which the inputs are research expenditures and OTT staff and expertise and the outputs are the results of the various stages of the technology transfer process. It is important to note that what is output from one stage of the process can be an

Table 6. Technology transfer in U.S. universities – miscellaneous data (data from interview material)

University	Policy wrt exclusive vs. non-exclusive licensing (*)	Patent ownership if institutional non-interest (*)	Sharing of license income (net) (*)	Monitoring of royalty agreements (*)	*Proposed success indicators* (*)
1	Industry sponsors get option on exclusive license	Univ. owns patent but licenses it back to inventor	50 % inventor(s), 25 % OTT, 25 % mgmt center or school	Needs improvement	Commercial use on the market. Local firms working w/ university. Faculty satisfaction
2	No policy	Assigns ownership to inventor	<$50K: 60 % inventor, 15 % dept, 5 % college, 20 % univ. $50–100K: 40 % inventor, 25 % dept. 15 % college, 20 % univ. >$100K: 30% inventor, 30% dept., 20 % college, 20 % univ.	Annual review, annual billing. Termination in case of non-payment	Still working on that
3	No policy	If sponsored research: univ. owns IP but may assign it to inventor. If not sponsored, inventor owns. In both cases, income is shared	4 basic distributions depending on how IP was created: Univ. 100%, inventor 0%; Univ. 0%, inventor 100%; 50/50; or 15/85. Univ. share is divided equally betw dept and univ. admin.	Licensees required to maintain records and make them available for annual audit	Number of technologies brought to market. Service to faculty. Regional economic development. Number of spin-off companies. Financial result for univ.
4	No, depends on circumstances	If govt. sponsored, rights are returned to govt. In other cases, OTT may license inventor	15 % of gross revenue to support OTT. Then direct expenses are deducted. Net revenue: 33 % inventor, 33% dept., 33% school	Licensing associates and OTT accounting monitor and enforce royalty agreements	Number of licences. New monies and royalties generated. High licensing rate of patents
5	No, but follow guidelines for federal funding	Reassignment to inventor	<$200K: 50% inventor, 25% dept., 25% mgmt center. $200–$2000K: 33.3% inventor, 33.3 % dept., 33.3% mgmt center. >$2000: 33.3% inventor, 66.7% mgmt center	Moving from manual to computerized monitoring and invoicing	Royalty income. Number of patents and licenses. Degree of success of transfers (not only monetarily)

Table 6 (continued)

University	Policy wrt exclusive vs. non-exclusive licensing (*)	Patent ownership if institutional non-interest (*)	Sharing of license income (net) (*)	Monitoring of royalty agreements (*)	Proposed success indicators (*)
6	No policy Income shared	Reassignment to inventor	<$75K: 50% inventor; of the remainder after expenses 33.3% inventor, 25% univ., 52.7% dept. >$75K: after expenses, 33.3% inventor, 25% univ., 52.7% dept.	Starting to implement computer program to monitor license payments. Have begun terminating non-compliant licenses	Service to faculty
7	Usually exclusive licenses	Reassignment to inventor	Non-medical: <$100K: 50% inventor, 10% dept. 30% school, 10% tech promotion. Non-medical >$100K: 40% inventor, 20% dept., 25% school, 5% tech promotion, 10% tech research. Medical <$100K: 50% inventor, 20% dept., 20% school, 10% tech promotion. Medical >$100K: inventor 40%, dept. 255, school 20%, tech promotion 5%, tech research 10%	An administrator monitors via a computer program.	Level of overall activity: patent applications, license income, royalties
8	Only exclusive licenses	Reassignment to inventor	Inventor 50%. Of remainder, 25% dept., 75% general fund not vice versa	Not effective. Companies usually contacts OTT	Number of invention disclosures
9	No policy	Licensed to inventor	40% inventor, 20% inventor's research, 10% dept., 10% school, 20% univ. adm.	Finance person within OTT monitors agreements	Activity level wrt number of disclosures. Restricted disclosure agreements, MTAs, license agreements, patent app., patents issued, lic. income, expenses reimbursed

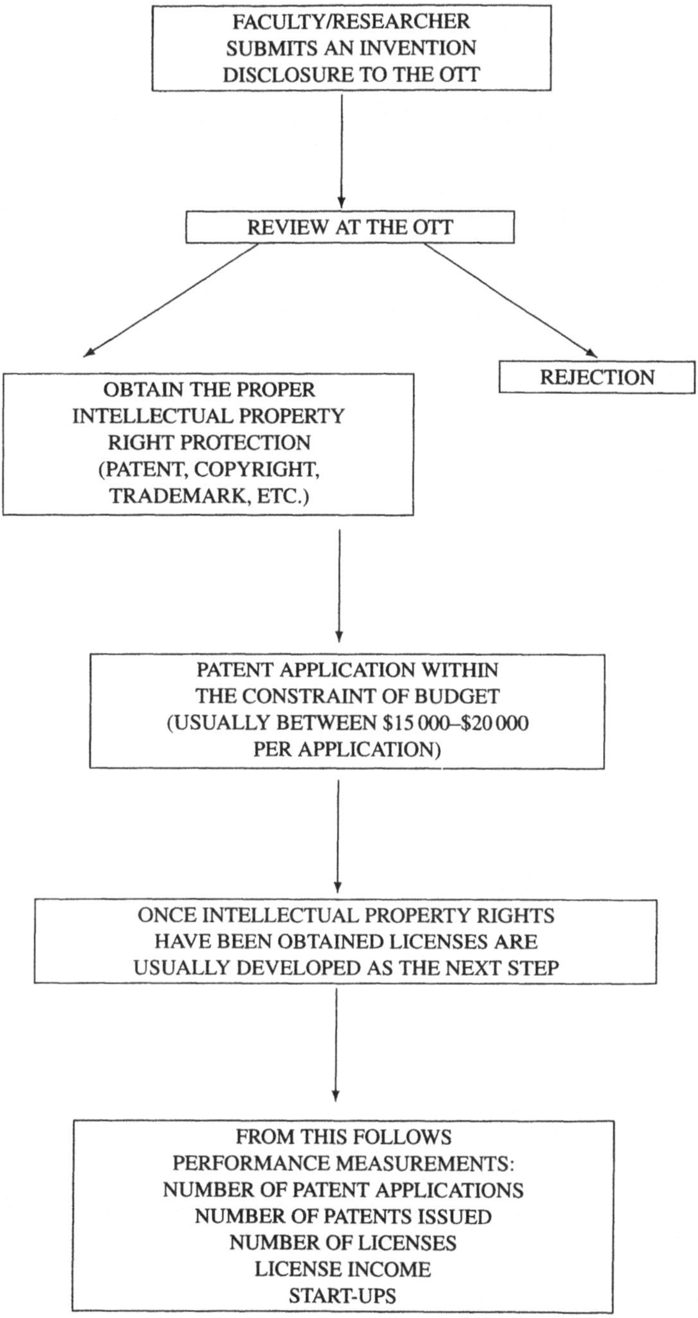

Fig. 1. The process of technology transfer and the most common performance measurements

input to another stage. For instance, invention disclosures may be regarded as an output but may also be used as an input in the patent application stage.[4] When describing the interaction between the industry and the university, a model of 'two-way' interaction is a more appropriate way of expressing the links between them. A study by Meyer-Krahmer and Schmoch (1998) shows that collaborative research and informal contacts are the most important interactions between universities and industry and that industrial firms are important producers of new knowledge which in turn is useful for academic research.

As indicated above, it may be expected that the more resources are spent on R&D, the greater the expected outcome. Therefore we expect a high correlation between total research expenditure and the number of licenses and patents, leading to license income.

Table 7. Total research expenditure, number of licenses and options, and total license income, FY 1996

Institutions	Total research expenditure ($M)	Cumulative active licenses and options	Total license income ($M)
5	441.3	142	1.1
4	395.5	903	43.8
10	328.0	94	0.8
11	250.0	218	9.4
8	218.0	N.A.	3.9
6	207.7	78	1.1
3	165.0	43	7.1
9	146.0	38	2.6
13	144.1	82	2.9
1	140.6	37	0.5
7	128.5	67	0.6
2	64.4	39	2.2
12	13.5	6	0.9

An interesting aspect is whether, in fact, those universities that spend the most money on R&D are also the ones that have the largest number of active licenses and patents and the highest royalty income. Table 7 shows the research expenditure and the cumulative active licenses per institution. The universities are ranked in descending order by the total research budget. The table gives an indication of the correlation between the input research expenditure and the output licenses and options in the technology transfer process.

The correlation between the two variables Total Research Expenditure and Cumulative Active Licenses and Options is high, 0.599. A simple linear OLS (Ordinary least squares) regression shows that Total Research Expenditure has a high

[4] It would also be possible to describe the process in a more complicated model with feedback but we would then lose simplicity and transparency without necessarily obtaining better models.

explanatory power, explaining as much as 35 percent of the variation in the number of licenses and options. It is also significant as indicated by the t-value (on the 5 % level) in explaining the number of licenses and options. If we look at all 212 universities in the AUTM survey, the relationship is even tighter: the corresponding figures are 0.99 and 99 percent, respectively. The correlation between the number of licenses and options and license income received is also high: 0.975 in our subset and 0.988 in the AUTM population.

3 Statistical analysis of the AUTM survey data 1991–1996

3.1 Data collected by AUTM

This section is based on two data sets, one for 1996 and one covering the period 1991–1995. Both data sets include both the entire population of universities surveyed by the AUTM and the subset of universities included in our own questionnaire survey.

The AUTM survey population for 1996 consisted of 300 institutions: 212 U.S. universities, 55 U.S. hospitals and research institutes, 28 Canadian institutions, and 5 third-party management firms. 58 percent (173 organizations) responded to the survey, including 131 U.S. universities, 26 U.S. hospitals and research institutes, 14 Canadian institutions, and 2 third-party management firms. The response rate from the universities, 62 percent, is somewhat higher than the overall response rate. There is a small over-representation of the top universities since the follow-up efforts were concentrated on the top 100 universities. This should not affect the results of the technology transfer function, which is the main focus of this report. For the purpose of our analysis, only the U.S. universities are included in the presentation of the data set, which we will refer to as N, the total sample.[5]

The AUTM data include information on more than 30 variables that all have to do with technology transfer. We have chosen to look more closely at variables emphasized as important by the universities included in our survey and what other researchers have pointed out as important in the literature (see Tables 8 and 9).

3.2 Results

As can be seen in the Tables 8 and 9, looking at the whole data set N, most of the technology transfer offices were started in the mid 1980s. This does not mean, however, that there was no such function before that time. Many universities engaged in technology transfer even without such an office. Some universities have a big office with more than 100 professional and support staff, but the mean is only 5.2 professional and support staff indicating a much more modest size of the technology transfer office in most universities. In the subset, the corresponding figures are 18 professional and support staff as maximum, while the mean is somewhat higher than for the whole group, 6.6 professional and support staff.

[5] The data material was processed with the SPSS software. The correlations were calculated using the Pearson correlation coefficient and the regressions using linear regression, OLS (ordinary least squares).

Table 8. Descriptive statistics n

	N	Minimum	Maximum	Mean	Std. deviation
PROGYEAR Year in which institution devoted 0,5 prof. FTE to technology transfer	13	1970	1992	1985	6
96PTTFTE Prof. FTEs in technology transfer office	13	1	7	3	2
96STTFTE Support staff FTEs in technology transfer office	13	1	11	3	3
96INDEXP ($ M) Research expenditures: Industrial sources	13	2.7	35.1	14.8	9.4
96FEDEXP ($ M) Research expenditures: Federal government sources	13	7.6	336.8	154.1	96.8
96TOTEXP ($ M) Total research expenditures	13	13.5	441.3	203.3	124.0
96LCEXEC Licenses/options executed	13	1	136	29	36
96ACTLIC Cumulative active licenses	12	6	903	146	245
96LIRECD ($ M) License income received	13	0.5	43.8	5.9	11.7
96LCGNLI Licenses/options generating licenses income	13	1	259	52	70
96LILIFE ($ M) Licenses income received in life science	13	0.2	36.1	4.6	9.7
96LIPHYS Licenses income received in physical science	13	0	7.7	1.4	2.7
96EXPLGF ($ M) Legal fees expended	13	0.1	2.3	0.9	0.7
96REIMLG ($ M) Legal fees reimbursed	13	0	941	363	347
96INVDIS Invention disclosures received	13	22	300	90	76
96TPTAPP Total patent applications filed	13	4	130	45	40
96NPTAPP New patent applications filed	13	3	90	33	28
96USPTIS U.S. patents issued	13	5	56	20	16
96STRTUP Start-ups initiated	13	0	14	3	5
Valid N (listwise)	10				

Table 9. Descriptive statistics N

	N	Minimum	Maximum	Mean	Std. deviation
PROGYEAR Year in which institution devoted 0,5 prof. FTE to technology transfer	120	1925	1997	1985	12
96PTTFTE Prof. FTEs in technology transfer office	130	0	60	3	6
96STTFTE Support staff FTEs in technology transfer office	130	0	41	2	
96INDEXP ($ k) Research expenditures: Industrial sources	130	59	115,497	11,771	14,447
96FEDEXP ($ m) Research expenditures: Federal government sources	131	1.4	869	94	124.7
96TOTEXP ($ m) Total research expenditures	131	5.4	1,518	142	180.3
96LCEXEC Licenses /options executed	128	0	137	17	25
96ACTLIC Cumulative active licenses	128	0	903	82	137
96LCEXSU Licenses executed to start-up companies	124	0	14	2	3

Table 9 (continued)

	N	Minimum	Maximum	Mean	Std. deviation
96LIRECD ($ M) License income received	131	0	63.2	2.8	7.8
96LCGNLI Licenses /options generating licenses income	130	0	513	38	63
96LILIFE ($ M) Licenses income received in life science	119	0	36.1	1.6	4.1
96LIPHYS ($ M) Licenses income received in physical science	119	0	7.7	0.4	1.1
96EXPLGF ($ M) Legal fees expended	127	0	18.0	0.6	1.7
96REIMLG ($ k) Legal fees reimbursed	126	0	7,090	227	686
96INVDIS Invention disclosures received	131	0	670	62	80
96TPTAPP Total patent applications filed	130	0	325	30	40
96NPTAPP New patent applications filed	129	0	207	21	27
96USPTIS U.S. patents issued	131	0	159	14	19
96STRTUP Start-ups initiated	129	0	14	1	2
Valid N (listwise)	80				

The total research expenditures vary strongly among the universities. The range is from \$5 million to as high as \$1.5 billion. The average expenditure is \$142 million. As these numbers indicate, the standard deviation is large, \$180 million. In the subset, the variation is somewhat smaller, from \$13 million to \$441 million. On average, the universities spend about \$203 million. The standard deviation in the subset is also smaller, \$124 million.

What can we say about performance? The number of active licenses represents one measure. This variable varies between 0 and 903 licenses. Overall, the mean is 82 licenses with a standard deviation of 137. Perhaps even more interesting is the license income which varies from 0 to \$63 million. On average, the universities have an income of about \$2.7 million from licenses. The standard deviation is \$7 million. In the subset, the number of active licenses ranges from 6 to 903, with a mean of 145 licenses and a standard deviation about 245. The license income varies between \$543,000 and \$43 million. The average income is \$5 million, and the standard deviation is as high as \$11 million.

It is even more interesting to look at the licenses and options generating license income. As shown previously, there are only a few very successful licenses that generate a large amount of income; most licenses do not. The number of licenses producing income ranges from 0 to 513. The average is 38 licenses per university and the standard deviation is 63. In the subset the number varies from 1 to 259. The average the number is 52 with a standard deviation of nearly 70.

Another output variable is the total number of patents, which varies between 0 and 325 applications filed. The average number is approximately 30, and the standard deviation is 40. For the subset, the number of applications varies between 4 and 130. On average, there are 45 applications per university and the standard deviation is 40.

The number of startups initiated varies from 0 to 14 with a mean of 1.4. The standard deviation is 2.3. In the subset, the maximum is the same but the mean is somewhat higher, 3.2. The standard deviation is also higher than for the whole group, 4.6.

Following the process of technology transfer as outlined in Figure 1 it becomes clear that the technology transfer process involves a sequence of events. It starts with the invention disclosure leading to a review; then the disclosure is either rejected or accepted, usually with some adjustments. The next step is to look for the right kind of property right protection, such as a patent. Once intellectual property rights have been obtained, licensing is likely to be the next step. In some cases, the end result is a start-up. Since it all depends on the previous step in the chain, we decided to build several models using the performance measures as our dependent variables[6]. The models are described in Matrix 1 below. One important factor that has to be incorporated in the models is time. Once a patent has been issued, it can take years before a license is executed. The same is true for license income. It depends on what has happened in the past; the path dependency is strong. The different steps in the technology transfer process can be described in several models. More models

[6] Looking at the correlations among some of the variables can give us some more information. See Table 10 for details. Overall, the correlations are very high, indicating that there are strong linkages between the variables.

Table 10. Correlation matrix N and n

Correlations on N

	PROGYEAR	96TOTEXP	96LCEXEC	96ACTLIC	96LIRECD	96LCGNLI	96INVDIS	96TPTAPP	96USPTIS	96STRTUP	96TOTFTE
PROGYEAR	1.000	−0.504	−0.641	−0.536	−0.378	−0.567	−0.559	−0.536	−0.594	−0.449	−0.463
96TOTEXP	−0.504	1.000	0.996	0.995	0.982	0.997	0.999	0.999	0.998	0.994	0.997
96LCEXEC	−0.641	0.996	1.000	0.998	0.984	0.998	0.997	0.997	0.997	0.995	0.993
96ACTLIC	−0.536	0.995	0.998	1.000	0.988	0.998	0.996	0.997	0.996	0.995	0.994
96LIRECD	−0.378	0.982	0.984	0.988	1.000	0.989	0.983	0.985	0.984	0.980	0.988
96LCGNLI	−0.567	0.997	0.998	0.998	0.989	1.000	0.997	0.998	0.998	0.993	0.997
96INVDIS	−0.559	0.999	0.997	0.996	0.983	0.997	1.000	0.999	0.998	0.995	0.996
96TPTAPP	−0.536	0.999	0.997	0.997	0.985	0.998	0.999	1.000	0.999	0.995	0.996
96USPTIS	−0.594	0.998	0.997	0.996	0.984	0.998	0.998	0.999	1.000	0.994	0.996
96STRTUP	−0.449	0.994	0.995	0.995	0.980	0.993	0.995	0.995	0.994	1.000	0.990
96TOTFTE	−0.463	0.997	0.993	0.994	0.988	0.997	0.996	0.996	0.996	0.990	1.000
PROGYEAR	1.000	−0.595	−0.819	−0.864	−0.734	−0.764	−0.599	−0.870	−0.723	−0.858	−0.670
96TOTEXP	−0.595	1.000	0.702	0.599	0.467	0.672	0.467	0.703	0.782	0.741	0.903
96LCEXEC	−0.819	0.702	1.000	0.980	0.930	0.991	0.362	0.760	0.770	0.801	0.810
96ACTLIC	−0.864	0.599	0.980	1.000	0.975	0.968	0.636	0.897	0.777	0.868	0.757
96LIRECD	−0.734	0.467	0.930	0.975	1.000	0.920	0.270	0.646	0.659	0.691	0.658
96LCGNLI	−0.764	0.672	0.991	0.968	0.920	1.000	0.255	0.690	0.717	0.723	0.790
96INVDIS	−0.599	0.467	0.362	0.636	0.270	0.255	1.000	0.827	0.631	0.784	0.344
96TPTAPP	−0.870	0.703	0.760	0.897	0.646	0.690	0.827	1.000	0.843	0.913	0.648
96USPTIS	−0.723	0.782	0.770	0.777	0.659	0.717	0.631	0.843	1.000	0.791	0.792
96STRTUP	−0.858	0.741	0.801	0.868	0.691	0.723	0.784	0.913	0.791	1.000	0.759
96TOTFTE	−0.670	0.903	0.810	0.757	0.658	0.790	0.344	0.648	0.792	0.759	1.000

Matrix 1. The technology transfer process step by step

Invention disclosure = b_1*program year + b_2*total expenditure + b_3*staff + e
Total patent application = b_1*program year + b_2*total expenditure + b_3*staff + e
Total patent application = b_4*invention disclosure +e
New patent application = b_4*invention disclosure + e
US patent issued = b_5*total patent app. + e
US patent issued = b_6*new patent app.+ e
Licenses = b_7*patent issued $_{(t-1)}$ + e
Licenses = b_5*total patent app. $_{(t-1)}$ + e
Cumulative licenses = b_5*total patent app. + e
License income = b_8*cumulative licenses + e
Start-ups = b_1*program year + b_2*total exp. + b_3*staff + e

t stands for time

than are being shown have been tested and analyzed, but only the most relevant and significant will be presented here.

As mentioned in the introduction, our main focus is on capturing the variables that can explain or give an indication of the universities' technology transfer process and activity. The most common output variables are the number of patents and the number of licenses. The output variables of greatest interest here are the number of total patents, the number of new patents, the number of U.S. patents issued, license income, the number of licenses, and the number of start-ups. These variables are our dependent variables and will be noted by Y. Concerning independent variables, i.e., variables that might explain the technology transfer activity, we assume that the more money is spent, the more activity can be expected. Examples of input variables are: research expenditures, number of invention disclosures, number of employees (i.e., number of staff), and the number of years that the OTT has been operating (we expect to find a learning curve). We will note the input variables with X.

The reason that we do not include the year in which the OTT was established (referred to as "program year"), total FTEs, and total research expenditures as input variables in all our models (they are included only in the first two and the last one) is the sequential nature of the process: the output of one step may be input in the next. Therefore, we expect the three variables mentioned above to be incorporated already. The correlation matrix (Table 10) shows that there are strong links between all the variables. We would therefore have to deal with multicollinearity (covariation among the independent variables) which could lead to incorrect estimation of the regression coefficients. The variables could even have the "wrong" signs. In the presence of multicollinearity, the effect of each variable on the dependent variable cannot be determined.

To estimate the models we used linear regression, OLS. The results from the regressions are shown in Tables 11 and 12. We have chosen to give the beta value (β) when dealing with multiple regression equations which could be helpful in analyzing what variable has the strongest impact in predicting the dependent variable. In all other cases the regression coefficient (b) is given.

Table 11. Regression on N

Y	b_1*X_1 program year	b_2*X_2 total exp.	b_3*X_3 Staff/ FTE	b_4*X_4 Inv. dis	b_5*X_5 total patent app.	b_6*X_6 new patent app.	b_7*X_7 US patent issued	b_8*X_8 Cum. licen-ses	R2 and R2 adj. *= sig. t-values
Inv. dis.	−.964	2.230 E-07	3.003						0.839 0.835
	β_1 =-0.139	$\beta_2 = 0.502$	$\beta_3 = 0.361$						*
Total patent	−0.347	1.313 E-07	1.206						0.839 0.835
	β_1 =-0.100	$\beta_2 = 0.593$	$\beta_3 = 0.291$						*
Total patent				0.468					0.888 0.888 *
New patent				0.306					0.847 0.846 *
Patent issued					0.436				0.812 0.811 *
Patent issued						0.707 (t-1)			0.807 0.805 *
License							1.234 (t-1)		0.702 699 *
Cum. Licenses					2.906				0.700 0.697 *
License inc.								45590	0.622 0.619 *
Start-ups	−3.775 E-02	5.428 E-09	2.157 E-02						0.398 0.382
	β_1 =-0.192	$\beta_2 = 0.429$	$\beta_3 = 0.091$ not sig. t –value						*

Table 12. Regression on n

Y	b_1*X_1 program year	b_2*X_2 total exp.	b_3*X_3 Staff/ FTE	b_4*X_4 Inv. dis	b_5*X_5 total patent app.	b_6*X_6 new patent app.	b_7Xx_7 US patent issued	b_8*X_8 Cum. licen- ses	R2 and R2 adj. *= sig. t-values
Inv. Dis.	−8.469 $\beta_1 = -0.685$	5.397 E-07 $\beta_2 = 0.886$ not sig. t-value	−13.733 $\beta_3 = -0.915$						0.509 0.345 *
Total patent	−5.229 $\beta_1 = -0.804$	2.188 E-07 $\beta_2 = 0.683$	−4.005 $\beta_3 = -0.507$						0.851 0.801 *
Total patent				0.435					0.684 0.655 *
New patent				0.316					0.739 0.715 *
Patent issued					0.337				0.711 0.685 *
Patent issued						0.547 (t-1)			0.410 0.357 *
License							1.688 (t-1)		0.818 0.802 *
Cum. Licen- ses					6.415				0.805 785 *
License inc.								48461	0.950 0.945 *
Start- ups	−0.482 $\beta_1 = -0.641$	1.254 E-08 $\beta_2 = 0.338$	2.171 E-02 $\beta_3 = 0.024$ not sig. t–value						0.819 0.759 *

As we expected from the correlation matrix, the overall R^2 values are high, indicating that we can predict the dependent variable fairly confidently. Invention disclosures can to 83 percent be explained by program year, total research expenditures, and number of staff. For the subset this is lower, 34 percent. The variable that has the strongest impact on invention disclosures is total research expenditures, but for the subset it is the number of staff, which has a negative (but statistically insignificant) impact on invention disclosure. Comparing the subset with the whole sample, it does not look like our sample is very different from its population. However, an exception is that in our sample, 95 percent of the variation in license income is explained by the number of cumulative licenses, whereas for the whole sample the corresponding figure is 62 percent. This means that adding one more license to the number of cumulative licenses has a rather large impact on license income. As indicated by the regression, it increases the license income by $45,590 and for the sample by $48,461 on an average basis. But we have not captured all the explanations behind license income. Licensing may not be appropriate for all inventions, and certainly chance (or luck) plays an important role.

One of the variables that is hard to predict is the number of start-ups. The only variables we can use as explanatory variables are program year, total research expenditures, and number of staff. Together they explain 38 percent of the variation in the number of start-ups. For our subset we can explain as much as 80 percent with the same model. These results show that there is more to start-ups than money and number of employees at the OTT. Earlier in the report we have mentioned that such things as culture and surrounding environment can have a big impact on entrepreneurial ability.

4 Conclusions

It is clear from this survey that technology transfer from universities to the commercial sector needs to be understood in its broader context. It is not simply a matter of maximizing income for the universities, even though in a few cases quite substantial income is generated. It is rather a matter of finding the proper balance between the basic functions of teaching and research within the universities on the one hand and providing service to the wider community on the other. The primary purpose of a technology transfer program is for the university to assist its researchers in disseminating research results for the public good. Success in this endeavor is only partially reflected in income generated for the university or the number of business start-ups. The degree of success depends not only on the nature of the interface between the university and the business community but also on the receptivity in the surrounding community as well as the culture, organization, and incentives within the universities themselves.

In the sample of universities studied here, the OTT is set up as a unit within the university. In most cases the director reports to the Vice President for Research or Provost, indicating that the activity is regarded as strategically important. The staffing varies from 2 to 18 and the annual budget for licensing and patenting varies from less than $400,000 to $4 million. The larger the OTT, the broader is the in-house expertise, and the more aggressive the pursuit of patents and licenses.

The steps to transfer or commercialize a technology are basically the same at all universities, as illustrated in Figure 1. In most cases, sign-off authority on Non-Disclosure Agreements, Material Transfer Agreements, and licenses resides within the OTT.

In universities with a medical school, medicine is usually the field most frequently served by the OTT, followed by engineering. Sometimes arts and sciences (particularly chemistry) are also engaged in technology transfer activities.

There are several stages in the technology transfer process, each associated with its own outcome: invention disclosure, patent application, patent issued, license sold, license income and/or business start-up. As a rule, only half of the invention disclosures result in patent applications; half of the applications result in patents; only a third of patents are licensed, and only a handful (10 – 20 %) of licenses yield substantial income. In our sample, the median gross license income in 1996 was $2.2 million and the median income per license was $80,000.

In our sample, industry-sponsored research maintained its share of about 7 percent of total research expenditures over the period 1991–1996. The share in 1996 varied between 4 % and 20 % in the institutions surveyed.

License income is shared among the inventor(s), the management center or school, and the OTT according to various formulae. The inventor's share generally varies between 30 and 50 %. Monitoring of royalty agreements is carried out routinely in some universities but seems to have received low priority in several cases. No general agreement seems to exist regarding indicators of success in technology transfer activities.

The policies with regard to start-ups vary considerably among the universities. It is well known that MIT and Stanford University have an excellent record of interacting with the local community to create new business ventures. MIT's role in the so-called 'Route 128 phenomenon' (Saxenian, 1996) is well known. An average of 25 companies spun off from MIT during each year in the 1980s. The same pattern is found in California's Silicon Valley originating from research carried out at Stanford University (Roberts and Malone, 1996).

It is interesting to note that while Stanford has a policy not to encourage start-ups, it has generated more start-ups than any other university; only MIT comes close. It almost looks like an inverse relationship: the more active policy the university has, the fewer start-ups can be observed. This may simply indicate that if there are few start-ups, the university needs to take more action. Conversely, if there are many start-ups, no university action is necessary. Thus, there are other things than university policies that explain start-ups. In the entrepreneurial literature culture, personality, and networks are often mentioned as conducive to new ventures. A recent report by Bank of Boston (1997) investigating companies founded by MIT graduates emphasized that MIT encourages its students to become risk-takers and to start their own business. It is also well established that children of entrepreneurs are more likely than others to become entrepreneurs – another indicator of the importance of culture and tradition in the surrounding community. As mentioned in the introduction, we are only investigating one side of the coin, the university side. In order to get the whole picture we would have to include the other part as well, which is beyond the scope of this report.

References

Abramson NH, Encarnacao J, Reid PP, Schmoch U (1997) Technology transfer systems in the United States and Germany. Fraunhofer Institute for Systems and Innovation Research, National Academy of Engineering, National Academy Press, Washington, DC

Adams JD, Griliches Z (1996) Research productivity in a system of universities. NBER Working Paper #6120

Association of University Technology Managers (AUTM) homepage (1998) http:autm.rice.edu/autm/

Audretsch DB, Stephan PE (1996) Company-scientist locational links: the case of biotechnology. American Economic Review 86 (2): 630–640

AUTM (Association of University Technology Managers, Inc.) (1996). AUTM five year Licensing Survey 1991–1995. AUTM, Norwalk, CT

AUTM (Association of University Technology Managers, Inc.) (1998) AUTM Licensing Survey 1996. AUTM, Norwalk, CT

Bania N, Eberts R, Fogarty MS (1993) Universities and the startup of new companies: can we generalize from Route 128 and Silicon Valley? Review of Economics and Statistics 76 (4): 761–766

Bank of Boston (1997) MIT: The impact of innovation. Economics Department, Bank of Boston, Boston, MA

Bessy C, Brousseau E (1999) Technology licensing contracts: features and diversity. Mimeo, CEE & ATOM, Université de Paris

Bremer HW (1999) University technology transfer evolution and revolution. http://web.mit.edu/osp/www/cogr/bremer.htm 2/18/99

Cooke I, Mayes P (1996) Introduction to innovation and technology transfer. Artech House, Boston

Council on Governmental Relations (COGR) (1993) The Bayh-Dole act: a guide to the law and implementing regulations. http:www.tmc.tulane.edu/techdev/Bayh.html 9/30/98, 1–7

Council on governmental relations (COGR) (1998) Homepage, http://www.cogr.edu/index.htm 9/30/98

Council on governmental relations, COGR (1993) Homepage, University technology transfer: questions and answers. http://web.mit.edu/osp/www/cogr/qa.htm 2/18/99

Fisher L M (1998) The innovation incubator: technology transfer at Stanford University. Strategy & Business 13: 76–85

Homepages of all the universities included in the survey

Jaffe AB, Trajtenberg M, Henderson R (1993) Geographic localization of knowledge spillovers as evidenced by patent citations. Quarterly Journal of Economics 108 (3): 577–598

Jensen R, Thursby M (1998) Proofs and prototypes for sale: the tale of university licensing. NBER Working Paper No. 6698

Meyer-Krahmer F, Schmoch U (1998) Science-based technologies: university – industry interactions in four fields. Research Policy 27: 835–851

Mowery DC, Nelson RR, Sampat BN, Ziedonis AA (1998) The Effects of the bayh-dole act on u.s. university research and technology transfer: an analysis of data from Columbia University, University of California, and Stanford University. Mimeo, Columbia University

Mowery DC, Sampat BN (1999) Patenting and Licensing of University Inventions: Lessons from the History of Research Corporation. Mimeo, Columbia University

Patent Laws (1998) United States Code: 35 U.S.C. 151–211, http:www.klusterlaw.com/lawrule/law2.htm 9/30/98, 1–28

Public Law 96–517 (1980) Text of the Bayh-Dole Act, http:maps.nemoline.org/techtrans/bayhdole.html, 9/30/98, 1–31

Roberts EB, Malone DE (1996) Policies and structures for spinning off new companies from research and development organizations. R&D Management 26 (1): 17–48

Rudolph L (1998) Overview of federal technology transfer. http:www.plc.edu/risk/vol5/spring/rudolph.htm, 9/30/98, 1–7

Saxenian A (1996) Regional advantage. Harvard University Press, Cambridge, MA

Siegel D, Waldman D, Link A (1999) assessing the impact of organizational practices on the productivity of university technology transfer offices: an exploratory study. NBER Working Paper #7256

Technology Transfer homepage, http:maps.nemoline.org/techtrans.html, 9/30/98

What is the systems perspective to Innovation and Technology Policy(ITP) and how can we apply it to developing and newly industrialized economies?*

Morris Teubal

Economics, The Hebrew University, Mount Scopus, Jerusalem, Israel
(e-mail: msmorris@mscc.huji.ac.il)

Abstract. Despite recent advances in the Evolutionary and Systems Perspectives to Economic Change (SI), confusion still exists about how to apply it to the design and implementation of Innovation & Technology Policy (ITP) in concrete settings. Since the 'Normative' aspects of SI are framed in terms so general to make them insufficient or inadequate as guides and tools for actual policymaking, a presumption exists that additional theoretical and conceptual *knowledge* is required. Thus a major objective of this paper is to contribute to the development of a realistic and 'grounded' theoretical framework for Technology and Innovation Policy which is particularly relevant both for the promotion of Business Sector R&D and of hi tech (especially IT) industries in Top Tier and other Industrializing Economies. A second objective is to contribute directly to the capability of successfully applying this conceptual framework in concrete policy settings. Rather than justifying ITP the paper focuses on characterising and applying "Salient *Normative* Principles or Themes" of the SI perspective to ITP. Several concrete examples are given and the notions of Policy Process, (Country) Program Portfolio Profile and Policy Environment are introduced.

Key words: Systems of innovation – Business sector – Innovation and technology policy – Policy process – Program portfolio profile

JEL Classification: H23, H50

---

* This is a shortened version of a paper with the same title (Teubal, 2001a). Thanks to Uwe Cantner for helpful comments and suggestions.

1 Background and objectives

Despite significant advances in the approaches to Innovation & Technology Policy (ITP) in the literature(see STI Review, 1998) confusion still exists about how to apply such a conceptual framework in specific settings. Most discussions during the 90s emphasise the increasing importance of Evolutionary and Systems Perspectives to Economic Change (see papers in Edquist, 1997) but these relate more to the processes of Innovation and Technological Change themselves rather than to *policy* directed to these areas. Moreover, whenever 'normative' aspects enters the analysis they are framed in terms so general to make it insufficient or inadequate as guides and tools for actual policymaking.[1] Underlying this paper is the view that additional theoretical and conceptual *knowledge* is required for an Evolutionary/Systems of Innovation (SI) perspective *on ITP* to be effectively applicable in a wide variety of settings.

A major objective of this paper is *to contribute to the development of a realistic and 'grounded' theoretical framework for Technology and Innovation Policy.* This is considered as a pre-condition for successful application of the SI perspective. Since the area is very broad I will be focusing on aspects which are relevant particularly but not exclusively in two contexts/ types of countries: a) the promotion of Business Sector R&D both in countries which recently have systematically initiated the introduction and diffusion of this activity and in the Top Tier Group of NIE (e.g. Korea and Israel) which have implemented such policies in the past b) the promotion of hi tech industries (particularly Information Technology and Software industries) which is becoming a major issue for both Top Tier and other Developing Countries(e.g India and China). A second aspect of our focus is our concern for the 'needs' of the business sector (direct and also indirect aspects). We will not be considering ITP directed to Health, Ecology, etc; nor that directed to the support of Science.[2] This does not mean that the so-called Business Sector 'Supporting Structure' will not be considered. It will, but only inasmuch as it supports the Business Sector. On the other hand I will not exclusively be dealing with Innovation policy narrowly speaking but also with aspects of Technology (and Industrial) policy as well.

A second objective is *to contribute directly to the capability of successfully applying this conceptual framework in concrete policy settings.* The focus will not be on providing justifications for Government support of Innovation and Technology, but in helping policy makers *apply SI principles.* Preciously few concepts and tools pertaining to ITP are readily available today, the most famous of all being two related and highly criticised notions: *market failure* as applied to R&D and Innovation, and '*market failure analysis*' (Arrow, 1962; Stoneman, 1988) which purports to provide the quantitative basis for estimating the subsidy to be given to R&D. That such concepts are either inconsistent with Innovation itself or inadequate as a guide

[1] One paper that has gone beyond others in this respect is B. Johnson. (see Johnson, 1997) (see also Teubal, 1999; Dodgeson and Bessant, 1996; Caracostas and Muldur 1998; Edquist, 1999).

[2] Some reference to Science of course will be made since it is part of Technology Policy as well as part of the Infrastructure for generating high level manpower. However I will not be directly and centrally be involved with Science Policy nor with the interface between Science and Technology.

to policy making.has been shown repeatedly by Evolutionary theorists (Nelson and Winter, 1982; Nelson, 1987, 1994, 1955; Dosi, 1988; Andersen, 1994; Metcalfe, 1996, Saviotti 1997 among others). Despite the emergence of substitute concepts such as *System Failures* (OECD 1997, Galli and Teubal, 1997; Smith 1991), *Technology Policy Cycle* (Teubal, 1996, 1997) *and a 'redefined' notion of market failure* (Teubal, 1998) critical ingredients in the overall Systems ITP framework seem to be missing. In this connection I will introduce and apply the notions of *Program Portfolio Profile* (PPP), *Policy Process* and *PPP Bias*- to name at least some of those missing elements (Teubal, 1999)[3]

2 The SI perspective to ITP: building blocks and general principles

A (National) Systems of Innovation perspective to Innovation/Technology (see Nelson, 1993; Lundvall, 1992; Edquist, 1997) is an 'intermediate' view concerning the role of market forces versus the State in conducting Innovation/Technology/Economic activities in general. It lies between a 'pure market' view of the operation of enterprises (with its emphasis on the individual enterprise facing input and product markets) and a 'planning or State-led' model or mode of operation/development. In the SI perspective individual enterprises operate within (or are embedded in) a set of institutions and non-firm organisations – a 'Supporting Structure' which determines to a large extent both behaviour and outcomes. It is an actor-based system rather than a 'firm' (and consumers) based system since active roles can be played also by decision making units within the Supporting Structure.

The SI perspective is less a framework for the allocation of resources than one for understanding innovation and innovation-based behaviour of firms and non-firm organisations. It emphasises the *Collective* nature of the Innovation Process– every entrepreneur is embedded in a set of institutions/ organisations supporting it; including social structure and culture. This paper's approach the SI perspective to ITP is based on a distinction between general principles and salient aspects of the "Positive" side of the perspective and those of the "Normative" or Policy side. Most of the literature refers to the former. Since "Policy" will not automatically flow from the SI perspective as it is usually presented (with a strong emphasis on 'Positive' General Principles) explicit consideration of 'Normative' principles is important. This will be the major task of this Section.

2.1 Building blocks

Component sub-systems. For our purposes we will consider a system of innovation as comprising *five (5) subsystems* or *meso-level components*: the *business sector* (BS); the *supporting structure* (SS); *interactions and links*; *institutions and markets*; and *culture and social structure* (Teubal and Andersen, 2000). For lack of space

[3] Pre-SI perspectives include Neoclassical (Arrow and Stoneman, op.cit), Structuralist (Lipsey and Carlaw, 1998a,b) and Evolutionary Perspectives(see above). These will not be systematically covered here. The SI approach includes all (or most) elements of the Evolutionary approach, as well as many of the Structuralist perspective. For short summaries of these approaches, (see Teubal 2001b).

our analysis will emphasise the first three. The distinction between the business sector on the one hand and the supporting structure on the other differs from that proposed by Lipsey and collaborators (e.g Lipsey and Carlaw, op. cit.). For them, the basic distinction is between technology and the supporting structure, where the latter would include both business firms and other organisations and institutions comprising our 'supporting structure'. My choice is based on the fact that the Business Sector generates a large and increasing share of GNP and of growth; and on the fact that – in the current era or innovation paradigm – "users" of new technology in this sector must to a large extend develop it themselves. Thus the Business Sector is the "backbone" of the process of SI transformation process.[4]

Business sector (BS). The *business sector* is the *backbone* of the system and its *restructuring* is the central axis of the process of transformation of systems of innovation. A Business Sector may have a wider or narrow *set of sectors* and it may have more or less *depth* that is sectors and firms linked among themselves through input/output market transactions and other links. The industrial base of small countries may be narrow;[5] and whenever the BS is heavily dominated by one sector the system of innovation would largely consist of the sectoral system which is specific to that sector.

From a SI transformation perspective full characterisation of the Business Sector would not be complete without considering *the capacity of firms to adapt to the changing internal and external environment they face*. The presumption of enterprise heterogeneity is crucial. There will be "innovators" which may also be called Schumpeterian entrepreneurs, "imitators", and "laggards"; and the *diffusion of R&D/Innovation (and more, of enterprise restructuring)* will proceed in phases, starting with the former group and ending with the latter (Teubal, op.cit.; Teubal and Andersen, op.cit.).

Supporting structure (SS). The *supporting structure* involves a number of organisations whose behaviour, in contrast to firms, does not follow (or strictly follow) market principles. These organisations indirectly or directly support companies and their 'restructuring' in response to changed conditions, thereby also contributing to the transformation of Systems of Innovation. The supporting structure includes Technology Centers, Universities, Government Laboratories; Government-owned Venture Capital (VC) companies; Business Associations (Saxenian, 1998) and organisations/institutions of Policy (Galli and Teubal, op.cit.). Needless to say that the borderline between the supporting structure and the wider system is not fixed but depends on the context and on the objectives of the analysis. Nor is the borderline between the business enterprise sector and the supporting structure. absolutely

[4] This is consistent with the view that the underlying knowledge and 'generic technologies' frequently are created within Universities and other institutions/organisations of the Supporting Structure. The statement about the business sector being involved in the generation of the innovations it uses includes both situations were there is little room–due to transactions costs and other reasons–for independent R&D companies and others where such companies can thrive. (see Teece, 1986).

[5] E.g. Israel's industrial sector is heavily biased towards hi tech industries and has a lower than OECD average weight of other sectors including a significantly narrow set of mid-high tech and low tech industries.

clear and immutable. Thus a Government-owned Venture Capital company would be part of the supporting structure if its objective was to effectively promote companies. But a privatised VC company which was formerly Government owned would be part of the Business Enterprise sector.[6]

Links and Interactions.[7] These, which include non-market interactions among actors of a SI, are frequently considered the central elements of a system of innovation, one which distinguishes the systems of innovation perspective from a conventional perspective to the business sector where each firm is considered more or less to be on its own, confronting an impersonal and neoclassical market.

Links are critical for SI and more specifically for clusters and industrial districts. Since interactions are very much associated with learning e.g. within user-producer networks (Lundvall, 1985; Johnson, op.cit.) there seems to be a tendency to state that *non-market links* are the essence of a system of innovation. While not denying the importance of non-market links, well functioning SI also require significant market links e.g. among firms in different stages of production, strategic partnerships or alliances due to technology or other complementarities; or inter firm links flowing from market-based processes of diffusion of new generic technologies. Moreover, a lot of the non-market interactive learning is related in some way to market links and market transactions(see Lundvall's "organized market").[8]

Focus: dynamics of BS restructuring and of SI transformation. For the Systems of Innovation perspective to be relevant to ITP today it should not be primarily interested in the *operation* of an existing system but, given its links with the Evolutionary Perspective, in its *transformation and transition* to a new system of innovation, presumably more adapted to the new international environment and internal context. By *successful operation* of a SI I mean not a static situation with no growth but rather a virtuous cycle of growth under more or less unchanged external and internal conditions. However the real challenge today is to sustain growth under fundamentally changed external and internal conditions. Most work on SI however deals

[6] This illustrates how Business Sector "depth" frequently will provide the support required for BS transformation or restructuring withoug the need for new elements of the supporting structure (see above story of Silicon Valley's restructuring during the 80s). In this connection a distinction should be made between two types of segments of the business enterprise sector: a segment whose activity *supports* companies through the provision e.g. of *non-tradable* specialised services and inputs (a largely *indirect* contribution to growth); and one that does not support such companies and whose contribution to valued added is *direct*. The latter would include domestic VC companies whose activities are directed to foreign SU companies. Many Israeli Venture Capital Companies are now establishing themselves in foreign markets(personal communication).

[7] In what follows I will not survey the notion of Institutions. Readers could consult various articles by Nelson and Edquist (ed) (op.cit.), and the survey by Greif (2000). The concept of Institutions is still under flux and two new approaches have recently been proposed (Nelson's social technologies and Greif's integrative analysis). For the purpose of this paper, institutions are 'rules of the game' or the organisations generating/overseeing these rules.

[8] Any characterisation of the SI links should not ignore the *External Links* of the system particulary when the SI is confronted with a new wave in the Globalization process. In the current wave of Globalization an important component of such links are links to Global Capital Markets (Teubal and Avnimelech, 2002).

with operation rather than with transformation of the system (e.g. most sections in Edquist, op.cit.).

System *transformation* is a *Cumulative Diffusion process fed by Collective Learning and System* Effects (Teubal, 1998; Teubal and Andersen, op.cit.). Learning about restructuring is "collective" especially in the early phases of system of innovation transformation.[9] 'System Effects' reinforce the effects on cumulativeness of the transformation process; they derive, e.g. from the activities of new or modified organisations/institutions supporting the Business Sector (BS).

BS restructuring involves both Technological and non-technological factors e.g. new Management Routines, Strategies and Organisational forms. It is not an automatic process. For this reason and because of its importance I view it as the *backbone and focus* of the analysis of SI transformation.

2.2 "Positive" general principles (compact presentation)

The General Principles underpinning a (Dynamic) System of Innovation Perspective to ITP are listed in Table 1.

Table 1. "Positive" SI General Principles

R&D and learning
New system components
Cumulativeness and co-evolution
Emergence of new demands (for new components, links, etc)
Fundamental uncertainty
Role of capabilities
Firm heterogeneity
Key agents
System embeddedness
Possibility of lock-in

I will very briefly expand here on the first five of the above items.

R&D and learning. The basis of the system of innovation perspective is that both R&D *and* Learning are important, and that there are many types of learning (as well as of R&D) including 'learning about R&D' or 'learning to innovate'(see Table 1). It is inconceivable that SI transformation take place without important learning (partly 'collective') processes taking place.

[9] Collective learning concerning 'whom' and concerning market and technology trends was emphasised by Saxenian in her book on the Silicon Valley hi tech cluster which is a 'regionally based' system of innovation. (see, Saxenian, op.cit.).

Table 2. Intrafirm 'learning about R&D/innovation'-early 'innovation phase' of countries

1. Learning to Search for Market & Technological Information
2. Learning to identify, screen, evaluate, select & generate new projects
3. Learned to generate and execute "complex" R&D projects
3. Learn about the importance of Marketing, thereby overcoming the previously held view that "my invention is so good that it will sell by itself"
4. Learn to manage the innovation process(linking Design to Production & Marketing; selection of personnel; budgeting; etc); etc

New sub-system components; new links. When changes in the environment are sharp it is unlikely that the transformation of systems of innovation could be undertaken without fundamental changes in the 'architecture' of the system, particularly the incorporation of new elements or components either into the business sector or the 'Supporting Structure'. Taiwan is a good example were the establishment of Government owned laboratories –a component of the 'Supporting Structure'– during the beginning of its 'Innovation Phase' (late 70s and 80s) enabled it both to successfully enter the Semiconductor industry *and* to upgrade more traditional industries.[10] Another example is the appearance of Venture Capital within the Supporting Structure of Israel's high tech industry during the 90s; and the simultaneous appearance of large numbers of new technologically based start up companies during the 90s. Both represented a radical transformation of the previous 'Electronics Industry' and 'Electronics'- oriented system of innovation of the 80s in that country.[11]

Changed architecture also relates to new links, both domestic links across subsystems (e.g. the reinforcement of University-Industry Links) and international ones. Thus the above-mentioned VC segment was very strongly linked from its onset with US Private Equity Investment Companies and other US organisations. More generally, globalisation raises the importance of new and more varied international links-not only standard product market links, but also links associated with cross border transactions in (or concerning) *assets.*

Cumulativeness and co-evolution. A major aspect of the dynamics of transformation is *cumulativeness* that is a process, which when initiated (due to positive feedback from learning and other processes) becomes self- reinforcing. This corresponds to the *post variation and selection* phase of Evolutionary Processes (*'reproduction').* It took place in Israel during the 90s. Once the process was triggered by changing external and internal circumstances (peace, immigration, new links with the US e.g. access to NASDAQ) and by policies (the enactment of three major new

[10] See the analysis of ITRI (Industrial Technology Research Institute) and one of its laboratories (ERSO-Electronic Research Service Organiaation) in Nelson (ed.) (1993, ch. 12).

[11] There are other examples of 'changed architecture', also concerning 'sectoral systems of innovation'. R. Nelson and Sampat (1998) describes how the growth and development of the German Chemical industry in the last decades of the 19^{th}C. depended closely on the creation of Chemistry departments at German Universities.

Government programs directed to VC, entrepreneurship and cooperative R&D) a cumulative process of change took place with success breeding more success. It seems that one critical aspect of this process was interactive learning between start up companies and newly formed Venture Capital companies; another was increasing mutual knowledge of both Israeli and relevant sets of Ucompanies. Thus as mentioned, cumulativeness should be viewed as the combined effect of collective learning and 'system effects' (Teubal, op.cit.). Cumulativeness could be related to *co-evolution*, e.g. between business sector and the supporting institutions/structure (Nelson, op.cit.) or among groups of agents in the Business Sector. Thus the hi tech cluster emerging in Israel during the 90s involved co-evolution between start up companies and Venture Capital.[12]

Emergence of new "demands". An important aspect of SI transformation is *emergence of new demands* – in our context, demands for the new products/services offered by the supporting structure(e.g. "financial and other services offered by VCs"); demand for new links; and demand for *BS* 'restructuring' itself. Coordinated growth of both, supply and demand, is part and parcel of successful transformations (Teubal, op. cit.). Demand should be distinguished from "Need" – there may be a Need for the various new elements of the SI but not demand, at least initially.

This view is important for policy, since there cannot be a presumption that demand for the critical new components of the system will be out there or will automatically appear. Policies–especially those inspired by the 'linear view of innovation'–have been very biased in the past towards the provision of incentives. They have traditionally not aimed directly at the creation of demand. This should and is changing. We know for example that both Supply and Demand policies might have had to be implemented for 'coordinated growth of supply and demand' of Venture Capital in Israel. Note that a major mechanism for enhancing 'demand' is interactive learning (Teubal, op. cit.).

Fundamental uncertainty. The basic insight is that Fundamental Uncertainty exists, not only risk that is translatable into 'states of nature' and corresponding probabilities. This is related to Metcalfe's view that the Capitalist System is an open system (Metcalfe, 1999). Therefore we cannot fully 'model' reality, the direction of change frequently cannot be predicted; and the basis for true 'maximisation' or 'optimisation' is not there. A policy implication is sthat *an element of judgement* is inevitable (Lipsey and Carlaw, op.cit.) although this judgement should be informed by theory and by data

2.3 "Normative" general principles of ITP

Given the multiplicity of SI components and even more of agents or agent types, and the potential complexity of SI transformation a normative perspective should

[12] Work in process. Saxenian op.cit., extensively discusses the collective learning and co-evolutionary processes spearheaded by Sun Microsystems and other companies which were part and parcel of the restructuring of Silicon Valley during the 80s. In the analysis of that period she does not focus , however, on adaptations of the supporting structure.

refer to a *set of policies* rather than with individual policies (Edquist, op.cit., Teubal, op.cit., Teubal and Andersen, op.cit.). System transformation to be effective requires addressing the whole system rather than individual components. Under these circumstances a successful impact of one program or policy will depend on existence or non existence of other policies; and the total impact of a particular policy will involve a significant 'indirect' component.[13] This implies that the policy effort is more complex than what would seem to be the case in a Neoclassical world; and that policy coordination (at a moment of time and through time e.g. in connection with program sequencing) is an important inter-program aspect of such an effort. Moreover a true system of innovation perspective emphasises rather than glosses over systemic differences among countries. This in turn implies that policies are context specific (Lipsey and Carlaw, op.cit.).

Most of the Normative features or general principles of the System of Innovation perspective to ITP are 'claims'; and a minority are normative 'Building Blocks'. A list of such features is shown in Table 3.

Table 3. 'Normative' SI General Principles

Policy objectives: learning and SI transformation
The nature of policy making: adaptive policy maker; policy as judgement; an explicit strategy; policies are context specific; policies are increasingly research and Knowledge Intensive; policy-business co-evolution; policy lock-in
Learning, demand and dynamics: learning in program implementation; stimulating demand for critical components; policy cycle and policy sequencing; policy learning and capabilities
Characteristics of policy set (or program portfolio): set of coordinated policies (programs); mix between top down and bottom up initiatives; mix between institutions and incentives; mix between different types of incentives' programs; focus on R&D *and* non-R&D factors; program profiles and policy biases

Some of the above flow directly from 'Positive' SI General Principles. Thus the "learning" objective is due to the fact that SI transformation requires the onset of a Cumulative process which is based based both on 'learning' and on overcoming System rigidities or failures. I will explain a few of the remaining 'Normative' General Principles.

2.3.1 Adaptive policy maker: The existence of fundamental uncertainty led Metcalfe to ascertain that policy makers are adaptive rather than optimisers (Metcalfe, op.cit.). An optimum allocation of resources' is difficult to conceive of (Metcalfe, op.cit.); and even if exists it would be unknown to policy makers.[14] The term 'Adap-

[13] It also implies that past failure of a particular program may imply, under certain circumstances, that such a policy-together with complementary policies-might have to be reinforced rather than be discontinued.

[14] As mentioned Fundamental Uncertainty also implies that there is an important element of Judgement in Policy so this will be a central feature of the Adaptive Policy Maker. Both concepts explain why market

tive' should not be construed to mean 'incomplete adjustment'; it rather means that neither the direction nor the extent of change can be fully ascertained or determined. Nonetheless, through learning and experimentation; search and research; and through a process of 'muddling through' – ITP should aim at a successful adaptation of the system of innovation to changed conditions.

The adaptive policy maker will act at the various levels or phases of policy making (see 'policy process' below) – strategy formulation, setting priorities, identifying programs and other policy actions. Strategy formulation in particular is fraught with the fundamental uncertainties suggested both by the Metcalfe and the Lipsey analysis.

2.3.2 An explicit strategy: *An explicit strategy* (see Lall and Teubal, 1997) would both establish priorities in certain areas and adopt a 'wait and see' and 'search and experimentation' posture in others. Strategy is justified given the impossibility of applying economic calculus (including risk analysis) to choose among all relevant SI transformation patterns or transition trajectories. Since optimisation cannot be applied policy making should strive a) to generate a small number of alternative future scenarios of the SI; and b) to select one or a very small number, which given the 'informed judgement' of policy makers, seem to be adequate for the country.

A strategic perspective involves at least three elements: a *vision* about the future of the system of innovation; a *strategy* to achieve it; and a set of *priorities* associated with such a strategy. It is not always possible to separate the three; therefore I shall consider *vision/strategy* on the one hand and *priorities* for Innovation and Technology Policy on the other. These priorities would then be translated into programs, institutional changes and other policy actions e.g. information.[15]

2.3.3 A set of coordinated policies: One reason why policy is serious business and a complex one as well is the likelihood that SI transformation requires a changing portfolio of policies with new coordination patterns among the new elements of the set. The accompanying box illustrates ITP program and policy co-ordination in Israel for the 1980–2000 period. These problems are likely to exist or to arise in other countries as well, both in those countries shifting from an 'imitation' to an 'innovation' phase and for top tier non-advanced countries facing the current winds of globalisation as pertains to high tech development (see below and Teubal, 2000).

---

failure *analysis* as propounded by the neoclassical perspective is not relevant for policy making today. This does not mean that a 'redefined' notion of market failure is not useful. It is (Teubal, 1998a, op.cit.) although its nature and role differs from those it has within a neoclassical perspective.

[15] One vision/strategy for many countries in Europe could be a business sector with an undergrowth of start up companies carrying new IT technologies which would facilitate entry into new areas of rapid growth. This would enhance the product innovation component of Europe. In Israel, a reasonable vision/strategy would be different. Beyond assuring the continued flow of SU it would involve a business sector populated by a larger subset of large global companies in IT high tech; a better 'Exploitation' of R&D results more generally speaking; and a thriving biotechnology cluster which may exploit the country's pool of skilled labor in the life sciences (Teubal and Avnimelech, op.cit.)

Table 4. Coordination of programs and policies-Israel 1980–2000

1. Support for Business Sector R&D should have been better coordinated with the supply of high level R&D personnel –engineers, scientists and technicians. Non-coordination caused R&D support to exacerbate R&D personnel shortages (depending on the institutional context). It may also have lead to a rise in compensation rather than to "increases in real R&D performed" (no-additionality).
2. Support for Business Sector R&D should be re-assessed and configured in light of increasingly dominant role which Venture Capital plays in financing Start Up companies (e.g. 3.4 B$ raised during 2000). Venture Capital emerged in 1993 as a direct result of a 'targeted' Government program–"ozma" Both types of finance are increasingly substitutes rather than being complementary (there seems to have been strong complementarities during the first phase of growth of the Venture Capital Industry); and t he marginal contribution of Government subsidies seems to be declining.
3. Enhanced importance of Institutional Changes (e.g. corporate law/governance, de-regulation of telecommunications; competition policy, etc); of general incentives e.g. capital gains taxation; and of other policies (e.g. strategy of the national telecom company and the process of its privatisation). Given that better institutions reduce the 'need' for Business Sector R&D support, changes in both institutions and in incentives must be coordinated.

2.3.4 Stimulating demand for "critical components": Traditionally the major role of policy was to create supply incentives e.g. to reduce enterprise costs of doing R&D, but gradually this was superseded by a more mixed perspective where both demand and supply stimulation was aimed at. The 'R&D/Innovation Learning' that flows from implementation of Horizontal Programs (see below) has a dual effect– first, it reduces cost & enhances the efficiency of R&D; second–it enhances the 'demand ' for R&D by business enterprises. (Teubal, 1996–1997, op.cit.). Similarly with respect to new components of the BS Supporting Structure which may be considered as critical for SI transformation: policies which are directed to stimulate their appearance or emergence must be designed with the view that a) the bottleneck to emergence of such an element may be lack of demand rather than issues of cost/efficiency in supply; b) success may well require enhanced demand as well as lower cost/increased efficiency in supply.[16]

Table 5 shows how a mix between Horizontal and Targeted Policies generated demand for VC within Israel's high tech sector.

2.3.5 The learning approach to program implementation: Given the 'Learning and SI transformation' objective mentioned above, there are a number of elements pertaining to implementation of specific incentives' programs (including 'creation of demand' From a study of Horizontal programs these would include

- Assuring a Critical Mass of Projects as early as possible during the Infant Phase of the program
- Creating a Policy Implementation Network–to assure learning by experience on the part of policy makers

[16] It is important to emphasise the link between learning or interactive learning on the one hand and "demand" for critical elements or components of SI transformation. Frequently one gets the impression from the literature that 'Demand side policies' are circumscribed to "Government Procurement".

Table 5. Learning and stimulation of demand for venture capital (Israel, 1990s)[17]

1. A pre-existing Horizontal R&D Support Program (the "Industrial R&D Fund") which supports R&D throughout the Business Sector. This enhances the flow of SU ('demanders' of VC services)[18]
2. Implementation of the Technological Incubators Program
3. The conditions of operation of 'Yozma'-a Government owned VC company created in 1993 to invest in private VC funds- involved partnerships with experienced foreign Investment Banks/Financial Institutions. This created the opportunity of 'learning from others' and better 'interactive learning' among local agents. Thus a measure of 'market building' was built -in into the program.

- Generating Policy- relevant typologies of R&D projects/Innovations, firms and areas
- Analysing, codifying and diffusing knowledge about the Learning & Growth Processes of successful companies
- Special Attention to promote *wide diffusion* of R&D (or other relevant activity promoted) during the mature phase of program implementation (including actions to avoid biases against SMEs)
- Explicit attention to developing Policy Capabilities
- Other: flexible budgets, use of grants rather than loans, bottom-up determination of projects, etc.

These factors have been analysed in Teubal op cit in the context of Horizontal programs. None of these factors can be assumed to be given, on the contrary–many would go contrary to the intuitions and to the routines of policymakers.

At the individual program level, proof of success in the adoption of a 'learning approach' to implementation, would be *program take-off* i.e a situation where the number of new projects applying for support increases fast and eventually outstrips the possibilities of support.[19] Under favourable conditions take off may happen a few years after initiation of program implementation Two other indicators of success are '*endogenization*' of the activity supported (e.g. R&D in an Horizontal Program); and achieving '*wide diffusion*' of such activity. The former would mean that an increasing number of projects would be implemented even without (or with reduced) Government Support.[20] Achieving wide diffusion, in my opinion, should at least substitute in part the objective of achieving a suitable rate of return on Government disbursements supporting the activity (this means that the 'return' would have a qualitative component-wide diffusion of a strategically important activity within the business sector- and a quantitative one- measured rate of return). Achieving all three objectives would be indicative of the cumulative, learning-induced process mentioned above.

2.3.6 Policy cycle and program sequencing: Application of a systems/evolutionary perspective to the implementation of a *specific* incentives program leads, in re-

[19] This might but need not coincide with the point of inflexion of the S-curve used in Diffusion Studies.

[20] The 'additionality' test justifying Government support to R&D would not succeed. In Neoclassical parlance 'market failure' would have disappeared.

sponse to collective learning and to the generation of new institutions/organisations serving the system, to a Technology Policy Cycle with Infant and Mature Phases. This parallels the Fluid and Rigid Phases of the Innovation/Product Life Cycle. This has been investigated systematically in a series of articles dealing with Horizontal Technology Policies that is policies oriented to the business sector as a whole (rather than to a specific sector or technology); and having a strong neutrality of incentives component, especially in the Infant Phase of implementation of the program.

Table 6. Technology policy cycle: dynamics of learning and market failure; and implicatons for incentives (infant and mature phases)

Stylised situation at the beginning of the infant phase
1. Pervasive learning externalities ('everybody learns from everybody')
2. Ignorance of policy makers (about location of most important market failures)
Implication: Neutrality of incentives
Dynamic/learning effects and their impact during the mature phase
1. Exhaustion of some types of learning (and associated externalities)–those associated with the emergence of a class of 'routine' projects
2. New opportunities for 'complex projects' and for programs promoting other technological activities e.g. generic, co-operative R&D
3. Enhanced knowledge of policy makers
Implication: Greater selectivity in incentives; and desirability of implementing new programs, either horizontal, or targeted or both.

Table 6 is an 'ideal' analysis of the Technology Policy Cycle: it is seldom applied fully in most contexts. Note that implementation according to the model leads to a measure of *program sequencing* i.e. success in the original program will create new opportunities for successfully implementing new programs. In Israel successful implementation of the 'backbone' "Industrial R&D Fund" (since 1969) led – 20 year later-to three new programs: Yozma (in support of Venture Capital, during 1993–1997), The Magnet Program (support of cooperative, generic R&D, since 1992); and the Technological Incubators Program. This program sequencing reflects both 'policy learning' and a changed internal and external environment e.g. the Oslo Peace Agreement, Russian Immigration, etc.

2.3.7 Policy learning and policy capabilities: Table 7 illustrates policy learning in Israel during the 80s which led to new programs in the early 90s.

2.3.8 Salient characteristics of the 'policy portfolio': Mix Between Incentives and Institutions. The view that 'institutions are paramount' (Edquist, op.cit.) seems sometimes to be taken to mean that there is no role for incentives in ITP directed

Table 7. Policy learning–Israel

Underlying the spate of new programs of the early 90s was a measure of both Experienced-Based Learning & research–induced knowledge acquisition
1) Policy- learning underlying Yozma: The high rate of failure of high tech Start Up companies during the late 80s led to a search for means to solve problem. It was found in a new financial institution prevalent in Silicon Valley–Venture Capital. The Chief Scientist, Ygal Erlich, considered alternative ways of stimulating Venture Capital. Some where implemented (the Inbal & Yozma programs); and Yozma was particularly successful
2) Policy-learning underling the Magnet Program: The consolidation of R&D/Innovation Capabilities in Israel during the first 20 years of implementation of the "ndustrial R&D Fund" created new opportunities for the promotion of Cooperative, Generic R&D. A systematic effort of assessing the "Strategic Need", of creating awareness, and of probing for alternative designs for the future program was undertaken by Academics at an independent policy-research institution (The Jerusalem Institute). Policy makers both at the Treasury and at the Office of the Chief Scientist 'learned' about this proposal, and approved its implementation in 1992.

to SI transformation.[21] This is not warranted despite the fact that in some contexts incentives could be largely ignored. Incentives e.g. R&D grants, conditional loans, or general tax policies may potentially play very important roles in system of innovation transformation. Moreover, incentives today could complement institutions today and may help generate patterns of behaviour that eliminate or reduce the need for incentives tomorrow.

Focus on R&D and non-R&D Factors. Enhanced R&D impact and enhanced Innovation during SI transformation in an era of globalisation require explicit attention to non-R&D factors such as organisation and management, marketing and production, and to institutions and other factors leading to the growth and consolidation of innovative companies beyond the R&D phase. Moreover, Saxenian has shown how new company organisational forms, business models & strategies have contributed to Silicon Valley's 're-configuration' during the 80s in response to the Japanese challenge to US dominance of the SC industry (Saxenian, op.cit., ch. 4). These considerations suggest that an ITP policy focusing only on R&D-the focus of the Neoclassical Perspective-could generate weak economy-wide impact due to widespread failure in the commercialisation of supported R&D projects. Related to this, support of R&D only (as in Israel and other countries) rather than support of Innovation as a whole reinforces the inherent technical biases of many technological entrepreneurs.[22]

[21] The reasons are not clear but the fact is that beyond "procurement policies" there has been very little consideration of incentives within the SI perspective (see Edquist, op.cit.)

[22] A balanced between R&D and non-R&D factors would seem to be even more important for small, peripheral economies like Israel who are strongly linked to world asset and capital markets (see Teubal and Avnimelech, op.cit.).

3 Applying general principles

3.1 Vision/strategy and policy process

I have already mentioned that application of General ITP Principles to concrete policy situations cannot be direct nor automatic. It requires not only familiarity with the specific setting within which the policy should operate, but also additional theoretical and practical 'knowledge' – about *policy phases/process* and the country's *policy environment*. This paper aims at providing a modest beginning to the first of these factors. The starting point is to view ITP as a *process* with *distinct phases and activities*- a view that parallels the view of Innovation itself as a process (Kline and Rosenberg, 1986). We have already mentioned a Strategic dimension which when activated ends up with a new set of country priorities for Innovation and Technology (and the Business Sector as a whole). *Upstream Phases* include generation of a *Vision* about the future SI, of a *Strategy* to achieve it, and translating Vision/Strategy into a *Set of Priorities* (Teubal, op.cit.). The policy process's *Downstream Phases* include *Identification of acceptable Program Portfolio Profile (PPP) & Institutional Changes for the country* which would advance the above-mentioned set of priorities, *Program Design,* [23] *Program Implementation* and *Impact Assessment and Feedback* (Table 8). The feedback from program assessment may influence both priorities and specific program design, that is either upstream and/or downstream phases of the policy process.[24]
The basis for initiating upstream phases is an *understanding of the trends in world markets, technology and competition; and the implications of this for the country in question* (Bartzokas and Teubal, 2001, op.cit.).[25] Thus globalisation of asset and capital markets and the emergence of the Internet created numerous new opportunities which only those countries who succeed in launching a new set of policies may exploit. In this context, formulating an explicit Strategy may lead to a recon-

[23] Program design would involve the identification of the activity to be promoted; the Business Sector focus; the program's company (or SS-organisation) orientation; the tools to be used (subsidies, conditional loans, etc); the effective incentive rate; and the criteria for project approval.

[24] The above view of the policy process further develops the one put out in Teubal (op.cit.). It differs from that put forth by Edquist (op.cit.) despite some correspondence between his categories and mine. His "roblems" come close to my "trategic priorities" although there is little discussion of the various levels of specificity in the formulation of 'problems' (I recognise that this is difficult and no good solution has yet been arrived at). A second point concerns the transition from 'problems' to policies: while Edquist makes it depend on the political system with very little further specification, the framework presented here tries to flesh out some of the elements of the process (e.g. a typology of policy processes, link with the policy environment, etc.–see later). This approach could shed light on the interactions between the professional aspects of policy and the political ones.

[25] This paper will not analyse in depth the Upstream Phases of ITP. Suffice to say at this point that a number of methods should be implemented such as Foresight and Prospective Studies, International Comparisons, Benchmarking exercises, etc. The Vision, Strategy and Priorities arrived at will also be influenced by Theory, Data and inevitably, Judgement.

Table 8. Policy process-phases (BS-directed incentives' programs)

Phase	Objective	Tasks-activity	Outcome
1. Upstream-strategy formulation	Formulate an explicit strategy	Search, research and interaction (stakeholders & experts), generating a vision/strategy	Set of (new) priorities in innovation, technology and for the business sector
2. Downstream-determining the program portfolio	Identifying an adequate program portfolio *Profile(PPP)*	Evaluation of alternative program profiles and mixes of programs; Preliminary design; trial implementation, final design	Determining a (new) set of programs & program designs which "it" priorities
3. Downstream- individual program design and implementation; assessment	Successful implementation and learning	Full implementation; operational adjustments; research on impacts and on success/failure.	Contribution to business sector restructuring; new information about 'policy needs'

sideration of new priorities and, indirectly, of new programs and of institutional changes.[26] [27] [28]

3.1.1 An integrated policy process. The above defines an 'Integrated' policy process that is one which involves upstream and downstream phases starting with a new vision/strategy; and almost inevitably– to new ITP priorities. It is triggered by radical changes in the environment facing the country (including its internal environment). The new priorities in turn would lead to policy activity in Downstream Phases which process leads to *a new portfolio of policies* (PPP, and institutions). The new PPP will generally involve a subset of new programs, the deletion of some old programs; and changes in the design / implementation of some of the remaining programs.

[26] As mentioned. in Israel a well functioning hi tech cluster may require a larger dose not of Start Ups (of which it has plenty) but of larger, global companies since it is believed that this could ensure the 'leveraging' of R&D in terms of value added and employment–both direct and indirect effects of R&D/SU & other "spillovers" to the national economy. There are two alternative Vision/Strategies: I & II. In I BS R&D is the strategic priority; and little or no explicitness about almost everything else in the system. Under II High Tech is assigned a leading role, but it also strives for a balance with mid/low tech and between the support (direct or indirect) of R&D and non-R&D factors.

[27] Whatever the final outcome of an explicit strategy it is clear that identifying the implications of the above-mentioned trends for the national economy requires significant additions to knowledge. In my opinion, collecting information is not enough, it is important to generate conceptual theory for integrating it into a coherent whole. This means that Academic research may have to play a role, side by side with data collection and commissioning work to consultancies.

[28] A major bottleneck in commissioning such research and conceptual work is the capacity of the policy making institutions to define objectives and to assess the true capabilities and the true outputs proposed by the organisations competing for the work.

3.2 "Reference" program portfolio profiles (PPP) based on a typology of programs

The next step in applying ITP General Principles is to identify a reasonable, acceptable and strategy-friendly Program Portfolio Profile for the country. This is not simply a list of programs–the term *profile* is supposed to indicate something about *structure or mix of programs of different types.* It suggests the need of a typology of programs for representing different PPPs. Beyond the PPP concept itself I will introduce two *Reference PPPs* to facilitate the subsequent analysis of a country's PPP-*Bias* (Sub-section 3.3 below). This procedure points out the way for 'changing' the actual PPP in the 'right' direction.[29]

3.2.1 A typology of incentives' programs. A major issue in ITP as a body of knowledge concerns the typology of incentives' programs, of institutional changes, and of other policy actions. Existing work on ITP following the SI perspectives is weak in this respect there being very little awareness that this is a crucial aspect of the knowledge base underlying policy[30]. While Edquist's latest work is useful and instructive, on this point he is very vague. His point of view is that policies are "selective" which is true but which does not carry us very far.[31] Due to historical reasons and to lack of background knowledge I will here focus on the typology of 'incentives' programs rather than that of institutions or institutional changes. Programs should *first* be classified according to their direct impact within SI structure. We thus have programs directed to one or more of the two subsystems–Business Sector or Supporting Structure *(Primary Classification);* or programs directed to System Links/Interactions; to Institutions and Markets or to Culture and Social Structure (*Secondary Classification*). For lack of space I will be considering here only the first group.

Primary classification: A distinction is made here according to *specific subsystem focus,* namely whether the program is directed to the Business Sector or whether it is directed to the BS's Supporting Structure (Justman and Teubal, 1986). Programs directed to the Business Sector Subsystem can be classified according to

- *The (Supported) Activity and its Characteristics (e.g. Breadth)* e.g. 'classical' or 'generic' R&D; Transfer/Absortion and/or Diffusion of Technology; a narrow activity like Design or a broad one like 'Technological Modernization'; etc.
- *Scope of BS Support-Horizontal Programs*-the whole BS; *or Targeted-* to a specific Sector, Technology or Region

[29] While this is not a fully operational method, it–together with the accompanying illustrations-represents a framework which policy makers can use as a starting point in their search for an 'appropiate' PPP.

[30] As mentioned, the underlying reason would seem to be that 'policy' is not really regarded as a field of knowledege in itself, but basically as "application".

[31] I mentioned that selectivity exists not only in Targeted programs but also in both the design and in the implementation of Horizontal programs. For a summary of some of the issues see Lall and Teubal, (op.cit.).

- *Company Focus:* Possibilities– focus on National Champions; focus on large or small companies; focus on incumbent companies or on SU; focus on Innovative/Schumpeterian or on Laggard companies; etc.
- *Link Emphasis:* whether or not programs directed to the BS *indirectly favour Links and/or Interactions* among or within Subsystems, or between Domestic and Foreign Agents.

A roughly similar distinctions could be made in relation to programs directed to the Supporting Structure.[32]

Horizontal versus targeted programs: For programs directed to the Business Sector, the notion of Horizontal program must involve 1) a *sufficiently wide scope of BS support and Company focus* of the program (e.g. the whole business sector rather than a single industry; a wide enterprise orientation rather than only directed to e.g. to National Champions); *and* ii) support of a *'wide-breadth' activity*. Thus, while support of 'classical' R&D throughout the business sector could be considered 'horizontal' such a term would be inappropriate to a program supporting 'classical R&D involving Co-operation between Israeli and US companies'. Despite the latter's orientation to the business sector as a whole, the breadth of R&D supported by the program is quite narrow so the term 'Horizontal' is inappropriate[33]. *Targeted programs* as far as the business sector is concerned can be directed either to a specific BS branch or technology or to a specific region; to specific links; to a small group of companies; or involve a a narrow activity being promoted (e.g. design or TQM even if it is directed to the BS as a whole). Table 9 briefly summarises the advantages and disadvantages of Horizontal and Targeted programs.[34] [35]

3.2.2 'Reference' program portfolio profiles. A country's (Program) Portfolio Profile (PPP) refers to the composition of its program portfolio and to the characteristics of the associated program set. Two major profiles come to mind: *PPP 1*

[32] For example, instead of Scope of BS Support we should refer to Scope of SS Support; and rather than Company Focus we should refer to SS Organisation focus. Note that Targeted Programs directed to the SS can be interpreted in two ways a) exclusive support of 'targeted' technologies and of the SS-organisations housing them; or b) support of only those SS-organisations which facilitate or induce R&D/innovation in 'targeted' BS branches.

[33] Such a program would in effect 'target' specific links between the Israeli and US Systems of Innovation

[34] Either Horizontal or Targeted programs could be launch to promote the *Supporting Structure* i.e. indirect support of the BS. An example of a 'Technological Infrastructure Policy' program which is Horizontal is Israel's Magnet program which operates since 1992. Taiwan's support of SC technology during the early 80s would be an example of a *Targeted* program (or set of programs and other policy actions) directed to a segment of the Support Structure /Technological Infrastructure supporting a particular technology and/or industrial sector. Other examples are the promotion of IT or biotechnology within Government Laboratories and Universities, or "targeted" support for Technology Centers belonging to priority industries. In same cases Targeted support of Innovation and Technology may be part of a coordinated set of programs and policies directed to generate comparative advantages in certain industries

[35] The Taiwanese example involved support of C-MOS technology absorption and development through ERSO (Electronics Research Service Organization) during the late seventies and early eighties. This in turn led to the development of that country's impressive SC industry. For a summary of Taiwan's policies towards the Semiconductor industry, see Mathews (1997).

Table 9. Horizontal versus targeted programs-advantages/disadvantages

Horizontal programs
1. Since there is no need to identify priority sectors, these programs are implementable without strong initial policy capabilities–as long as the Government Agency has a disposition and a capability to learn .
2. Can effectively promote unpredictable, random "variety" in the Business Sector–an increasingly important component of successful growth and evolution.
3. Frequently they represent an important component of policies directed to Business Sector Restructuring (promotion of R&D/Innovation & of Organizational Change) since there is a strong Strategic Need in a number of contexts: NIC's opening up, peripheral countries of Europe including Eastern European countries, support of SMEs and of Start Ups, and Advanced countries wishing to promote sophisticated forms of R&D cooperation, etc.
Targeted programs.[36]
1. Promote 'systemic' variety into the economy.
2. Required in the presence of strong sectoral/ technology specificities.
3. Feasible when there are clear sectoral or technology priorities which are known or are easily identifiable by policy makers.

which directly focuses on the Business Sector through implementation of (mostly) Horizontal Programs or a mix of programs, with relatively few incentives given to the supporting structure. *PPP 2* on the other hand provides incentives to the Supporting Structure (*indirect* support of the BS) through use of Targeted programs which support only institutions/organisations related to a narrow group of business sectors. Each one of these might have variants of their own depending on specific circumstances and depending on other factors such as the particular activities emphasised. We may have two extreme types of *activity emphasis*: *R&D Orientation* and *Diffusion Orientation.* Thus *PPP 1* may be either R&D oriented (variant 1a) or Diffusion Oriented (variant 1b). This distinction could also be relevant in relation to *PPP 2*. Thus 2b may be relevant for industrialising countries or for countries in the "Imitation" Phase; while 2a (R&D orientation) would be relevant for advanced countries. The outcome is summarised in Table 10.

3.3 Ascertaining PPP bias and some illustrations

The approach to determine the PPP Bias of a country would involve two stages a) *identify a reasonable/acceptable Strategy* and associated priorities; b) *determine the "Fit" between Strategy and actual PPP.* Thus, under the SI perspective, a country PPP 1a would or would not be biased towards R&D depending on whether or not the 'reasonable/acceptable strategy' involves or does not involve overwhelming support of R&D over diffusion. Needless to say, Strategy (and indirectly, potential PPP bias) will depend on the stage of development of a country. For simplicity

Table 10

Program Portfolio Profile (PPP)	Sub-system focus	Type of Program	Type of Activity
PPP 1	Business Sector (BS)	Horizontal (direct support of all sectors and technologies)	1a: R&D oriented 1b: Diffusion oriented
PPP 2	Supporting Structure (SS)	Targeted(indirect support of a specific sector or technology)	2a: R&D oriented 2b: Diffusion oriented

we consider two: the *Imitation Phase* – where the developing country focuses on absorbing foreign technology and in building the basic Science, Technology and Innovation institutional set up (Korea during 1982–1980; see Lee, 2000; Kim, 1997; Westphal, 2001); and the *Innovation Phase* (e.g. Korea, 1980–1997; Israel, 1970–1990) where the focus is on empowering the Business Sector and on Business Sector R&D.

3.3.1 Illustrations of PPP bias: imitation phase For the imitation phase I consider two extreme cases of program portfolio profiles–Korea (Lee, 2000)[37] and Indonesia (Dodgeson, 2000). It is assumed that thc *Strategic Priority* is to build manufacturing industry through an explicit policy of transferring and absorbing technology from abroad. It is also recognised that, due to the weakness of the business enterprise sector during that phase of development, that a measure of 'supply push'; 'top down' and SS-oriented policies is inevitable. A PPP reflecting this features 'with moderation' could be considered 'balanced' or adequate for a typical country in this phase.

3.3.2 PPP bias: illustrations for the innovation phase (Korea, Israel) I assume that a major strategic priority for ITP is the introduction and diffusion of R&D and Technological Innovation within the Business Sector; and this objective would ideally imply strong direct support to business enterprises. A PPP reflecting this and which also does not ignore the other branches of the BS could be considered 'balanced' or adequate for a typical country at this phase in its development.

Comments: Israel's high level of consistent support for Business Sector R&D during 20 years and its success in generating (towards the end of the 80s) strong increases in Business Sector R&D expenditures is a reflection a Program Portfolio Profile with a clear Business Sector focus. Its weak performance in diffusion of technological development (including R&D) within mid/low tech industry is a reflection both of its Type of Activity focus–R&D–and of its *bias* against those elements of the Supporting Structure most oriented to mid/low tech. e.g. sector specific Technological Centers.

[37] See also Westphal op.cit.

Table 11. Policies and PPP: korea (imitation phase 1962–1980)

1. Creation and support of basic science and technology institutions(KIST, MOST,GRI)

2. Support of technological infrastructure through a network of Government Research Institutes (GRI). These and the other institutions also attracted Korean Scientists and Engineers working abroad

3. Presumed relative absence of direct, horizontal support of technological transfer/absortion in the business sector[38]

Policy and PPP: Indonesia (imitation phase)

1. Extreme targeting favouring one sector (Aircraft Industry) and even one firm/national champion(IPTN-a government owned company)

2. No systematic policies promoting the transfer/absorption of technology (and diffusion policies) oriented to the business sector as a whole

3. Weak support of human resource development.

A clearly biased PPP:

- no systemic promotion of the SS (including human resources)
- absence of broad based direct support of the business sector (excepting aircraft industry)
- bias against technology transfer & diffusion policies

4 Summary and conclusions

The major objective of this paper was to systematically analyse the Systems of Innovation perspective to Innovation and Technology Policy (ITP) with a view of applying it to concrete policy settings. Systems of Innovation research on policy has focused until now on the Positive rather than on the Normative Side. This coincides with the view prevailing in quarters within Academia that policy is simply application (an "artifact" in the language of Layton, 1987, and Vincenti). It is also one of the reasons why–beyond the incorporation of 'systems of innovation', 'learning' and 'interactions/links' into Policy Makers language–not much progress seems to have been made.

The approach followed is to see the SI perspective to ITP first and foremost as an area of *theoretical knowledge whose general principles* and *other knowledge components* may contribute to the design and implementation of policies in different national settings. Beyond the general principles of the *positive* side of the SI perspective (most of which have appeared in the literature and are only briefly summarised here), I analysed ten ITP "Salient *Normative*\Policy Principles or Themes" classified into four groups: *ITP Objectives* (Learning and SI transformation); *The Nature of Policy Making* (Adaptive Policy Maker, An Explicit Strategic Dimension; Policy as Judgement; and the Context Specificity of Policy); *Learning, Demand & Dynamics* (the importance of: New SI Components, explicit consideration of Demand; Learning during Implementation; Policy Learning, etc); *Characteristics of the Policy Set (e.g.* mix between Targeted and Horizontal programs, etc). These

Table 12

a) Korea (1980–1997)
Korea's PPP-'innovation phase'(1980-1997)
1. Direct support of R&D/innovation at business enterprises (through horizontal and probably targeted programs providing tax incentives to R&D; tax credits for human capital development, loans for technological development etc)
2. Targeted infrastructural support of industrial technologies of strategic importance (including the HAN project) – a component of the Supporting Structure
3. Targeted support of venture capital (and/or other mechansims of financial support for R&D/innovation of enterprises
A balanced PPP- involving substantial direct, (presumably) horizontal support to R&D/innovation at business enterprises; and targeted support of relevant components of the 'Supporting Structure'

b) Israel (1970–1990)
Israel's PPP-innovation phase (1970–1990)
1. Strong, consistent R&D support of the BS (use of grants/conditional loans)
2. Bias against the supporting structure
3. Bias against Mid/Low Tech Industry
A Balanced PPP–with respect to priority of developing high tech industry; a somewhat biased PPP from the overall economy perspective

principles are amply illustrated with examples, mostly from Israel (the country whose ITP I know best). I have also proposed a typology of programs which follows a SI perspective, and suggested two "Reference" Program Portfolio Profiles (PPP) which would help focus the analysis of PPP Biases in industrialising countries at the "imitation" (Korea, Indonesia) and "innovation"(Korea, Israel) phases. The analysis also considers other components of the SI perspective to ITP more directly related to Application in specific country settings. These include *The Policy Process* and *The Policy Environment.*

The central emphases of the paper are a) presenting the Normative Principles of ITP as an integrated whole; b) elevating the Strategic Dimension of ITP (including its integration into a view of the 'Policy Process' which parallels that of the 'Innovation Process') to the status it deserves in the present world semi-chaotic environment and c) dealing with knowledge items relevant for ITP Application–Reference PPPs based on a SI-based program typology; the notion of PPP Bias which is a step forward towards determining a Strategy-Friendly PPP; etc.

The goal of policy making is to ensure adoption by the country of a reasonable and acceptable PPP. Sometimes this is relatively easy since no fundamental changes in the policy process and in policy making 'routines' are required. In other contexts it may be difficult due to the need of inducing a shift in the way the formulation and implementation of policy is done. This shift in the policy process is rarely automatic. Policy makers must, to be successful, take account of what could be termed the "Policy Environment" of the country. This may be of central importance for example in the transition from the imitation to the innovation phase of a country's economic development. Other wise the policy process itself may become trunkated or locked in into the old mode of operation.

A successful transition to the integrated mode of policy making will require significant amounts of information and probably new institutions of policy such as an 'interactive stakeholder forum' common in many countries; or a specialized ITP institution such as Israel's Office of the Chief Scientist at the Ministry of Industry or Trade.

One additional area of knowledge required for successful application of ITP General principles concerns the above *Policy Environment* and its characterization. This will be left for future work.[39]

References

Andersen E (1994) Evolutionary economics: post schumpeterian contributions. Pinter, London

Arrow K (1962) Economic welfare and the allocation of resources to inventions. In: Nelson R (ed) The rate and direction of inventive activity 609–625. Princeton University Press, Princeton, NJ

Bartzokas A, Teubal M (2001) A framework for policy-oriented innovation studies in industrialising economies. Introduction to special issue on: the political economy of technology policy in developing countries. The Economics of Innovation and New Technology (forthcoming)

Caracostas P, Muldur U (1998) Society, the endless frontier. European Commission, Directorate General XII

Dodgeson M (2000) Policies for science, technology and innovation in asian NIE. In: Kim L, Nelson R (eds) Technological learning and economic development: the experience of the Asian NIEs. Oxford University Press, Oxford

Dodgeson M, Bessant M (1996) Effective innovation policy: a new approach, International Thomson Business Press, London

Dosi G (1988) The nature of the innovation process. In: Dosi C, Freeman C, Nelson R, Silverberg R, Soete L (eds) Technical change and economic theory. Pinter, London

Edquist C (ed) (1997) Systems of innovation: technologies, institutions and organizations. Pinter, London

Edquist C (1999) Innovation policy-a systemic approach. Linkoping University, Department of Technology and Social Change

Galli R, Teubal M (1997) Paradigmatic changes in national systems of innovation. In: Edquist (ed) Systems of innovation: technologies institutions and organizations. Cassel, London

Greif A (2001) Historical institutional analysis. Typescript

Johnson B (1997) Implications of a systems of innovation perspective on innovation policy in Denmark. Paper presented at the International Symposium on RTD Policies in Europe, Jerusalem

[39] The Policy Environment plays two main roles: a) it mediates between ITP *Knowledge* (e.g. ITP Normative Principles) on the one hand and the specific ITP applied in a particular country; and b) it influences the set of actions (and their impact)taken to induce the
transition towards an integrated policy process. See Teubal 2001a op. cit., for a preliminary analysis of how the Policy Environment may influence the means by which a desirable shift to an integrated policy process mode would take place.

Justman M, Teubal L (1995) Technological infrastructure policy: generating capabilities and building markets. Research policy. Reprinted in: (1996) Teubal M et al. (ed) Technological infrastructure policy (TIP): an international perspective. Kluwer, Dordrecht
Kim L (1997) From imitation to innovation: the dynamics of Korea's technological learning. Harvard Business School Press, Boston
Kim L, Nelson R (eds) 2000) Technological learning and economic development: The experience of the Asian NIEs. Oxford University Press, Oxford
Kline SJ, Rosenberg N (1986) An Overview of Innovation. In: Landau, Rosenberg N (ed) The positive sum strategy. harnessing technology for ecoomic growth. National Academy Press, Washington
Lall S, Teubal M (1997) A framework for market stimulating industrial and technological policy. World Development 26 (8): 1369–1385
Layton ET (1987) Through the looking glass or news from lake mirror image. Technology and Culture 15: 594–601
Lee Won-Young 2000) The role of s and t policy in korea's industrial development. In: Kim L, Nelson R (eds) Technological development and economic development: the experience of east Asian newly industrialising economies. Oxford University Press, Oxford
Lipsey R, Carlaw K (1998) Technological policies in neo-classical and structuralist-evolutionary models. Special issue on best practices in technology and innovation policy. STI Review No. 22
Lipsey R, Carlaw K (1998) The implications of knowledge-based growth for micro-economic policies. The Industry Canada Research Series. The University of Calgary Press, Calgary
Lundvall BA (1985) User producer interaction. Aalborg University Press, Aalborg
Lundvall BA (ed) (1992) National systems of innovation: towards a theory of innovation and interactive learning. Pinter, London
Mathews J (1997) A Silicon Valley of the east: creating Taiwan's semiconductor industries. California Management Review 39 (4)
Metcalfe S (1996) The economic foundations of technology policy: equilibrium and evolutionary perspectives. In: Stoneman P (ed) Handbook of the economics of innovation and technological change. Blackwell, Oxford, UK, Cambridge, MA
Metcalfe SJ (1999) Lecture in international symposium technology policy and developing countries. UNU/INTECH, Sussex
Nelson R (1987) Roles of government in a mixed economy. Journal of Policy Analysis and Management 6 (4): 541–557
Nelson R (ed) (1993) National systems of innovation: a comparative study. Oxford University Press, Oxford
Nelson R (1994) The co-evolution of technology, industrial structure and supporting institutions. Industrial and Corporate Change 3
Nelson R (1995) Recent evolutionary theorizing about economic change. Journal of Economic Literature 23: 48–90
Nelson R Sampat B (1998) Making sense of institutions as a factor in economic growth. Typescript
Nelson R Winter S (1982) An evolutionary theory of economic change. Harvard University Press, Boston, MA
OECD (1997) National innovation systems. Paris
Saviotti P (ed) Evolutionary theories of economic and technological change, pp 256–275. Harwood, London
Saxenian A (1998) Regional advantage: Silicon Valley and route 128, 2 edn. Oxford University Press, Oxford
Smith K (1991) Innovation policy in an evolutionary context. In: Metcalfe S (ed), Saviotti P (eds) Evolutionary theories of economic technology change. Harwood, London
STI Review (1998) Special issue on new rationale and approaches in technology and innovation policy. OECD, Paris
Stoneman P (1988) The economic analysis of technical change. Oxford University Press, Oxford
Teece D (1986) Profiting from technological innovation: implications for integration, collaboration, licencing and public policy. Research Policy 15: 285–305
Teubal M (1996) R&D and technology policy at NICS as learning processes. World Development 24: 449–460

Teubal M (1997) A catalytic and evolutionary approach to horizontal technology policy. Research Policy 25: 1161–1188

Teubal M (1998a) Policies for promoting enterprise restructuring in national systems of innovation: cumulative learning and system effects. STI Review. Special Issue on: New rationale and approaches in technology and innovation policy. OECD, Paris

Teubal M (1999) Towards an R&D strategy for Israel. The Economic Quarterly (Hebrew), December

Teubal M, Andersen E (2000) Enterprise restructuring and embeddedness: a policy and systems perspective. Industrial and Corporate Change 1 (9): 87–111

Teubal M, Avnimelech G (2002) Company growth, acquisitions and access to complementary assets in Israel's data security sector. European Planning Studies (forthcoming)

Teubal M (2001a) What is the systems of innovation perspective to innovaton and technology policy (ITP) and how can we apply it to developing and newly industrialised economies. Presented at June 2000 ISS Meeting, Manchester; and at the June 2001 DRUID Meeting

Teubal M (2001b) Pre-systems of innovations perspectives to innovation and technology: a summary. Prepared for the EU sponsored ECLA (Economic Commission for Latin American and the Caribbean, United Nations), Course Module on Innovation & Technology Policy, Santiago de Chile

Westphal L (2001) technology strategies for economic development in a fast changing global economy. Special issue on: The political economy of technology policy in developing countries. Economics of Innovation and New Technology (to appear)

Knowledge production and distribution and the economics of high-tech consortia

Maurice Cassier[1] **and Dominique Foray**[2]

[1] CNRS, CERMES, Paris, France (e-mail: cassier@ext.jussieu.fr)
[2] CNRS, IMRI, Dauphine University, Paris, France (e-mail: dominique.foray@oecd.org)

Abstract. This paper deals with the issue of management of knowledge and intellectual property in research consortia. We develop a framework based on three axes describing the processes and procedures of collective invention (the collective production of knowledge, the appropriation and distribution of results, the composition of the group and the mode of managing externalities). We, then, use this framework to analyse various tensions and conflicts between the degree of distribution and collectiveness of the production and circulation of knowledge, the degree of collectiveness of the appropriation of results, and the scope and level of organisation of final dissemination. Based on case studies carried out in the field of biotechnology, the paper develops some policy implications with a particular focus on European policy.

Key words: Collective invention – Knowledge – Intellectual property rights – Research and development

JEL Classification: L20-O31-O33

1 Introduction[1]

It is increasingly recognised that the knowledge generated for innovation is not solely a process of intra-organisational conversion, combination, adaptation and extension, but also a process that is collectively organised by industries and other

[1] The authors are especially thankful to Ed Steinmueller for a number of useful comments and suggestions. They are glad to Stan Metcalfe for his encouragements to finalize the final draft of this paper. Financial support from the European Commission under the TSER programme (contract n°SOE1 - CT97-1062) is gratefully appreciated.

larger domains of inter-firm or government-industry relations. The collective nature of innovation is in the first instance a recognition that the domain of innovation is larger than that of the organisation, and that the unit of analysis must therefore be larger as well.[2]

The growing appreciation that it is appropriate to adopt a broader unit of analysis based upon understanding of particular research communities and networks has been augmented by efforts to identify and to enumerate the institutional forms of various inter-organisational relations supporting knowledge creation. Technology development agreements in the form of joint ventures, technology exchange agreements and university-industry linkages provide persuasive evidence that collective invention is a rapidly growing component of national innovation systems.

All of this can be characterised as an advancement in knowledge of how the innovation process works. It reflects the fact that the innovation process is itself evolving.

There is, nonetheless, a gap in our understanding. The "forms" of collaborative arrangements do not adequately describe their content or process. Moreover, when we return to examine specific organisations in the original innovation studies tradition, we find firms engaged in collaborative processes that are not implemented within the formal structures of joint ventures or strategic alliances. This suggests that the collaborative invention process is even greater than some of the indicators would suggest. What happens in the process of collective invention is therefore becoming a more pressing issue. Before we can develop broader and more inclusive measurements of the activity and the determinants of its successes, we need to develop deeper empirical knowledge of the varieties and purposes of collective invention. Conceptual advancement and empirical knowledge generation must be developed simultaneously and interactively. This is what we have done in the *Collective Invention (Colline)* project supported by the EC under the TSER programme (Foray and Steinmueller, 1999).

2 A study on biotechnology consortia

The concept of an R-D consortium is an important tool and institutional mechanism for technology policy.

- It creates spaces for sharing knowledge, in which there is a break from technological secrecy and the retention of knowledge by private agents. It generates a new economic category of knowledge called collective or pooled knowledge, which is shared among participants during the period of research.
- It allows agents to develop concerted actions by organizing the division of labour to explore a certain domain and by providing an institutional framework to assemble divided and dispersed knowledge.
- It enables agents to create a more consistent and coherent initial endowment of intellectual property rights, which does not fragment the knowledge base.

[2] See for instance, among many others, Dosi (1988); Freeman and Soete (1997); Lundvall (1992); Metcalfe (1995); Mowery and Rosenberg (1989); Teece (1986) and von Hippel (1988).

When the knowledge is initially fragmented ("anti-commons property"), the consortium provides a space in which rights can be exchanged at a low cost, because partners are well identified and collective learning can occur.

All these good properties of consortia prove to be extremely useful for the regulation of the knowledge-based market economy. They show that consortia have the potential to "regenerate" the two main mechanisms which traditionally organize research activities: private property and public organization.

The *Colline* project was initially based on the perception of a high rate of creation of R&D consortia in Europe. It was also based on the assumption that beyond a sort of generic form of consortium, there are numerous particular constructions, and it might be useful to explore thoroughly the mechanisms of collective invention and to try to identify some good practices.[3]

To investigate the mechanisms of collective invention, we use a framework which articulates three dimensions: the production of knowledge (division of labour, access to the common pool of resources); the attribution of results (intellectual property rights policy); and the internalization of externalities and management of spillovers.

3 The framework

Collective invention has been the subject of much research.[4] It nevertheless seems to us that few analyses attempt to grasp the processes and procedures through which communities are formed, collective goods produced, and the social benefits of the activity acknowledged. *Colline* has, by contrast, tried to gain insight into these processes and procedures, i.e. the constituent elements of collective invention. Its work is based on three axes describing the processes and procedures of collective invention:

- Production of knowledge (division of labour, technical co-ordination, organisation of resources and knowledge access)
- Appropriation and distribution of results (intellectual property)
- Composition of the group (internalisation) and dissemination beyond the group (externalities)

– First, collective invention is a way of producing knowledge, based on the possibilities afforded by the division of labour when a number of entities or agents come together to pursue a particular goal. This first dimension is that of the "technical" organisation of the circulation and production of knowledge, which uses collective invention as a solution to problems of dispersion and division of knowledge, and of excessive incentives to invest in R&D (tragedy of commons). Technical organisation concerns, first, the distribution of roles, and thus the division of labour, and secondly, access to a pool of shared resources. While the former is characteristic of

[3] The analysis presented here draws on 8 monographs on biotechnology consortia (6 are European consortia supported by Community programmes and 2 are international consortia).

[4] See Cassier and Foray, 1999a, for a survey.

explicit forms such as consortiums (although repeated informal trading can spawn primitive forms of division of labour), the latter characterises most forms of collective invention, i.e. consortiums but also client-supplier relations and informal exchange between members of the same community.

– Secondly, collective invention is a mode of appropriation and distribution of the results of research and innovation, based on the creation of a community. This involves problems that are neither simple nor easy to solve. A multitude of possible "rules" can be identified, from the establishment of collective property rights to the maintenance of private rights. This dimension is important in the context of galloping privatisation, which is having the effect of fragmenting the knowledge base and producing monopolies that preclude any subsequent development by a third party. From this perspective, collective invention may be seen as an instrument for producing a more coherent logic for the attribution of property rights, or for providing a framework for trading rights at the lowest cost.

– Lastly, collective invention produces a new boundary, an original partition between a set of co-ordinated agents and the rest of the world. The question of the dissemination of results and thus of the social returns to collective research represent the third aspect of collective invention. This aspect raises the question of the "composition" of the group, that is, the internalisation of knowledge externalities. Is the group composed of all the members of a set (an industry, for example)? – in which case the question of dissemination is less relevant. Or does it consist of a significant part of this set? – in which case collective invention may become an obstacle to the entry of new actors into the industry. Or, lastly, is it limited to a very small number? – in which case the question of dissemination beyond the circle is raised. One important issue to be considered here is the elaboration of rules about "rights of entry" for newcomers. Another is the consortium's intention to promote broad use of the invention, as in the case of standards, or to appropriate gains from invention within the consortium. If the latter, it is important to know whether there is a clear 'decision-maker' who can choose to negotiate exchanges with non-consortium members.

Having identified these three axes, an object of analysis emerges: the communities which support collective invention are not all accomplished in the same way, in terms of the following criteria: degree of distribution and collectiveness of the production and circulation of knowledge; degree of distribution of the appropriation of results; and the scope and level of organisation of final dissemination. Moreover, these three criteria are not independent of one another, and there may be tensions between them. For example, the integration of new partners (composition of the group) may occur at the expense of a change in the collective nature of the production and distribution of knowledge. A balance between these three requirements will then have to be defined. These balances are at the centre of our questioning.

4 Two generic models

The consortium naturally offers an appropriate framework for implementing complex forms of division of labour and circulation of data and materials (production of knowledge), for designing sophisticated mechanisms for the attribution of results, and for defining controlled procedures of diffusion towards society. The problem here is the creation of a balance between the composition of the group and the collective production of knowledge. We have identified two models in the particular case of European biotechnology consortia.

In a first model (model A), there is little diversity of institutional actors. Consortia are mainly academic, which means that externalities are seldom internalized. On the other hand, the processes of knowledge production and the attribution of results are highly collective:

- There is a formal and strong process of organization of the division of labour;
- The pool of resources (data, tools) is highly collective and the resources circulate freely among the members of the consortium;
- Some consortia have established a regime of collective property. In these consortia, each partner agrees that any patentable result obtained during the project will be the joint property of all partners and that each partner's share will depend on their participation in the work. Because the system of collective property can pose particular management problems, some consortia create an ad hoc institution for managing their collective property.

These consortia (model A) are therefore very close to what is seen as a "perfect community", i.e. a community that is strong enough to resist private incentives generated by patent and publication races. They are, however, relatively weak as regards the internalisation of externalities, for industrial partners remain on the outside. That is certainly why the community is so "perfect" in its ability to share knowledge and tools, with little concern for opportunistic behaviour. However, the main economic issue in this type of consortium is the transferability of knowledge to society (to particular classes of users), which is the realisation of a social return of the consortium. This return must come about through explicit "outputs" from the consortium rather than "spillovers" from the processes of knowledge generation occurring within it.

In a second model (model B) there is a wider diversity of partners. Industrial partners are integrated into consortia, although at the expense of the collective nature of both the knowledge generation process and the attribution of results. Consortia are characterised by very little sharing of research and by smaller knowledge pools. The results are ultimately appropriated separately by participants. Why is this the case? Industrial companies working within a consortium are reluctant to disclose confidential information. The circulation of data is therefore strictly limited to confined areas, so that rival firms participating in the same consortium work in different sub-parts of the general project and are never in contact. Finally, private partners cannot release their rights to a collective body. These consortia nevertheless fall into the category of collective invention because there is also some pooling of resources, knowledge sharing, and concerted decisions to organise the division of labour.

We can thus differentiate between these two models, in which the main problematic is different in each case (see Cassier and Foray, 1999b, for a detailed presentation of the empirical results of the study).

We have here a clear illustration of our assumption that the development of the three dimensions of collective invention corresponds to conflicting objectives: externalities can be internalized (which means integrating industrial partners) at the expense of the collective nature of the enterprise which will be altered, and the generation of strong communities requires that industry be left out.

For each model the problematic is different. We shall now consider each one successively.

4.1 Model A: Issues of the industrial transferability of knowledge

In the first "purely academic consortium" model, the ultimate problem is one of spin-offs and the transferability of results to industry and users. An important practice here is the building of an industrial platform, which includes certain privileged users. In this design, however, the industrial partners do not participate in the consortium's agenda.

Another solution is for the industrial platform to become a contracting party within the consortium. This form of quasi-integration aims at strengthening interactions between laboratories and industrial users, who participate at least in the definition of the programmes and the discussion of results, if not in the research itself.

We thus have two different schema. It immediately becomes apparent that the second scheme is intended to achieve a fragile equilibrium between keeping the collective nature of the process at a very high level, and integrating the industry. This is an interesting type of practice.

Another mechanism is absolutely critical in the production of such an equilibrium: partial dissemination of knowledge. This is a remarkable innovation used both to reserve the results for a certain period (which preserves academic institutions' ability to file for patents)[5] and to inform industry immediately about certain interesting results. It is based on the differentiation between the effective diffusion of knowledge and the sending out of signals. Signalling has a twofold advantage: the consortium immediately publicises information on its results while maintaining access rights and confidentiality; and, secondly, the firms receive early signals on research underway, and can undertake direct negotiation to enter into contractual relationships with the laboratories that own the data.

To sum up, it is clear that a "perfect community" with weak performance in terms of the transferability of knowledge would have no impact on society. That is why such a model has to be equipped with specific mechanisms aimed at promoting knowledge transferability. The combined system of (quasi-integrated) industrial platform and partial dissemination of knowledge is in this sense valuable.

[5] In Europe one cannot patent a knowledge, which has been disclosed through a scientific publication, while in the United States a grace period keeps the option to patent during one year following the publication.

4.2 Model B: How to produce a strong community with heterogeneous actors?

Model B raises quite different issues. In this model the internalization of externalities is the strong aspect of collective invention. Industries are within the consortia, so that the transferability of knowledge is not an issue. But pushing industry into a consortium implies explicit policies and mechanisms to protect the industrial entities within the community of partners. Two possibilities exist. The first is simply to redirect the research agenda towards "non critical" projects (this was the case, for instance, of SEMATECH; see Grindley et al., 1996). The alternative is to reinforce the protection of intellectual property. This is the case of most of the biotech consortia. However, such an option has the potential to deteriorate the collective and multilateral character of partnerships. The problematic of model B is thus to find an equilibrium between the need for providing better protection to private companies within the consortium, and the need for preserving some kind of collectiveness. Thus, problems here concern the production of knowledge and the attribution of results.

There are various difficulties in pooling data when industry is involved. Partners may bring in private materials that they want to exploit with the help of academic partners, without disclosing information on these materials to potential rival companies that are also members of the consortium. Such difficulties are partly overcome through the building of complex systems for circulating data, which are strictly compartmentalised within small teams consisting of only one firm each. The same kind of system is used for the knowledge and data produced during the collective research activity. The basic principle here is the "controlled" dissemination of data, which can lead to very complex patterns of concentric circles and gradual processes of dissemination.

The attribution of property rights also raises difficult issues. There are different possibilities:

- Disjoint property rights: in some consortia, while participants agree to exchange research results, they set up a system of separate ownership in which each firm retains control over its confidential material, technical know-how and invention. For example, one consortium is divided into five sub-projects, with one firm only working on each sub-project. Because firms are spread out among separate contracts, they are not forced to grant licenses to their rival partners, as in a typical EC contract. Each firm controls its territory and files patents on the inventions developed within its own sub-project.
- Temporary property rights: the other alternative is to grant participants temporary ownership, lasting for the duration of the research, on fragments of knowledge (e.g. particular chromosomes they receive and decrypt). During that period, the participants can publish or patent data concerning their fragments. The right to do so lapses once the research has been completed.

To sum up, although the second type of consortium provides a more explicit set of mechanisms for the industrial use of knowledge (because industry is involved in the actual activity), there is a need for some institutional innovation in order to establish a balance between the collective nature of the process and the protection of private interests. These innovations have to be developed to create such a balance at two critical points: the circulation of data and knowledge during the process, and the attribution of results. Depending on the mechanisms at work at both levels, a continuum can be drawn from the least to the most collective arrangement. Here again, a compromise appears to be difficult to find.

4.3 *In search of good practice*

Good practices in model A concern the transferability of knowledge (platform, partial dissemination). In model B, they concern the maintenance of some collectiveness, while the entry of the industry into the consortium is promoted (organization of the circulation of knowledge and the attribution of property rights).

In the absence of good practice in model A, there is a risk of funding collective research with a very weak impact on the economy. Without good practice in model B, there is a risk of having communities that are not resistant enough to withstand private incentives (patent races). This was the case of the International Breast Cancer Consortium. This does not mean, however, that the IBC consortium had no impact as a form of collective invention (see our policy implication, last section).

We can roughly draw a continuum pertaining to the different criteria set out above: circulation of data and organization of work; attribution of results; and management of externalities. This continuum must not, however, mask a gap between:
– On the one hand, forms of collective invention producing strong communities, which propose real alternative solutions – both local and temporary – to the mechanism of market coordination (we mean that they propose clearly different coordination mechanisms to those used in a private market, based on clearly defined and relatively codified forms of self-regulation).
– And, on the other hand, forms of collective invention producing far weaker communities – situations in which collective invention seems to be subjected to market mechanisms (we mean that collective constraints are weakened greatly and that the community does not seem to be strong enough compared to patent race mechanisms). It is remarkable to note that a consortium does not automatically fall into the category of "weak communities", even though a private property regime precedes its creation (participants initially possess private rights).

Admittedly, the solutions of collective property observed are applied by consortia composed essentially of academic laboratories or small private sequencing laboratories, with no direct interests in the industrial exploitation of knowledge. These consortia define collective property regimes to manage indivisible knowledge or to facilitate transfers toward industry. Industrial users can benefit from this form owing to lower transaction costs (the yeast industrial platform upholds the principle of a trust and of the collective property of the EUROFAN consortium). These solutions of collective property are therefore perfectly articulated to the market in so far as the trust sells access rights to industrial users who will subsequently develop proprietary innovations or take out application patents. However, these solutions are disputed by those who, within the consortia, have special agreements with industrial firms or strategies for exploiting the results (e.g. the Pasteur Institute is bound by exclusivity agreements to three firms in the field of diagnosis and therapeutics).

5 Shifting from model A to model B: European consortia at the crossroads

In this respect, the European model of collective invention in biotechnology is at a crossroads. The new principles (implemented within the 5th framework programme) aimed at increasing the industry's participation in consortia by reinforcing the protection of intellectual property within the community of partners, are positive and certainly unavoidable. These new principles concern, firstly, limitations to scientific publication (the scientist has to send a copy to his industrial partner, who has a period of 60 days to make a "motivated opposition"), and secondly, a requirement that the academic laboratory secure an authorisation of its initial partner if it wants to continue a research programme with a new partner. These principles should not, however, lessen the collective and multilateral character of such partnerships. A fragile equilibrium might be found between the need to provide better protection to private companies within the consortia, and the need to promote strong knowledge communities. There is, however, a risk of these new principles pushing consortia to form far weaker communities, which have proved unable to withstand private incentives generated by the patent and publication races. These developments create a paradox. Although the purpose of strengthening intellectual property rights is to increase incentives for the generation of new knowledge, incentives created by strong protection may prove detrimental to knowledge generation and distribution mechanisms.

6 Policy implications

Three policy implications can be drawn from our study:

A first implication concerns the tension between the uniformity of intellectual property regimes at the level of the European RTD consortia, and the variety of knowledge creation tasks – as described in our case studies. A "single rule" may be inconsistent with the diversity of situations. It is, however, clear from our case studies that the uniformity of intellectual property regimes mainly characterises the general conditions of intellectual property and confidentiality – that is to say, the legal infrastructure for cooperation and networks – while there is ample leeway for devising specific practices and rules under those general conditions. The general framework includes the definition of various classes of knowledge and the legal status of participants; the attribution of property rights to those who have produced the knowledge; the sharing of knowledge under reciprocity rules among the participants; and the granting of non-exclusive access rights to non participants. Those conditions express a certain policy orientation in favour of knowledge dissemination and exploitation. But this general framework also includes the possibility of devising specific contractual agreements (the so called "consortium agreement") in order to allow participants collectively to generate ad hoc modes of circulation of data and knowledge and of intellectual property rights attribution. It is, moreover, in the "spirit" of the actions of the DG XII scientific officers to leave networks with extensive leeway for finding original modes of regulation. This principle of a complementary design of institutional practices and norms is now formally recognised in the rules of the 5th Framework Programme. Thus, there is most often more than one solution to deal with problems of data circulation, attribution of property rights or dissemination of knowledge beyond the boundaries of the network. The important policy question now is what kind of practices – as elaborated within one particular context – could fit into a large class of activities? This leads us to another fundamental implication.

Our study has identified and documented practices and rules dealing with the issues of attributing property rights, organising access to a pool of resources and knowledge, and managing the dissemination of knowledge beyond the boundaries of the community. Thus, high-tech consortia are areas of high levels of institutional creativity where the actors have to devise rules for sharing and appropriating knowledge. These rules are used to manage the multiple tensions between individual priorities and collective invention, members and non-members, academic participants and industrial partners. We have documented an abundant production of rules and institutional innovations. The reason for such creativity is the somewhat unusual context of collective invention in some of the fields investigated. Moreover, two specifically "European constraints" necessitate creativity in the design of consortium agreements for supplementing European standard contracts:

– The absence of a grace period implying the need for deferred or partial access;

– The thrust toward knowledge sharing (produced by the European standard contract) which implies the need for reinforcing individual protection.

These local arrangements are mainly produced by researchers. Very frequently such rules and practices emerge spontaneously from the search by participants for "better organisations". Some are merely guidelines; others appear as appendices to the contract. For lawyers, guidelines and local rules are private arrangements between actors, organised under the principles of "self-discipline of a professional partnership". These rules express a tendency to the decentralisation of economic regulation. In this respect, they serve as experiments that can be influential in the development of persistent patterns of inter-organisational behaviour. That means that some of these rules are now codified and replicated in various contexts. But most of them, while performing very well, remain local, largely non codified and therefore very difficult to transfer and to generalise to other contexts. There is, thus, a risk of under-performance in institutional learning. This first point suggests a need for more in-depth institutional experiments. "Institutional experiments" does not necessarily mean a search for and diffusion of "best practices". It can mean, instead, processes of institutional learning through which actors and participants in various contexts are informed about original practices developed in other contexts, and are encouraged to test them, at least intellectually.

A third implication deals with the fact that the collective invention model – under its particular organisational form, the consortium – has been chosen as a crucial coordination mechanism to reinforce competitive advantages in some fields of high technology (and in particular in biotechnology). Now, an important policy question is whether the consortium remains the appropriate and efficient scheme as research advances towards the discovery of new knowledge (a new molecule), when patents races are likely to occur. While the value of collective invention seems to be indisputable during the first phases of basic research (when the pay-off is essentially informational), the question remains open for the latter phase of the process.

The case of the international breast cancer consortium is interesting (Cassier et Gaudillère, 1999): participants have effectively shared resources and data to narrow the gene research field (enabling them to target their own work better so as to remain in the race, and excluding groups not participating in the consortium). But the field of breast cancer genetics was so competitive that once the area of investigation had been defined, strategic knowledge was kept secret while each group negotiated with industrial firms and a patent race started within the consortium itself. The analogy with minerals prospection or deep-sea fishing is obvious. With the sharing of "geographic" information between a few agents, a first selection can be made among the groups, before the real competition starts.

Should we conclude here that the constraints of collective invention and the assignment of collective property rights can slow down the process, while racing has the potential to accelerate it? This cannot be taken for granted, for gene hunting has the potential to impede the cumulative production of knowledge. Genes would have been discovered earlier if all non-disclosed information had been diffused and shared among the participants.

It is, however, obvious that there is no constant best form of co-ordination during the whole process of research, development and innovation. We might think, instead, in terms of finding a best policy flux, an optimised path and rate of movement between promoting forms of organisation that support the collective production of public goods at some stages, and forms of market coordination that can support the acceleration of research at some further stages.

These consortium-related issues immediately lead us to a last issue – the need for evaluating the impact of the evolution of EU policy in terms of intellectual property, as described in section 4 of this paper. The extent to which the number of industrial partners is increasing as a result of the establishment of new rules (aimed at providing a better protection of intellectual property within the consortium), and the way in which these changes affect the "quality" of collective invention, are some of the issues that warrant further research.

References

Cassier M, Gaudillère JP (1999) Les relations entre science, médecine et marché dans le domaine du génome: pratiques d'appropriation et pistes pour de nouvelles régulations: le cas de la génétique du cancer du sein. Rapport de Recherche, Programme Génome du CNRS, Working Paper IMRI, Université Paris Dauphine, 2000/04

Cassier M, Foray D (1999a) The sharing of knowledge in collective, spontaneous or collusive forms of invention. Colline WP 01, IMRI, University of Paris Dauphine

Cassier M, Foray D (1999b) Les modes de régulation de la propriété intellectuelle dans les consortia de haute technologie. Economie Appliquée Tome LII, 2, pp 155–182

Dosi G (1988) Sources, procedures and microeconomic effects of innovation. Journal of Economic Literature vol.36, 1126–1171

Foray D, Steinmueller E (1999) Collective invention and European policies, - Executive Summary. Colline WP13, IMRI, University Paris Dauphine

Freeman C, Soete L (1997) The economics of industrial innovation. (third ed.), Pinter
Grindley P, Mowery DC, Silverman B (1996) Sematech and collaborative research: lessons in the design of high-technology consortia. In: Teubal M, Foray D, Justman M, Zuscovitch E (eds.) Technological Infrastructure Policy 173–211
Lundvall BA (1992) User-producer relationships, national systems of innovation and internationalization. In: Foray D, Freeman C (eds.) Technology and the Wealth of Nations, Pinter
Metcalfe S (1995) Technology systems and technology policy in an evolutionary framework. Cambridge Journal of Economics 19, 25–40
Mowery D, Rosenberg N (1989) Technology and the pursuit of economic growth. Cambridge University Press
Teece D (1986) Profiting from technological innovation: implications for integration, collaboration, licensing and public policy. Research Policy 15 (6), 285–305
Von Hippel E (1988) The sources of innovation. Oxford University Press

The manufacturer's authorised representative in the EU is Springer Nature Customer Service Centre GmbH, Europaplatz 3, 69115 Heidelberg, Germany. If you have any concerns regarding our products, please contact ProductSafety@springernature.com

Printed and bound by CPI Group (UK) Ltd, Croydon, CR0 4YY
22/09/2026
02220988-0006